**New Results in Numerical and Experimental Fluid Mechanics**

Edited by
Horst Körner and
Reinhard Hilbig

Notes on Numerical Fluid Mechanics (NNFM) Volume 60



- Volume 59 Modeling and Computation in Environmental Sciences. Proceedings of the First GAMM-Seminar at ICA Stuttgart, October 12–13, 1995 (R. Helmig / W. Jäger / W. Kinzelbach / P. Knabner / G. Wittum, Eds.)
- Volume 58 ECARP – European Computational Aerodynamics Research Project: Validation of CFD Codes and Assessment of Turbulence Models (W. Haase / E. Chabut / E. Elsholz / M. A. Leschziner / U. R. Müller, Eds.)
- Volume 57 Euler and Navier-Stokes Solvers Using Multi-Dimensional Upwind Schemes and Multigrid Acceleration. Results of the BRITE/EURAM Projects AERO-CT89-0003 and AER2-CT92-00040, 1989-1995 (H. Deconinck / B. Koren, Eds.)
- Volume 55 EUROPT – A European Initiative on Optimum Design Methods in Aerodynamics. Proceedings of the Brite/Euram Project Workshop „Optimum Design in Aerodynamics“, Barcelona 1992 (J. Periaux / G. Bugeda / P. K. Chaviaropoulos / T. Labrujere / B. Stoufflet, Eds.)
- Volume 54 Boundary Elements: Implementation and Analysis of Advanced Algorithms. Proceedings of the Twelfth GAMM-Seminar, Kiel, January 19-21, 1996 (W. Hackbusch / G. Wittum, Eds.)
- Volume 53 Computation of Three-Dimensional Complex Flows. Proceedings of the IMACS-COST Conference on Computational Fluid Dynamics, Lausanne, September 13-15, 1995 (M. Deville / S. Gavrilakis / I. L. Ryhming, Eds.)
- Volume 52 Flow Simulation with High-Performance Computers II. DFG Priority Research Programme Results 1993-1995 ( E. H. Hirschel, Ed.)

Volumes 1 to 51 are out of print.
The addresses of the Editors are listed at the end of the book.

# New Results in Numerical and Experimental Fluid Mechanics

Contributions to the
10$^{th}$ AG STAB/DGLR Symposium
Braunschweig, Germany 1996

Edited by
Horst Körner and
Reinhard Hilbig

Die Deutsche Bibliothek – CIP-Einheitsaufnahme

**New results in numerical and experimental fluid mechanics :**
Braunschweig, Germany, 1996 / ed. by Horst Körner and Reinhard Hilbig. –
Braunschweig : Vieweg, 1997
(Contributions to the ... AG STAB/DGLR symposium ... ; 10) Notes
on numerical fluid mechanics ; Vol. 60)
ISBN-13: 978-3-322-86575-5 e-ISBN-13: 978-3-322-86573-1
DOI: 10.1007/978-3-322-86573-1


Softcover reprint of the hardcover 1st edition 1997

Vieweg is a subsidiary company of Bertelsmann Professional Information.

http://www.vieweg.de

Produced by W. Langelüddecke, Braunschweig
Printed on acid-free paper

ISSN 0179-9614
ISBN-13: 978-3-322-86575-5

This book is dedicated to the memory of

Dr. H.-G. Knoche

# FOREWORD

This volume contains the contributions to the 10th DGLR / AG STAB- Symposium held at the German Aerospace Research Establishment (DLR) - Braunschweig Research Centre - November, 11 to 13, 1996. AG STAB is the German Aerospace Aerodynamics Association, founded at the end of the 70', while DGLR is the German Society for Aeronautics and Astronautics (Deutsche Gesellschaft für Luft- und Raumfahrt, Lilienthal Oberth Gesellschaft).

In the AG STAB German scientists and engineers from universities, research-establishments and industries are involved, who are doing research and project work in numerical and experimental fluidmechanics and aerodynamics for aerospace and other applications.

About 20 years ago it became obvious for this community that a joint effort of members of universities, the DLR and industry was necessary to counter-act declining budgets in the field. It was decided to approach high-level persons in industry, ministries and the parliament for help to shift the trend with its negative effects for research and industry. From the begin it was clear that an effort should be built around a central theme. "Flow with Separation" became the topic of the AG STAB (Arbeitsgemeinschaft Strömung mit Ablösung), which developed fast into a lively association, with, however, a larger scope than just flow with separation.

One of the general guidelines of STAB is to concentrate resources and know-how in the institutions involved and to avoid duplication in research work as much as possible. Today, this is more then ever necessary. The experience made in the past makes it easier now, to obtain new knowledge for solving today's and tomorrow's problems.

Strongly involved in STAB from the beginning has been Dr. Hans-Georg Knoche, who held leading positions in several German companies. His enthusiasm as well as his managing qualities were very important. He was the first Speaker of the AG STAB. Because of his merits and his 70th birthday in 1996 the $10^{th}$ STAB Symposium was dedicated to him. Dr. Knoche was expected to join the Symposium. He died two weeks before. It is the AG STAB who is very grateful to Dr. Hans-Georg Knoche.

Since 1986 the Symposia are organized every two years at different locations in Germany. In between STAB workshops are held. It is now for the first time that the contributions to the proceedings are reviewed. Many of the contributions are giving first results from the "Luftfahrtforschungsprogramm der Bundesregierung (German Aeronautical Research Programme) 1995-1998". Some of the papers report on work sponsored by the Deutsche Forschungsgemeinschaft (German Research Council), DFG. Therefore, the volume gives a broad overview over the ongoing work in this field in Germany.

The Review-Board, which is almost identical with the Program-Committee, consisted of J. Ballmann (Aachen), U. Dallmann (Göttingen), B. Ewald (Darmstadt), R. Friedrich (München), P. Hennig (München), R. Hilbig (Bremen), E.H. Hirschel (München), H. Körner (Braunschweig) - Chairman -, W. Kordulla (Göttingen), D. Kröner (Freiburg), G.E.A. Meier (Göttingen), G. Redeker (Braunschweig), P. Thiede (Bremen), S. Wagner (Stuttgart) and H.B. Weyer (Köln). Nevertheless, the authors sign responsible for the content of their contributions.

The Symposium was sponsored by the Bundesministerium für Bildung, Wissenschaft, Forschung und Technologie, BMBF, (Federal Ministry of Education, Science, Research and Technology), which is gratefully acknowledged.

The editors are also grateful to Prof. Dr. E.H. Hirschel as the general editor of the „Notes on Numerical Fluid Mechanics“ and to the Vieweg-Verlag for the opportunity to publish the results of the symposium.

The collection of the contributions was organized by Dr. H.-J. Heinemann, DLR.

H. Körner, Braunschweig; R. Hilbig, Bremen June 1997

CONTENTS

Page

Invited Papers:



Contributed Papers:

CONTENTS (continued)

CONTENTS (continued)

CONTENTS (continued)

# COOPERATION PROJECT AERONAUTICAL RESEARCH

H. Diehl

Federal Ministry of Education, Science, Research and Technology

Heinemannstr. 2, 53170 Bonn, Germany

## SUMMARY

The recent mergers in the US aircraft industry are confronting the European aeronautical industry with new challenges. US funding on the American aircraft industry is several times that spent in Europe. The German aeronautical research programme attempts to compensate for this disadvantage by greater efficiency. For this purpose, a close network is being established within which companies, universities and the national aerospace agency cooperate on research projects.

## 1. THE COMPETITORS

Since the beginning of 1997, competition in the world market for large commercial aircraft with more than 100 seats has become keener. McDonnell Douglas was forced to give up the competition with Boeing and Airbus. Due to its limited range of products, it was no longer able to keep up with the other competitors. When the company tried to embark on the production of wide-body aircraft, the operating funds available were already so scarce that relevant plans could no longer be realized. And after McDonnell Douglas failed to be awarded either of two large strategic defence contracts by the Pentagon, the merger with Boeing became inevitable.

The merger of McDonnell Douglas with Boeing has created the world's largest aircraft group with a staff of almost 200000 and an expected turnover of $48 billion in 1997. In comparison, the turnover of Airbus Industrie is far lower with about $9 billion. It has, however, to be considered that the entire turnover of Airbus Industrie is achieved in the field of civil aviation.

Publications of the White House and the Pentagon leave no doubt that it is of the greatest importance to the US Administration that the United States be the leader in worldwide aeronautical competition. This policy is supported on the one hand by the goal of the world power to be superior with its airforce everywhere and at any time. In addition the aeronautical industry is of great economic significance because it is the largest industrial foreign exchange earner of the United States.

After the end of the Cold War, the connection between military and civil aviation has become even closer. As defence budgets shrink, civil aviation gains significance in relative terms. Emphasis is placed increasingly on dual-use aspects of technological development. As a result of the concentration in the aircraft industry, the Pentagon expects savings in development and procurement to the tune of several billion dollars.

The US Administration, with its research agency NASA, and the aircraft industry are working closely together to consolidate the supremacy of American aviation.The government provides a great variety of aids to the aeronautical industry, ranging from research grants to the procurement of military aircraft to assistance with acquisition abroad through diplomatic channels.

In the United States, aeronautical research receives about ten times the support that is available to German aeronautical research. In comparison with Europe as a whole, the United States' aeronautical research still receives four times as much in support.

Support in terms of procurement contracts is difficult to state in exact amounts, but it is certainly of major significance. To name an example: the large commercial aircraft of the post-war period were modifications of military transport aircraft and thus practically free of development costs.

## 2. EUROPE

By the early 1960's, the United States had succeeded in establishing a monopoly for large commercial aircraft. Great Britain and France had, in a joint effort, launched the supersonic commercial aircraft Concorde, which, although a great technological achievement, was an economic failure.

With the AIRBUS project the Europeans started towards the end of the 1960's to break open the monopoly of the United States and to establish a competitive aircraft industry of their own. Airbus has meanwhile succeeded in conquering 30% of the world market for large commercial aircraft. Airbus is an outstanding success in the history of European cooperation. And this success was made possible by the family concept pursued by Airbus as well as by the use of the most advanced technology.

With its 37.9% share in Airbus Industrie, the German Daimler Benz Aerospace Airbus has, formally, the same strong position as the French Aerospatiale; however, it produces the rather less attractive workshares in the field of aircraft manufacturing. One of the reasons for this is the lag of the German aircraft industry resulting from the ban on activity during the post-war years. With the Jumbo, the B747 and the young B777, the United States succeeded in consolidating its monopoly in the field of wide-bodied aircraft with more than 350 seats. This monopoly guarantees certain profits and a lasting competitive edge for Boeing over Airbus. The position of Airbus is, in addition, weakened by the structure of Airbus Industrie as a so-called Groupement d'Intérêt Economique. This rather loose corporate group results in more complex and slower decision making processes which obstruct consistent cost optimization.

The Airbus group - DASA, Aerospatiale, British Aerospace and CASA - is determined to eliminate the above two weaknesses. In the course of 1997, the prerequisites will be established both for the merger of the Airbus group into a single corporate entity and for a wide-body aircraft, with a seating capacity approximately the same as that of the B747, or even larger. Transatlantic competitition has thereby gained a new dimension. This new move is hardly less significant than the foundation of Airbus.

## 3. THE NEW AERONAUTICAL RESEARCH PROGRAMME

The German aeronautical research programme 1995 - 1998 aims to strengthen the German aircraft industry for the European merger by means of technological excellence and to assist it in holding its own in transatlantic competition on the basis of European cooperation.

The four-year programme is endowed with DM 600 million, with 80% coming from the BMBF budget and 20% from that of the BMWi. The BMBF has central responsibility for the programme.

The programme has a few features different from other research and development programmes. As its goal is to strengthen industrial efficiency, it is oriented to project concepts for the year 2010. These project concepts are called lead concepts, a term which is meant to express that there is a concept on which the research and development activities of all those participating in the programme are required to be based. The lead concept indicates the direction for system firms, for the supplies industry, for the DLR and for the universities. Such lead concepts are

- megaliner,
- helicopter,
- propulsion system.

Two of the lead concepts planned at the start of the programme, the Regioliner and the Eurojet, have lost some of their significance as a result of the changes in the German aircraft industry over the last two years.

The lead concepts underline the fundamental intention of the aeronautical research programme. On the basis of the realization that the German aeronautical industry is weak compared with its American competitor and that relative funding is lower, all forces are to be joined and oriented to relevant industrial development goals.

The aeronautical research programme needs to become the cooperative project of all those working in the field of aeronautical technologies. Development niches which do not contribute to the overall goals will not receive public funding. The programme funds will be used with a view to strengthening existing cooperative activities and creating new ones. Ultimately, the goal is to design the aeronautical research environment in an optimal manner, to establish a close network of cooperative activities, with a view to using synergies and making efficient use of scarce human and financial resources.

## 4. THE LEAD TOPICS

While lead concepts indicate directions, the relevant lead topics provide for concrete cooperation, define common goals, determine interfaces, and fix agreed schedules. The lead concept of the megaliner has a number of lead topics assigned to it. The megaliner includes, for example,

- high-lift systems,
- passenger systems,
- new technology fuselage,
- electronic flight control systems,
- flexible aircraft,

to name only a few. In the case of the majority of lead topics, the system leader determines the framework within which the activities of the supplies industry, the DLR and the universities have to fit. These projects serve to extend existing cooperative activities and create new ones. The research network is becoming closer and more efficient.

The two surveys on electronic flight control systems (EFCS) and on passenger systems indicate the large number of partners contributing to these topics in a coordinated manner.

More or less the same is true of the lead concepts for propulsion systems and helicopters, although the range of cooperative activities is naturally more narrow there.

## 5. COLLABORATIVE PROJECTS

The DLR is an integral part of the aeronautical research programme. It makes available a substantial part of its capacities for application-oriented research. This may result in gaps in basic research which will then have to be closed again by means of close cooperation with the universities. In this way, the DLR is increasingly assuming the role of mediator between and turntable for industry and basic research.

Quite a number of collaborative projects have been initiated in the meantime. They include

- megaflow,
- flexible aircraft,
- CFC damage tolerance,
- Keromix (combustion chamber),
- cockpit,
- noise.

With the exception of the Cockpit collaborative project, which is coordinated by the TH Darmstadt, all other collaborative projects are coordinated by the DLR.

In these collaborative projects, several university institutes and institutes of the DRL work together on application-oriented research projects. In most cases, companies directly participate in the activity; where this is not the case, research is at least closely coordinated with industry.

Special mention should be made of the collaborative activity entitled "Megaflow". Under this activity, a computer code is being elaborated for aerodynamic design recognized by the entire aeronautical research community. In this way an efficient standardized tool for aerodynamics is being created. Also, the standard will contribute to preventing computer codes from becoming useless because of interface problems, thus resulting in the loss of man-years and funding.

## 6. THE UNIVERSITIES

The German universities hold a major potential for aeronautical research. According to a BMBF survey, about 800 persons work in the field of aeronautical research. However, this research potential is spread over some 80 institutes. The aeronautical research programme provides the opportunity for this research potential to be focused on application-oriented fields and, by means of cooperation, creates the possibility of mastering tasks which are beyond the capabilities of individual institutes. The collaborative projects are one means of achieving this goal, another are the priority programmes and special research programmes of the DFG. In the engineering sciences, the DFG seeks application orientation. Every year, about DM 20 million are provided for aeronautical research - quite a hefty sum for work done at universities.

Outside the universities, the DLR participates in the DFG priority programme entitled "Transition"; the BMBF also supports this priority with funds of its own. Special research programmes are in place, for example, on the topics of

- adaptive structures, and
- flow-structure interaction.

The DFG thus makes an essential contribution to the aeronautical research programme. Both the BMBF and the DFG plan to extend existing concepts and intensify cooperative activities.

The setting-up of the AG Stab underlined the necessity of close coordination in the important field of aerodynamics entitled "Separated flow". It anticipated today's efforts towards cooperation, and its unbroken tradition proves just how productive the cooperation is between industrial aerodynamics, the DLR and the universities.

## 7. THE NEXT FEW YEARS

In order for Germany to be able to hold its own in European and global competition, the existence of an efficient German scientific and technical infrastructure will be crucial. With regard to the Airbus and in the field of military projects, the aircraft industry is already organized at the European level. Companies will grow together to a greater extent, and a single corporate entity for the Airbus is forthcoming. In this way, German locations will be put to the test and they will need to prove that they are more efficient than others. In the research-intensive aeronautical industry, they will be able to do so only by joining forces and by resorting for contributions to the DLR and the universities. Existing resources need to be used in an optimal manner and the Cooperation Project Aeronautical Research must create the innovative network that consolidates the German locations with their highly skilled jobs. The AG Stab is part of this network.

With a view to further developing this network, two things are important:

- the transition from basic research to application, and
- the establishment of centres of competence.

Many studies have shown that Germany holds key positions in basic research while commercialization of innovations leaves something to be desired. This is where targeted measures will have to be taken. It is not necessary to do more research where Germany holds a leading position worldwide anyway without being able to turn its excellence into marketable products. The emphasis on aeronautical research must be shifted further towards application. A beginning has been made with the lead concepts and the collaborative projects. The DLR is an important mediator and turntable for basic research and industry. An attempt should be made to integrate the activities of big science with industrial companies' predevelopment work, wherever research topics lend themselves to such an approach. This process presupposes changes on both sides. Companies will be relieved of development costs, but they will become dependent on external supplies. The research establishments, on the other hand, will have to introduce industrial structures that guarantee that research goals will be implemented within a reasonable time and at reasonable cost. Staff exchanges as well as mixed industry/DLR teams appear to be important. Such close interlinkage binds workshares to German locations.

Such linkage will of course have a future only if it is based on research and development excellence. It is not enough to have a great research potential; what is also necessary is a breakthrough to European and worldwide leadership positions. It will be necessary to establish centres of competence that are based on industrial products or capable of establishing new product strategies for German locations.

The establishment of centres of competence must not stop at universities either. The research potential of universities is too big and too valuable to be neglected. Mediocrity as a result of fragmentation to a - for today's tasks - subcritical size of working groups must stop. The universities' teaching mission and the responsibility of the German Länder constitute the limits. However, modern communication techniques can, via collaborative projects, help break down these frontiers and create a virtual centre of competence that includes several sites.

The European aircraft industry will have to collaborate more closely. There is the 3E programme, a development cooperation of the Airbus group; in addition the national research centres have jointly formed the AEREA, and further promising approaches are the German-Netherlands Wind Tunnel (DNW) and the European Transsonic Wind Tunnel (ETW). Despite all the practical difficulties resulting from still existing competition and the preponderance of national problems, the integration process will continue. Germany can foster such integration and secure its national interests if it redesigns its research and development structures on the basis of cooperation among all those involved in relevant R&D in industry, big science and the universities.

**Survey: Lead topic Passenger systems**

Project participants:

| | | | |
|---|---|---|---|
| Daimler-Benz Airbus | Hamburg/Bremen | Bavaria Avionik | München |
| Dornier | Friedrichshafen | Hella | Lippstadt |
| Blücher | Erkrath | FhG | Dortmund |
| Illbruck | Leverkusen | ESW | Wedel |
| Liebherr | Lindenberg | TU Hamburg-Harburg | Hamburg |
| Nord-Micro | Frankfurt | Uni Münster | Münster |
| AOA | Gauting | TU Berlin | Berlin |

System firms, university institutes and the entire supplies industry work for passenger comfort in future wide-bodied aircraft. Government funding: currently DM 60.8 million

**Survey: Lead topic Electronic Flight Control System**

Project participants:

| | | | |
|---|---|---|---|
| Daimler-Benz Airbus | Hamburg/Bremen | DLR | Braunschweig |
| BGT | Überlingen | TU Hamburg-Harburg | Hamburg |
| Liebherr | Lindenberg | TU Berlin | Berlin |
| ZFL | Kassel | TU Darmstadt | Darmstadt |

Electronic flight control covers the whole range of knowledge in the field of aircraft. The project holds a key position in the aeronautical research programme.
Government funding: currently DM 77.3 million

# REDUCTION OF AERODYNAMIC DRAG (RaWid)
# in the national Research and Technology Programme for Aeronautics
# STATUS AFTER THE FIRST YEAR OF THE PROGRAMME

Josef Mertens
Daimler-Benz Aerospace Airbus GmbH
D-28183 Bremen

## SUMMARY

The technology programme "Reduction of aerodynamic drag (RaWid)" for high speed aerodynamics at Daimler-Benz Aerospace Airbus is sponsered by the German ministry for education, research and technology since July 1, 1995. Connected to this industrial programme are the cooperation programmes "MEGAFLOW" under leadership of the DLR and "Transition" by the DFG, and several contributions by DLR and universities.
The programme is oriented towards technologies required for a MEGALINER which gains momentum by the ambitious plans for a new large Airbus A3XX.
In the first year new technological steps were undertaken in theory, design and experiment. Some critical steps were verified by wing designs checked in wind tunnel tests.

## INTRODUCTION

The programme "Reduction of aerodynamic drag (RaWid)" is oriented towards a MEGALINER which gains momentum by the ambitious plans of Airbus Industry for the A3XX. The A3XX can only compete against Boeing's planned B747-500/600, if it becomes much more efficient for the airlines (15-20%). Such a big improvement in economy is impossible to reach only by reduction of the aircraft's price -especially in Europe-, but it requires application of new cost reducing technologies, as it was all the time with Airbus.

RaWid provides aerodynamic technologies for transonic speeds. Industrial goals are

- performance improvement for large civil aircrafts via
  reduced fuel burn and operating costs, increased operating flexibility (payload/ range capability) and lowered emissions;
- productivity improvement of the development process via better tools providing
  secured geometry definition and warranted aerodynamic data at go-ahead.

Technologies investigated to reach these goals concern

- aerodynamic optimisation of the whole aircraft incl. interferences of components,
- variable camber including load control for wing development,
- reduction of friction drag,
- tools required for aerodynamic design like CFD, transition prediction, numerical optimizers, measurement and model techniques

and include design exercises and experimental validation.

The -then still planned- technology programme RaWid was presented on the 9$^{th}$ STAB-Symposium in Erlangen [1]. Since July 1, 1995, the programme leader Daimler-Benz Aerospace Airbus (DA) is sponsored by the German ministry for education, research and technology (BMBF). RaWid is coordinated with the parallel aerodynamic programmes "High Lift Concepts (HAK)" and "A320 Laminar Fin (ELAS)" [2]; the latter being the first flight test of hybrid laminarisation on large aircraft in Europe.

In the framework of RaWid, the BMBF funds investigations at universities. Several and RaWid contributions by DA are included in the cooperation programme MEGAFLOW; its objective is the development of a unified CFD code under leadership of the DLR. This cooperation was established with start of RaWid; it was publicly presented in a kick-off meeting on Jan. 30, 1996. Many other activities of the DLR are oriented towards the goals of RaWid, especially the DLR programme "Transition" [3] and new or improved measurement techniques. The German research foundation (DFG) supports new methods of cooperation between universities, research labs and industry in the priority research programme (SPP) "Transition"; it is supportet by DA with RaWid and DLR and started in spring 1996 [4].

## THEORY

Emphasis of the theoretical work was on contributions to MEGAFLOW and to the SPP "Transition". For both several presentations are given in this symposium. Here only some examples -mainly DA contributions- are referenced:

First test cases of ATTAS flight tests were investigated using the PSE code COPS. Cases dominated by Tollmien-Schlichting (TS) waves [5] agreed well with $e^N$-predictions. PSE analysis promises improvements in the understanding of recepticity, e.g. concerning free stream disturbances. Cases dominated by cross flow (CF) waves [6] did not yet allow for a satisfactory evaluation.

With DLR Göttingen the design of the A320 hybrid laminar fin was investigated using the NOLOT/PSE codes [7]. Analyses of TS waves show acceptable agreement with linearized theory; an explanation for the unsatisfying effect of curvature in local theory can be given. Results for CF waves are not conclusive; tendencies seem to be contradictory to experiments. DLR investigated many other topics in the "Transition" programme contributing to RaWid goals, see [3, 8] and many symposium presentations.

In MEGAFLOW, first test cases for the Navier-Stokes and Euler solver FLOWer were performed. The first version of a postprocessor was provided allowing the analysis of the physical drag components: induced, wave and friction drag. It is the starting point for the MEGADRAG development at TH Darmstadt. Large improvements in computing speed have been achieved by parallelisation, which for numerical optimisation is handled by the MEPO system at TU Braunschweig [9].

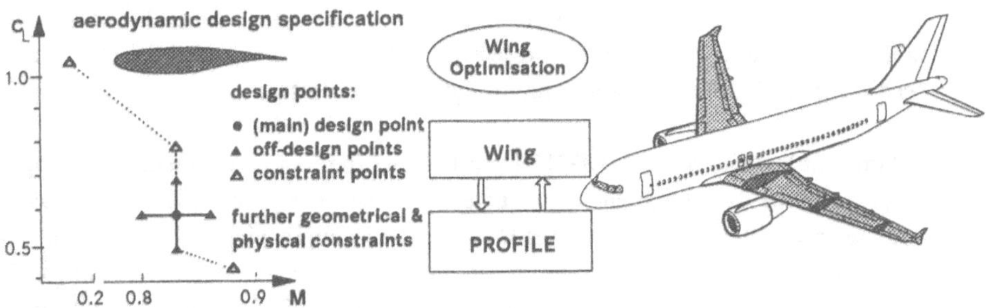


Figure 1: Aerodynamic Wing Design Optimisation AEROPT

The DA knowledge on numerical optimisation of airfoils was compiled in the code AEROPT (AERofoil OPTimisation) by A. Van der Velden with extensive documentation. A special study was devoted to geometry generation suited for optimizers. Multipoint design of airfoils for transonic viscous infinite swept or tapered wing flow is standard at DA; buffet, dive and low speed ($C_{Lmax}$) constraint points are included. It allows advanced wing designs we were unable to realize before, so e.g. higher design Mach numbers at lower sweep angles, and improved variable camber (VC) and laminar airfoils. A complete wing design was performed using multipoint profile optimisations coupled by a simpler 3D calculation for the induced effects and load distribution (fig. 1); the MEPO parallelisation speeds up turn around in a work station environment [9].
Inverse and partially inverse designs (including constraint handling) are possible using this method. Next step is introduction of *transonic aerodynamic* design goals into the DA predesign environment MIDAS (Multi Disciplinary Integration of Deutsche Airbus Specialists) [10].

## AERODYNAMIC DESIGN AND WIND TUNNEL TESTS

In application of the new technologies, several wings were designed for testing in wind tunnels:

- A first laminar glove **"HLFC"** (HLFC = Hybrid Laminar Flow Control) for testing of hybrid laminar wing flow was designed. The glove is positioned between the engines on the starboard side of an A340. Aerodynamic high speed design was checked in a transonic test at ARA, still without suction. Aim of this test was the verification of the pressure design, especially with respect to the fairing areas at both sides of the glove, which have to shield the laminar area against the turbulent pressure distributions of the A340 wing and the influences of the nacelles [11].
  The same **"HLFC"** glove with Krüger high lift devices was tested in the DNW to estimate performance of the high lift devices and handling qualities compatible with the unchanged design of the port side wing [12].
  Both wind tunnel tests confirmed the selected approach and provided the information needed for final glove design in a future European cooperation.

- A turbulent/laminar wing **"HLFC1"** for a MEGALINER was designed. This wing is to start service as a turbulent wing. Later on, when operational HLFC technology becomes available, it will offer HLFC performance improvements by application of a new suction nose without change of the wing box structure. ARA tests (still without suction) demonstrated limits of turbulent/laminar wing design: For the given conditions ($M_{Cr}$ = 0.85, $M_{Mo}$ = 0.88, MEGALINER size) and using the present knowledge on HLFC physics, the laminar designed wing provided insufficient performance in the turbulent mode, which will limit performance in the most important first years of service. But performance gains were confirmed for an ab initio laminar wing designed for the a.m. high Mach numbers, especially for smaller aircraft. For lower design Mach numbers, even a turbulent/laminar wing may be designed. Because requirements for a MEGALINER wing refurbishment cannot be met at this time, further HLFC wing designs were postponed until flight test data of the A320 HLF Fin programme (ELAS) are available.

- An optimized turbulent fixed camber wing **"TFC2"** for a MEGALINER was designed. This is a study for wing sweep reduction (TFC2 = 30°), compared to DA's wing improvement study "TFC1" (33°) (fig. 2). The TFC2 wing was tested in the ARA transonic wind tunnel and verified the predicted improvement potential.

- A turbulent variable camber (VC) wing **"TVC1"** was designed, with same planform as TFC2. Wind tunnel tests will be in November/December 1996.

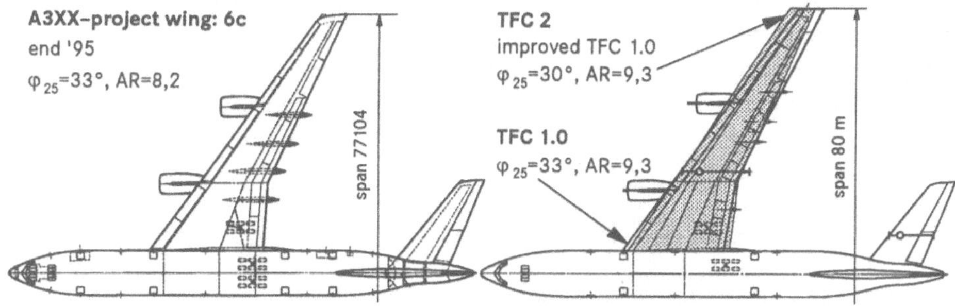


Figure 2: MEGALINER wing planforms

Both wings (TFC2, TVC1) will be followed next year by a further improvement step (TFC3, TVC2) to exploit the possibilities of nonlinear lofting and compare optimized fixed and VC wings with suited wing size adaption.

The left and right wings in fig. 2 result from different design concepts. Left wing "6c" is much like the Boeing concept:

large wing area, large sweep, high taper, lower aspect ratio, thin wings.

In contrast, classical Airbus concepts like the TFC2 have

high aspect ratio, lower taper, lower sweep, smaller wing area, thicker wings.

Both concepts show their own benefits and drawbacks, like:

- Fuel volume depends on wing area and thickness. It can set lower wing size limit, but not for a MEGALINER. Because volume rises by 3rd power of scale, lift and wing area only by 2nd, a high performance MEGALINER is not fuel limited.
- Wing weight rises with span, area, sweep, and decreases with thickness and taper.
- Aerodynamic performance increases with span (induced drag) and decreases with wing area (friction drag) and thickness.
- Speed performance increases with sweep and decreases with thickness.
- Stall characteristics become dangerous with high taper, because the outboard wing is highly loaded at high lift conditions.
- Low speed performance increases with span and decreases with sweep.
- Tailplane areas (i.e. weight and drag) rise with wing area and sweep.
- Landing gear integration becomes simpler at reduced sweep.
- Aeroelastic deformations decrease with wing area, thickness and increase with span, sweep and camber.

The Airbus concept provides better aerodynamic performance which becomes most important for long range. But it requires a highly sophisticated airfoil technology to enable wings with lower sweep and increased thickness. Especially the huge MEGALINER with a span of 80 m requires new technologies to limit wing loads and control deformations. This is the Variable Camber (VC) technology (fig. 3). It provides performance improvements, but more important are expansion of the flight enveloppe, load alleviation, correction possibilities for deformation and future adaptions like new engines.

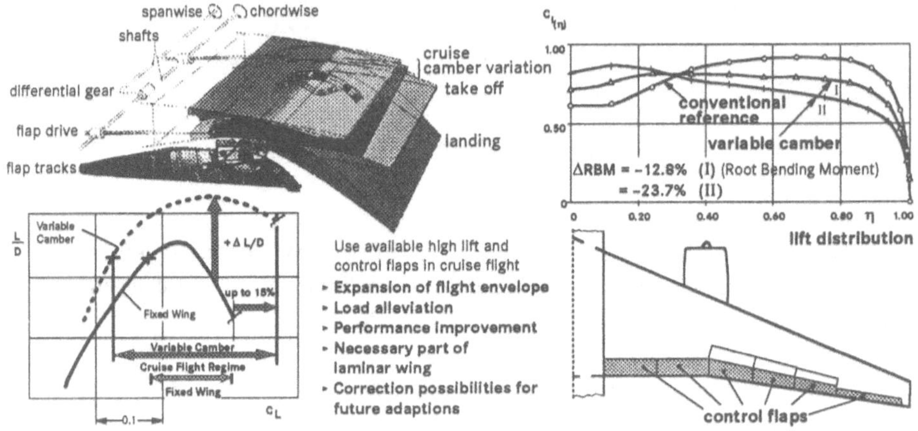


Figure 3: Variable Camber (VC) Technology

In order to reduce shock-boundary layer interference, several concepts were investigated. Best solution seems to be a wing contour bump at the shock position to enable weakening of the shock with only limited thickening of the boundary layer [13]. First tests in the wind tunnel were promising (fig. 4).

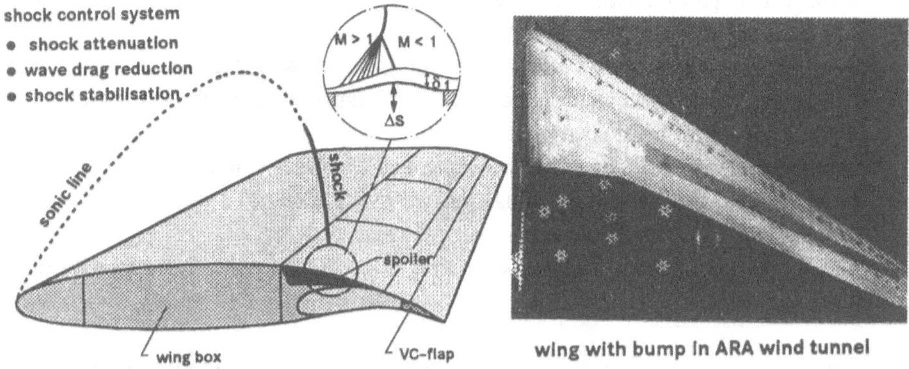


Figure 4: Shock-Boundary Layer Interference Control by a Bump

## INTERDISCIPLINARY CONTROL SYSTEM STUDIES

New aerodynamic technologies like HLFC or VC require interdisciplinary cooperation in the whole development process. In RaWid these investigations started in parallel with the aerodynamic development.

For the turbulent/laminar wing "HLFC1", the requirements of all disciplines were collected and a first balancing towards a solution was performed. The wing nose has to provide: suction for laminar flow including suction energy; deicing during climb, descend and on the ground; store Krüger flaps used in low speed flight; prevent contamination.

- Aerodynamics defines the suction flow, suction distribution, surface quality and high lift (Krüger) sections.
- System departments determine energy and space required for several solutions of the suction system and Krüger flaps; maintainability and reliability are considered.
- Structure departments investigate limits for blowing pressures and define the required space for the suction structure, system mounting and wing structure implications. New structural concepts for suction systems are envisaged.

It was shown, that a MEGALINER with suction on the upper wing surface can be realized, although space allocation for high lift system and suction ducts remains a challenge.

## REFERENCES


[1] Mertens, J.: "Reduktion aerodynamischer Widerstand (RaWid) im Luftfahrtforschungsprogramm des BMFT". DGLR-Bericht 94-04 "Strömungen mit Ablösung", 9. DGLR-Fach-Symposium, 4.-7. Oktober 1994, Erlangen, S. 323-327.

[2] Henke, R.: "Das Programm 'A320 HLF Fin': Eine multidisziplinäre Herausforderung neuer Qualität". DGLR-Bericht 94-04 "Strömungen mit Ablösung", 9. DGLR-Fach-Symposium, 4.-7. Oktober 1994, Erlangen, S. 317-322.

[3] Hein, S., Bertolotti, F.P., Gordner, A., Stolte, A., Koch, W., Bippes, H., Lerche, T., Wiegel, M., Dallmann, U.: "Status of DLR's 'Non-Empirical Transition Prediction' Project". DLR-Report IB 223-96 A 37.

[4] Wagner, S.: "Das Verbund-Schwerpunktprogramm 'Transition'". DGLR-JT96-102, Jahrestagung DGLR 1996, Dresden, Jahrbuch DGLR 1996 II, pp. 697-705.

[5] Schrauf, G., Herbert, Th., Stuckert, G.: "Evaluation of Transition in Flight Tests Using Nonlinear Parabolized Stability Equation Analysis". Journal of Aircraft, **33**, 3, May-June 1966, pp. 554-560.

[6] Herbert, Th., Schrauf, G.: "Crossflow-Dominated Transition in Flight Tests". AIAA-Paper 96-0185, 34th Aerospace Sciences Mtg, Jan. 15-18, 96, Reno, NV.

[7] Hein, S., Bieler, H., Gordner, A., Simen, M.: "NOLOT/PSE Transitionsanalysen zur Entwicklung von HLFC-Absaugesystemen". DGLR-Fachaussch. T 2.2 'Experimentelle Aerodynamik', 2.-3.5. 1996, TU Berlin, ILR Mitt. 311 (1996), Ed. W. Nitsche, R. Henke, pp. 79-89.

[8] Bippes, H., Fischer, M., Wiedemann, A., Bertolotti, F.P.: "Experimente zur Beeinflussung der Querströmungsinstabilität mit Hilfe von Absaugung durch perforierte Wände". DGLR-Fachaussch. T 2.2 'Experimentelle Aerodynamik', 2.-3.5.1996, TU Berlin, ILR Mitt. 311 (1996), Ed. W. Nitsche, R. Henke, pp. 43-50.

[9] Axmann, J.K., Hadenfeld, M., Frommann, O.: "Parallel Numerical Airplane Wing Design". 10. DGLR/AG STAB-Symposium, 11.-13. Nov. 1996, Braunschweig.

[10] Van der Velden, A., von Reith, D.: "Multi-Disciplinary SCT Design at Deutsche Aerospace Airbus". 7th European Aerospace Conference EAC'94 "The Supersonic Transport of Second Generation", Toulouse, Oct. 25-27, 1994, pp. 449-464.

[11] Schmid-Göller, S., Hansen, H.: "Design of a Laminar Glove for the A340 and First Results of a High Speed Wind Tunnel Test". 10. DGLR/AG STAB-Symposium, 11.-13. Nov. 1996, Braunschweig.

[12] Gramlow, R.: "Design of a Krüger for the A340 Laminar Glove and First Low Speed Wind Tunnel Results". 10. DGLR/AG STAB-Symposium, 11.-13. Nov. 1996, Braunschweig.

[13] L. Gildemeister, G. Dargel, H. Hansen: "Theoretische Grundsatzuntersuchungen zum Thema 'Adaptive Hautstrukturen im Verdichtungsstoß'". DGLR-Fachaussch. T 2.2 'Experimentelle Aerodynamik', 2.-3.5.1996, TU Berlin, ILR Mitt. 311 (1996), Ed. W. Nitsche, R. Henke, pp. 69-78.

# National CFD Project MEGAFLOW
# - Status Report -

Norbert Kroll
DLR, Institute of Design Aerodynamics, Lilienthalplatz 7
D-38108 Braunschweig, Germany

## Summary

In the framework of the national air transport research program the CFD project MEGAFLOW was initiated. Its goal is the development and validation of a dependable and efficient numerical tool for the aerodynamic simulation of complete aircraft in cruise as well as in start and landing configurations, which meets the requirements for industrial implementation. This requires a concentrated cooperation with the aircraft industry, DLR and universities. The project started mid 1995 and will be finished at the end of 1998. This paper shortly outlines the plan of action and presents the current status of the project.

## Introduction

In the last decade, considerable progress has been made in the development and validation of numerical simulation tools for aerodynamic applications. As a result, next to the traditional tools such as wind tunnel and flight testing, Computational Fluid Dynamics (CFD) is widely accepted as an integral part of the aerodynamic design procedure of new aerospace vehicles. Numerical flow simulation does not only offer reduction of turn-around times and costs, but it also enables flow and geometry parameter variations which are not possible with wind tunnel testing.

Despite the recent advances, CFD still suffers from deficiencies in accuracy, robustness and efficiency for complex applications, such as complete aircraft flow predictions. From industry's point of view, numerical simulation tools are expected to deliver detailed viscous flow analysis for complete configurations at realistic Reynolds numbers, prediction of aerodynamic data with assured high accuracy, fast response time per flow case at acceptable total costs as well as aerodynamic optimization of the main aircraft components. In order to meet these requirements in earlier and earlier stages of the design process, considerable improvements of the CFD methods currently available in industry are necessary. One of the largest problems in computational fluid dynamics is the generation of appropriate computational meshes for complex configurations. Another problem for many industrial applications relates to the insufficient modeling of turbulent flows and the associated high numerical costs. State of the art in industry is the calculation of inviscid flows around complete aircraft configurations based on the solution of the Euler equations. The simulation of viscous flows by solving the Reynolds-averaged Navier-Stokes equations is limited to basic configurations such as isolated wings or wing-body configurations using a rather simplified modeling of turbulence. Consequently, current computational methods do not allow a critical evaluation of the aerodynamic potential and performance limits of realistic aircraft designs with sufficient accuracy.

Within the framework of the national air transport research program, the project MEGAFLOW was initiated with the objective to enhance the capabilities of current CFD methods and to support the establishment of numerical simulation as an effective tool in the industrial design process. The goal of the project is the production of a dependable, efficient and quality controlled program system for the aerodynamic simulation of complete transport aircraft in cruise as well as take-off and landing configuration. The software system, which is based on block-structured grids and contains of the grid generator MegaCads and the Reynolds-averaged Navier-Stokes solver FLOWer, should meet the requirements of industrial implementation. The performance of the complete system will be validated using industry relevant applications. The extensive development and validation work is carried out in concentrated cooperation with aircraft industry, DLR and universities. The MEGAFLOW project is partially funded by the German Federal Ministry for Education, Science, Research and Technology (BMBF). It started mid 1995 and will be finished at the end of 1998. The founding partners are Daimler-Benz Aerospace Airbus, DLR and the universities of Berlin, Braunschweig, Darmstadt and München. Since the start of the project, additional partners (TZN, GMD, Universität Stuttgart, RWTH Aachen and Hochschule Bremen) have been associated with the MEGALOW project.

The MEGAFLOW project consists of 5 major activities [1]

- development of the block-structured grid generator MegaCads
- improvement and enhancement of the block-structured Navier-Stokes solver FLOWer
- validation of the software system for industrial applications
- software quality management
- alternative methods.

This paper presents the current status of the project. In some areas, detailed presentations of the achievements have been presented at this conference by individual papers.

## Results

### Grid Generator MegaCads

The interactive grid generation system MegaCads was developed for the construction of multi-block grids for complex configurations [2]. The complete grid-generation sequence is performed in a parametric manner and is stored in a protocol file using a simple script language. In combination with interactive editing, the sequence can be reused for similar grid generation tasks with a minimum of user interaction. Implemented basic CAD techniques enable the design of nearly arbitrary block topologies in space and on curved surfaces.

In the first phase of the MEGAFLOW project, emphasis was put on the improvement and enhancement of techniques for surface and volume grid generation for complex applications. A specific projection technique was developed and implemented which allows the generation of grids on arbitrary surfaces. Furthermore, different algorithms for the determination of the surface/surface intersection of aircraft components were integrated. Efficient two and three-dimensional algebraic grid-generation methods are available, which enable the construction of initial grids of good quality. In order to smooth the algebraic grids, elliptic techniques were implemented, allowing different controls for the point distribution at grid boundaries. Activities concerning the integration of multigrid techniques were started, in order to reduce the computational costs of elliptic grid-generation techniques. A two-dimensional β-version clearly demonstrated the potential of the multigrid acceleration for elliptic grid generation.

The graphical user interface and the visualization techniques were improved, in order to reduce and simplify the necessary user interaction. In addition, a code was developed which almost automatically detects the block topology and interblock connections for the generated grid. This tool closes the interface between grid generator and flow solver.

Besides the algorithmic and data-structure development of MegaCads, three-dimensional Navier-Stokes grids around various MEGAFLOW related configurations were developed. Figure 1 shows the grid around the DLR-F6 wing/body/pylon/nacelle configuration. This Navier-Stokes grid will be used for the validation of the MEGAFLOW software system for the cruise flight configuration. The grid consists of 55 blocks and 3.2 million grid points. In order to accurately resolve the boundary layers, embedded C-type grids were selected and constructed around the wing and nacelle.

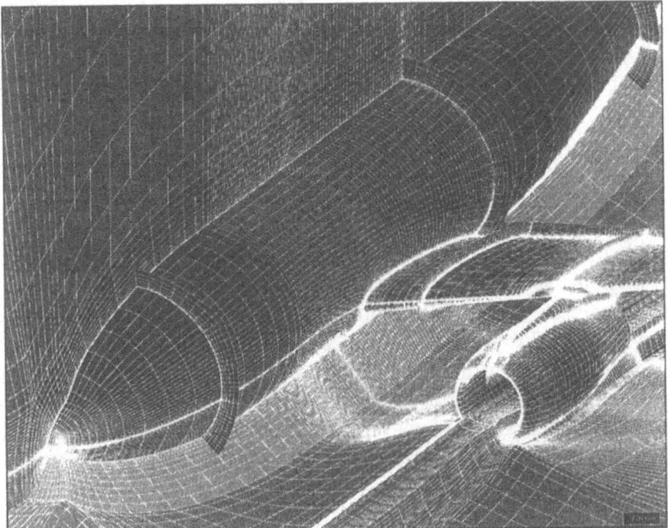

**Fig. 1** Navier-Stokes grid for DLR-F6 configuration

**Navier-Stokes Solver FLOWer**

The FLOWer code solves the compressible, three-dimensional Reynolds-averaged Navier-Stokes equations in thin-layer form. It uses a cell-vertex, finite-volume, central-difference formulation on block-structured meshes and is designed to simulate flows around complex aerodynamic configurations. FLOWer has been directly evolved from the DLR CEVCATS code [3], and includes special features from the DASA Airbus MELINA code and the DASA-LM IKARUS code. In the framework of the BMBF project POPINDA [4], in close cooperation with the German aeronautical industry and German National Research Center for Information Technology (GMD), it has been extended to a parallel, fully portable code.

Within the MEGAFLOW project, the development activities for the flow solver are concentrated on the improvement of efficiency and accuracy for viscous flow simulations. For large aerodynamic applications, parallel computing is an important ingredient of the numerical simulation. It not only enables the reduction of turn around times and computational costs, but also offers the possibility to conduct simulations with large CPU and memory requirements. Since the effective use of parallel computers requires load balancing of the computation, a tool was developed which splits an arbitrary block structure based on the number of processors available and modifies the block topology accordingly. Figure 2 shows an inviscid calculation of a wing/body/pylon/nacelle configuration on the IBM SP2 parallel computer. It is clearly demonstrated that a reorganization through block splitting of the basic block-structured grid created by the grid generator leads to a significant reduction of the computational time.

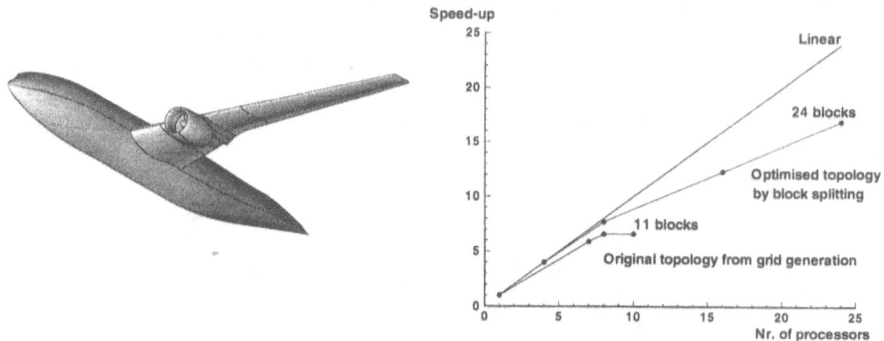


**Fig. 2** Performance increase through parallelization for a generic transport configuration

In addition, based on the parallel FLOWer Version 113, a technique for self-adaptive local-grid refinement was implemented. Those parts of the grid will be refined where high discretization errors of the flow quantities occur. The communication library CLIC-3D, developed by GMD in the POPINDA project, clusters the detected cells, creates a new block structure and distributes it in a load-balanced way to the allocated processors. The new block structure is embedded as a finer level in the multigrid structure. The first results obtained for a wing/body configuration are very encouraging. However, the fine/coarse grid interface treatment has to be improved before the local grid refinement technique can be made operational in the central FLOWer version.

The multiblock grids for complex configurations often contain geometrical and topological singularities. In a cell vertex code, a special treatment of singular points or lines is necessary in order to avoid accuracy and efficiency losses. Therefore, a procedure was developed and implemented which detects singularities in a given grid and discretizes consistently the corresponding inviscid, viscous and dissipative fluxes.

For the calculation of incompressible flows, a preconditioning of the compressible flow equations was implemented [5]. This technique allows an improved accuracy and efficiency for low Mach number cases, such as high-lift flows.

With respect to unsteady calculations, the FLOWer code was extended to simulate time-accurate flows around a rigid body in arbitrary motion. The motion is taken into account by a transformation of the governing equations. In order to bypass the severe time-step restriction associated with explicit schemes, a simple implicit time-stepping method, known as the dual-time stepping approach, was implemented. In combination with multigrid acceleration, this scheme allows very efficient calculations of viscous time-accurate flows. For oscillating wings a speed-up of about three orders of magnitude compared to the basic explicit scheme was obtained. It could be demonstrated that the combination of implicit dual time-stepping method and parallelization is a very promising approach for predicting unsteady aerodynamic loads [6]. Furthermore, the code was extended to allow grid deformation. This enables the FLOWer system to be used for aeroelastic simulations.

The improvement of turbulence modeling is one of the major activities of MEGAFLOW, since the prediction of aerodynamic performance of aircraft or aircraft components requires an accurate simulation of the viscous, turbulent effects. Based on the experience gained with the DLR CEVCATS code, the k-ω turbulence model of Wilcox was implemented as a standard two-equation transport model. The model was integrated in a generalized way, in order to allow an effective implementation of other, improved models which will be developed in the framework of the MEGAFLOW project. Furthermore, a point implicit treatment of the source terms of the turbulence transport equations was implemented resulting in an improved efficiency of the solution method. Investigations for a realistic wing/body configuration showed [7], that in contrast to the algebraic model, the k-ω transport model delivers results which are independent of the number of blocks and the block topology used in the calculation (see Figure 3). The current implementation of the k-ω model forms the basis for the complex applications planned in the MEGAFLOW validation phase.

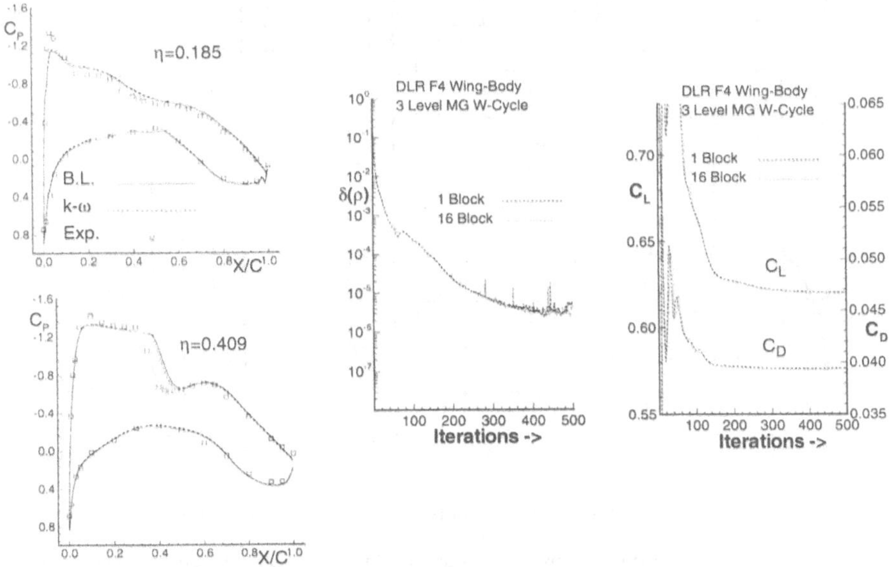


**Fig. 3** DLR F4 wing/body, M=0.75, α=0.93°, Re=3x10⁶, results with k-ω model

With respect to aerodynamic optimization, several activities were carried out. Based on existing optimization strategies available at Daimler-Benz Aerospace Airbus and Technical University of Braunschweig, an environment for parallel optimization (EPOS) was designed and implemented [8]. The FLOWer code was integrated as an analysis tool. The system was successfully tested on a workstation cluster by the aerodynamic shape optimization of airfoils. Currently EPOS is being extended to the MEPO system (Modular Environment for Parallel Optimization), which will allow parallelization on different levels of the optimization procedure, including enhanced load balancing. Further activities were concentrated on the investigation of different procedures for geometry parametrization and of optimization strategies for drag minimization of transonic airfoils [9]. In addition, work was devoted to the development of an efficient optimization procedure based on the solution of the adjoint Euler equations [14].

**Validation**

The MEGAFLOW system was already applied to several industry relevant configurations. Airbus put much effort in the comparison of the capabilities of the FLOWer code and the in-house code MELINA for viscous flow simulation. For this purpose a detailed test matrix was defined, which includes various relevant configurations and different flow conditions covering important aspects of transport aircraft. Where available, the comparisons were supplemented with experimental data. These investigations showed that FLOWer delivers results which are at least of the same quality those produced by MELINA. In many cases, the FLOWer results compare better with experimental data (Figure 4). Moreover, FLOWer shows less sensitivities with respect to numerical parameters and the code is almost 40% more efficient on the NEC SX-4 supercomputer. As a consequence, Airbus replaced MELINA with FLOWer for production use.

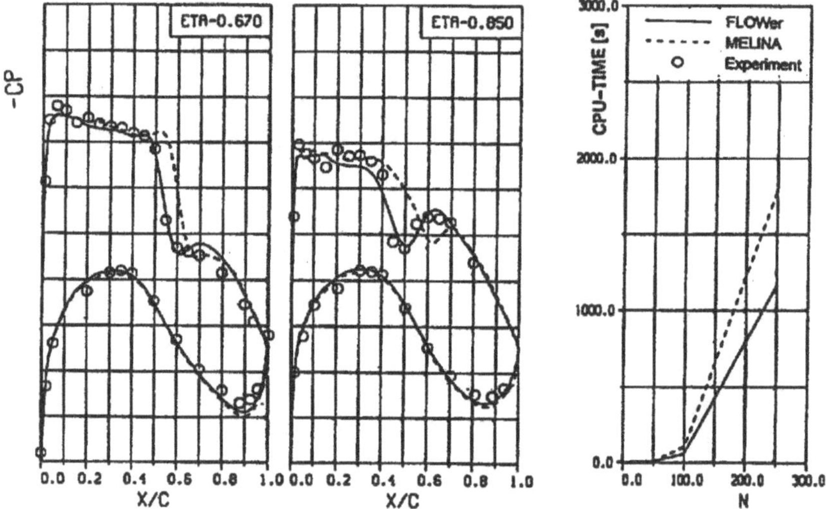


**Fig. 4** Navier-Stokes computation for wing/body configuration

With respect to high lift configurations, in the past the MEGAFLOW system was successfully applied to several airfoils using two-equation transport turbulence models. As a first step towards the validation of three-dimensional high-lift configurations, a generic configuration consisting of wing and flap was calculated [10]. The grid generated by MegaCads has 13 blocks and about 2.2 million points. The calculation was carried out with the k-ω turbulence model. Figure 5 presents the pressure contours and the stream lines on the upper surface of wing and flap. The vortex which is generated at the side edge of the flap is also shown. The comparison of the calculated and measured surface pressure in two wing sections near the side edge of the flap demonstrates the good agreement of prediction and experiment.

The validation process also includes the improvement of turbulence modeling. Activities are devoted to the assessment of linear two-equation transport models for modeling complex aerodynamic flow phenomena, such as shock induced separation and large flow separation at maximum lift condition [11]. Various transport models were investigated for relevant two dimensional flows. The results indicated that a low Reynolds number k- ω model is superior to the other usually used linear models. The investigations of the high lift configurations showed, that improved results can be obtained using a locally controlled modification of the production of the turbulence energy. In addition to the work related to the two-equation turbulence model, the one-equation model of Spalart-Allmaras was tested and prepared for the implementation into the FLOWer code. For this purpose a generalized procedure for the determination of the wall distance was developed, which is independent from the block topology of the grid. Furthermore, the capability of algebraic models to simulate high lift flows was investigated.

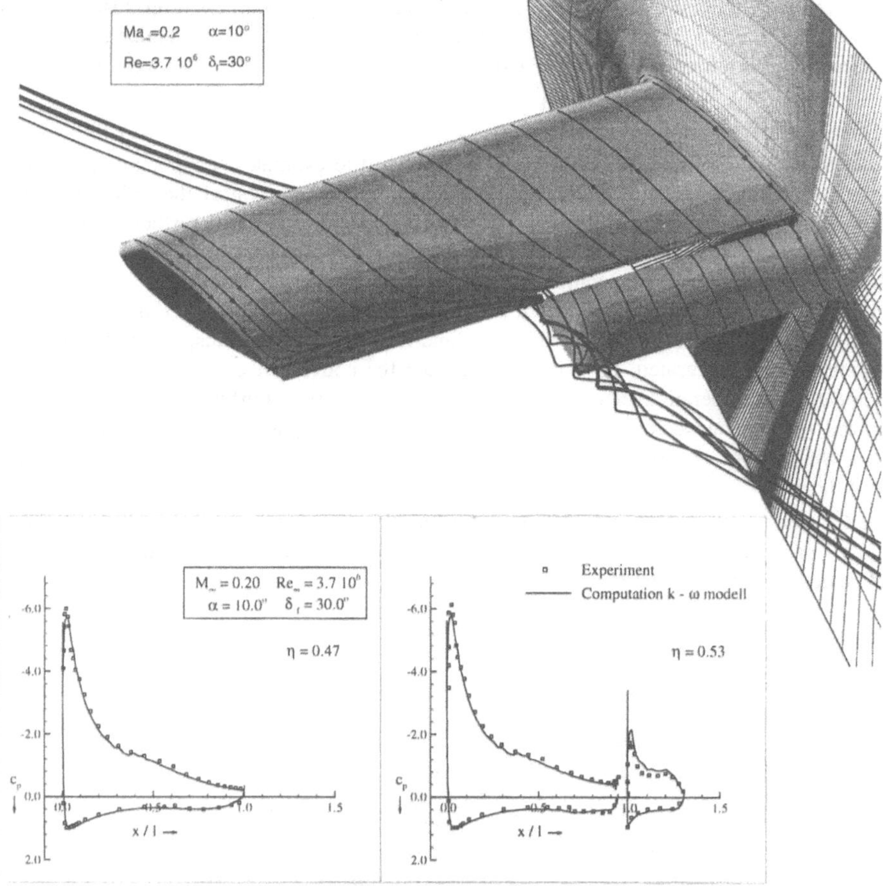


**Fig. 5** Navier-Stokes computation for wing/flap configuration

Based on experience gained at Airbus, the software package MEGADRAG was developed, which determines the contributions to the drag from the computed flow solution. The modules for calculating the wave drag based on the entropy rise through the shock and the friction drag using the entropy flux balance in the flow field were tested for airfoil flows [12]. Since the code uses the same data structure as the flow solver FLOWer, it can be directly coupled as a postprocessing tool.

## Software quality management

In order to insure the supportability, dependability and further expendability of the MEGA-FLOW system, modern quality management and software-engineering methods will be implemented. So far, activities were mainly focused on the flow solver FLOWer. In close cooperation with the DLR software quality assurance department, mandatory procedures were established to integrate the software components developed by the different partners into the central version. Guidelines were developed for programming, documentation, testing, error and problem reporting. For software version control a configuration management tool was introduced. For both MegaCads and FLOWer installation and user handbooks were written.

## Alternative methods

In order to be open to other scientific developments and to estimate their potential, research activities are being devoted to alternative approaches for the solution of the Euler and Reynolds-averaged Navier-Stokes equations. The work is concentrated on the further development of the unstructured hybrid DLR $\tau$-code for transport aircraft applications [13]. The code uses hybrid grids, which consist of a combination of prismatic and tetrahedral grids. Various spatial discretization schemes were implemented and tested. For the simulation of turbulent flows, the algebraic turbulence model of Baldwin-Lomax and the one-equation transport model of Spalart-Allmaras were integrated. In order to accelerate the convergence, a multigrid procedure based on the agglomeration of control volumes was developed. Furthermore, the code was optimized for use on vector and parallel computers. An adaptation procedure was implemented, which in the case of viscous flows allows a directional grid refinement in the prismatic layer of the grid.

Figure 6 shows results for the simulation of the turbulent flow around the ONERA M6-wing using the one-equation turbulence model. A considerable improvement of the convergence rate could be obtained with the multigrid method.

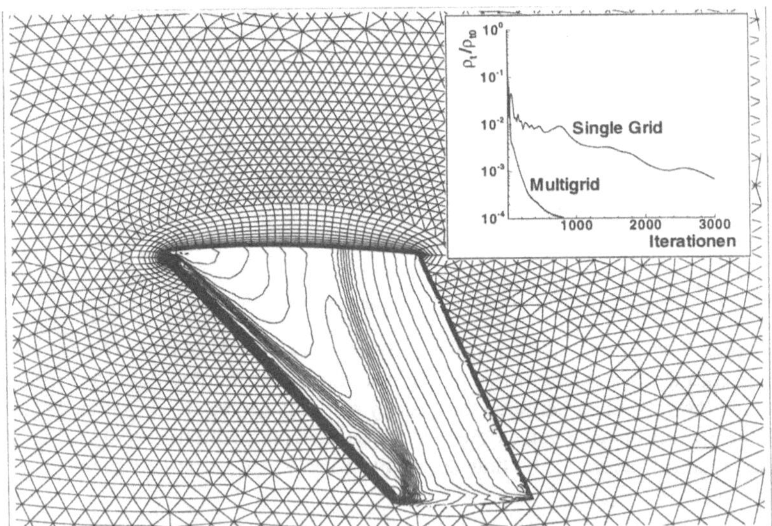


**Fig. 6** Results of hybrid $\tau$-Code for ONERA M6-wing, M=0,84, $\alpha$=3,06$^0$, Re=$1.1 \times 10^7$

## Conclusions

The objective of the BMBF project MEGAFLOW is the production of a dependable, efficient and quality controlled program system for the aerodynamic simulation of complete aircraft. The project is a close cooperation of aircraft industry, DLR and universities which has run smoothly and has been very productive. So far, activities were concentrated on the development and improvement of the grid generator MegaCads and the flow solver FLOWer, with the next major goal of meeting the July 97 milestone computation of a complete aircraft under cruise conditions.

## Acknowledgment

The author would like to thank all colleagues of the MEGAFLOW project who provided him material for this status report.

## References


[1] Bartelheimer W.: *Projektbeschreibung MEGAFLOW.* DLR-IB 129-96/8, 1996.

[2] Brodersen, O.; Hepperle, M., Ronzheimer, A.; Rossow, C.-C.; Schöning, B.: *The Parametric Grid Generation System MegaCads.* Proceedings of 5th International Conference on Numerical Grid Generation in Computational Field Simulations, Ed. Soni, B.K.; Thompson, J.F.; Häuser, J.: Eiseman, P.R., NSF Center Mississippi, pp 353-362, 1996.

[3] Kroll, N.:, Radespiel, R.; Rossow, C.-C.: *Accurate and Efficient Flow Solvers for 3D Applications on Structured Meshes.* Lecture Series 1994-05 of the van Karmann Institute for Fluid Dynamics, Brussels, March 1994.

[4] Eisfeld, B.; Bleecke, H.-M.; Kroll, N.; Ritzdorf, H.: *Parallelization of Block Structured Flow Solvers.* AGARD R-807, pp 5.1-5.20, 1995.

[5] Radespiel, R.:, Turkel, E.; Kroll, N.: *Assessment of Preconditioning Methods.* DLR-FB 95-25, 1995.

[6] Heinrich, R.; Bleecke, H.-M.: *Simulation of Unsteady, Three Dimensional, Viscous Flows Using a Dual Time Steping Method.* This Volume.

[7] Monsen, E.; Rudnik, R.; Bleecke, H.: *Flexibility and Efficiency of a Transport-Equation Turbulence Model for Three-Dimensional Flow.* This Volume.

[8] Axmann, J.K.; Hadenfeld, M.; Frommann, O.: *Numerical Airplane Wing Design with the MEPO Parallel Adaptive Optimization System.* This Volume.

[9] Bartelheimer W.; Orlowski, M.; Wild, J.: *Investigations on Aerodynamic Optimization of Airfoils.* This Volume.

[10] Rudnik, R.; Ronzheimer, A.; Schenk, M.; Rossow, C.-C.: *Berechnung von 2- und 3-dimensionalen Hochauftriebskonfigurationen durch Lösung der Navier-Stokes Gleichungen.* Tagungsband der DGLR Jahrestagung, Dresden, 1996.

[11] Rung, T.; Thiele, F.: *Computational Modelling of Complex Boundary-Layer Flows.* Ninth International Symposium on Transport Phenomena (ISTP-9) in Thermal-Fluids Engineering, Singapore, 1996.

[12] Kreuzer, P.; *MEGADRAG a Postprocessing Tool for Drag Analysis of Flow Fields.* This Volume.

[13] Gerhold, T.; Friedrich, O.; Evans, J.; Galle, M.: *Calculation of Complex Three-Dimensional Configurations Employing the DLR τ-Code.* AIAA 97-0167, 1997.

[14] Jameson, A., *Optimum aerodynamic Design using CFD and Control Theory,* AIAA Paper 95-1729, AIAA 12th Computational Fluid Dynamics Conference, San Diego, CA, June 1995.

# Laminar-Turbulent Transition Research at DLR

Uwe Ch. Dallmann

DLR – Institut für Strömungsmechanik,
Bunsenstr.10
D-37073 Göttingen

## 1 Summary

DLR institutes and divisions in Braunschweig, Göttingen and Cologne are presently involved within several European projects on transition research and Hybrid Laminar Flow Control. DLR also contributes to transition research within FESTIP (Future European Space Transportation Investigation Programme) and ESA/ESTEC–TRP hypersonic flow programmes. The following review, however, concentrates on recent results on the physics of laminar–turbulent transition and on non–empirical transition prediction for subsonic and transonic flows. Such investigations were performed at DLR Göttingen.

The main goal of this laminar–turbulent transition research is the construction of a reliable, non-empirical transition prediction method for transonic flows over wings of commercial airplanes. This transition–prediction method should aid industry in handling cases that are intractable using any simpler, but empirical prediction techniques. Experimental identification and theoretical / numerical modeling of both the receptivity mechanisms that generate disturbances inside the boundary layer and the subsequent mechanisms that drive transitional flows into turbulence are indispensable to reach the goal of a non–empirical transition prediction. The transition-research plan at DLR Göttingen rests on a physical insight into these processes not only as a necessary prerequisite of new and more reliable transition predictions for industrial applications but also for testing efficient laminar-flow control concepts. The present article is a shortened and updated version of the paper [1] "Status of DLR's non–empirical transition prediction project" by St. Hein et al.. The reader is kindly asked to check that paper [1] for references on any detailed work. In addition to the results presented below we refer to the book of abstracts of the Second European Forum on Laminar Flow Technology [2], where DLR's contributions to the wind tunnel and flight experiments utilizing so called Natural and Hybrid Laminar Flow Control technologies have been presented.

Each of DLR's instability and transition experiments has been designed to concentrate on certain aspects of laminar–turbulent transition. This reduces the number of phenomena present in the individual experiment and thereby allows a much deeper insight into the physical mechanisms. New theories and new numerical codes had to be developed to deal with such nonlinear and nonlocal flow instabilities as well as disturbance amplifications and interactions. Moreover, some of the experiments have been supplied with disturbance generators that allow controlled excitation of disturbances within the boundary layer. Together with some advanced measurement techniques this further improves the possibilities of analysing the phenomena observed and helps to compare with theory.

Investigations of receptivity issues have considered the disturbances generated by small, steady and time-varying, localized surface roughness. Downstream, detailed measurements of the nonlinear stages of transition were made. Data on the effect of free-stream turbulence on transition were also collected [6]. Such data were used to verify and improve the theoretical and numerical instability on transition models. Due to the relatively high turbulence levels in wind-tunnels, however, the experimental results are not directly applicable to the free-flight transonic flows of interest. Thus, once the theoretical and numerical tools have been tested and proven reliable, the tools can be applied to the "quiet" flows found in actual flight.

# 2 Numerical tools for non–empirical transition prediction and analyses of transition control: NOLOT/PSE

Transition prediction methods used in industry usually require numerous flow–instability studies to be done in succession, varying parameters such as the frequency and the wavelength of the disturbance in such a way as to blanket all possible cases of relevance. Therefore, usage has been practically limited to fast instability–analysis codes that use little computer time. Up to now, the most sophisticated methods used are based on classical local linear stability theory. Fastness has been achieved at the cost of precision by simply neglecting the "nonlinear-stages" of transition that are responsible for the onset of turbulence, and by neglecting all streamwise varying properties of the geometry and the meanflow (i.e. "local" instability analyses).

Due to the loss of modeling power caused by the use of the linear codes, empirical correlations have to be introduced for the prediction of the transition location. Following the philosophy of the $e^N$–method, the growth rates provided by linear local theory are integrated in downstream direction for transition prediction following various solely empirical integration strategies.

Besides the uncertainty about the 'correct' N-factor, the shortcomings of the $e^N$-method are obvious. Within the framework of linear stability theory, "receptivity" of the boundary layer to outer-flow disturbances like turbulence or sound and nonlinear mechanisms are out of scope of this transition prediction method. Furthermore, local stability theory does neither properly take into account the effects of flow nonparallelism, i.e. boundary–layer growth and streamline curvature, nor the effects of surface curvature. There is no rational criterion to ensure the reliability of the $e^N$-transition prediction method presently used in industry.

Reliable non-empirical transition prediction requires numerical methods that allow a physical modeling of all mechanisms that affect the transition process significantly. Therefore, DLR in cooperation with FFA/KTH is developing nonlocal nonlinear methods for transition analysis i.e. the NOLOT-code utilizing parabolized stability equations PSE for convectively unstable flows, like boundary layers [10]. NOLOT/PSE includes nonparallelism, i.e. boundary–layer growth, streamline curvature etc., and surface curvature also for three-dimensional, i.e. swept wing flows, in a mathematically consistent manner. In contrast to local methods a straightforward extension to nonlinear problems is possible. Moreover, NOLOT/PSE models some of the known receptivity mechanisms due to distributed roughness, distributed suction, sound etc. by using appropriate inhomogeneous boundary conditions [3]. Hence, it covers a much wider range of the laminar–turbulent transition process, i.e. from the generation of disturbances, their linear and nonlinear development up to the final rise of the skin–friction coefficient dependent on the disturbance environment. NOLOT/PSE covers the whole Mach–number range from subsonic to hypersonic flow regimes. The present review is focused on subsonic to transonic flows, however.

# 3 Experimental set-ups for instability and transition research with disturbance control

Due to the appropriate design of each of the DLR experiments it is possible to concentrate on certain aspects of the laminar–turbulent transition, which helps to analyse and understand the physics of interest in detail. Moreover, the conception of the experiments tries to meet the requirements from simplifying assumptions of the theory.

Although the main emphasis of the DLR work is on three-dimensional (wing-like) boundary layers the transition scenarios in two–dimensional flows are still of some interest due to several reasons. First of all, there are basic transition mechanisms and different transition scenarios, involving different modes of instabilities (like in the oblique wave breakdown), which are not well understood yet. In order to clearly select between the different transition scenarios and to provide well–defined initial conditions for theory, novel devices for controlled excitation of disturbances had to be developed. Whereas sound waves through a single spanwise slot are used to initiate two–dimensional Tollmien–Schlichting waves, a second disturbance generator has proven to be suitable for initiation of a pair of oblique waves. (Results are discussed in Section 5.1.)

Particle Image Velocimetry (PIV) can not only be used for quantitative measurements but also provides a fast qualitative overview on the characteristics of the laminar–turbulent transition process. Having qualitative information from flow visualization using laser light sheet techniques and quantitative information from the subsequent image processing of the PIV-method, it is possible to restrict the time–consuming pointwise measurements with hot-wire anemometry to some hand–picked regions.

With respect to three-dimensional wing-like flows the DLR experiment of a swept flat plate with a displacement body on top (fig.1) proposed by U.Dallmann and E.H.Hirschel already in 1974 and substantiated by H.Bippes and coworkers since 1980 was originally designed to study transition scenarios dominated by crossflow instabilities with instability modes generated solely by the small environmental disturbances always present under natural conditions. The displacement body prescribes an almost constant favourable pressure gradient in downstream direction. This pressure gradient suppresses Tollmien–Schlichting instabilities but causes crossflow instabilities due to the inflectional crossflow profiles of the resulting three–dimensional boundary–layer flow. Hence, the situation is very similar to that on the front part of a wing without having any surface curvature affecting the transition process, however. End plates with potential streamline curvature are used for approximation of infinite swept wing conditions necessary for comparison with theory. The swept flat plate experiment has been supplied with a disturbance generator driven by loudspeaker signals which are controlled in such a way that the phase shift between two neighbouring membranes is used to fix the spanwise wavelength of generated traveling modes [12],[13],[14]. Stationary crossflow vortices of desired spanwise wavelength are introduced by equally spaced roughness elements glued onto the surface. A huge amount of important data have been collected under these controlled conditions from

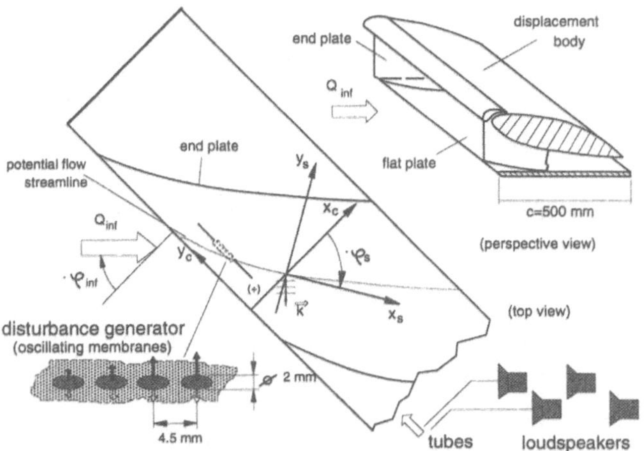


Figure 1: Experimental set–up used for investigations of instability and transition phenomena in three–dimensional boundary–layer flows excluding surface–curvature effects explicitly.

which a lot has been learned about laminar–turbulent transition in three–dimensional flows and which highlights the dependence of the transition process on individual wind-tunnel characteristics. (Results are summarized in Section 5.2.)

Many problems of transitional flow control have appeared in the course of BRITE/EURAM–projects which deserve rigorous tests and basic physical understanding. In order to validate Hybrid Laminar Flow Control devices, like suction panels, these have been adapted to the experimental set-up as well [5]. Whereas the swept flat plate experiment excluded effects of surface curvature an ongoing experiment using a HQ26 profile is intended to clarify laminar–turbulent transition dominated by crossflow instabilities on a cambered wing profile. Last but not least, the flow along the leading edge of a swept wing is of central importance with respect to the boundary layer's receptivity to external disturbances.

# 4 Receptivity studies

Sustained efforts, both experimental and theoretical, have been made at DLR Göttingen to study the generation of flow–disturbances inside the boundary layer by disturbances in the free–stream or by roughness and waviness on the surface itself. This disturbance–generation phenomenon is commonly called "boundary–layer receptivity".

To date, three classes of receptivity phenomena have been investigated at DLR Göttingen:

1. Receptivity of Tollmien–Schlichting waves in the Blasius boundary layer to free–stream sound and distributed wall roughness (fig.2),

2. Receptivity of streamwise vortices and steady or low–frequency "streaky structures" in the incompressible boundary layer to free–stream vorticity (fig.3),

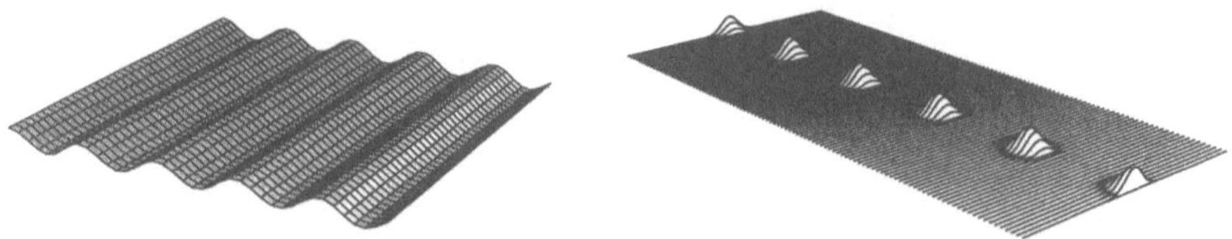

Figure 2: Examples for non–localized and localized surface roughness.

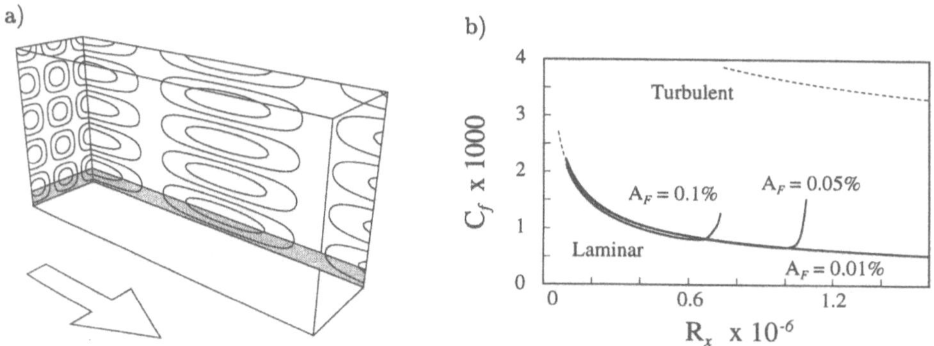


Figure 3: Receptivity to freestream vorticity for a Blasius boundary layer:
a) Iso–vorticity contours of a streamwise aligned vortical mode for modeling free–stream turbulence. The boundary–layer region is marked by the gray band.
b) Calculated skin–friction coefficient for three different amplitudes $A_F$ of the free–stream vortical modes indicating dependence of laminar–turbulent transition on turbulence level.

3. Receptivity of crossflow vortices in incompressible three–dimensional boundary layers to surface roughness (fig.4), and to unevenness in wall–suction distribution.

For case 1) theoretical models have been proposed by various authors and the nonlocalized case has been incorporated in the PSE formulation. The comparison with experimental data, obtained in the Blasius boundary layer, is quite good.

In case 2), the free–stream motions are rotational like free-stream turbulence, and travel with the free–stream. Our experience with some of the published models has indicated that these models are not very useful since they either fail to agree with, or are practically impossible to compare with experimental data. With this fact in mind, a new physical receptivity has been developed at DLR Göttingen, that successfully predicts many of the experimental measurements [4]. The key step in the PSE analysis is the construction of physically accurate initial conditions. As a result the free–stream vortical modes (fig.3a) generate steady "streaks" in the boundary layer, and when these "streaks" combine with Tollmien–Schlichting waves, the flow may go rapidly through transition. Fig.3b shows the calculated skin friction as function of streamwise Reynolds number, for three different strengths of the free–stream vortical modes (e.g. the free–stream turbulence level).

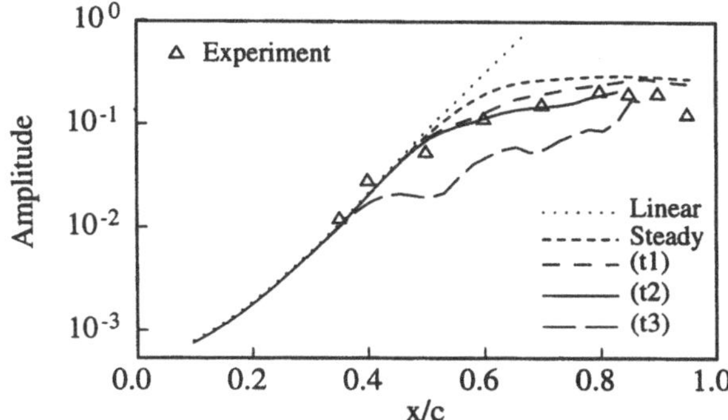


Figure 4: Calculated development of a stationary crossflow vortex in an incompressible three–dimensional boundary layer generated by localized surface roughness according to linear theory (Linear), nonlinear theory without any traveling wave present (Steady) and nonlinear theory with traveling waves of increasing amplitudes present (t1 – t3). The measured data from DLR–experiment are given for comparison.

In case 3) the stimulation of flow disturbances is created by surface roughness, surface undulation, or variations in the wall suction distributions that are fixed in space and steady in time. The stimulation is particularly effective in three–dimensional boundary layers because steady crossflow vortices are naturally amplified in such flows. The computed results [3] are in agreement with the measurements of Bippes and colleagues (for the experiment see fig.1), as shown in fig.4. The initial amplitude of all the computed curves is given by the receptivity model.

# 5 Transition scenarios

## 5.1 Oblique breakdown in a 2D boundary layer

Making use of the disturbance generators mounted on the flat plate experiment three different transition scenarios have been forced to appear inside the two–dimensional boundary layer. Since within the last decade both the fundamental and the subharmonic type (i.e. K-type and C/H-type) have been studied frequently in experiment (though using other devices for artificial excitation) and in theory, DLR has focused on the oblique breakdown as well [17]. This transition scenario could be of importance under natural flow conditions where localized disturbances such as roughness elements or suction holes initiate the transition process. So far, the oblique breakdown has been known from theory only. In contrast to the two previous scenarios, the oblique breakdown does not rest on a two–dimensional Tollmien–Schlichting wave as primary disturbance but is initiated by a pair of oblique waves only. Nonlinear interaction between these two traveling oblique waves excites rapidly growing streamwise vortices, which in turn generate large amplitude low– and high–speed streaks due to the so–called lift–up effect. Secondary instability of the resulting new profiles finally leads to transition. Results from PIV measurements are shown in fig.5. Strong streamwise aligned "streaky structures" are clearly visible. Flow direction was from left to right. Corresponding results from direct numerical simulations

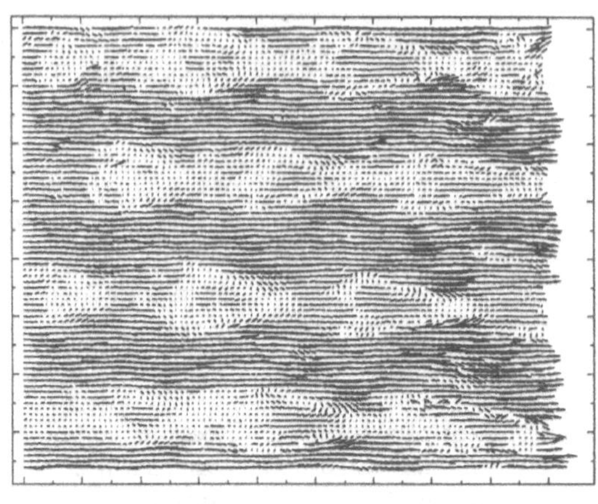

Figure 5: Instantaneous spatial scenery of flow phenomena at the oblique breakdown in a surface–parallel plane. Flow direction is from left to right. PIV-measurement.

performed at KTH Stockholm are in very good agreement with the DLR experimental results.

## 5.2 Transition in a 3D boundary layer dominated by crossflow instabilities

Within the last few years the amplification of disturbances developing under conditions of natural transition in the unstable three–dimensional flow of the DLR swept–plate experiment (fig.1) has been studied in detail. It was found that the characteristics of the disturbance development strongly depend on the different turbulence levels of the different used wind–tunnels [6]. Furthermore a mechanism that might explain the onset of turbulence in cases where both stationary and traveling modes have reached a saturation state could not be clearly identified yet.

The improved experimental set–up renders it possible now to control both the initial amplitudes of the stationary crossflow vortices and those of the oblique traveling waves independently by means of fixed surface roughness elements and the newly developed unsteady disturbance generator, respectively. Additionally, it does allow the identification of both amplitude distribution and propagation direction of individual modes using the phase-locked signals of a single hot–wire probe [14].

Fig.6a shows the measured amplification curves for a freestream velocity of $Q_\infty = 16.3m/s$, where the onset of laminar–turbulent breakdown can be observed near the end of the model. A stationary and a traveling mode have been excited artificially in this case and both saturate further downstream. In the nonlinear saturation region a high–frequency secondary instability has been detected. The frequency spectrum (fig.6b) does not only display sharp peaks for the excited traveling mode of $f = 82Hz$ and its higher harmonics, but also clearly indicates a second frequency band of amplified modes located around 2000 Hz, which is in qualitative agreement with the predictions of secondary theory [7],[8] for similar flow conditions. A method based on nonlinear bifurcation theory

a)

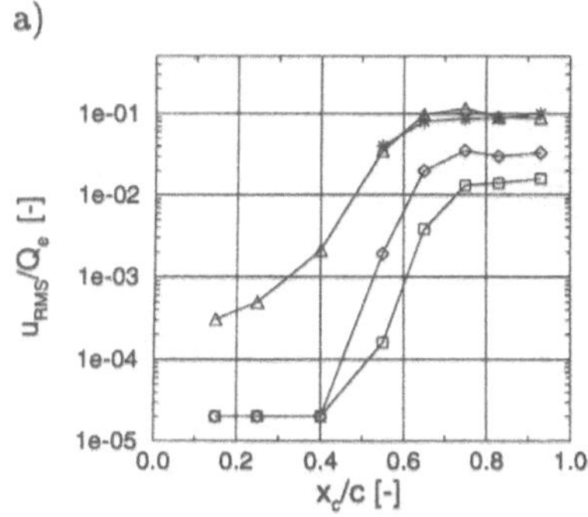


b)

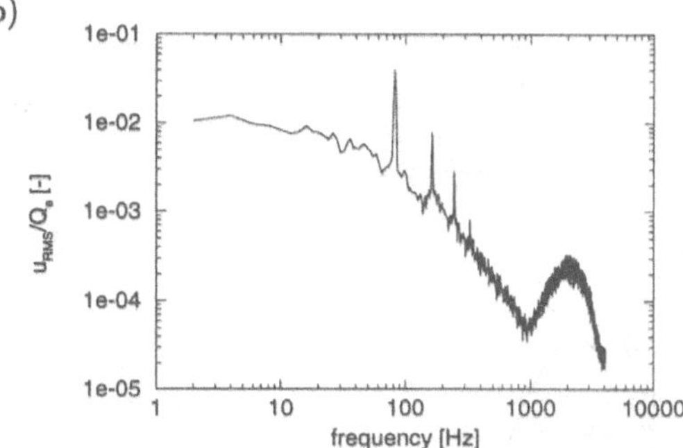


Figure 6: DLR swept flat plate experiment:
a) Measured amplitudes as a function of chord position $x_c/c$ for the stationary mode ($*$) initialized by roughness elements, the traveling mode ($\triangle$) initialized by the disturbance generator and its higher harmonics ($\diamond, \square$).
b) Measured Fourier spectrum of the streamwise velocity component showing the appearance of high–frequency secondary instabilities.

is currently being developed, which provides directly nonlinear saturation solutions for the primary mode and therefore allows meaningful quantitative comparisons with experiments [11].

## 5.3 Nonlocal linear instability analysis for Hybrid Laminar Flow Control on the A320 fin

The preliminary results of our studies on the A320 fin with suction using linear nonlocal theory will provide the prerequisites for future nonlinear transition investigations. Conditions measured in experiment were chosen for these investigations ($Re_\infty = 23.87 \cdot 10^6$, $Ma_\infty = 0.777$ and a sweep angle of $\phi = 41.78°$). Various suction distributions were prescribed including failure of a certain suction chamber [9].

As a first step the instability characteristics with and without suction were compared (fig.7). According to linear nonlocal theory, suction at the leading edge area stabilizes crossflow instabilities significantly, whereas it only causes a downstream shift of the instability region for Tollmien–Schlichting waves. Latter observation might be of some interest for future nonlinear transition analyses and transition predictions to evaluate Hybrid Laminar Flow Control concepts. Moreover, we have proven that treatment of surface curvature effects within the framework of a local theory as suggested within ELFIN tasks was wrong. The ongoing BRITE/EURAM project EUROTRANS helps to clarify this issue further.

## 5.4 Transition analysis of the ATTAS flight experiment

The transition process on the laminar glove of the ATTAS flight experiment is considered for this study [15]. An up to now unknown transition scenario has been identified (fig.8a), which originates from a stationary crossflow vortex (denoted as mode (0,-1,1)) and a traveling Tollmien–Schlichting wave (denoted as (1,1,0)) with wave front parallel

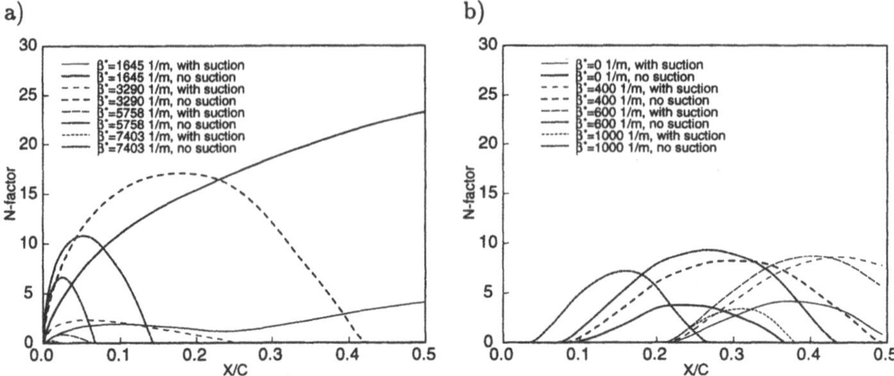


Figure 7: Effect of suction on the calculated N-factors for the A320–fin:
a) for stationary crossflow vortices with different spanwise wavenumbers $\beta^*$,
b) for Tollmien–Schlichting waves of $f^* = 5kHz$ and different spanwise wavenumbers $\beta^*$.

to the leading edge. These two modes are responsible for the nonlinear excitation of a spanwise traveling mode (1,0,1) together with a nonlinear distortion of the mean flow (mode (0,0,0)). Explosive growth of the modes sets in, which suggests a highly energetic interaction between these modes and which finally leads to laminar–turbulent transition as indicated by the skin–friction rise plotted in fig.8b. Changing the initial amplitudes of the modes would shift the downstream location of the calculated skin–friction rise. Hence, receptivity models (Section 4) will have to be incorporated in the future.

# 6 Conclusions

The reliability of any transition prediction method can now be tested via its application to the DLR instability and transition experiments performed at subsonic speeds. One such test case is under investigation between various European partners within GARTEUR–AG27. The application of NOLOT/PSE with additional expert's knowledge on receptivity and transition scenarios on transonic Laminar Flow Control experiments like A320 fin, ATTAS or F100 data provides validation tests of the DLR/FFA–code via direct comparisons within the BRITE/EURAM project EUROTRANS. Apart from such international collaborations transition research within performed at German Universities and at DLR is linked to laminar flow control technology developments at DASA (Daimler Benz Aerospace) in collaboration with DLR and others via the "Verbund–Schwerpunktprogramm Transition" [16] sponsored by the DFG (Deutsche Forschungsgemeinschaft) and the BMBF (Bundesministerium für Bildung, Wissenschaft, Forschung und Technologie).

The goal of non–empirical [1] transition prediction can be reached by close cooperation between theory and experiment only. An overview of the theoretical and experimental tools developed and used at DLR Göttingen in order to reach this goal was given above.

[1] "Non–empirical" herein is used to indicate that a rational definition of upstream and outer disturbance flow conditions (given either experimentally or via receptivity theory) is all that is required and no further "calibration" of N–factors etc. is necessary for transition prediction.

a)

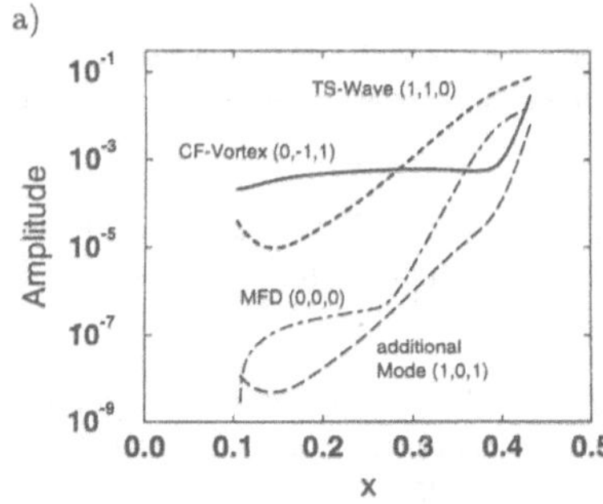


b)

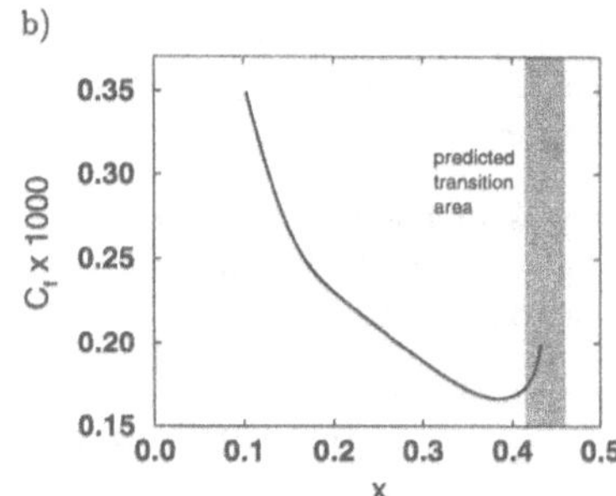


Figure 8: Transition analysis for the three-dimensional boundary layer of the ATTAS flight experiment:
a) Transition scenario originating from the interaction of a stationary crossflow vortex and a Tollmien–Schlichting wave.
b) Skin friction rise due to laminar–turbulent transition as predicted by nonlinear nonlocal theory.

Examples have shown, how theory, basic experiment and prototype free–flight applications influence and benefit from each other.

Due to the application of disturbance generators it is now possible to make measurements under controlled conditions. Together with novel measurement techniques it allows a much clearer and deeper insight into the laminar–turbulent transition process. By that means further details and new phenomena have already been and will be further identified.

NOLOT/PSE, i.e. a nonlocal nonlinear transition analysis method based on the parabolized stability equations, has demonstrated its potential for non-empirical transition prediction provided that the relevant modes together with their initial amplitudes are known. However, the predicted location of laminar flow breakdown not only depends on the modes taken into account but also on their initial amplitudes. These uncertainties about the relevant modes and their initial amplitudes have to be removed. Hence, successful modeling of the receptivity mechanisms and the knowledge about the nonlinear interaction mechanisms which finally lead to transition are required. DLR's future experimental and theoretical work will furthermore concentrate on these mechanisms also in order to minimize the risks to be taken in extrapolating wind tunnel transition data to free–flight prototype applications, including any Hybrid Laminar Flow Control on transonic wings.

*Cordial thanks to the many colleagues who have contributed with enthusiasm, sagacity and inventiveness to the substantial progress made in our transition research project and to those who have supported the present team.*

# References


[1] Hein, S. ; Bertolotti, F. P. ; Gordner, A. ; Stolte, A.; Koch, W. ; Bippes, H. ; Lerche, T. ; Wiegel, M. ; Dallmann, U. : *Status of DLR's "non-empirical transition prediction" project.* DLR-IB 223-96 A37.

[2] *Abstracts of the Second European Forum on Laminar Flow Technology*, organized by the Association Aéronautique et Astronautique de France (AAAF), June 10-12, Bordeaux, France.

[3] Bertolotti, F. P. : *On the birth and evolution of disturbances in three-dimensional boundary layers.* Proc. IUTAM Symp. on Nonlinear Instability and Transition in three-dimensional Boundary-Layers, Manchester, UK, 1995.

[4] Bertolotti, F. P. : *Response of the Blasius boundary layer to free-stream vorticity.* Submitted to: Physics of Fluids

[5] Bippes, H. ; Fischer, M. ; Wiedemann, A. ; Bertolotti, F. P. : *Experimente zur Beeinflussung der Querströmungsinstabilität mit Hilfe von Absaugung durch perforierte Wände.* DGLR-Fachausschußsitzung "Beeinflussung von Tragflügelumströmungen", 2./3. Mai 1996, TU Berlin.

[6] Deyhle, H. ; Bippes, H. : *Disturbance growth in an unstable three-dimensional boundary layer and its dependence on environmental conditions.* Journal of Fluid Mechanics, vol. 316, pp. 73–113, 1996.

[7] Fischer, T. M. ; Hein, S. ; Dallmann, U. : *A theoretical approach for describing secondary instability features in three-dimensional boundary-layer flows.* 31st Aerospace Sciences Meeting & Exhibit, January 11-14, 1993, Reno, NV, AIAA-93-0080.

[8] Fischer, T. M. : *Ein mathematisch-physikalisches Modell zur Beschreibung transitioneller Grenzschichtströmungen. I. Die linearen und nichtlinearen Störungsdifferentialgleichungen. II. Lokale und nichtlokale Eigenschaften sekundärer Instabilitäten.* DLR-FB 95–06, 1995.

[9] Gordner, A.; Simen, M. ; Hein, S. ; Stolte, A. ; Dallmann, U. : *Nichtlokale Instabilitätsanalyse kompressibler Seitenleitwerksumströmungen mit Absaugung.* DLR-IB 223-96 A50.

[10] Hein, S. ; Bertolotti, F. P. ; Simen, M. ; Hanifi, A. ; Henningson, D. : *Linear nonlocal instability analysis — the linear NOLOT code —.* DLR-IB 223-94 A56.

[11] Koch, W. : *Nonlinear crossflow saturation in three-dimensional boundary layers.* IUTAM-Symposium on Nonlinear Stability and Transition in Three Dimensional Boundary Layers, July 17-20, 1995, Manchester, U. K. , Kluwer.

[12] Lerche, T. ; Bippes, H. : *Experimental investigations of cross-flow instability under the influence of controlled disturbance excitation.* Colloquium on Transitional Boundary Layers in Aeronautics of the Royal Netherlands Academy of Arts and Science, 1995.

[13] Lerche, T. : *Experimental investigation of nonlinear wave interactions and instability in three-dimensional boundary-layer flow.* 6th European Turbulence Conference - ETC6 , Lausanne, Switzerland, July 1996.

[14] Lerche, T. : *Experimentelle Untersuchung nichtlinearer Strukturbildung im Transitionsprozeß einer instabilen dreidimensionalen Grenzschicht.* Dissertation Universität Hannover, Fakultät für Maschinenwesen, 1996. Also appeared as: Fortschrittsberichte VDI Reihe 7, Nr.310, 1997.

[15] Stolte, A. ; Bertolotti, F. P. ; Hein, S. ; Simen, M. ; Dallmann, U. : *Nichtlokale und nichtlineare Instabilitätsuntersuchungen an kompressiblen Strömungen.* DLR-IB 223-95 A54.

[16] Wagner, S. : *Das Verbund-Schwerpunktprogramm "Transition".* Vortrag auf der Jahrestagung der DGLR in Dresden, 24.–27. Sept. 1996, DGLR-JT-96-102. Jahrbuch der DGLR II, S. 697-705.

[17] Wiegel, M. : *Experimentelle Untersuchung von kontrolliert angeregten dreidimensionalen Wellen in einer Blasiusgrenzschicht.* Dissertation Universität Hannover, Fakultät für Maschinenwesen, 1996. Also appeared as: Fortschrittsberichte VDI Reihe 7, Nr.312, 1997.

# The Development of Advanced Internal Balances for Cryogenic and Conventional Tunnels

B. Ewald
K. Hufnagel

Technical University of Darmstadt, Germany
Department for Aerodynamics and Measuring Technique
Faculty of Mechanical Engineering

Petersenstr. 30
D-64287 Darmstadt

## Summary

The measurement of the aerodynamic forces is the most important task in the wind tunnel at least for aircraft development work. The accuracy and reliability of the balance is the key factor in this test technology. The urgent requirement for more and more accuracy of force testing leads to a demand for more and more balance accuracy. The most urgent demand in this field comes from the Cryogenic Tunnels like the new European Wind Tunnel. Funding from the German Ministry for Research and Technology made possible about 12 years of uninterrupted research in internal wind tunnel balances, especially in cryogenic balances, at the Technical University of Darmstadt in co-operation with Daimler Benz Aerospace Airbus GmbH. This work resulted in an advanced technique of internal balance design, construction ,instrumentation and calibration methods. The outcome of this effort is not only a cryogenic balance technology, which allows transport performance measurements in the ETW with a repeatability of less than one drag count but also a considerable improvement of balances for conventional tunnels. For this result all aspects of the balance technology had to be treated.

- For the balance design a computerised method was developed, which allows an optimisation of the structural design in a short time. Principal aspects of the design were studied with Finite Element analysis for optimised solutions.
- The technique of the electron beam welded balance was established successfully. This construction method gives considerable advantages with respect to design for optimum structure stiffness and low interference.
- The difficult problem of strain gaging and wiring for cryogenic environment, which normally includes severe moisture problems, was solved as a result of lengthy investigations.
- For cryogenic balances a novel arrangement of the axial force measurement system was developed, which minimises the problem of temperature gradient induced error signals. For residual errors of this type numerical correction methods are proven.
- For balance calibration a new strategy is used. A novel mathematical algorithm extracts a third order measuring matrix (no matrix inversion necessary) from the calibration data set. In a mathematical sense this is the best possible closed solution. In co-operation with Deutsche Airbus and the Carl Schenck Company a fully automatic calibration machine was developed for ETW. A smaller and simplified version of this machine is under construction at the Technical University of Darmstadt.

Finite element analysis turned out to be a powerful tool in the development of optimised structures for internal balances. Novel balance structures with minimised linear and non-linear interference and with minimised sensitivity against temperature gradients have been developed and will be demonstrated especially in the paper of Mr. Zhai and Mr Hufnagel.

Four balances for the Cologne Cryogenic Tunnel (KKK) at the DLR and three balances for the European Transsonic Wind Tunnel have been constructed and delivered during this programme.

## 1. INTRODUCTION

The successful design and development of commercial transport aircraft depends (among many other problems !) on excellent aerodynamics. Especially the flight performance reacts very sensitively on aerodynamics. Since flight performance must be guaranteed to potential future customer long before the first flight of the prototype, the success of the aircraft depends heavily on wind tunnel tests with the utmost accuracy. This ever rising requirement for accuracy in wind tunnel testing and especially the challenge of precise force testing in cryogenic wind tunnels gave a strong impetus for strain gage balance research in the recent past. Since accuracy limits for conventional strain gage balances are set mainly by thermal effects, the target to achieve at least the same or possibly even better accuracy with cryogenic balances in cryogenic tunnels is an extremely difficult task. For the research work on cryogenic balances the ambitious target of one drag count repeatability for transonic transport performance testing was set.

To achieve considerable improvements compared to balances known and used today, a single clever balance design idea respectively a single successful detail improvement is not sufficient. A systematic search through all parts and aspects of balance technology and the improvement of all details of this technology to the limits of the available technology is necessary. The important parts of the technology are :

- Design philosophy
- Design computation and optimisation
- Selection of spring material for the balance body
- Material heat treatment
- Balance fabrication methods
- Strain gage selection and wiring method
- Moisture proofing respectively cryogenic environment proofing
- Data acquisition electronic hardware
- Mathematical calibration algorithm
- Calibration equipment
- Strategy of balance use in the wind tunnel

Research funds from the German Ministry for Research and Technology put us in a position to do concentrated balance research and development for about twelve years. The aim of this research was to improve each of these partial aspects of balance technology to the scientific limits available today. Due to this research in Cryogenic Balances many improvements also for balances for conventional wind tunnels resulted. Part of the work was done in close cooperation with the Deutsche Aerospace Airbus GmbH at Bremen with some contributions of the DLR.

## 2. BALANCE DESIGN PHILOSOPHY

For a successful balance design some essentials must be fulfilled :

1. ***Choose the balance ranges as close as possible to the actual measuring task. In defining the ranges include the consideration, that ranges of the balances can be overloaded, if other ranges are not fully used in the tests. This overload capacity of a balance normally is defined by the 'load rhombus'.***
2. ***Choose the geometric dimensions of the balance as large as allowed by the available space in the model.***
3. ***Design the balance structure for maximum stiffness.***

The first essential requires the design of dedicated and tailored balances for the different tasks of a wind tunnel. As an example for a typical transport configuration model matched in scale to the test section dimensions in a transonic wind tunnel at least three different balances are required for high accuracy testing :

- Very sensitive balance for cruise condition L/D optimisation work.
- Less sensitive balance for cruise condition work including buffet tests, maximum lift tests and $M_{DIVE}$ tests.
- Envelope balance for stability and control tests up to $M_{NE}$ including full control surface deflections and large angles of attack and yaw.

This requirement results in a numerous and expensive balance equipment of a wind tunnel but improves tunnel accuracy very much.

The maximum load capacity of a balance design within a fixed diameter D is limited even if an ultra high tensile strength steel (High Grade Maraging Steel) is used. In our balance design method we introduced a balance load capacity parameter S.

$$S = \frac{Z \cdot l^* + M_Y}{D^3} \quad [N / cm^2]$$

The characteristic length $l^*$ of the balance is defined as the distance from the reference centre to the end of the active part of the balance, see Figure 1. So this „Balance Capacity Parameter“ is a simplified measure of the bending stress in the balance body close to the balance connection to model or sting, which may be a cone or a flange. In most balance designs this is the critical position with respect to stress.

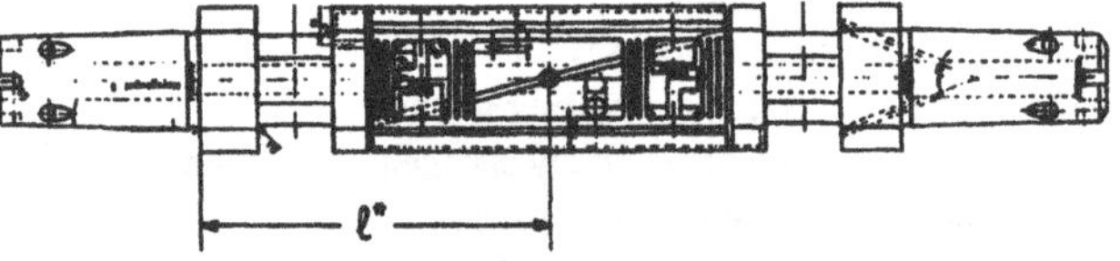

Figure 1 Characteristic Length

- Beyond a value of S = 2000 N/cm$^2$ the design of a precise balance including an axial force system is not possible.
- For a transport performance high precision balance the load capacity parameter should not exceed S = 500 N/cm$^2$.

Even lower load capacity parameters are recommended for optimum precision in drag measurement, if the space in the model allows for the larger balance diameter.

The third essential mentioned above - high stiffness of the balance body - is difficult to achieve with the conventional balance fabrication process by EDM (Electric Discharge Machining). With this method all internal cuts in the balance body must be accessible for the electrode from the outer side of the balance body. This compromises the stiffness requirement. So the fulfilment of the stiffness requirement is mainly a question of the fabrication method.

The ultimate solution of this problem is the Electron Beam Welded Balance concept, which was developed by the author at VFW (now Deutsche Airbus) more than fifteen years ago. The balance is fabricated from four pieces, which are prefabricated to the final dimensions of all internal surfaces and welded together by electron beam welding. All external machining including opening of the flexure systems is done after welding. The production steps are clarified by the Figures 2.

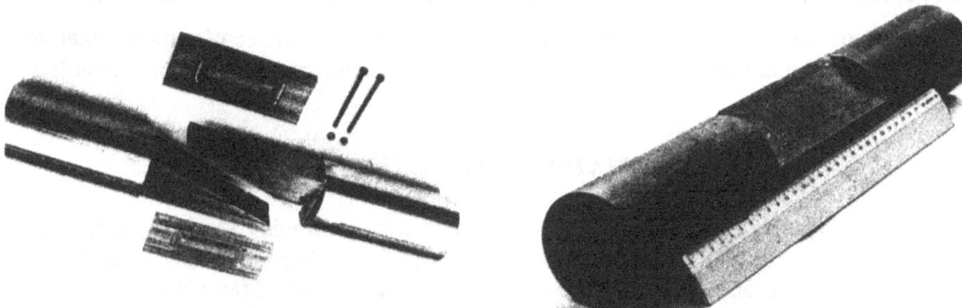

Figure 2a. Prepared Balance Parts

Figure 2b. Welded Balance Body

Figure 2c. Finished Balance Body

Provided that a proper material is selected and a sophisticated heat treatment after the welding process is done, full material strength is restored in the welding zone and the finished balance *is* a one piece balance and - with respect to strength and hysteresis - definitively behaves like a one piece balance. In a polished cut through the welding seam the welding zone is hardly visibly.

The concept of the Electron Beam Welded Balance turned out to be highly successful and was used since the invention for all balances constructed by the Deutsche Airbus GmbH and by the Technical University of Darmstadt. This fabrication method gives complete freedom in the internal design of the balance structure and allows a much stiffer design of the balance.

## 3. Balance Design Computation and Optimisation

At the Technical University of Darmstadt the design computation is done with the interactive computer programme „SEKOWA". With each step this programme completely computes the stress situation at all critical positions of the balance body and some additional characteristic parameters. All results are printed. The designer checks the results and according to his experience with the design process he modifies one or several geometric dimensions. Each step is designated as a "RUN". An experienced balance designer needs about 40 to 60 runs for a final satisfying result or for the understanding, that a good balance with the specified ranges can not be designed within given dimensions. The computation is based on basic stress and strain formula for short bending beams and short torsion beams.

The use of finite element analysis for routine balance design is not possible, since the discretisation of the complicated structure with many modifications for the optimised design is to laborious. Nevertheless for principal optimisation of strain gage balance designs finite element analysis proved to be an extremely valuable tool, this was demonstrated by the work of Junnai Zhai [22] at the Technical University of Darmstadt. Work on balance optimisation with the tool of finite element analysis is continued at the Technical University of Darmstadt.

The analysis of balance structures by the Finite Element Method demonstrated, that the computation and the minimisation of linear and non-linear interference effects is possible by this method.

## 4. Material Selection

The conventional material for strain gage balances is either maraging steel or precipitation hardening steel like PH 13.8 Mo (1.4534) or 17.4 PH (1.4548). For the welded balance concept we use Maraging 300 (1.6354) for conventional balances resp. Maraging 250 (1.6359) for cryogenic balances. Maraging steel is excellent for electron beam welding; the precipitation hardening steels should be good for welding as well, but no experience was gathered up to now with welded balances from precipitation hardening steels.

A very comprehensive study on force sensor spring materials was performed at the Technical University of Darmstadt. One important result of this study was a detailed insight in the large influence of the heat treatment on hysteresis.

An excellent material for force sensors may be the titanium alloy Ti Al Mg 4 (3.7164). Hysteresis is almost non existing with this material. Nevertheless more experience especially in electron beam welding and in gage application must be gathered before application of titanium as a strain gage balance material. Another promising material for conventional and cryogenic balances is Copper Beryllium (2 % Beryllium), if the load capacity factor allows for the lower tensile strength of this material compared to maraging steel. Hysteresis is extremely low and electron beam weldability is good. The excellent heat conductivity of copper beryllium will considerably reduce the temperature gradient problems with cryogenic balances. A cryogenic balance for the ETW from copper beryllium was designed and constructed at the Technical University of Darmstadt.

## 5. Strain Gaging and Wiring methods.

Up to now we used strain gages exclusively from Micro Measurement (Vishay). From the available range of gages types can be selected, which are very well suited for the cryogenic range and as well for conventional tunnel conditions. For the extreme temperature range of cryogenic balances misadaptation of the STC-Factor is recommended. We use SCT-Factors of 11 or 13 for balances constructed from maraging steel.

For a very low zero drift over the temperature range of cryogenic balances misadaptation of STC-factor, close coupled arrangement of the gages of one bridge etc. is not sufficient. Even the gages from one pack of five show considerable scatter in thermal behaviour. Gage matching improves this situation very much and was first proposed by Judy Ferris (NASA Langley). Since the thermal behaviour of gages can be evaluated only from the applicated gage, each individual gage is applicated to a common maraging steel sample by cyano cryalate bond. After a measurement of the zero drift of each gage in a cryogenic chamber the arrangement is heated beyond the stability of the cyano cryalate bond and the gages are carefully cleaned. From the results of this process the gages for each bridge are individually selected for minimum bridge zero drift. This procedure is time consuming but reduces bridge zero drift very much.

For final gage application on strain gage balances epoxy hot bonding is used exclusively. Preparing the surfaces, preparing the gages and the bonding procedure must be done with the utmost care, patience and perfect observance of the manufacturers instructions. Even the utmost care is not sufficient, it must be combined with years of experience in the art of strain gage application.

The internal wiring of the bridge circuits is carefully designed for symmetric length and symmetric temperature on all internal bridge wire connections. All bridges are wired separately for excitation lines, excitation voltage sensing lines and signal lines. All circuits are connected to the tunnel data system via a high quality miniature connector mounted at the sting end of the balance. Normally 80 pin connectors are used.

## 6. Moisture Proofing resp. Proofing for Cryogenic Environment

To achieve excellent zero point stability, moisture proofing is most important. For conventional balances a very careful observance of strain gage manufacturers instructions may be sufficient. For cryogenic balances moisture proofing is perhaps the most difficult detail of balance construction. Strain gage manufacturers give no sufficient instructions and offer no sufficient materials for these environmental conditions. A very careful application of multiple thin layers of nitril rubber is the best moisture proofing method we found up to now.

## 7. Data Acquisition Equipment

It stands to reason, that for balance signal acquisition top quality equipment is used only. Nevertheless between wind tunnel instrumentation experts there is a certain disagreement on the basic type of equipment. In most tunnels DC measuring techniques are used in form of specially designed signal conditioning and digitising units or in form of high quality digital multimeters.

In recent years some commercial developments, especially the DMC data acquisition unit (600 Hz carrier frequency) of the German company Hottinger has brought the AC measuring technique back into the field again. This AC equipment is equivalent and in some cases even superior to the best of DC equipment and has the big advantage of blocking any thermal voltage signals. In the case of cryogenic tunnels with their large temperature differences in the test region this may be essential. The disadvantage of the AC measuring method is the limited frequency range, which may cause concern, if dynamic balance stresses shall be monitored. Nevertheless dynamic signals up to 200 Hz can be monitored with this equipment satisfactorily.

The Hottinger measuring system DMC 9012 resp. its successor DMCplus is used successfully in the Aerodynamics Lab and the Wind Tunnel of the Technical University of Darmstadt, the Aerodynamic Department and Wind Tunnel Department of Deutsche Airbus, the Cologne Cryogenic Tunnel (KKK) and the ETW Calibration Machine. The system can be equipped with up to 28 data channels. The maximum speed of this system is 100 000 measurements per second and the resolution is up to 300 000 parts. The system is fully computer controlled; several systems may be managed in parallel by one PC. The system provides also the excitation for the strain gage bridges.

## 8. Mathematical Method of Calibration

The field of calibration perhaps includes the largest improvement potential of the balance technology. The first item in this field is the mathematical description of the balance behaviour. The generally used method is the so called second order calibration. Since many years we extended this to a third order approximation of the balance behaviour :

$$S_i = R0_j + \sum_{j=1}^{6} A_{ij} F_j + \sum_{j=1}^{6} \sum_{k=1}^{6} B_{ijk} F_j F_k + \sum_{j=1}^{6} C_{ij} F_j^3 .$$

In this description for the direct component calibration terms a third order term is taken into account. The advantage of this description compared to the conventional second order calibration was often questioned by other experts, nevertheless the use of the third order approximation is simply logical.

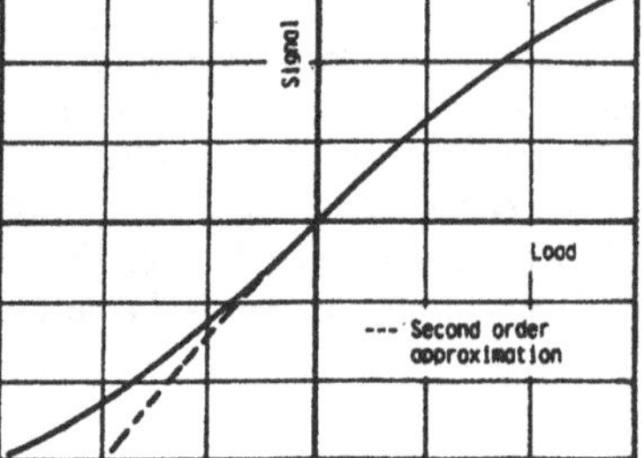

Figure 3. Second/Third Order Approx.

Certainly there are physical reasons for a non-linearity of the characteristic line of one component of a strain gage balance (or other force sensor) as shown in the positive quadrant of Figure 3. Since a strain gage balance is a symmetrical structure, almost certainly in the third quadrant the non-linearity of the characteristic line should be mirror inverted to the line in the positive quadrant as shown in Figure 3 by the continuous line. There is no reason to expect a monotonic curvature as shown by the dotted line. The non-linearity of the continuous line in Figure 3 can be described in a polynom by the third order term only. This is the only reason why we use the third order description of the balance behaviour. Applied to actual calibration data the comparison of second and third order calibrations shows that in the case of the third order approximation the third order coefficient have a considerable size and the second order terms come out smaller than in the case of a second order approximation. Nevertheless the quadratic terms should not be neglected. Very often a strain gage force sensor has a slightly different sensitivity in the positive and the negative quadrant. This behaviour is approximated by the quadratic term. Since all the work is done very fast by the computer, the higher mathematical complexity of the third order approximation is no argument against this algorithm.

# 9. Calibration Equipment

## 9.1 The Need for Automatic Calibration

The man power consumed for calibration is a large part of the total cost of a balance. This problem is aggravated in the case of cryogenic balances. The temperature is an additional parameter, so the calibration effort is two to five times higher compared to the case of a conventional balance. Also for higher accuracy the compilation of a larger data base for the evaluation of the calibration matrix is beneficial. So calibration efforts will increase even more.

The other reason for the search for automatic calibration methods is the sensitivity of the conventional calibration method against human errors. The utmost care and attention is necessary for the manual calibration. Especially with respect to the reference point accuracy errors occur easily since the balance is hidden in the loading sleeve, since the balance must be rotated during calibration and since the loading sleeve must be readjusted after rotation. A fully automatic machine avoids all these sources of human errors in the calibration.

So in several places of the world (ARA and DERA in Great Britain, FFA in Sweden, IAI in Israel) and in Germany the possibilities for automatic calibration were studied. A general target of these studies was to perform a complete six component calibration including all single loads and all pairs of two single loads in one working shift (8 hours).

A unique design principle was invented by the author and realised by the Carl Schenck Company and Deutsche Airbus together with the Technical University of Darmstadt. With this machine the functions *'calibration load generation'* and *'calibration load measurement'* are separated totally. The Internal Balance is connected with its model end to a six component force measuring device similar to an external wind tunnel balance. This device (the 'measuring machine') measures the calibration loads applied to the balance precisely. The 'Master Calibration Matrix' of this measuring machine allows for the small misalignment resulting from the elasticity of the connection between balance and measuring machine.

The loads are generated by push-pull pneumatic load generators acting on a loading frame connected to the sting end of the balance. Since the loads are measured precisely at the model end, load generation can be done rather crude and fast.

Figure 4 demonstrates the design of the machine. The 'Measuring Machine' is accentuated by dots on the left side; the force generating system is located on the right side of the Internal Balance. The perfect separation of calibration load generation and measurement of the component loads contributes largely to the excellent accuracy of the machine.

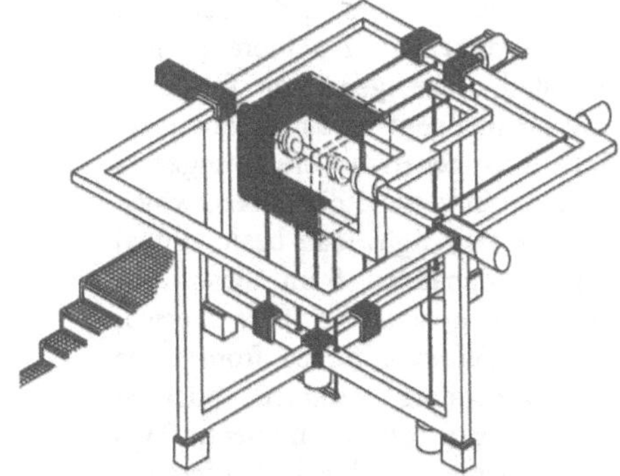

Figure 4. Design of Automatic Calibration Machine

The prototype of this machine was constructed for the ETW (European Transonic Wind Tunnel) and is successfully in operation at ETW. Figure 5 shows the ETW machine in the ETW Balance Calibration Lab. Figure 4 also demonstrates another big advantage of this machine. The internal balance itself is not hidden by a loading sleeve or surrounded by loading levers. So balance may be easily enclosed by a climate chamber for perfect temperature conditioning of the balance. For the accuracy of cryogenic balance calibration a perfect temperature conditioning between 100 K and ambient temperature is a must. In the case of the ETW machine the balance is connected to the measuring machine and to the load generating system by thin walled titanium tubes, which transfer the forces and form a nearly perfect blockage for the heat flow into the chamber. So temperature gradients in the balance, which are disastrous for the calibration accuracy, are prevented.

Figure 5. ETW Balance Calibration Machine

The machine is fully computer controlled. The fast operation allows a complete six component third order calibration including all single loads and all pairs of two loads in one working shift. So balance calibration costs are considerably reduced compared to the conventional procedures. This allows frequent recalibrations, a provision which improves accuracy and reliability of wind tunnel testing very much. The improved reliability is especially important in expensive tunnels like cryogenic transonic tunnels, where faulty test results cause big money losses.

## 9.2 Second Generation Automatic Calibration Machine

The success of the ETW Automatic Calibration Machine motivated for further development of this principle. A comprehensive analysis was done at the Technical University of Darmstadt on all lessons learnt with the ETW machine. The outcome was a machine with the same basic principles. The design was simplified and some minor imperfections of the first prototype were avoided. The main difference is, that the principle of a more or less dedicated force generator for each load component was abandoned. Three force generators are arranged in a triangle in vertical directions, which act on a very light load beam. Equal forces commanded from these load generators generate a pure normal force. Differential forces from these load generators produce pitching moment and rolling moment. Two force generators acting in Y-direction generate side force and yawing moment; only the axial force is

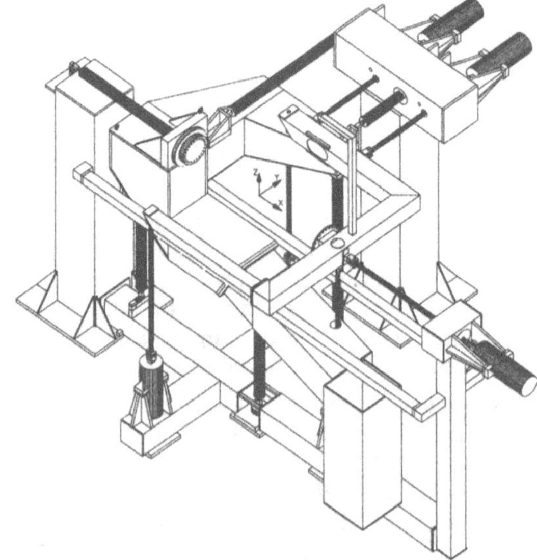

Figure 6 : Second Generation Calibration Machine Design

generated by a single and dedicated force generator.

The targets of this second generation design was a simplified design to save costs and an improved dynamic behaviour for increased speed.

A prototype of this machine is under construction at the Technical University of Darmstadt. Also this machine will be available on a commercial basis.

## 10. Cryogenic Balance Design

A standard design philosophy for cryogenic balances has not yet been established; among the cryogenic community there is even no agreement if unheated or heated balances are to be preferred. The majority of cryogenic balance designs is unheated up to now but there are also strong promoters of the heated balance.

Nevertheless the author is pessimistic with respect to the heated balance. The massive joints on model and sting end of the balance will cause considerably large heat flows, so a lot of local heating power will be required to condition the balance to ambient room temperature with no spatial temperature gradients. The result most probably will be even worse temperature gradients in some regions of the balance body. From the authors point of view the more promising solution is a special balance design which tolerates temperature gradients without unacceptable deterioration of the accuracy especially in the axial force measurement.

This was achieved successfully with the concept of the tandem axial force elements, which are integrated in the front and aft flexure groups of the axial force system (see Figure 7 for example). The predominant part of temperature gradient generated axial force errors is proportional to the mean temperature difference in the upper and lower cantilever beams of the axial force system. With the conventional central position of the axial force bending beam the error signals are a function of the arbitrary temperature distribution in the cantilever beams.

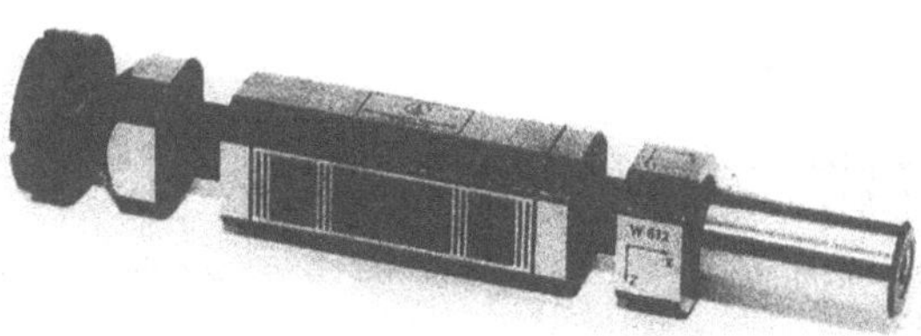


Figure 7. Cryogenic Balance W 612 (KKK) with tandem axial force elements

With the tandem axial force system the error signals due to temperature gradients in the front and in the aft bending beam element have the same magnitude but opposite signs. By adding the signals of the front and the aft sensor the signals due to temperature gradients are cancelled. The unavoidable tolerances in bending beam dimensions and gage position result in a small residual error signal due to temperature gradients. Nevertheless these residual errors may be removed by a simple numerical correction.

The concept of the tandem axial force elements is very successful. For temperature gradients of 5 degrees centigrade along the balance length the gradient induced error of the axial force signal without additional numerical correction is less than 1 µV/Volt in the case of the ETW balance W 618.

## 11. Future Developments

Some plans for future developments of the balance technology were already indicated in the previous pages. Our most important plans for future developments are :

- Black Box Balance Concept
- Further evaluation of Copper-Beryllium Balance Concept
- Evaluation of Titanium Balance Concept
- Integrated Balance Bridge for signal lines, power lines and pneumatic lines
- On Board Measurement for pressure distribution or other data with light transmission of data into the earth system
- Further optimisation of balance design by finite element analysis
- Further development of the Automatic Calibration Machine

## 12. Conclusions

The extensive research on strain gage balances done at the University of Darmstadt in co-operation with Deutsche Airbus demonstrated, that a substantial improvement of the wind tunnel force testing technology requires engineering progresses in any detail of balance design concepts, actual balance designs, material selection, balance fabrication method, gaging methods and calibration equipment and calibration algorithms. So all these details where included into our balance research efforts and any detail was improved to the technological limits available today. The outcome is a balance technology, which leads to much improved balances for conventional tunnels and to cryogenic balances which up to now (this development is not jet finally finished) bring the target of less than one drag count repeatability for transport configuration performance measurements within reach.

## REFERENCES


[1] E. Graewe, *"Development of a Six-Component Balance for Cryogenic Range"*. Forschungsbericht W 84-022 BMFT 1984.

[2] B. Ewald, *"Grundsatzuntersuchung zum Temperatur-Verhalten von DMS-Axialkraftteilen"*, BMFT LVW 8420 10, Nr. 10/85 1985.

[3] B. Ewald , *"Development of Electron Beam Welded Strain-Gage Wind Tunnel Balances"*, Journal of Aircraft Volume 16, May 1979.

[4] B. Ewald, G. Krenz, *"The Accuracy Problem of Airplane Development Force Testing in Cryogenic Wind Tunnels"*, AIAA Paper 86-0776, Aerodynamic Testing Conference, March 1986.

[5] B. Ewald, E. Graewe, *"Entwicklung einer 6-Komponenten-Waage für den Kryo-Bereich"*, 3. BMFT-Status-Seminar, Hamburg, May 1983.

[6] B. Ewald, E. Graewe, *"Development of Internal Balances for Cryogenic Wind Tunnels"*, 12th ICIASF, Williamsburg, VA, June 1987.

[7] B. Ewald, *"Balance Accuracy and Repeatability as a Limiting Parameter in Aircraft Development Force Measurements in Conventional and Cryogenic Wind Tunnels",* AGARD FDP Symposium, Neapel, September 1987, AGARD CP 429.

[8] B. Ewald, P. Giesecke, E. Graewe, T. Balden, *"Feasibility Study of the Balance Calibration Methods for the European Transonic Wind Tunnel"*
Report TH Darmstadt A 37/88, January 1988.

[9] B. Ewald, Th. Balden, *"Balance Calibration and Evaluation Software",* Proc. Second Cryogenic Wind Tunnel Technology Meeting, ETW, Cologne, June 1988.

[10] B. Ewald, T. Preusser, L. Polanski, P. Giesecke, *"Fully Automatic Calibration Machine for Internal Six Component Wind Tunnel Balances Including Cryogenic Balances",* ISA 35th International Instrumentation Symposium, Orlando, Florida, May 1989.

[11] B. Ewald, P. Giesecke, E. Graewe, TH. Balden, *"Automatic Calibration Machine for Internal Cryogenic Balances",* Proc. Second Cryogenic Wind Tunnel Technology Meeting, ETW, Cologne, June 1988.

[12] B. Ewald, T. Preusser, L. Polanski, P. Giesecke, *"Fully Automatic Calibration Machine for Internal Six Component Wind Tunnel Balances Including Cryogenic Balances",* ICIASF Congress, September 1989, Göttingen.

[13] Alice T. Ferris, *"Cryogenic Strain Gage Techniques used in Force Balance Design for the National Transonic Facility",* NASA TM 87712, May 1986.

[14] B. Ewald, L. Polanski, E. Graewe, *"The Cryogenic Balance Design and Balance Calibration Methods",* AIAA "Ground Testing Conference" July 1992, Nashville, Bericht A 99/92.

[15] B. Ewald, K. Hufnagel, E. Graewe, *"Internal Strain Gage Balances for Cryogenic Wind Tunnels",* ICAS-Congress, Sept. 92, Peking.

[16] B. Ewald, E. Graewe, *"The Development of a Range of Internal Wind Tunnel Balances for Conventional and Cryogenic Tunnels",* European Forum on Wind Tunnels and Wind Tunnel Test Techniques, Sept. 92, Southampton.

[17] H.F. Rush, *"Grain Refining Heat Treatment To Improve Cryogenic Toughness of High-Strength Steels",* NASA TM 85816, 1984.

[18] F. Schnabl, *"Entwicklung eines numerischen Algorithmus und eines Rechnerprogramms zur Auswertung der Eichversuche an 6-Komponenten-DMS-Waagen.",* Technical University of Darmstadt, Diploma-Thesis A-D-69/87, 1987.

[19] T. Balden, *"Ein neues Konzept zur Kalibration von Kryo-Windkanal-Waagen",* Deutsche Airbus Bremen, Proceedings of DGLR Jahrestagung 1993, Göttingen.

[20] B. Ewald, *"Theory and Praxis of Internal Strain Gage Balance Calibration for Conventional and Cryogenic Tunnels"*, 18 AIAA Ground Testing Conference, June 1984, Colorado Springs.

[21] G. Viehweger, B. Ewald, *"Half Model Testing in the Cologne Cryogenic Tunnel (KKK)",* 18th AIAA Ground Testing Conference, June 1984, Colorado Springs.

[22] J. Zhai, B. Ewald, K. Hufnagel, *„An Investigation on the Interference of Internal Six-Component Wind Tunnel Balances with FEM",* ICIASF '95, Dayton, Ohio.

[23] B. Ewald, *„Advanced Force Testing Technology for Cryogenic and Conventional Wind Tunnels",* ICAS Congress 1994, Anaheim, California.

# PARALLEL NUMERICAL AIRPLANE WING DESIGN

J.K. Axmann[1], M. Hadenfeld[1], O. Frommann[2]
1. Technische Universität Braunschweig, IFR
Hans-Sommer-Str. 5, D-38106 Braunschweig
2. Daimler-Benz Aerospace Airbus GmbH, EFV
Huenefeldstr. 1-5, D-28183 Bremen

## SUMMARY

A parallel numerical optimization method for the airplane wing design has been established. The extension of traditional Evolution Strategies by new and alternative methods combined with a load management for parallel processes leads to the MEPO optimization system. The concept of the system allows beneficial usage of parallel processing features for running the optimization algorithms as well as the application-specific simulation codes. In this way unacceptable turn-around-times in the design process of airplane wings can be reduced significantly. The low amount of communications between computer processor units makes it possible to employ workstation clusters efficiently. Coupled with simulation codes of DASA and DLR, a powerful program system for optimizing airplane wing designs has been created. Highly promising results in the calculation of wing shapes for the future Airbus A3XX now lead to a commercially usable program.

## INTRODUCTION

Technical and economical considerations require the use of numerical optimizations in the design process of new airplanes. For the determination of aerodynamically improved wing shape configurations computational amounts of several days on powerful modern workstations have to be spent. However, without the inclusion of the experience of qualified design engineers sufficient results can not be obtained by numerical optimization methods in use [1]. Methods capable of global optimizations require several hundreds of simulation runs. The high computational expenses lead to unacceptable turn-around-times in todays sequential computational manner. Parallelism has not yet been established in the industrial design process.
Funded by the German ministry of education and research (BMBF), an adaptive parallel numerical optimization system has been developed since 1995. Based on features of the optimization code EVOBOX, developed at the Institute for Flight Mechanics and Spaceflight Technology (IFR, former IfRR) at the TU Braunschweig [2-5], a modular program has been coded. One aim of this re-engineering task was to achieve higher degrees of parallelism and modularity within the newly structured program. The MEPO system (Multipurpose Environment for Parallel Optimizations) takes into account additional levels of parallelism in the design process of airplane wings as well as improved load management and communication structures [6]. By use of parallel computer processors on workstation clusters the numerical optimization of wing designs now can be carried out with significantly shorter turn-around times.

## AIRPLANE WING OPTIMIZATION

A wing profile in two-dimensional description can be given with 5 to 21 design variables in a chordwise definition. For the spanwise definition 5 to 10 intersections are used to simulate approximatively the airflow around the 3D wing [7]. Viscous-inviscid flow simulations are based on different codes like VICWA [8], XLS6 [9] or EPPLER [10] in dependence of the airflow conditions. They are combined by use of a control program, which manages the calls of the different routines and evaluates the results with an objective function defined by the design engineers.

For the determination of transonic aerodynamic wing shapes several numerical optimization methods are in use. Apart from algorithms based on inverse surface techniques [11,12] and methods using control theory [13] most practical optimization methods evaluate gradient information from search steps for the calculation of mathematical optima [14,15]. Therefore, the application of such direct optimization techniques lead to complex iterative processes for the design of new, aerodynamically optimized airplane wings. Numerous restrictions have to be taken into account. Beside aerodynamical requirements - high lift and low drag values for different Mach numbers - also geometrical and technical constraints have to be considered. The objective function, the mathematical formulation for the quality of a examined wing configuration in dependence on the input design variables, arises from a combination of numerous evaluation functions with additional terms taking into account practical demands.

By use of direct optimization methods the objective function forces variations step by step toward improved configurations. Without any parallelism the design engineers are faced with days of computing time for a single wing optimization. To reduce the turn-around times for an optimized wing design a parallelization can be introduced.

## MULTI-LEVEL PARALLELISM IN THE WING DESIGN

Three levels of parallelism in the design methodology can be distinguished, as shown in Figure 1. A fine grain parallelism is based upon block structured solvers for the flow simulation. A so-called medium grain parallelism takes into account the possibility to determine the airflow from the wing intersections and the different flight conditions in parallel. On the highest level of parallelism different designs are suggested by the optimization algorithms.

Fine grain parallelism has been realised in the FLOWer code [16]. The code is under development at DLR and DASA and can be used as a design code in the future in German aircraft industry. Consequently, additional work in parallelizing today used codes was avoided. Alternatively a medium-grain parallelism was established in the methodology. The concept of the MEPO system allows the usage of parallel processing features for the simulation codes, too. Therefore, the design of wing was split up in parallel tasks for the calculation of intersections and so-called *multipoints*.

Usually a wing is separated into 5 to 10 intersections with varying wing shapes along the span. The chordwise definitions of the profiles can be determined independently by optimization procedures. However, for each shape calculation a flow simulation must be done. Each quasi 3D boundary layer and shock wave calculation needs about 30 seconds by use of the XLS6 code. The more sophisticated code VICWA takes about 3 minutes of calculation time. Additionally, different flight conditions have to be taken into account. A wing design must be optimized for a wide range of Mach-numbers and angles of attack. To fulfil this requirement several design points - the so-called *multipoints* - have to be tested. In the numerical optimization process a complex objective function takes into account the qualities of a wing design for important flight conditions. The multipoint calculations can be carried out in parallel, too. In this way the complete wing design problem can be parallelized into 25 to 50 independent tasks without a high amount of computational work. Only some additional structures in the control and evaluation program had to be coded.

Additionally, coarse grain parallelism can be established without any expense in the design methodology by the appropriate choice of an optimization code. 25 to 80 variables have to be optimized in dependence to the number of intersections and shape describing parameters. Several optimization algorithms are usable [17-23]. However, their degree of parallelism differs in principle and may vary during the optimization process, too. Several optimization methods have been combined in the MEPO program system [6].

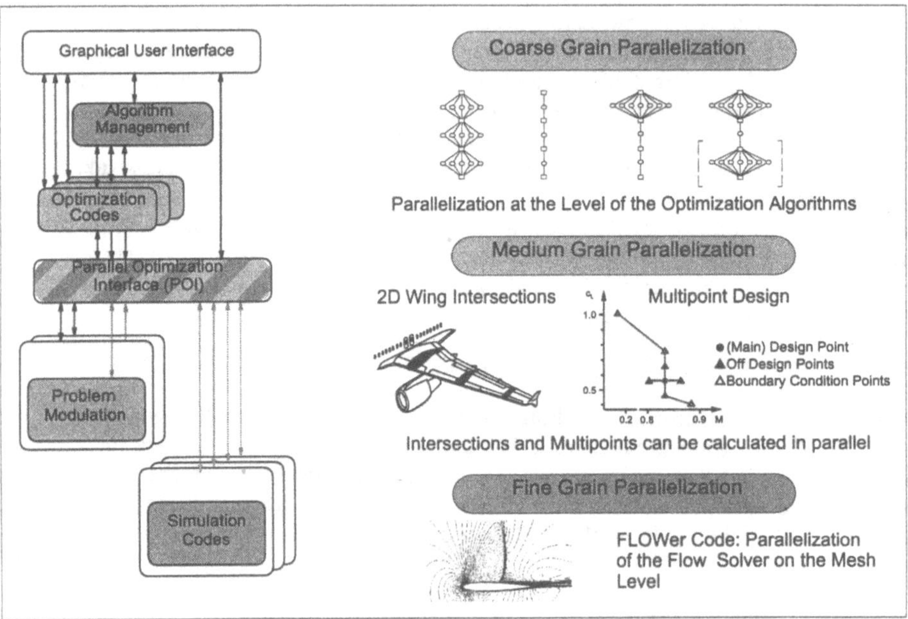


Figure 1: The structure of the optimization program system MEPO and the levels of parallelism in the design methodology of airplane wing designs

## THE STRUCTURE OF THE PROGRAM SYSTEM

Based on the knowledge gained during the development of the first parallel optimization system EVOBOX a newly structured code with a higher degree of modularity has been coded since 1995. The structure of this code is given in Figure 1. Different optimization codes can be selected by help of a graphical user interface or by an algorithm management system concurrently under development. All parallel applications on the different levels are managed by the Parallel Optimization Interface (POI). This module supports the fine, medium or coarse grain parallelism as well as the interconnection with the optimization modules. The parallelism is introduced by PVM (Parallel Virtual Machine), a widely used message passing software. In contrast to the former EVOBOX code, which was based on a uniformly structured *Master-Slave* concept, MEPO creates independent parallel processes.

The workstation pools in Bremen and Braunschweig consist of different types of machines offering different computing power of the processors for the parallel optimizations. As a consequence heterogeneous virtual parallel computers can be temporarily installed with the help of the message passing software. The efficiency of such clusters can be increased significantly by use of a resource management system [24]. For the MEPO system the integration of a management tool into the optimization code was chosen without any dependence to the operating system.

## LOAD MANAGEMENT

Coarse and medium grain parallelism lead to a high number of parallel processes in the established wing design method. Parallel tasks are produced not only the optimization algorithms, on a different level multipoints and intersections can be determined in parallel, too. This allows a precise load balancing. With the help of hardware dependent scaling factors the partitioning of the parallel tasks will be started. In the following generations the parallel tasks will be distributed basing of measured runtimes of the individual processes and the workload during the last minutes on the respective processors. With the help of these measured values a sufficient partitioning can be calculated in an approximative manner. With this method the individual workload and the power of each processor can be taken into account without any modifications on the operating system.
Additionally the inclusion and the release of further workstations can be managed by the POI depending of the time of the day. Especially at night times workstations are often without any or only with little load. Therefore they can be added to and release from an optimization cluster without an influence on the online activities during working hours. With the help of the management tool additional workstations can also be included in the case of hardware failures or shutdowns caused by users. In an analogy manner the restart of lost or stopped parallel processes in an optimization run can be handled, too.

## ADAPTIVE ALGORITHM MANAGEMENT

The optimization of an airplane wing design using only one classical optimization method leads to relatively high numbers of iterations. An adaption of the variation routine to the best fitted algorithm based on artifical intelligence, heuristics or additional information can decrease the necessary number of iterations significantly and thus the computational amount. The inclusion of an algorithm management module in the MEPO code will be completed in the near future. A first version is in use in the EVOBOX program demonstrating the capability of such an approach [25]. The basic idea is to correlate adaptively the convergence of the quality function for an optimization problem with the use of several specific optimization routines.
In MEPO several different optimization methods can be used for the variation of the parameter set describing the shape of the wing profile as the initial values for the airflow simulation and thus for the calculated objective function. However, the improvement in quality caused by a specific variation algorithm can not be predetermined without additional information. Therefore a meta-routine has been introduced.
The meta-routine controls and manages the calls of the different alternative optimization methods by so-called *call probabilities*. For each routine in use a call probability is calculated by its contribution to the quality improvements in former iterations. All values are normalized and related to the worst quality calculated after the first variation. Optimization algorithms which generate repeatingly parameter sets with improved qualities

get higher call probabilities. The corresponding contributions of others will be decreased toward smaller values, however never reaching zero. In this way routines which probably offer improvements are called more often without neglecting others totally. Modules showing the same call probabilities will be selected by equally distributed random numbers.

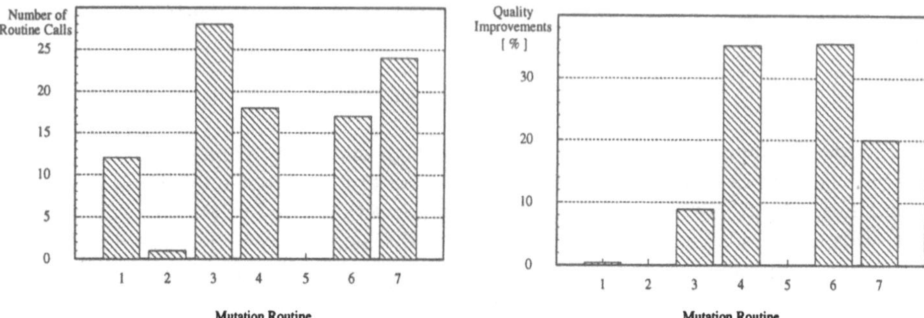


Figure 2: Statistics of the number of mutation routine calls: Number of routine calls (left) and improvements in quality caused by the different mutation routines (right)

The algorithm management routine in the EVOBOX code was able to reduce the computational amount of an arrangement problem with the same computational expense as the wing design by a factor of 10 to 15. Figure 2 shows a typical statistics of the number of routine calls. 6 different variation algorithms were in use. Routine number 5 was disabled. 380 different parameter sets were calculated and evaluated with a simulation code during 95 iterations. The generated quality improvements were normalized and subsumed for each optimization module. Routine number 4 and 6 were used 18 respectively 17 times during this run (left) with an overall quality improvement of about 35% for each routine. Routine 7 caused an improvement of about 20% during 24 calls. The other routines were not very successful in this run but were called randomly by the meta-routine. About 99% of the overall quality increase was generated during 82 of 95 calls. Thus, only 13.7% of the optimization routine calls produced no improvements.
Statistical results for the same problem without any use of the meta-routine and randomly varied optimization algorithms showed 46% - 78% routine calls with no improvements. Management routines with fixed call schemes produced no better convergences. However, these results may depend on the optimization problem, but the tests indicate the potential of an algorithm management. Further routines currently under development will utilise Neural Networks to determine sensitive and insensitive values in the parameter sets of the optimization problem. The intention is to learn the structure of the search space. This information correlated with successful applications of specific variation methods can be stored in an Expert System for similar optimization tasks.

## OPTIMIZATION RESULTS

At DASA the parallel 3D wing optimization method for the Airbus A3XX has been established since August 1996 and first results have been produced with the new program system. Without any parallelization up to 15 hours for regular single point optimizations and up to 60 hours for complete designs are necessary for the iterative airplane design on a HP 9000/700 workstation. With the help of the parallel method results can be produced within a few hours in dependence of the number of processors involved.

At DASA as well as at the TU Braunschweig the optimization jobs run on configurable clusters as low-priority batch jobs during everyday computer operations. This parasitic parallelism is only acceptable on powerful workstations, because several users share memory and processor power. Under these conditions measurements of a speed-up values are only of academic interest.

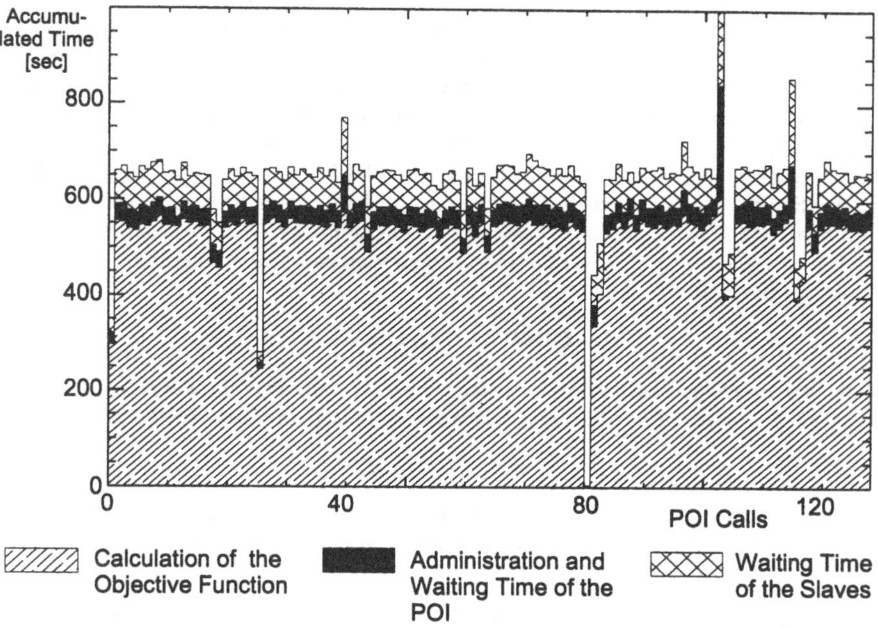


Figure 3: Test run of the optimization program system MEPO for a wing optimization of the A3XX: Parallelization of four multipoint determinations on four HP9000-7xx processors

With the help of the statistical output of the POI the real time partitions for calculations, communication and waiting can be analysed. Fig. 3 shows a test run result of the MEPO system for a design calculation of an Airbus A3XX wing. The accumulated time values for the determination of the objective functions, administration and waiting related to the number of POI calls are shown. Most of the available processor power could be used for the calculations of the objective function. For several hours the multipoint design code combining the different flow analysis codes XLS6, VICWA and EPPLER for the determination of the flow conditions ran on four HP 9000-7xx in a parasitic parallel mode. Only small time portions were used for active waiting or for communication and administration. A few peaks were caused by short online activities causing high processor loads. Depressions in the time history arise from illegal parameter combinations which lead to instant stops of the simulation codes. The optimization algorithms vary the parameters within predetermined boundaries. However, they can not test any interdependencies. Such checks are done in the design code leading to immediate stops with bad objective function values for the forbidden parameter value combinations. In this way time-wasting determinations of unnecessary flow solutions can be avoided.

## CONCLUSION

Based on the optimization algorithms and management modules of the EVOBOX code of the IfRR an improved and reengineered parallel optimization code system MEPO has been designed at the TU Braunschweig since 1995. The extension of classical optimization methods not only by new and combined algorithms but also by the inclusion of management methods based on artifical intelligence will lead to a widely usable optimization system. Parallel algorithmic structures coupled with a low amount of communication between computer processor units in use make it possible for workstation clusters to be employed efficiently. In this way the turn-around times for optimization calculations with hundreds of airflow simulations depend only on the number of disposable processors. There are no restrictions for an extension to other parallel hardware.
A first version of the code has been installed at the DLR in Braunschweig and at DASA in Bremen. Several optimization runs for the Airbus A3XX have been carried out at DASA demonstrating the capability of the established methodology. With the help of the MEPO system the development of a new airplane wing design can now be realized in significantly reduced turn-around times.

## ACKNOWLEDGEMENT

*The authors gratefully acknowledge the contributions and comments of colleagues from the Institute for Flight Mechanics and Spaceflight Technology, from DASA and from the DLR*

## REFERENCES

[1] Becker, K.; *Paralleles Rechnen in der Transportflugzeugentwicklung - Status und Ausblick*, in: Matthies, H.; Schüle, J. (Herausgeber); *Paralleles und Verteiltes Rechnen*, Shaker Verlag, Aachen (1996).

[2] Voegt, S; Lehmann, S; Axmann, J.K.; *Wiedereintrittsbahnen und ihre Optimierung durch einen Evolutionsalgorithmus*, DGLR-Fachausschuß-Sitzung: Flugleistungen und Bahnen, Ottobrunn (1991).

[3] Axmann, J.K.; *Die Optimierung von Raumflugmissionen mit Evolutionsalgorithmen auf Parallelrechnern*, DGLR-Fachausschuß-Sitzung: Flugleistungen und Bahnen, Braunschweig (1993).

[4] Axmann, J.K.; *Raumflugtechnische Optimierung mit adaptiven Evolutionsalgorithmen auf Parallelrechnern*, Jahrbuch der Deutschen Gesellschaft für Luft- und Raumfahrt, 94-E1-074, Band I (1994).

[5] Axmann, J.K.; *Parallelrechner-Nutzung zur mathematisch-technischen Optimierung mit Evolutionsalgorithmen*, in: Rönsch, W.; Schüle, J. (Herausgeber); *Parallelisierung im wissenschaftlichen Rechnen*, Informatikberichte 93-04, TU Braunschweig (1993).

[6] Hadenfeld, M.; Axmann, J.K.; *Parallele Tragflügelprofil-Optimierungen auf Workstation-Clustern*, in: Matthies, H.; Schüle, J. (Herausgeber); *Paralleles und Verteiltes Rechnen*, Shaker Verlag, Aachen (1996).

[7] Forbrich, D.; Van der Velden, A.; Dargel, G.; personal communication (1996).

[8] Dargel, G.; *Ein Programmsystem für die Berechnung transsonischer Profil- und konischer Flügelumströmungen auf der Basis gekoppelter Potential- und Grenzschichtlösungen*, STAB 92-07, 8. STAB-Fachsymposium, Köln (1992).

[9] Dargel, G.; Thiele, P.; *Viscous Transonic Airfoil Flow Simulation by an Efficient Viscous-Inviscid Interaction Method*, AIAA-87-0412, Reno, Nevada (1987).

[10] Eppler, R.; Somers, D.M.; *A Computer Program for the Design and Analysis of Low-Speed Airfoils*, NASA Technical Memorandum 80210, National Technical Information Service, Springfield, Virginia (1980).

[11] Sobieczky, H.; *Progress in inverse design and optimization in aerodynamics*, AGARD 1989, paper 1 (1989).

[12] Greff, E.; Mantel, J.; *An engineering approach to the inverse transonic wing design problem*, Communication in Applied Numerical Methods 2, pp.47-56 (1986).

[13] Reuther, J.; Jameson, A.; *Aerodynamic Shape Optimization of Wing and Wing-Body Configurations Using Control Theory*, AIAA 95-0123 (1995).

[14] Hicks, R. M.; Henne, P. A.; *Wing design by numerical optimization*, AIAA 79-0080 (1979).

[15] Van der Velden, A.; *Aerodynamic Shape Optimization*, AGARD R 803, paper 3 (1994).

[16] Kroll, N.; Radespiel, R.; Rossow, C.C.; and Eisfeld, B.; Bleecke, H.-M.; Kroll, N.; Ritzdorf, H.; *Structured Grid Solvers I and II - Accurate and Efficient Flow Solvers for 3D Applications on Structured Meshes*, AGARD Report R-807 (1995).

[17] Nelder, J.A.; Mead R.; *The Downhill Simplex Algorithm*, Computer Journal, Vol.7, S.308 (1965).

[18] Kirkpatrick S.; Gerlatt C.D.; Vecchi M.P.; *Optimization by Simulated Annealing*, Science, Vol. 220 (1983).

[19] Rechenberg, I.; *Evolutionsstrategie: Optimierung technischer Systeme nach dem Prinzip der biologischen Evolution*, Frommann-Holzboog Verlag, Stuttgart (1972).

[20] Schwefel, H.-P.; *Numerische Optimierung von Computer-Modellen mittels der Evolutionsstrategie*, Birkhäuser Verlag, Basel und Stuttgart (1977).

[21] Holland, J.; *Adaption in Natural and Artificial Systems*, University of Michigan Press, Ann Arbor (1975).

[22] Goldberg, D. E.; *Genetic Algorithms in Search, Optimization and Machine Learning*, Addison-Wesley, Reading, Massachusetts (1989).

[23] Fogel, L.J.; Owens, A.J.; Walsh, M.J.; *Artifical Intelligence through Simulated Evolution*, John Wiley & Sons, New York (1966).

[24] Axmann, J.K.; Krüger, S.; *Lastadaptiver Einsatz eines parallelen Optimierungscodes auf Basis Evolutionärer Algorithmen auf heterogenen Workstation-Clustern*, in: Cap, C. (Herausgeber); *Workstations und ihre Anwendungen*, vdf-Hochschulverlag, Zürich (1996).

[25] Axmann, J.K.; Van de Velde, A.; *Nuclear Fuel Management Optimization Using Adaptive Evolutionary Algorithms with Heuristics*, Proceedings: Int. Conference on the Physics of Reactors, Physor 96, Mito (1996).

# Experiments on Active Control of Tollmien–Schlichting Waves on a Wing

M. Baumann & W. Nitsche

Technische Universität Berlin, Institut für Luft- und Raumfahrt, Sekr. F2
Marchstr. 14, 10587 Berlin, Germany

## Summary

The paper describes experiments aimed on active control of Tollmien–Schlichting (TS) instabilities on airfoils. Natural 2D instabilities (TS-Waves) of a laminar boundary layer were reduced by superposition with artificially excited cancelling-waves to achieve a transition delay. The experiments were carried out in a laminar boundary layer of an unswept wing (airfoil: NACA 0008 mod. $c = 1300$ mm, $b = 600$ mm) at a Reynolds number of $Re_c = 1.3 \cdot 10^6$. A sensor-actuator system was flush mounted in the wing surface and operated by an adaptive controller to perform an automated closed loop control of the wave cancellation process. The actuator (spanwise suction/blowing-slot, located at $x/c = 0.46$) was used to produce the cancelling wave. The controller based on a digital signal processor (DSP) executing an LMS–adaptive FIR–filter algorithm (LMS — Least Mean Square, FIR — Finite Impulse Response). Hot-wire measurements were conducted to analyze the downstream development of the wave cancellation process. The averaged attenuation rate was 50 % in the TS-frequency range. Attenuation of natural disturbances of up to 90 % could be observed at individual frequencies.

## 1 Introduction

A delay of the laminar–turbulent boundary layer transition can be achieved by traditional methods which modify the mean velocity profile in order to stabilize it, such as a negative pressure gradient, distributed suction or wall cooling in air. Furthermore, it is well known that a transition delay can also be achieved by a direct dynamic reduction (active control) of unstable oscillations of the laminar boundary layer using a wave superposition principle. The actually old idea of TS-wave cancellation can more easily realizable with modern electronic devices. In the present work similar control techniques like in recent ANC–applications (ANC — Active Noise Control) are used for active wave cancellation (AWC) of TS-waves in windtunnel experiments. The principle of AWC on the wing surface is shown in figure 1. For a reduction of natural occurring TS-waves it is first necessary to detect them by a disturbance sensor. Then, the controller derives a driver signal for the actuator from the disturbance signal to generate the appropriate cancelling wave in order to minimize the residual disturbances, measured by the downstream control sensor.

Already Schubauer & Skramstad [11] (1943) probably knew that it is possible to reduce instability waves by superposition. Also the linear stability theory suggests the possibility of linear wave superposition, because of the linearized disturbances. However, in experimental work it is difficult to generate the appropriate canceling wave, especially in the case of natural broadband disturbances.

As already referred in [2] or described in more detail in the survey by THOMAS [3], the first known experiments were conducted under simplified test conditions, all indicating the feasibility of TS-wave cancellation of artificially excited sinusoidal TS-waves in flat plate experiments, e.g. by MILLING [9] (1981) or by LIEPMANN, NOSENCHUCK & BROWN [7, 8] (1982). Advanced approaches to cancel broadband disturbances performing a closed loop control of the cancellation process were made by LADD [5] (1988), [6] (1990) and by PUPATOR & SARIC [10] (1989). LADD already used adaptive filter techniques (LMS–adaptive FIR–filter) to cancel artificially excited random TS-waves.

In our previous work [1] (1995), [2] (1996), an advanced closed loop control of natural TS-wave cancellation was demonstrated on a wing. A sensor-actuator arrangement similar to that of the presented work was integrated in an unswept wing (NACA 0012, $c = 800$ mm, $Re_c = 0.8 \cdot 10^6$) and operated by a powerful controller based on an LMS–adaptive FIR–filter. An amplitude reduction comparable to results of the presented work was already obtained, but no detailed research was performed concerning the wave cancellation. A better cancellation performance was observed for pre-triggered disturbances, following the idea that the natural development of instability waves can be forced to a less random and phase stabilized 2D development by a very small sinusoidal excitation, which is introduced by an actuator located upstream in the stable region. This lock-in effect seems promising for future use of active wave control and will be researched in detail in the present project. Natural disturbances in this context generally means cases with no artificial excitation of TS-waves, but under the disturbance environment of the windtunnel used.

# 2 Experimental Apparatus

For the cancellation experiments a new test wing was designed to allow a more detailed study of sensor-actuator systems including boundary layer measurements in the present research project. The wing has a modified NACA 0008 airfoil with a chord of $c = 1300$ mm and a span of $b = 600$ mm. The contour was modified to obtain a flat surface in the rear region, i.e. instability region to enable also test cases with almost zero pressure gradient. The curvature of the NACA 0008 is continuously reduced beginning at $x/c = 0.25$ until the flat part is reached at $x/c = 0.36$. The wing is especially designed for the testsection of the boundary layer wind tunnel at the ILR (testsection: $h = 400$ mm, $b = 600$ mm, $l = 1350$ mm) and offers easy mounting conditions for various sensor-actuator systems.

A sensor-actuator system was integrated in the wing surface, as shown in the principle sketch of figure 1. The actuator consists of a spanwise slot with five small loudspeakers arranged in a cavity beneath, in a similar design like in [2]. The two-dimensional perturbations are induced in the boundary layer by suction and blowing through the slot, resulting in small periodical velocity fluctuations $v'$ perpendicular to the wall. Small surface hot-film arrays (5 sensors per array) are located upstream and downstream of the slot. One of the upstream sensors is used as disturbance sensor and one from downstream as the control- or error sensor. The dimensions of the arrangement are shown in figure 2.

The adaptive controller was almost the same as in the previous work [2], based on an adaptive filter algorithm, following the principle of figure 3 and also known as 'filtered–x–LMS' [4]. This control algorithm was programmed in assembly language for a floating point DSP (Analog Devices ADSP 21020). The algorithm consists of an LMS–adaptive FIR–filter ($FIR_1$) to model the transfer function between the disturbance sensor and the actuator (path 1) and an additional error correlator to ensure the convergence of the adaptation by the LMS. For this purpose a second FIR–filter ($FIR_2$) is used to model the error path (path 2) between the actuator and the control or error sensor. The disturbance signal is also filtered by this $FIR_2$ and multiplied with the real error signal to get a correlated error signal for the stable adaptation of the first FIR by the LMS. The second FIR can be adapted off-line in a pre-training while sending a random noise signal to the actuator and measuring the response with the error sensor. In regular operation, the first FIR–Filter is continuously adapted by the LMS–algorithm in order to minimize the error signal and therefore maximizing the cancellation effect. The sampling rate of the controller is 8 kHz, $FIR_1$ has 100 coefficients and $FIR_2$ has 200 coefficients. The described control algorithm is running very stable and is already fully adaptive in its latest version, because of the additional implemented continuous adaptation of $FIR_2$.

# 3 Results of Active Wave Control

The measurements of the TS-wave cancellation effects were carried out at a moderate adverse pressure gradient along the flat area of the upper wing surface, where the sensor-actuator system is located. The angle of attack was chosen to $\alpha = -1.46°$ and the freestream velocity was $U_\infty = 17$ m/s, resulting in a Reynolds number of $Re_c = 1.3 \cdot 10^6$. The active cancellation is first indicated by the signal of the downstream control or error sensor, when the case with AWC is compared to the case without control. Figure 4 shows typical time traces of the sensors and the actuator in streamwise order. In the left case no control was applied, indicated by the missing actuator signal. Now, the control sensor shows typical TS-wave signals in the linear range. In the case of active wave control the cancelling actuator signal can be observed, which is the adaptively filtered signal of the disturbance sensor. The control sensor shows a significant reduction in amplitude compared to the case without control. Averaged power spectra of the control sensor (figure 5) are indicating the best amplitude reduction in the frequency range of the TS-waves (300 Hz – 500 Hz) characterized by the bump and also for higher harmonics between 700 Hz and 900 Hz. Typical amplitude reductions in the TS-frequency range are around 12 dB, with a attenuation of single peaks in the spectra by up to 20 dB.

To analyze the cancellation effect downstream of the actuator, hot-wire measurements were conducted in the boundary layer. Mean velocity profiles ($u$) and their fluctuations ($u'$) were measured at different $x/c$-positions on the wing surface, comparing the case with AWC and without control. First it was tested whether the cancellation effect is homogenous in spanwise direction. Only a small spanwise area behind the control sensor could be reached with the actual 2D traversing unit. Across the possible spanwise range of $z = -10$ mm up to $z = 26$ mm relative to the control sensor ($z = 0$) the $u'$ distribution was almost constant, showing only a slightly better cancellation directly downstream of the control sensor.

The streamwise development was measured starting at a position 10 mm downstream of the actuator-slot in the centerline of the wing, which has a spanwise distance of 10 mm to the control sensor (figure 2). The mean velocity profiles (figure 6) indicate almost no change when control is applied. For a further downstream location (last two profiles) they are identical. The small variations in the first four profiles should be discussed on the basis of more data to be measured in the present project. The corresponding $u'$-profiles are clearly indicating the effect of active wave control. The data are averaged only over the relevant TS-frequency range (200 – 450 Hz) to visualize the effects of primary interest. At the first profile (10 mm downstream the slot) the cancellation effect takes place only in the near wall region, but already indicating an amplitude reduction of approximately 50% at the wall ($y$ = 0.2 mm). While developing downstream the cancellation spreads up to greater wall distances until the whole boundary layer is affected, beginning at $x$ = 650 mm (50 mm behind the slot). The TS-wavelength is approximately 25 mm. Further downstream the amplification of the TS-waves is clearly indicated by the strong growth of the $u'$–profiles. For the controlled case a significant amplitude reduction can be observed. The $u'$–maximum found at $y$ = 1.44 mm of the last profile is reduced from $u'$ = 2.4 % down to 1.2 %, showing an attenuation of the local turbulence level by 50 %. The power spectra of the local turbulence level, measured at the $u'$–maximum of the last profile (figure 8) shows again the typical TS-bump across the marked frequency range used for the data evaluation of $u'$–profiles. Higher amplitude differences than in the spectra of the control sensor (figure 5) are found and the amplitude reduction is more distributed across a wide frequency range, which is a comparable result to our previous work [2].

# 4 Conclusions

Active cancellation of natural 2D TS-waves was successfully investigated in experiments on a wing. A sensor-actuator system was employed in the instability region of the wing surface and operated by a controller to perform an automated closed loop control of the TS-wave cancellation. The controller based on an LMS–adapted FIR–filter algorithm running on a digital signal processor. The downstream development of the cancellation process was studied with hot-wire measurements. Compared to the case without control, amplitude reductions of the $u'$–fluctuations between 50 % (averaged over the TS-frequency range) and 90 % at individual peaks in the spectra were obtained with AWC (Active Wave Control). Further studies with an improved sensor-actuator system are planned, also employing the pre-trigger technique of our previous work [2]. The investigations of 2D control are an essential base for more advanced studies of 3D control in the future.

# 5 Acknowledgments

The work was conducted in a research project with the financial support of the DFG (Deutsche Forschungsgemeinschaft)

# References

[1] Baumann, M., Nitsche, W. — Aktive Grenzschichtbeeinflussung laminar–turbulenter Profilströmungen. ILR–Mitteilung 293 (Jan. 1995), TU–Berlin.

[2] Baumann, M., Nitsche, W. — Investigation of Active Control of Tollmien–Schlichting Waves on a Wing. In: Transitional Boundary Layers in Aeronautics, Eds: Henkes, R.A.W.M., van Ingen, J.L., KNAW, Vol. 46, 89–98, Amsterdam 1996.

[3] Thomas, A.S.W. — Active Wave Control of Boundary–Layer Transition. Edrs: Bushnell, D.M., Hefner, J.N., Viscous Drag Reduction in Boundary Layers. Progress in Astronautics and Aeronautics Vol. 123, AIAA 1990.

[4] Elliott, S.J., Nelson, P.A. — Active Noise Control. IEEE Signal Processing Magazine, October 1993, 12–35.

[5] Ladd, D.M. — Active control of 2–D instability waves on an axisymmetric body. Experiments in Fluids, 6, 69–70.

[6] Ladd, D.M. — Control of Natural Laminar Instability Waves on an Axisymmetric Body. AIAA Journal, 28, 367–369.

[7] Liepmann, H.W., Brown, G.L., Nosenchuck, D.M. — Control of laminar–instability waves using a new technique. J. Fluid Mech., 118, 187–200.

[8] Liepmann, H.W., Nosenchuck, D.M. — Active control of laminar–turbulent transition. J. Fluid Mech., 118, 201–204.

[9] Milling, R.W. — Tollmien–Schlichting wave cancellation. Phys. Fluids, 24, 979–981.

[10] Pupator, P.T., Saric, W.S. — Control of random disturbances in a laminar boundary layer. AIAA Paper 89-1007.

[11] Schubauer, G.B., Skramstad, H.K. — Laminar–Boundary–Layer Oscillations and Transition on a Flat Plate. Report No. 909, National Bureau of Standards, Washington D. C., April 1943.

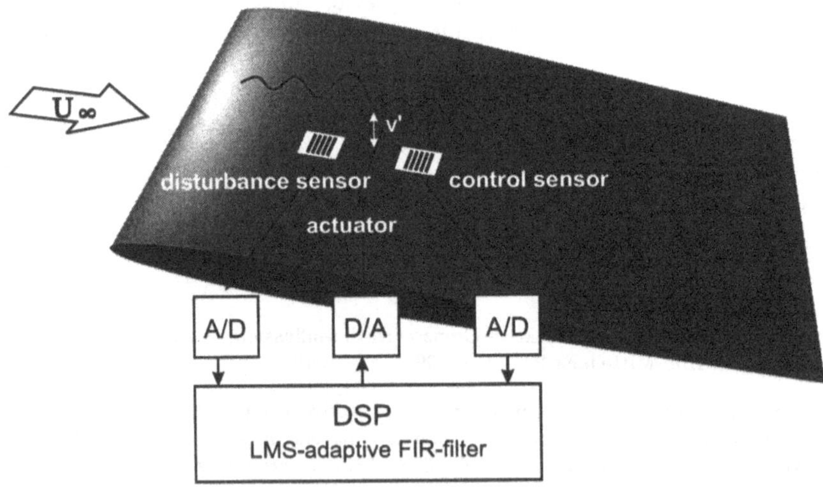


Figure 1: principle of active wave control of TS-waves (2D instabilities) on the test wing

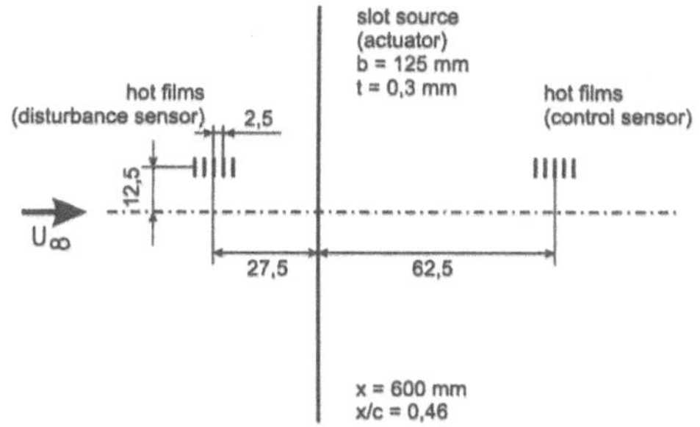


Figure 2: Layout of the sensor-actuator system

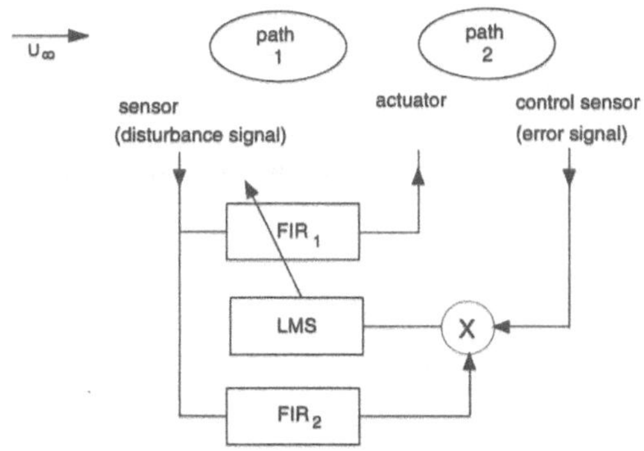


Figure 3: Principle of the adaptive controller ('filtered–x–LMS') [2]

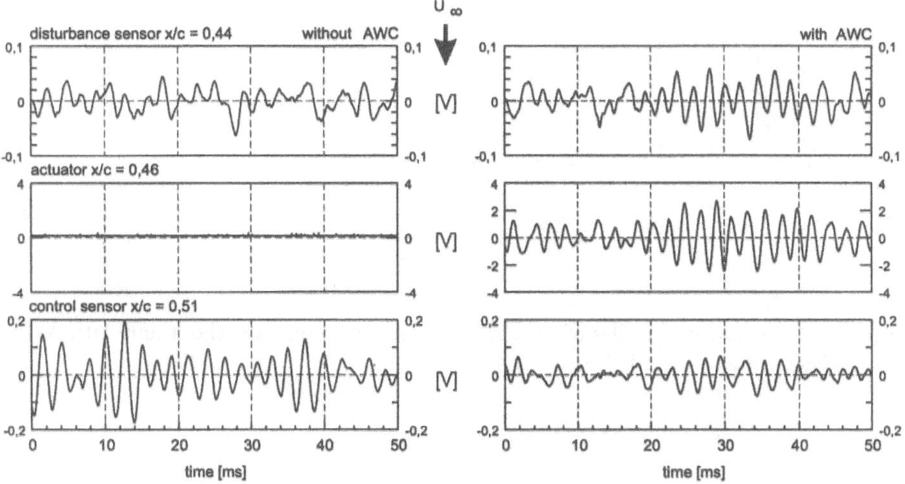


Figure 4: Signals of the sensor-actuator system without and with AWC ($U_\infty = 17$ m/s, $Re_c = 1.3 \cdot 10^6$, $\alpha = -1.46°$)

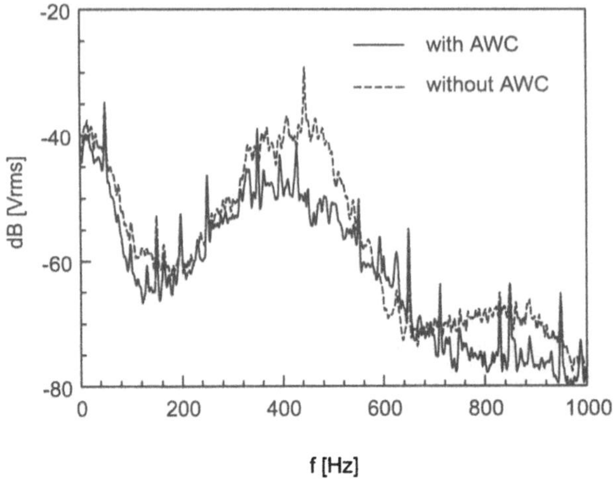


Figure 5: Power spectra of the control sensor without and with AWC ($x$ = 665 mm)

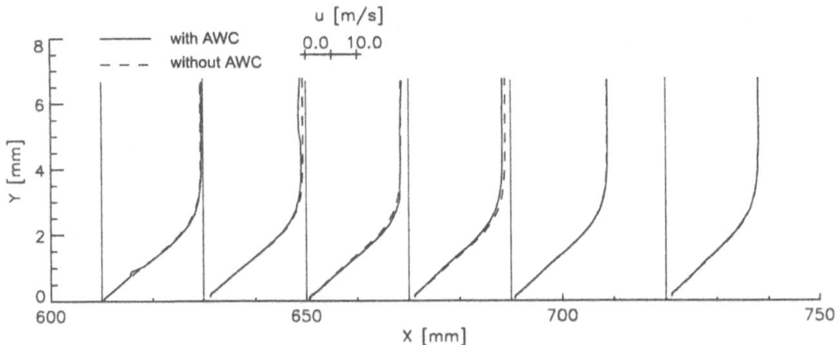


Figure 6: Mean velocity profiles of the donwstream development without and with AWC, ($U_\infty = 17$ m/s, $Re_c = 1.3 \cdot 10^6$, $\alpha = -1.46°$)

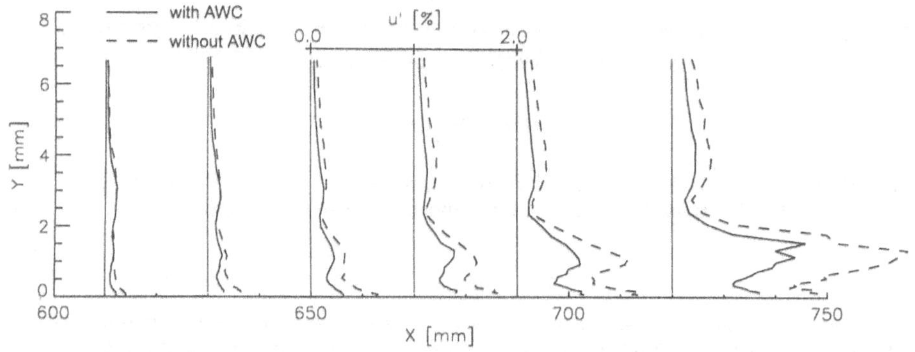


Figure 7: $u'$–profiles of the downstream development without and with AWC (actuator at $x = 600$ mm, $u'[\%] = 100 \cdot u'/U_\infty$, frequency range: 200 - 450 Hz)

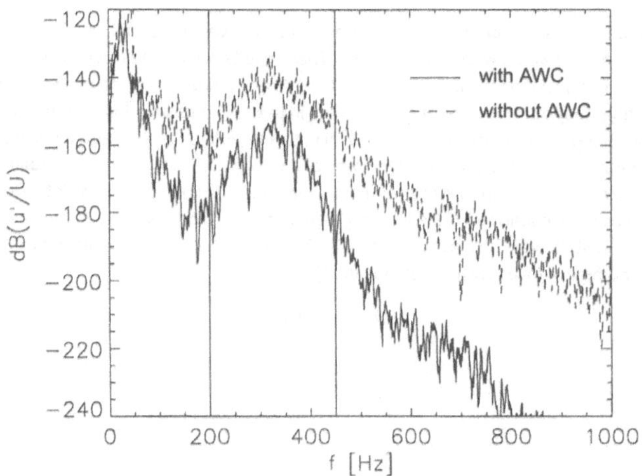


Figure 8: Power spectra of the local turbulence $u'$ without and with AWC ($x = 720$ mm, $y = 1.44$ mm)

# Stability Analysis of two- and three-dimensional Boundary Layer Flows with varied Wall Temperatures

F. P. Bertolotti#, H. Bieler *

# Institut für Strömungsmechanik, DLR Göttingen, Bunsenstr. 10, D-37073 Göttingen
* Daimler-Benz Aerospace Airbus GmbH, EFV, D-28183 Bremen

## 1. Summary

In this work, performed in cooperation between DLR and Daimler-Benz Aerospace Airbus, we have investigated the effect of heating strips on the stability of boundary-layers in air. Our work is based on numerical analysis, and is motivated by previous experimental observations that have shown a net stabilizing effect in subsonic two-dimensional boundary layers when thin heating strips are introduced near the stagnation region. Our focus is specific to flow conditions existing over transonic commercial aircraft, with particular attention given to the two-dimensional flow over a nacelle, and the three-dimensional flow over an Airbus A340 wing. Accordingly, our presentation is divided into two main parts. In the first part we show that a single heating strip roughly 4 centimeters wide and placed near the leading-edge of a nacelle can reduce by half the amplitude of the most unstable waves, and, hence, it is an effective method for laminar-flow control.

In the second part we show that localized heating does not lead to any appreciable gain in stability when applied to an Airbus A340 wing. However, localized cooling over the wing-box region does have a strong stabilizing effect, and may be a potentially useful laminar-flow-control technique in future aircraft employing cryogenic fuel.

## 2. Previous Work

One of the first studies of the effect of local heating on transition was performed in 1987 by Landrum and Macha [1]. They used a NACA 0012 airfoil with a heated nose. Their results are inconclusive, however, because over most of the chord there was an adverse pressure gradient, whose strong destabilizing effect overshadowed the more delicate effects of heating. They observed only a slight change in the transition location. In contrast, the results of Maestrello and Nagabushana [2] were more impressive. With high amounts of heating at the leading edge of a flat plate, they could relaminarize a fully turbulent boundary layer at a position significantly downstream of the heating strip. However, Maestrello's results are limited to two-dimensional flows. Experimental work by Dovgal, Levchenko and Timofeev [3] in Novosibirsk confirmed the stabilizing effect of localized heating in 2D flows, but also showed that the effect of heating in 3D flows is no longer beneficial. Their results for the two-dimensional boundary-layer were reproduced in the numerical investigations of Masad [4].

## 3. The Nacelle

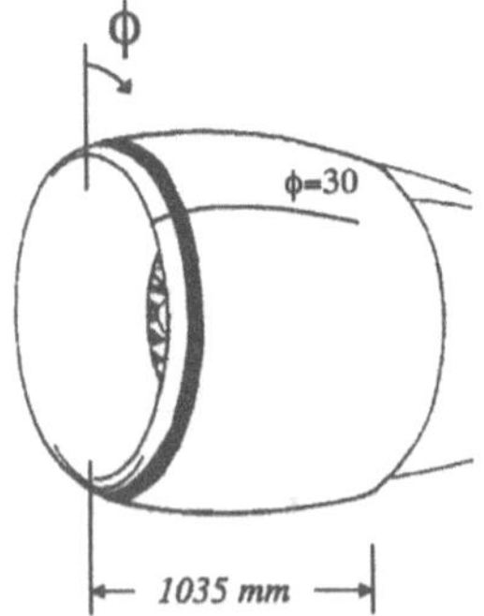


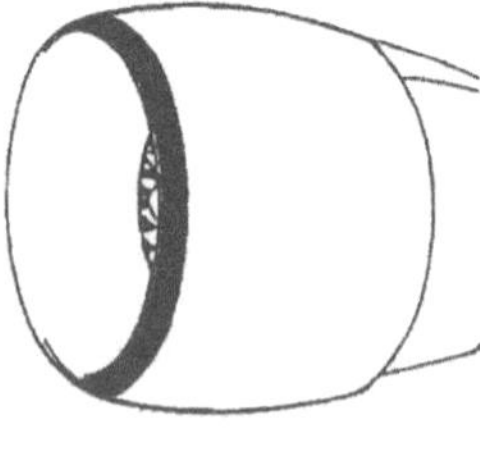

Fig. 1: Geometry of the nacelle. The black stripe shows the location of the heating strip. Left, cases C and D; right cases A and B.

The geometry and pressure distribution for the nacelle were provided by the Institut für Entwurfsaerodynamik at DLR Braunschweig. The nacelle was mounted on top of the ATTAS wing, flying at 6500 altitude at Mach 0.56. Fig. 1 shows relevant geometry. The angle $\Phi$ measures the angular position around the leading edge, and herein we investigate the stability properties of the boundary layer along lines of constant $\Phi$, at $\Phi$=30, 140, 220 and 330 degrees. The reference quantities used to non-dimensionalize all the variables in our mathematical formulation are L = 0.391 millimeters, V = 150 m/s and T=250 Kelvin. The chord length c of the nacelle is c=1035 millimeters. The plots that form the left column in Fig. 2 show four heating-strip configurations investigated. In cases A and B a 4-cm wide strip is placed near the stagnation line. Case A uses 250 deg Celsius of overheat, case B uses 125. The temperature downstream from the strip is specified to be equal to the stagnation temperature, which, in view of the low Mach number, remains close at all downstream locations to the adiabatic wall temperature in the reference case without heating. In cases C and D a 2-cm strip is placed a little downstream of the stagnation line. The only difference between case C and D is that in case D the surface downstream of the strip is insulated (adiabatic). In all four cases the heating is applied in the stable region. The plots on the right column of Fig. 2 show the heat-fluxes at the wall. Negative values signify heat added to the flow, positive values heat subtracted from the flow. As is well known, the relative cooling that takes place downstream of the strip in cases A, B and C has a stabilizing effect on the flow. However, that's not the whole story. In case D there is no heat-flux downstream of the strip, yet a stabilizing effect exists, as will be shown shortly. Local heating also delays separation, but we will not speak about this effect in this paper. An additional benefit of wall heating not considered herein, but which should be considered in a global analysis, is the reduction of wall shear stress in a turbulent flow, as discussed in [5].

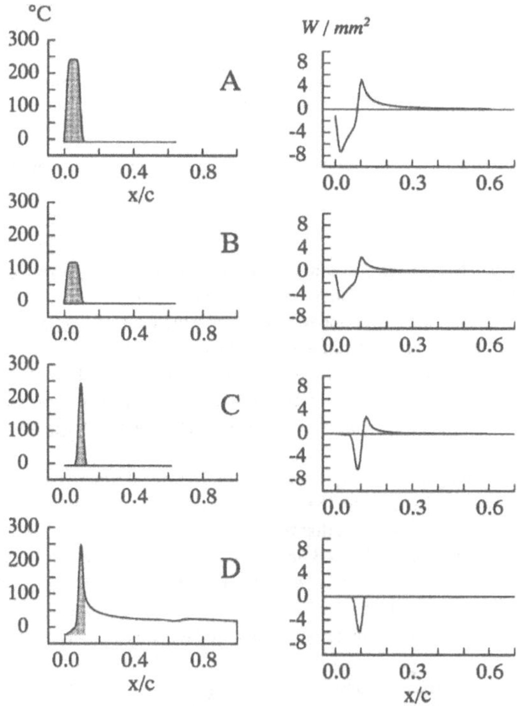


Fig. 2: Heating-Strip data: Left, temperature distribution along the wall; Right, heat-flux, in W/mm$^2$, along the wall.

In Fig. 3a we show the N factor for a TS wave at frequency F=30 (where $F = 2 \times 10^6 \, \pi \, f \nu / L^2$) for the four heating cases plus the reference case without any heating. The difference between the amplitude in the reference case and the case D shows the stabilizing effect of localized heating in a flow without relative cooling along at the wall. The cause of this stabilizing effect is not yet understood. The difference between the amplitude in case D and C indicates the additional stabilizing effect that heat transfer at the wall brings. Case A displays the strongest stabilizing effect. The N factor is just about halfed. The stabilizing effect of local heating, case A, on all unstable TS-wave frequencies is visible through the reduction in size of the neutral stability curve, as shown in Fig. 3b. Heating has no stabilizing effect in the trailing part (x/c > 5) because there the flow is inflectional there, prior to laminar separation.

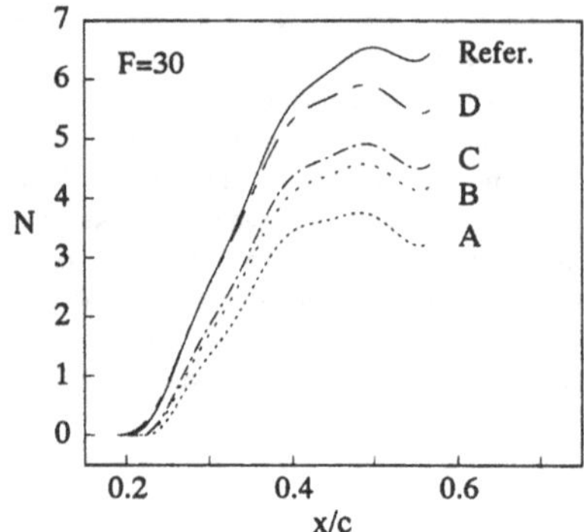


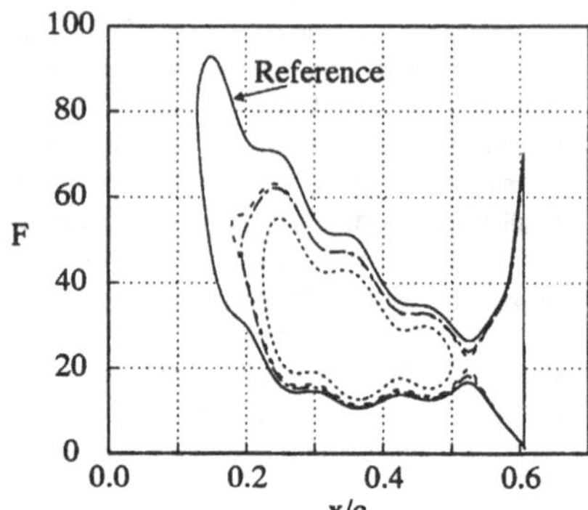


Fig. 3: (a) N-factors for the four heating cases. (b) neutral stability curves, $\Phi$=30.

## 3. A340 Glove

In order to investigate the effect of heat fluxes into and out of the wall for a boundary layer flow around a swept wing, the A340 (equipped with HLFC glove) at cruise conditions has been chosen.

Because our study is concerned with laminar boundary layers only, the laminar type pressure distribution (upper side) of the glove at the mid-span position is a suitable case. Our investigation is motivated by the fact, that for sufficient high Reynolds numbers and sweep angles some type of boundary layer flow control might be necessary in the box region of wings or tails (where suction techniques are difficult to apply due to space limitations). Additionaly, if projects like the hydrogen driven aircraft will be realized in the future, an energy reservoir for cooling purposes is already available.

Fig. 4 gives an overview of the heating and cooling arrangements. The wall downstream from the heating strip and cooling strip is adiabatic. Like the heating case for the nacelle described above, heating has been applied close to the nose of the glove; i.e. in the stable region. Note that the axis goes from 0 to s/c of only 5 %. This might produce possible benefits from relative cooling as the heated flow passes over a "relative cold" wall directly downstream of the strip. This strip - however - is very narrow because the neutral curve is almost located at the attachment line (about 0.2 % of arc length). The heating levels $T_w/T_{inf}$ = 1.3 and 1.5 were considered. A wider heating strip with $T_w/T_{inf}$ = 1.3 was applied to check the effect of disturbance amplification on travelling modes only, which get amplified at about 2 % of arc length.

The adiabatic wall temperature is for the considered Mach number (M = 0.82) approximately 10 % above the free stream temperature ($T_\infty$ =216 K for the cruising altitude of about 39000 ft). This means $T_{w,adiabat}$ = 240 K. The heat levels correspond therefore to $T_w$ = 280 K and 324 K, respectively. These settings were chosen to illustrate - within a basic study - the effects. The consequences for the size of the energy source and power requirements, which lead to a certain drag penalty, have to be evaluated in the future.

The pictures on the right hand side of Fig. 4 indicate the cooling strip arrangement. It is known from early flat-plate experiments that cooling in unstable regions lowers the amplification rates. In the box region cooling was applied between 20 and 30 % of arc length; a region where stationary and travelling modes are present (see the neutral points in the middle pictures). The cooling level of $T_w$ / $T_\infty$ =0.9 corresponds to a wall temperature of 195 K, which is roughly 40 K below the adiabatic value.

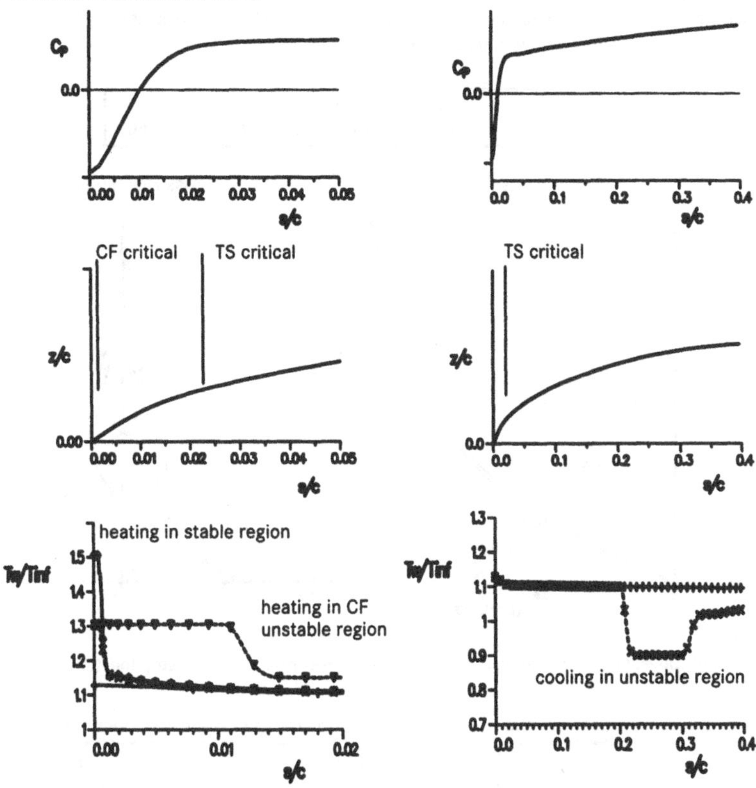


Fig. 4: A340 Glove at cruise conditions (M = 0.82)
Heating and cooling strip arrangements

As it was done in a prior work [6], a blending of the wall temperature at the beginning and ending of the heating and cooling strip with the neighboring adiabatic wall temperature is used to obtain a smooth temperature distribution. This is necessary to avoid numerical problems and to obey the assumptions which reduce the Navier-Stokes equations to the boundary layer equations. In practice, a certain heat flux in streamwise direction exists anyway.

## 4. Stability Analysis of the Heating Cases

An inspection of the streamwise momentum equation in the vicinity of the wall reveals the balance that exists between the streamwise pressure gradient, $\partial p/\partial x$ , the wall normal derivative of the viscosity

$$\frac{\partial \mu}{\partial z}\frac{\partial U}{\partial z}$$

and the second wall normal derivative of the velocity,

$$\frac{\partial^2 U}{\partial z^2} \, .$$

A change in the wall temperature has basically two effects. First, heating in air makes $\frac{\partial \mu}{\partial z}$ negative, hence increases the value of $\frac{\partial^2 U}{\partial z^2}$ .

If the pressure gradient is zero, then heating introduces an inflection point in the velocity profile, and the flow is made less stable. Second, a heated wall lowers the density close to the wall but leaves the driving force (i.e. the pressure gradient) unaffected, producing a fuller velocity profile.

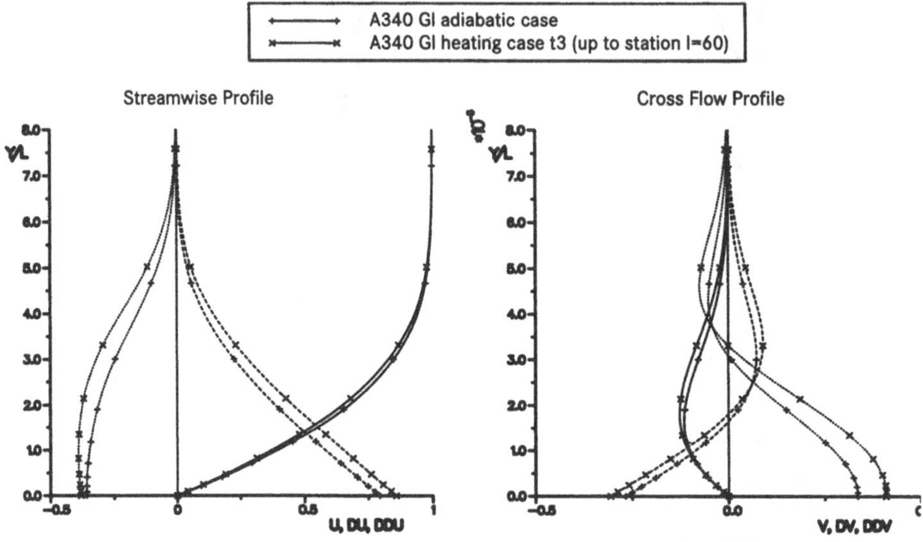


Figure 5: A340 Glove: Effect of heating strip on boundary layer profile; i = 50 (strip location i=1 to i=60)

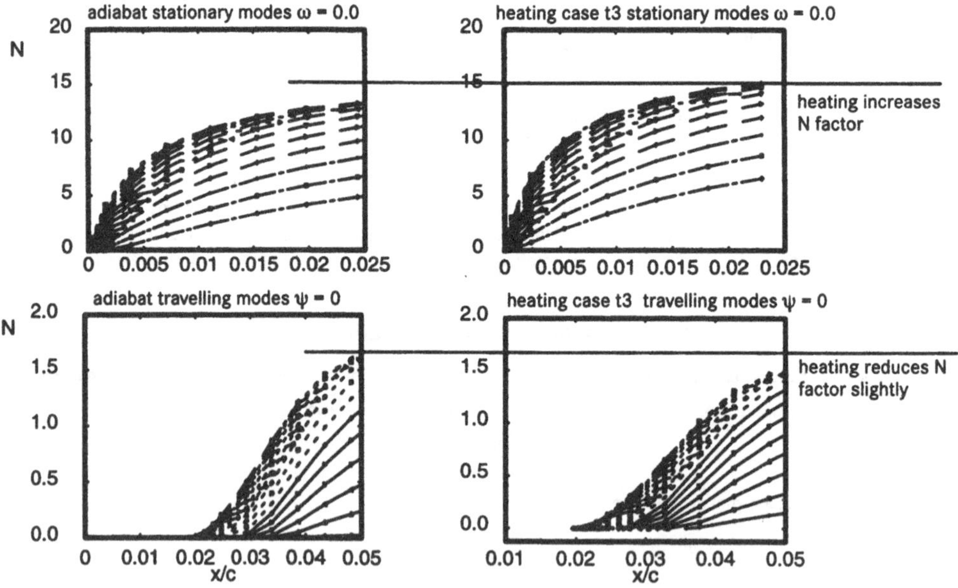


Fig. 6: Compressible N-Factors (No curvature) for CF- and TS-type disturbances

This effect makes the flow more stable.Thus, the net effect of heating depends on the actual values of the different terms. We can infer, furthermore, that localized heating in the vicinity of the stagnation flow should provide the most beneficial effect, since in this region the pressure gradient is strongly favorable, while cooling in air is most beneficial in a region wherein the pressure gradient is weaker (e.g. over the wing-box).

Fig. 5 shows the streamwise and the cross flow profiles and their wall normal derivatives at a specific location inside the heating strip as a function of the wall normal. The streamwise profile (left picture) is less full in case of heating compared to the reference case (adiabatic flow conditions). The cross flow maximum velocity (right picture) increases due to heating and the inflection point moves upwards from the wall.

From these figures one might expect an increased amplification rate due to heating. Fig. 6 gives an overview of the N-factor developmentfor the case with $T_w / T_\infty = 1.3$. Plots are shown for stationary (upper plots) and travelling (Tollmien-Schlichting type disturbances, lower plot) in comparison with the reference case (adiabatic). Heating reduces the N-factor for Tollmien-Schlichting type modes very slightly. But, heating increases the N-factor for stationary modes significantly.

## 5. Stability Analysis of the Cooling Case

Cooling has been applied inside the wing box region; the idea being that in future cryogenic-fuel aircraft, there will be an ample source of a cold sink.

Fig. 7 indicates that cooling acts in the opposite direction, as far as the changes of the streamwise and cross flow profiles are concerned. The data correspond to a position inside the cooling strip. The streamwise profile becomes fuller due to cooling (left plot) and the maximum cross flow velocity drops slightly and the inflection point moves a little bit downwards (right plot). This is quite understandable, since cooling reduces the temperature near the wall, hence the viscosity. The density is increased, but at this s/c location the pressure gradient is quite small.

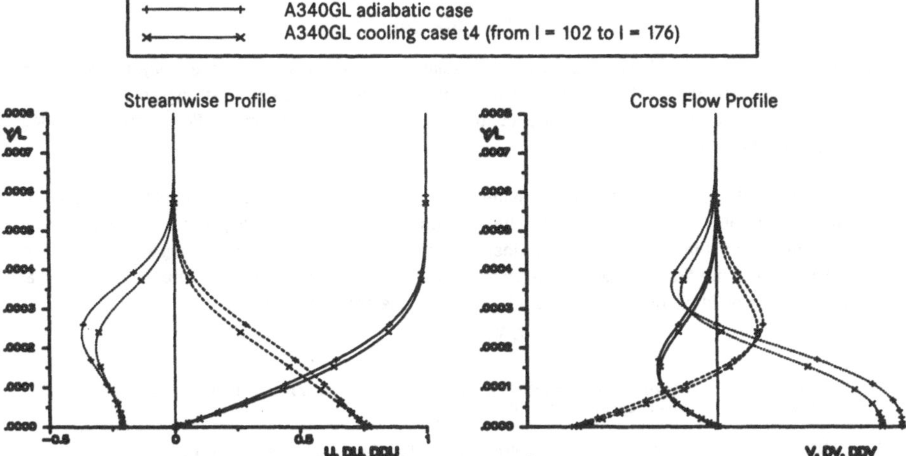


Fig. 7: A340 Glove: Effect of cooling strip on boundary layer profile
i = 155 (strip location i=102 to i=176)

A further effect (not shown here) is a reduction of the displacement thickness, which leads to smaller shape factor and a lower local Reynolds number.

The applied cooling level affects the disturbance development for the travelling modes strongly (Fig. 8 lower pictures). The reference case (adiabatic flow conditions) contains amplified modes downstream of the neutral point. With cooling applied, the amplification drops to zero in the strip region (between 0.2 and 0.3 of chord) and downstream of the strip the disturbances grow again. The cross flow amplification rates are practically unaffected.

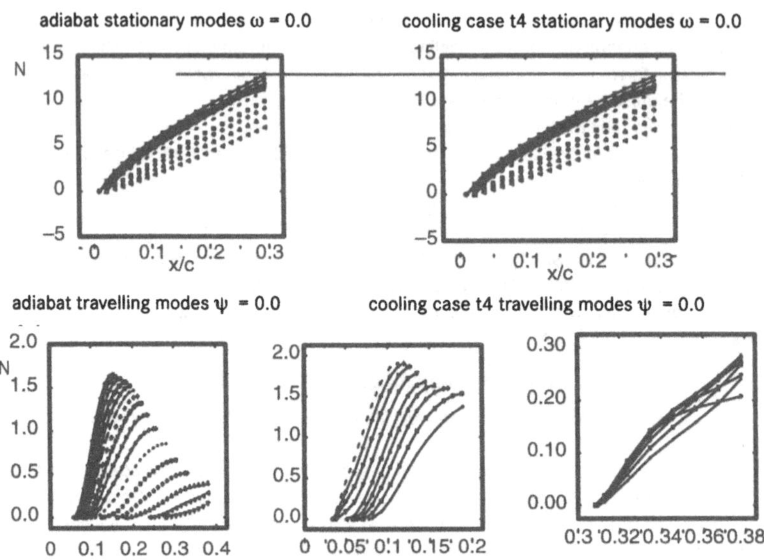


Fig. 8: Compresible N-Factors (No curvature)
for CF- and TS-type disturbances

## 6. Conclusions

Our investigation into the effect of heating strips on the stability of a 2-dimensional boundary-layer flow over a nacelle at Mach 0.56 has shown that the amplitude of Tollmien-Schlichting waves can be strongly reduced. The heating strips, thus, may offer an alternative form of control to suction for hybrid-laminar-flow designs, provided a "free" source of heat is available.

Our investigation into the effect of heating strips on the stability of the 3-dimensional boundary-layer flow over the Airbus A340 wing with HLFC glove, has shown that the amplitude of Tollmien-Schlichting waves is only weakly reduced, while the amplitude of the steady cross-flow instabilities is weakly amplified. The heating strips, thus, do not seem to be beneficial in this flow. A cooling strip, on the other hand, has a clear stabilizing effect on TS waves, and may prove to be an interesting option for future aircraft employing cryogenic fuel.

## 7. References


1. Landrum D.B., Macha J.M., Influence of a heated leading edge on boundary layer growth,stability, and transition. AIAA Paper 87-1259, 1987.

2. Maestrello L., Nagabushana K.A., Relaminarization of turbulent flow on a flat plate by localized surface heating. AIAA Paper 89-0985, 1989.

3. Dovgal A. V., Levchenko V. Ya., Timofeev V. A., Boundary layer control by a local heating of the wall. IUTAM Symposium on Laminar-Turbulent Transition, Toulouse, France 1989, D. Arnal, R. Michel (Eds.), Springer-Verlag.

4. Masad, J. A. Transition in flow over heat-transfer strips. Phys. Fluids 7 (9), 1995, pp. 2163, 2174.

5 Hirschel, E. H. Review of surface temperature effects on vehicle aerothermodynamics in high-speed flow. Daimler-Benz Aerospace Report No. DASA/LMLE3/S/STY/209/A, September 1996.

6 Bieler, H., Bertolotti, F.P. Einfluß lokaler Heizung auf die Stabilität dreidimensionaler Grenzschichtströmungen. Presented at 7. STAB-Workshop; 14.-16.11.1995; Göttingen.

# HEAT LOADS DURING HYPERSONIC -FLIGHT-CONDITIONS AT LOW ALTITUDES: COMPUTATIONS AND FREE-FLIGHT INVESTIGATIONS OF SEVERAL TESTDOMES

F. St. Boller, P.G. Fisch, A. Merten, K. Moldenhauer
Bodenseewerk Gerätetechnik GmbH, Postfach 101155, D-88641 Überlingen, Germany

## SUMMARY

This paper presents the results of numerical analyses and free-flight experiments to investigate aerokinetic heating of testdomes during accelerated hypersonic flight at low altitudes. A combination of a finite-volume CFD program and a FEA structure program was used to compute the aerokinetic heating of the testdomes. Within the German hypersonic missile program HFK, one realistic free-flight has been carried out. During this free-flight, several temperatures on the dome have been measured. At the present status of the project, an easy integration of the test dome into the nose of the cone of the HFK is desirable. Therefore, in this paper only centrally located test domes are considered. The results of the simulation and of the free-flight measurements are compared and discussed. The measurement technique and the influence of the modelling of the transition region are described.

## INTRODUCTION

The German Ministry of Defence (BMVg) together with the Federal Agency of Defence Technology and Procurement (BWB) initiated the technology project Hypersonic Missile (Fig.1) for concept studies of a new generation of short range air defence guided missiles. For missile flights at high supersonic and low hypersonic speeds ($2 < Ma < 6$) the problem of aerodynamic heat loads becomes important and has to be taken into account in the development phase [1]. The induced thermal stresses can affect the structural integrity of the missile or parts of the missile. In general, heat loads cover [2]:

- amount of heat :Q [J]
- heat flux: Heat transported per unit area A and time Δt: $\dot{q} = \frac{Q}{\Delta t\, A}$ $\left[W/m^2\right]$
- temperature T [K].

Bodenseewerk Gerätetechnik (BGT) projects a missile with autonomous guidance using an infrared seeker. The infrared seeker technology is based on target identification via infrared radiation. An infrared transmitting seeker window is needed, which has to be resistant to all aerodynamic and aerothermodynamic loads. A critical issue for seeker windows is the danger of window fracture as a result of thermal residual stresses generated during and after the acceleration phase of the missile. In order to understand the nature of the problem that we face in predicting the aerokinetic heating of the Hypersonic Missile it is necessary to mention the characteristics of this missile: The Hypersonic Missile (HFK) accelerates extremely fast while operating at low flight altitudes (with high air density) and hence at very large Reynolds numbers. The missile accelerates in a time interval of 0.8 seconds to a Mach number of 5.3. Due to the given combination of Reynolds number and Mach number, both laminar and turbulent boundary layers occur. The transition from a laminar to a turbulent boundary layer over the dome is investigated in several studies [3-5]. The transition region depends on Reynolds number, geometry, curvature effects, pressure gradients, propulsion oscillations, wall roughness and other boundary conditions. In Germany, existing ground facilities for high supersonic and low hypersonic heat transfer experiments do not provide realistic combinations of Mach number, Reynolds number, total temperature and transient velocity profiles. Moreover, windtunnel and shock tube experiments are not suitable to be directly applied to accelerating missiles. Consequently, an engineering tool for calculating aerokinetic heating and thermal residual stresses of arbitrarily shaped IR-windows or missile parts had to be established. To accomplish this task, two commercial software tools were coupled: an explicit finite-volume Navier-Stokes solver and a FEA-structure code. The Navier-Stokes solver calculates the heat loads and the FEA program computes the resulting heating and stresses in the structure. The tools run on standard workstations. Regarding the limiting hardware, the two software codes were not linked completely. The two codes are interrelated via the pressure and the heat loads (or via the corresponding coefficients respectively). Each program runs separately to avoid interactive iterations (no closed software loop).

## AEROKINETIC HEATING

The high kinetic energy of an hypersonic missile leads to an extreme heating of the surface. Aerokinetic heating is caused by a transport of energy from the airflow into the structure as a result of the temperature gradient in the boundary layer surrounding the missile. Among other effects, the actual amount of the heat flux into the wall depends on the wall temperature, i.e., aerodynamic heating is a combined fluid mechanics and structural mechanics problem. The aerodynamic heating will increase the surface temperature of the structure and the heat will be conducted through the structure with time. In the present simulations no radiation-cooling effects are taken into account, because at low flight altitudes and at usual external temperatures the aerokinetic heat flux is three orders of magnitude larger than the radiation-cooling flux. For practical calculations it is not suitable to operate directly with the heat flux as variable, because it is too strongly dependent on the wall temperature. In general, heat transfer data are presented in the form of the heat transfer coefficients $\alpha$, defined as:

$$\alpha = \frac{\dot{q}}{T_{aw} - T_w}$$

or by the Stanton number St, defined as:

$$St = \frac{\alpha}{\rho_\infty u_\infty c_p}$$

where $T_w$ is the wall temperature, $T_{aw}$ is the adiabatic wall temperature and where $\rho_\infty$, $u_\infty$, $c_p$ are the reference density, velocity and the specific heat at constant pressure of the free stream. The Stanton number is a dimensionless coefficient nearly independent from the actual wall temperature [1]. The Stanton numbers for different wall temperatures are presented in Fig. 2 versus the Reynolds number for a flat plate for the Mach number 5. Fig. 2 shows also the Stanton number for a cold wall and for an adiabatic wall. Comparing the Stanton number (Fig. 2) with the corresponding temperature boundary layer profile (Fig. 3) we observe only a small influence of the temperature profile or wall temperature on the Stanton number. For design purposes this influence is not very important. Regarding this, a practical way to calculate the aerokinetic heating is to compute data-arrays of Stanton numbers for constant wall temperatures and to operate with these numbers for all wall temperatures. This results in a decoupling of the problem:

The Navier Stokes program calculates steady-state solutions for various Mach number points along the missile trajectory. The total number of all steady-state results leads to an array of characteristic values. In this array the pressure and the heat loads are depending on Mach number, wall temperature and geometrical coordinates. The content of this array represents the input data for the FEA-program and characterizes the boundary conditions for the FEA-program. Only in the FEA-program the physical time and time dependence are considered.

The approach to exchange Stanton number data via data tables (or arrays of characteristic values) is well suited to handle the problem of aerokinetic heating. When the connection is established via data tables there is no feedback from the FEA-simulation to the CFD-calculation, accordingly this procedure is an „one-way principle“ simulation. This procedure has three significant advantages as long as quasi-steady approach is valid at all:

1. The acceleration of the missile does not influence the CFD-simulation. One array of characteristic values is used for all missile accelerations.
2. A time-dependent CFD solution process is not necessary. Only a few steady-state solutions lead to an array of characteristic values. This leads to a significant CPU-time reduction.
3. The FEA simulation runs detached from the CFD simulation. Studies with different material parameters (e.g. thermal conductivity) can be performed very quickly.

With the present knowledge in fluid mechanics no exact prediction of the transition area is possible. In accordance with [5] in all laminar-turbulent computations a constant location of the transition has been assumed at a radial coordinate of $r = 8$ mm (Fig. 4). Corresponding to [3, 5] we expect for a generic dome a continuously increasing Stanton number versus the y-coordinate starting from the laminar boundary layer and continuing via the transition area to the turbulent boundary layer. Over the fillet-item we expect a rather diminishing Stanton number (Fig. 4). The expected Stanton number curve leads to a non-regular heating of the surface. The incoming axial heat flux, resulting from the aerokinetic heating, and the radial (or cross ) heat-flux, resulting from the non-regular heating over the y-coordinate, create a three-dimensional heat transport problem in the structure. To describe this heating-problem, it is necessary to solve the time-dependent multi-dimensional heat transfer equations (e.g. with a FEA-tool).

## NUMERICAL MODELING

The following commercial codes were used for numerical modelling:

- Rampant© -CFD-solver, version 3.1.35, distributed by Fluent.Inc®,
- Ansys®-FEA-solver (version 5.2), offered by Ansys.Inc®. and
- P-Cube© grid-modeller, supplied by ICEM®.

The Rampant© code solves the Favre-averaged Navier-Stokes equations using structured or unstructured grids. All studies presented in this paper have been performed with the RNG-$k\varepsilon$ turbulence model for an ideal gas. Rampant allows to use multigrid convergence acceleration, typical values for gridcoarsing are 2-4 stages. The CFD-mesh was generated with the ICEM-software P-Cube©. Ansys© is a general purpose Finite Element-Analysis code to solve problems like structural and thermo-mechanical design tasks. For the presented results, only the time-dependent multi-dimensional or axisymmetric heat transfer equations are solved. Special attention is given to temperature dependent material properties of the structure. The data exchange and the integration process of the heat transfer equations are established as follows: Load data i.e. all data necessary for calculating the time and space dependent heat transfer, are stored in form of multi-dimensional tables, depending on Mach numbers ($M_1...M_m$) and surface locations. The required values are interpolated automatically when being called. The tables for Stanton numbers (St) and adiabatic wall temperatures ($T_{aw}$) for each node ($N_1...N_n$) of the FE-mesh are structured in principle as follows:

Tab.1: Structure of Load Data Tables

| | $M_1$ | $M_2$ | $M_3$ | ... | $M_m$ |
|---|---|---|---|---|---|
| $N_1(x,y,z)$<br>$N_2(x,y,z)$<br>$N_3(x,y,z)$<br>...<br>$N_n(x,y,z)$ | | | St | | |

| | $M_1$ | $M_2$ | $M_3$ | ... | $M_m$ |
|---|---|---|---|---|---|
| $N_1(x,y,z)$<br>$N_2(x,y,z)$<br>$N_3(x,y,z)$<br>...<br>$N_n(x,y,z)$ | | | $T_{aw}$ | | |

The flight profile is part of the data set, too. For a given flight time the appropriate Mach numbers are calculated and are used to extract the appropriate heat rates and adiabatic wall temperatures from the tables. The heat input into the structure is calculated for each time step, taking into account the actual temperature of the structure. The heat input per time increment causes a certain temperature rise. The integration of the total heat load is performed by summing up all heat fluxes to and from the structure when stepping through the flight profile.

## SELECTION OF A SUITABLE TEST DOME

To select a suitable test dome shape, fully-turbulent computations for several shapes have been performed at Mach 5.3 and at a constant wall temperature of Tw = 300K. With this fully-turbulent study, the maximum heat flux of the different shapes could be compared. Some typical results for three test dome shapes are presented in Fig. 5. Test dome 1 is a classically designed IR-seeker shape. Test dome 2 represents a simple integration of a test dome into the nose. Test dome 3 represents the most advanced shape with a sophisticated integration into the missile configuration. The three test domes are sketched in Fig. 5 together with the computed fully-turbulent heat flux versus the radius, or y-coordinate, respectively. Test dome 3 has a monotonous heat flux over the radius, because the recirculation domain in front of the window balances the heat-flux (Fig. 6). The heat flux is limited over the recirculation domain. The steady-state CFD-calculation did not converge, because the bow shock oscillated in front of the test dome 3. In Fig. 7, the grid was adapted for three arbitrarily selected bow shock positions. Test dome 3 was not selected for the experimental program because the oscillating bow shock represents additional fluctuating loads on the structure and leads to a possible failure of an IR-window. Test dome 2 dropped out due to the high drag compared to dome 1. However, test dome 1 was selected for the free-flight test.

## MEASUREMENT TECHNIQUE

Due to the rough environment for the test missile during free-flight, only conventional robust measurement techniques and sensors are reliable for the special application of free-flight measurements. The radially positioned Ni-Cr/Ni-thermoelements (Typ K, DIN586) are welded into the surface skin (Fig. 3). These thermoelements measure the surface temperature. The reference point of the thermoelements is the absolute temperature inside the missile, measured by a thermo-diode. After sampling using a combined multiplexer anlog/digital converter, the signals from the thermoelements are stored in the flight data recorder. The sampling frequency was 250 Hz per data channel. The flight data recorder had a storage capacity of 2 MB.

## EXPERIMENTAL PROGRAM

In the autumn of 1995 the free-flight was executed successfully in the bay of Meldorf at the German North Sea. After a total flight time of approximately 1.5 seconds, the missile was intentionally destroyed by an explosive charge. The distance from the launch position to the impact point of the missile cone was 3000 meters. The missile cone with the inertial measurement unit, telemetry-equipment and the flight data recorder was salvaged by the test team for a post flight analysis. All data are analysed at the ground station after the free-flight. Further free-flights are projected over a flight distance of 12000m.

## COMPARISON OF NUMERICAL AND EXPERIMENTAL RESULTS

The Stanton numbers for different Mach numbers and wall temperatures, computed by Rampant©, are displayed in Fig. 8 with special attention to the laminar-turbulent transition. Principally the shape of the curves of the computed Stanton numbers (Fig. 8) agrees with the expected curves (Fig. 4). For high super- and hypersonic conditions all curves in Fig. 8 are very similar and of the same order of magnitude. Only the curve for Ma=1.2 deviated considerably from the other curves, because it is a transonic Mach number. Ansys calculated the time-dependent temperature curves with the given Stanton numbers. The results are compared with the corresponding temperature curves measured by the thermoelementes during the free-flight. Presented are the difference between calculated temperature and measured temperature normalized by the measured temperature (relative deviation of computed and measured temperatures) via the flight time. Fig. 9 presents the relative deviation of computed and measured temperatures versus the flight time for a laminar-turbulent computation with a transition point at r = 8mm. Fig. 10 shows the deviation for a fully turbulent computation. The fully turbulent calculation generates heat loads which are too low on the outer radii. For the correct design of IR-windows it is important to assure that the calculated heat loads are not lower than expected in free-flight conditions. Therefore, the laminar- turbulent heat transfer calculation was chosen, which leads to a slight over-estimation of thermal stresses. The radial temperature gradient in the laminar-turbulent calculation is larger than measured.

## CONCLUSION

The „one-way principle“ simulation is a suitable engineering tool for calculating aerokinetic heating and thermal residual stresses of arbitrarily shaped IR-windows or missile parts with a high grade of flexibility in geometry and material changes. Further improvements in the prediction of aerokinetic heating are expected with a better definition of the laminar-turbulent transition. To verify the presented initial results it would be desirable to perform additional free-flight tests.

## REFERENCES

[1] E.R. Van Driest, Turbulent boundary layers in compressible fluids, J. Aero. Sciences, Vol 18, No. 3, 1951.

[2] E.H Hirschel, Heat loads as key problem of hypersonic flight, ZFW, Vol 16, 1992.

[3] G.F. Wildhopf, R. Hall, Transitional and turbuelent heat-transfer measurements on yawed blunt conical nosetips, AIAA Journal, Vol 10, No 10, Oct. 1972.

[4] K. F. Stetson, Boundary layer transition on blunt bodies with highly cooled boundary layers, Journal of the aero/space-science, Vol27, Feb 1960.

[5] E. Herpfer, Aerothermodynamic aspects for the design of missile domes with infrared seekers, ZFW, Vol 22; 1974.

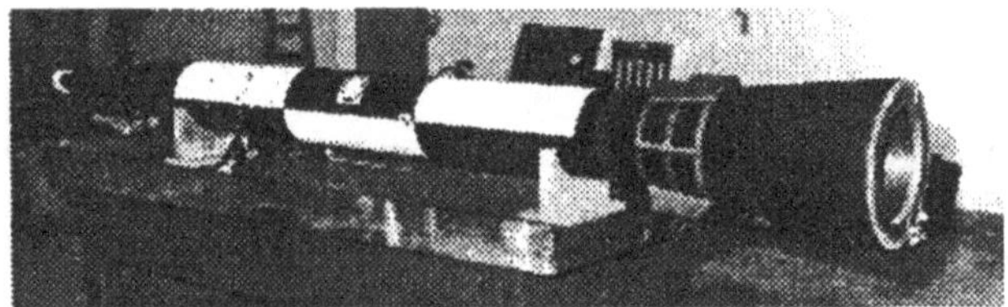

Fig. 1:Hypersonic Missile (HFK)

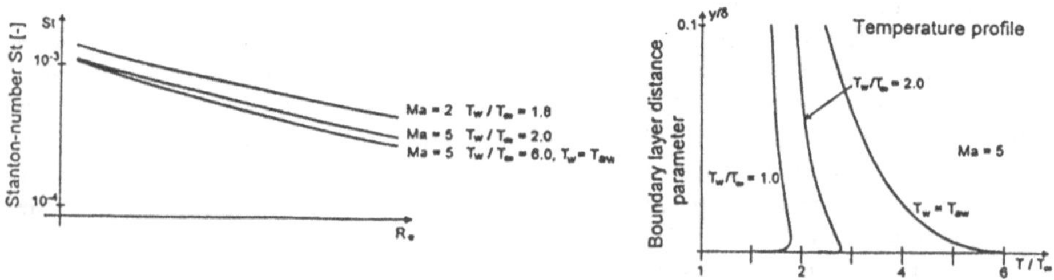


Fig.2: Turbulent Stanton number for a flat plate

Fig. 3: Temperature boundary layer for a flat plate

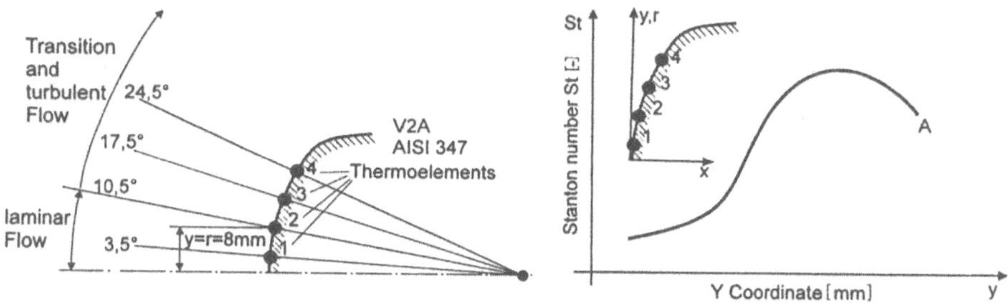


Fig 4: Position of thermoelements(1,2,3,4) and expected Stanton number curve (schematical)

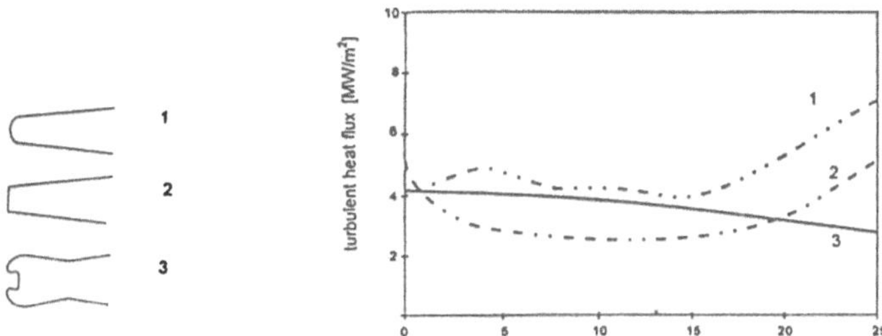


Fig. 5: Investigated test domes and heat flux at Ma=5.5 and Tw=300K

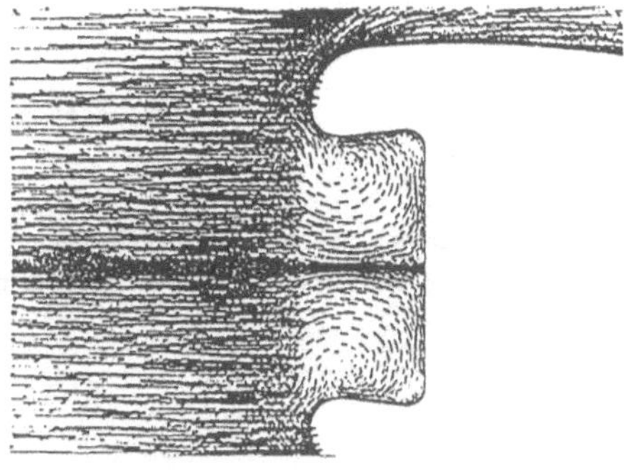

Fig.6: Recirculation domain in front of dome 3

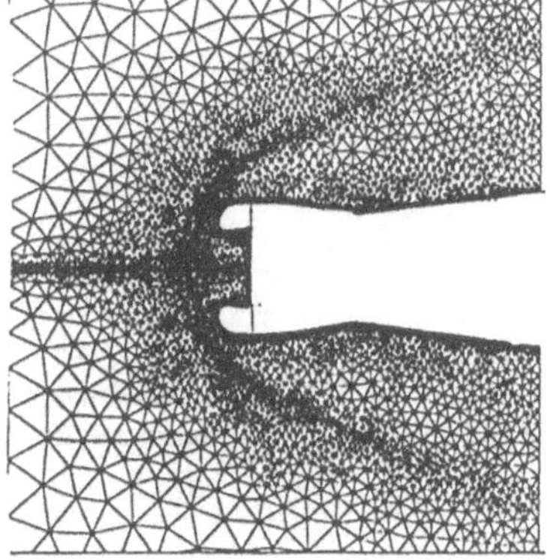

Fig. 7: Adapted grid

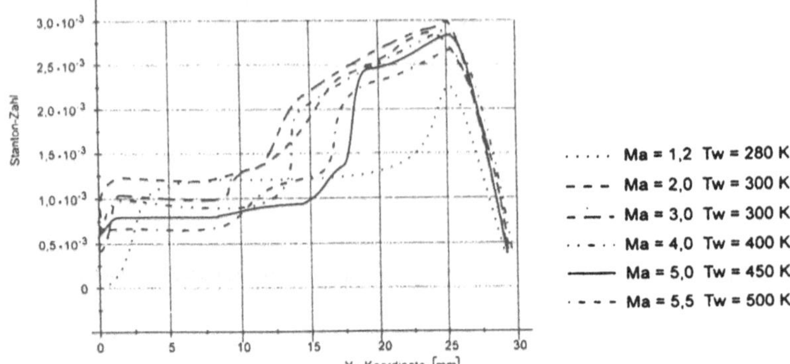


Fig. 8 : Stanton-number curves

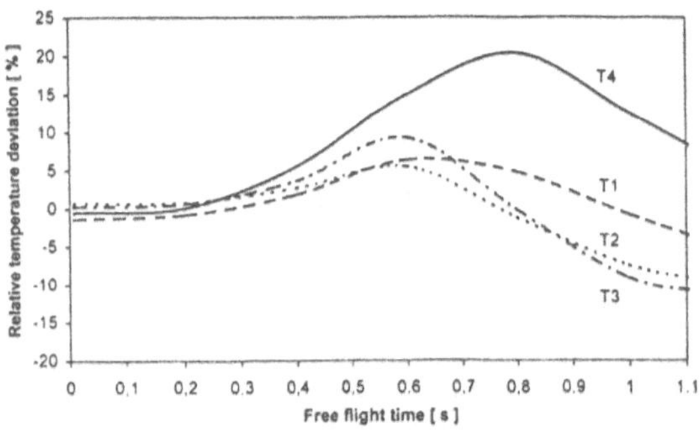


Fig. 9: Comparison of temperatures resulting from a laminar-turbulent calculation with measured temperatures. T1-T4 are the deviations for the corresponding thermoelements see Fig. 4.

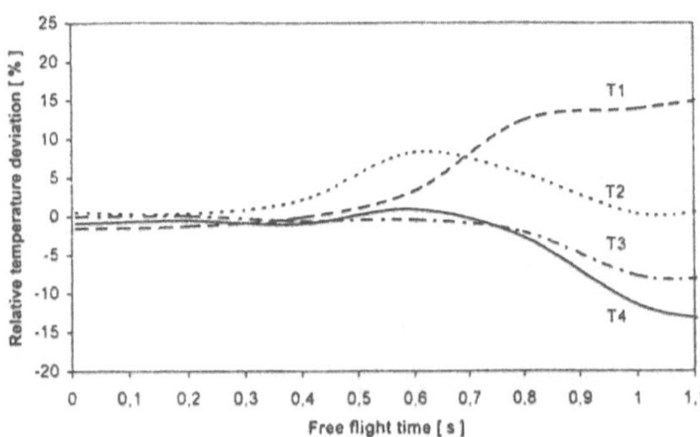


Fig. 10: Comparison of temperatures resulting from a full-turbulent calculation with measured temperatures. T1-T4 are the deviations for the corresponding thermoelements see Fig. 4.

# Experimental Studies of the Turbulent Flow Structure of Leading–Edge Vortices

Chr. Breitsamter

Lehrstuhl für Fluidmechanik,
Technische Universität München,
Boltzmannstr. 15, 85748 Garching, Germany

## SUMMARY

This paper presents selected results from an extensive experimental investigation on velocity and surface pressure fluctuations caused by leading-edge vortices, in particular, when breakdown occurs. The models used include delta- and delta-canard configurations. The vortical structures are clearly shown by rms and spectral distributions. Downstream of bursting the velocity fluctuations are the strongest in an annular region around the vortex center. The related spectra indicate quasi-periodic oscillations arising from a helical mode instability. When breakdown moves over the wing to the apex there is a strong increase in surface pressure fluctuations. They exhibit dominant frequencies corresponding to those of the velocity fluctuations. A frequency parameter based on the local semi-span and the sinus of $\alpha$ can be used to estimate the frequencies of dynamic loads due to bursting.

## NOMENCLATURE

| | |
|---|---|
| $c_{p_{rms}}$ | rms value of pressure coefficient |
| $\hat{c}_p$ | amplitude spectra of pressure fluctuations, $\sqrt{2\,S\,\Delta k\,U_\infty/l_\mu}$ |
| $c_r$, $c_c$ | wing root chord, canard root chord, $[m]$ |
| $f$, $f_D$ | frequency, dominant frequency, $[Hz]$ |
| $k$ | reduced frequency, $f\,l_\mu/U_\infty$ |
| $l_\mu$ | wing mean aerodynamic chord, $[m]$ |
| $Re_{l_\mu}$ | Reynolds number based on $l_\mu$, $U_\infty\,l_\mu/\nu$ |
| rms | root mean square value |
| $S$ | spectral density, $[1/Hz]$ |
| $s$, $s_l$ | wing semi-span, local wing semi-span, $[m]$ |
| $U_\infty$ | freestream velocity, $[m/s]$ |
| $u, v, w$ | axial, lateral and vertical velocity (wind tunnel-axis system), $[m/s]$ |
| $u_{xz_{rms}}$ | sum of axial and vertical rms velocity, $\sqrt{1/2\,(\overline{u'^2}+\overline{w'^2})}$ |
| $Y, Z$ | nondimensionalized coordinates of the measurement planes, related to $s_l$ |
| $x, y, z$ | wind tunnel axis system or model coordinate system, $[m]$ |
| $x_B$ | chordwise coordinate of burst location, $[m]$ |
| $\alpha$, $\beta$ | aircraft angle of attack, aircraft angle of sideslip, $[°]$ |
| $\varphi_W$, $\varphi_C$ | wing leading-edge sweep, canard leading-edge sweep, $[°]$ |
| $\Lambda_W$, $\Lambda_C$ | wing aspect ratio, canard aspect ratio |
| $\nu$ | kinematic viscosity, $[m^2/s]$ |
| $\Delta f$, $\Delta k$ | frequency resolution of power spectral density |
| $\bar{..}$, $..'$ | time-averaged, fluctuation part |

## INTRODUCTION

Leading–edge vortices have received much attention to provide enhanced maneuverability for present and future fighter aircraft [1]. Due to improved performances the aircraft will be required to operate at high–$\alpha$ for longer periods. However, in this flight regime the unsteady flow arising from bursted leading–edge vortices cause problems such as severe buffeting on fins [2,3] and wing planforms [4]. The induced narrow–band aerodynamic loads excite the natural frequencies of the structure [3]. It constitutes a threat to the structural integrity and corrective actions may be required to ensure both vibration free control and long fatigue life [5]. Therefore, a detailed analysis of the buffet flow environment is of distinct utility for the design of future fighter aircraft.
An extensive research program has been undertaken at the Lehrstuhl für Fluidmechanik of the Technische Universität München to determine the characteristics of the turbulent flowfield produced by leading–edge vortices in pre– and post–breakdown stage. Thus, light is shed on the very complex flow phenomenology resulting in prediction parameters for vortex–induced buffet excitation [6].

## EXPERIMENTAL TECHNIQUE AND TEST PROGRAM

### Description of Model and Facility

The wind tunnel models used include a 76° swept delta wing and a detailed fighter aircraft model of canard–delta wing type, Fig. 1. The sharp–edged carbon–fibre delta wing is composed of a flat plate and a triangular cross section. Fast–response pressure transducers are embedded in the right part of the upper surface positioned at $x/c_r = 0.3$ (9 transd.), 0.5 (9 transd.), 0.7 (11 transd.) and 0.8 (10 transd.). At $x/c_r = 0.9$ 25 transducers are located along the whole local span (Fig. 1a).
The delta–canard model consists of nose section, front fuselage including rotatable canards and a single place canopy, center fuselage with delta–wing section and a through–flow air intake underneath, and rear fuselage with nozzle section. All pieces are made of stainless steel. For the model tested, the leading– and trailing–edge flap deflections as well as the canard setting angle were set to zero degree.

The experiments were conducted in the Göttingen type low–speed wind tunnel facilities 1 and 2 of the Lehrstuhl für Fluidmechanik of the Technische Universität München. Dimensions of the open test sections are 1.5 $m$ × 3 $m$ and 1.22 $m$ × 2.06 $m$, respectively. Maximum usable velocity is 55 $m/s$ and 70 $m/s$. Turbulence intensity is less than 0.3% - 0.4% over the speed range of interest.

### Measurement of Fluctuating Velocity and Surface Pressure

Miniature triple– and dual–sensor hot–wire probes were used to measure the time–dependent velocity components. The sensors consist of 5–$\mu$m–diam. platinum–plated tungsten wires forming a measuring volume of about 1 $mm^3$. In order to achieve best angular resolution the sensors are arranged perpendicular to each other. The probes were operated by a multi–channel constant–temperature anemometer system (DISA C). Bridge output voltages were low–pass filtered at 1000 Hz before digitization and simultaneously sampled with 12–bit precision at 3000 Hz over 26.4 sec. The sampling parameters were arrived at by preliminary tests to ensure that all significant flowfield phenomena are detected. The method to evaluate the instantaneous velocity vector is based on look-up tables derived from the full velocity and flow angle calibration of the sensors [7].

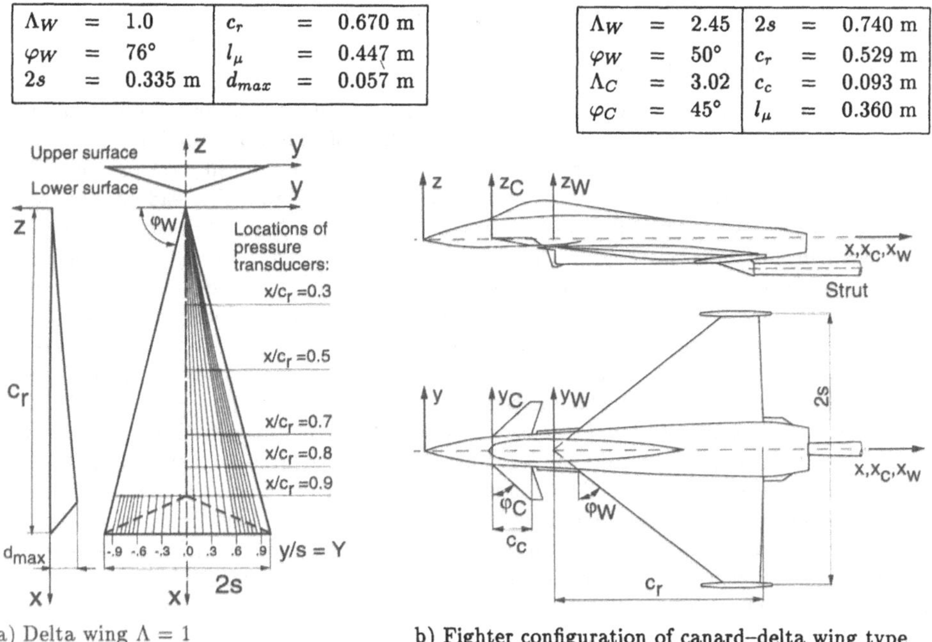

| | | | | | |
|---|---|---|---|---|---|
| $\Lambda_W$ | = | 1.0 | $c_r$ | = | 0.670 m |
| $\varphi_W$ | = | 76° | $l_\mu$ | = | 0.447 m |
| $2s$ | = | 0.335 m | $d_{max}$ | = | 0.057 m |

| | | | | | |
|---|---|---|---|---|---|
| $\Lambda_W$ | = | 2.45 | $2s$ | = | 0.740 m |
| $\varphi_W$ | = | 50° | $c_r$ | = | 0.529 m |
| $\Lambda_C$ | = | 3.02 | $c_c$ | = | 0.093 m |
| $\varphi_C$ | = | 45° | $l_\mu$ | = | 0.360 m |



a) Delta wing $\Lambda = 1$

b) Fighter configuration of canard–delta wing type

**Fig. 1:** Delta wing and Delta–Canard–Configuration

The pressure fluctuations were detected by transducers of active differential type (Kulite XQC-107-093-5D) giving a maximum resolution of 2 Pa. The sensors are connected to computer–controlled data acquisition system containing 8 assembly groups each fitted with 16 differential amplifier modules, Butterworth low–pass filters, and sample&hold units linked to the multiplexer of the 14 bit A/D–converter. Thus, 128 analog inputs are supplied with a total sampling frequency of 256 kHz. For the present tests, the sampling rate for each channel was set to 1000 Hz and the low–pass filter frequency to 256 Hz. The sampling time is 30 sec.

## Description of Tests

The delta wing tests were made at angles of attack ranging from 0° to 60° at symmetric freestream. The reference velocity $U_\infty$ was hold constant at 37 $m/s$ giving a Reynolds number of $Re_{l_\mu} = 1.07 \times 10^6$. The velocity measurements were performed in planes normal to the wing surface located directly above the pressure transducer sections. The surveys cover an area of $-1.1 < Y < 1.1$, $0 < Z < 1.2$ where the measurement points are evenly spaced with a distance of 0.045 related to the local semi–span (Fig. 2).
Test conditions of the delta–canard investigations correspond to those of the delta wing studies: $U_\infty = 40\ m/s$, $Re_{l_\mu} = 0.97 \times 10^6$, $\alpha = 28°$, $\beta = 0°$. There, the wing flowfield was traversed at 18 planes chordwise each of them with 33 points in spanwise and 16 points in vertical direction (Fig. 4). Over the wing the increments of the measurement grid were set to 0.027 laterally and to 0.042 vertically based on the local semi–span. At all tests turbulent boundary layers were present at wing and control surfaces proved by shear–stress sensitive liquid crystal measurements.

## RESULTS AND DISCUSSION

### Turbulence Intensity Distribution

For the delta wing, Fig. 2 shows contours of turbulence intensity obtained from axial and vertical rms velocities at 90% root chord and $\alpha = 25°$, 30°, and 35°. At fully developed vortices, $\alpha = 25°$, regions of moderately increased turbulence intensity are present, Fig. 2a. They clearly depict the rolled-up shear layers separated from the leading-edge and the secondary vortices. Absolute rms maxima of 15% - 22% indicate the viscous subcores. The local rms-maxima detected in the shear layers represent the formation of coherent vortical structures due to a instability of Kelvin-Helmholtz type. Because of bursting at $x/c_r = 0.86$ the vortex core structure is significantly changed at $\alpha = 30°$ showing an asymmetric pattern of severe velocity fluctuations, Fig. 2b. With increasing incidence, $\alpha = 35°$, breakdown takes place at $x/c_r = 0.49$. Downstream an annular region of maximum turbulence intensity is formed which approaches the wing surface.

Particularly, the axial turbulence intensity at $\alpha = 35°$ can be taken from carpet plots of Fig. 3. The adverse pressure gradient reaching a critical level leads to a stagnation of the axial flow near the vortex axis. Thus, next to the original jet-like core flow a region of strong decelerated flow occurs. The steep velocity gradient due to the change from jet to wake flow evokes an overall maximum in streamwise turbulence intensity at the vortex center, Fig. 3a. Downstream, the region of maximum turbulence intensity expands rapidly with the region of retarded axial flow while the largest velocity fluctuations arise within a limited radial range relative to the vortex axis, Fig. 3b.

A similar distribution is shown by the core profiles of axial rms velocity measured over the wing of the delta-canard configuration, Fig. 4. The highest turbulence level exists at the remaining swirling vortex sheet corresponding approximately to the point of inflection in the radial profile of decelerated axial velocity.

### Surface Pressure Fluctuations

As shown above, the impingement of bursted vortices is a source of buffet excitation on an aircraft experienced on the wing surface or on other surfaces such as the fin. This is substantiated by the corresponding surface pressure fluctuations, Fig. 5. When breakdown moves to a mid-chord position pressure fluctuation intensity increases strongly, especially beneath the vortex axis, where the distance to the region of maximum turbulence intensity is the smallest, Fig. 5a. A tremendous situation exists at $\alpha = 45°$ where the pressure fluctuations reach rms levels above 30%, Fig. 5b. The pressure fluctuation intensity decreases when the breakdown position is located very close to the apex from which the vortex axes are detached from the surface, Fig. 5c.

### Spectral Content and Dominant Frequency

Within the radial range of maximum turbulence intensity pronounced quasi-periodic oscillations are found, Fig. 6. Depending on the growth of the vortex core the wing surface is influenced such that also the surface pressure spectra exhibit peaked distributions tuned by the flowfield velocity ones. Downstream of bursting a dominant frequency can be observed, until breakdown has reached the apex. Then, the shear layer separated from the leading-edge will not be able to form a swirling flow with axial motion.

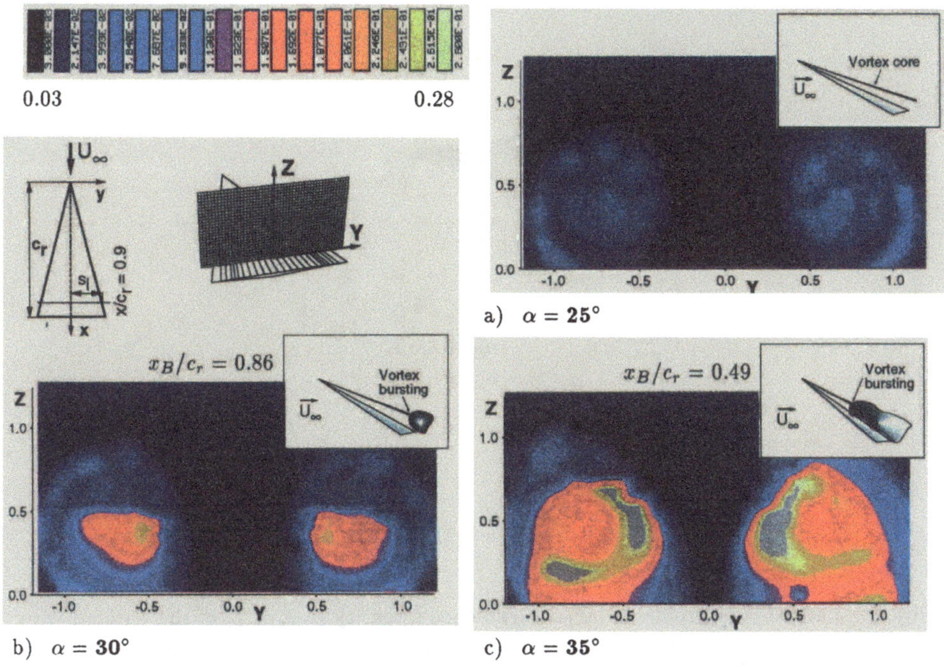


a) $\alpha = 25°$

b) $\alpha = 30°$

c) $\alpha = 35°$

**Fig. 2:** Contours of axial and vertical turbulence intensity $\mathbf{u_{xz_{rms}}/U_\infty}$ at the delta wing $\mathbf{\Lambda = 1}$ surveyed in a plane normal to the wing surface at $x/c_r = 0.9$ and $\alpha = 25°$, 30°, and 35°; $U_\infty = 37$ m/s, $Re_{l_\mu} = 1.07 \times 10^6$, $\beta = 0°$.

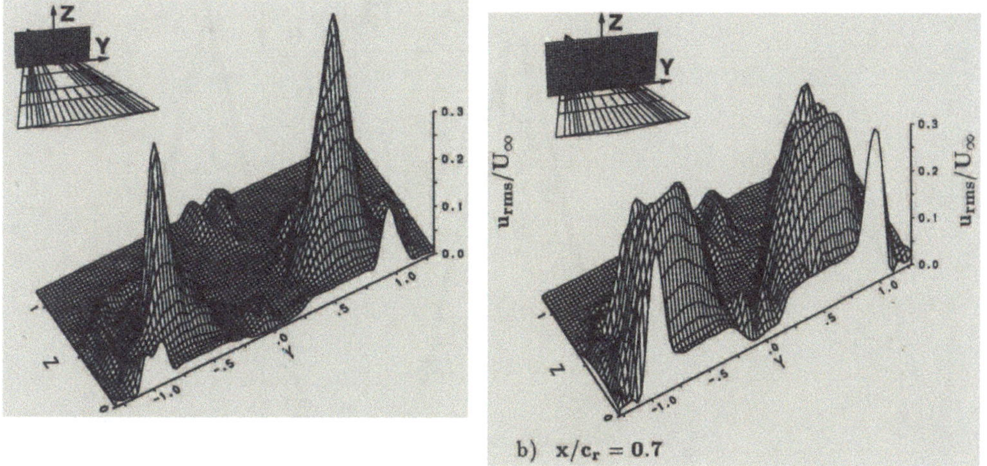


b) $\mathbf{x/c_r = 0.7}$

**Fig. 3:** Carpet plots of axial turbulence intensity $\mathbf{u_{rms}/U_\infty}$ at the delta wing $\mathbf{\Lambda = 1}$ surveyed in a plane normal to the wing surface at $x/c_r = 0.5$, 0.7 and $\alpha = 35°$; $U_\infty = 37$ m/s, $Re_{l_\mu} = 1.07 \times 10^6$, $\beta = 0°$.

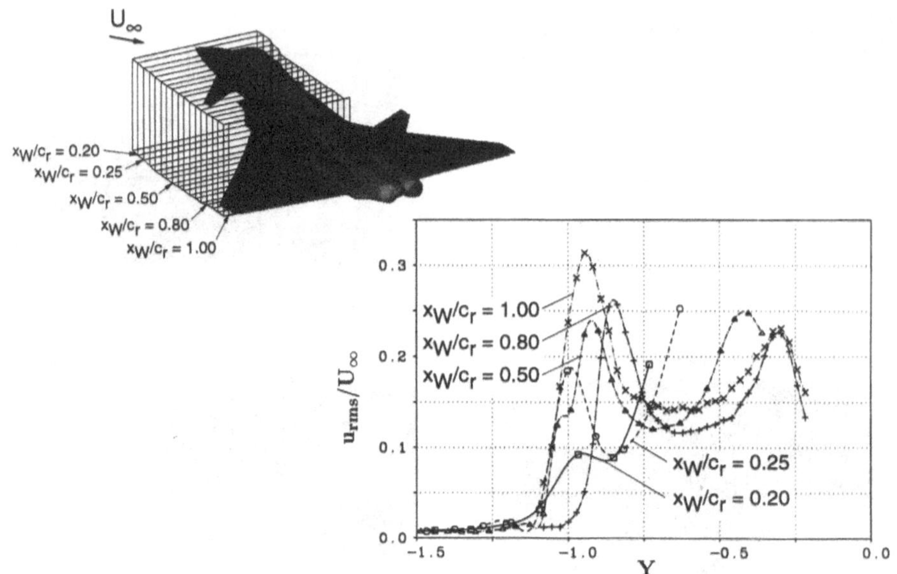


**Fig. 4:** Lateral core profiles of axial turbulence intensity $\mathbf{u_{rms}/U_\infty}$ of the bursted wing vortex at the delta–canard–configuration at $\alpha = 28°$; $U_\infty = 40$ m/s, $Re_{l_\mu} = 0.97 \times 10^6$, $\beta = 0°$.

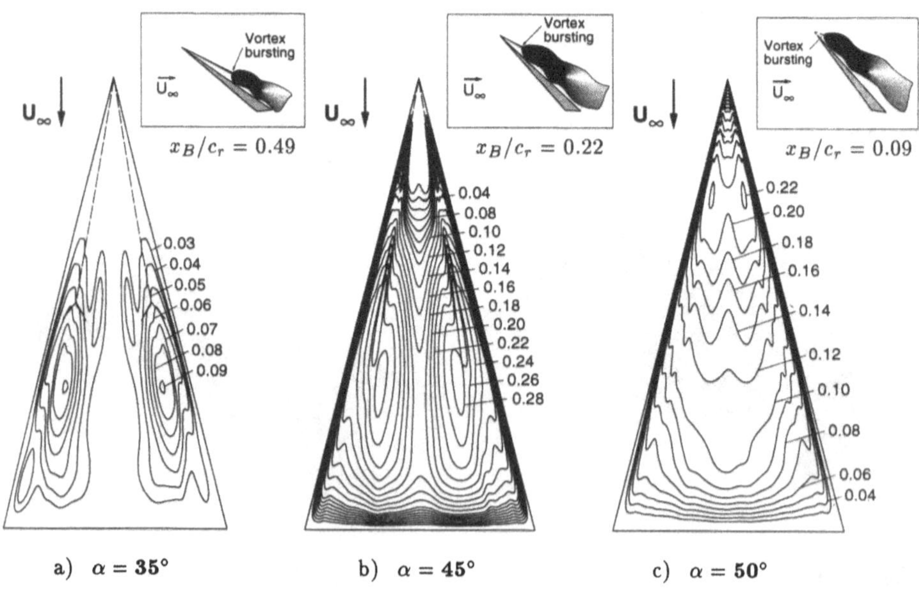


a) $\alpha = \mathbf{35°}$ b) $\alpha = \mathbf{45°}$ c) $\alpha = \mathbf{50°}$

**Fig. 5:** rms surface pressure distribution at the delta wing $\mathbf{\Lambda = 1}$ at $\alpha = 35°$, $45°$, and $50°$; $U_\infty = 37$ m/s, $Re_{l_\mu} = 1.07 \times 10^6$, $\beta = 0°$.

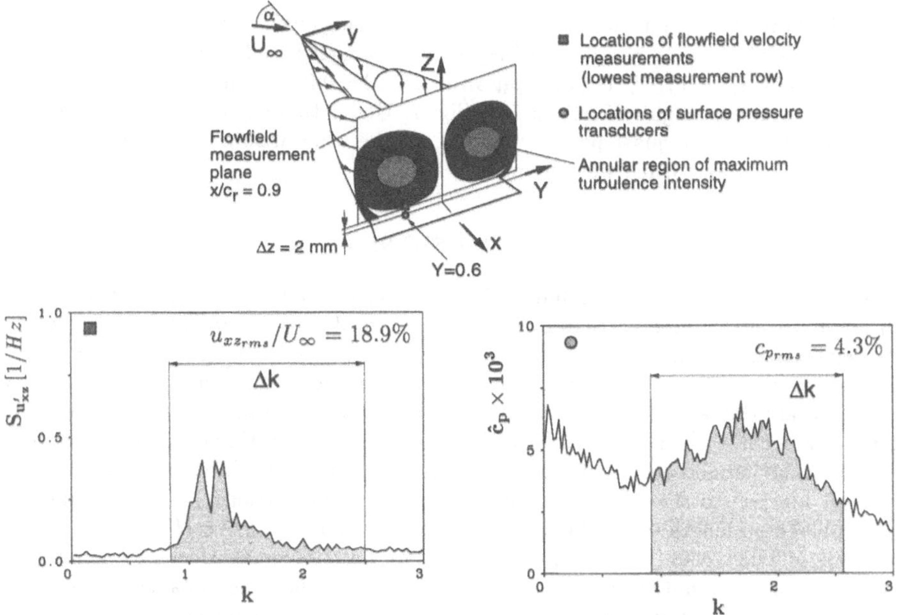


**Fig. 6:** Power spectral density of axial and vertical velocity fluctuations $\mathbf{S_{u'_{xz}}(k)}$ obtained above the wing surface and amplitude spectrum of surface pressure fluctuations $\hat{\mathbf{c}}_\mathbf{p}(\mathbf{k})$ at the delta wing $\mathbf{\Lambda} = \mathbf{1}$ at $x/c_r = 0.9$, $Y = 0.6$ and $\alpha = 35°$ ($\Delta f = 2.93$ Hz); $U_\infty = 37$ m/s, $Re_{l_\mu} = 1.07 \times 10^6$, $\beta = 0°$.

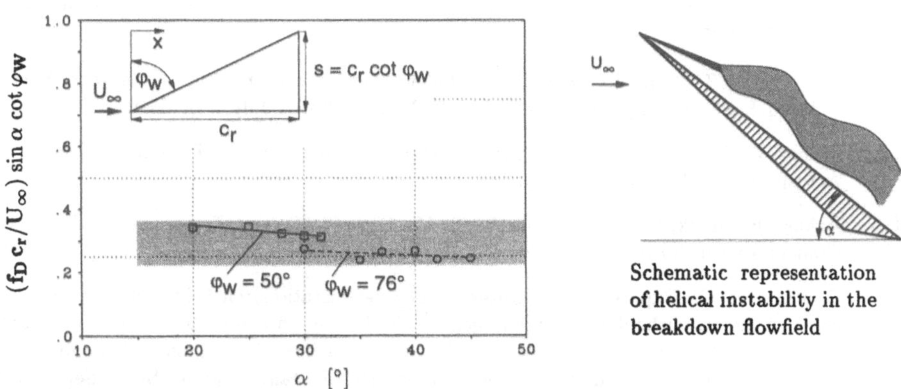


**Fig. 7:** Dominant reduced frequency ($f_D\, c_r/U_\infty$) scaled with the local semi-span and the sinus of $\alpha$ for the considered delta wing configurations; $Re_{l_\mu} \approx 1 \times 10^6$.

The dominant reduced frequencies decrease with increasing angle of attack and chord position. The correlation found results in a frequency parameter based on the local semi-span and the sinus of $\alpha$, i.e. $(f_D c_r/U_\infty)$ $\cot\varphi_W\, sin\alpha = 0.28 \pm 0.05$ which gives an appropriate correlation for sweep angles of 50° to 76°, Fig. 7. Under these conditions breakdown is the only mechanism for quasi-periodic loading on the wing arising from a helical mode instability with an azimuthal wave number of 1 [6].

## CONCLUSIONS

Detailed investigations of the turbulent flowfields associated with fully developed and with bursted leading-edge vortices have been conducted on a 76° swept delta wing and on a delta-canard-configuration. At the burst position the turbulence intensity shows an overall maximum near the vortex center whereas in the breakdown wake the velocity fluctuations are the largest in a restricted radial range. There, pronounced quasi-periodic fluctuations occur corresponding to a helical mode instability. When breakdown moves to the apex the annular region of maximum turbulence intensity comes close to the wing surface giving rise to strong coherent surface pressure fluctuations. The dominant frequencies based on length scales due to the vortex expansion, i.e. $(f_D c_r/U_\infty)$ $\cot\varphi_W\, sin\alpha$, give values of $0.23 - 0.33$. This result can be used for design studies covering sweep angles from 50° to 76°. The buffet excitation increases strongly when the distance between the vortex axis and the airframe component becomes smaller than one core diameter.

## REFERENCES


[1] Herbst, W. B.: Future Fighter Technologies. Journal of Aircraft, Vol. 17, No. 8, Aug. 1980, pp. 561–566.

[2] Lee, B. H. K., Brown, D., Tang, F. C., and Plosenski, M.: Flowfield in the Vicinity of an F/A-18 Vertical Fin at High Angles of Attack, Journal of Aircraft, Vol. 30, No. 1, Jan.-Feb. 1993, pp. 69–74.

[3] Breitsamter, C., and Laschka, B.: Turbulent Flow Structure Associated with Vortex-Induced Fin Buffeting. Journal of Aircraft, Vol. 31, No. 4, July-Aug. 1994, pp. 773–781.

[4] Mabey, D. G.: Some Aspects of Aircraft Dynamic Loads Due to Flow Separation. Prog. Aerospace Sci., Vol. 26, 1989, pp. 115–151.

[5] Ashley, H., Rock, S. M., Digumarthi, R. V., Chaney, K., and Eggers, A. J., Jr.: Active Control for Fin Buffet Alleviation. Wright-Patterson Rept. WL-TR-93-3099, Jan. 1994.

[6] Breitsamter, C.: Experimentelle Untersuchung der instationären Feldgrößen und Oberflächendrücke bei wirbeldominierter abgelöster Strömung an einem Deltaflügel. - In: Jahrbuch 1995 I der Deutschen Gesellschaft für Luft- und Raumfahrt - Lilienthal-Obert e.V., Deutscher Luft- und Raumfahrtkongreß 1995/ DGLR-Jahrestagung, Bonn, Sept. 26-29, 1995, pp. 163–175.

[7] Breitsamter, C., and Laschka, B.: Velocity Measurements with Hot-Wires in a Vortex-Dominated Flowfield. AGARD-CP-535, Wall Interference, Support Interference and Flow Field Measurementes, Brussels, Belgium, Oct. 4-7, 1993, pp. 11-1–11-13.

# Density Measurements on a Delta Wing by Means of Planar Intensity Distribution of Rayleigh Scattered Laser Light (PIRS) in the 1m x 1m Transonic Wind Tunnel Göttingen (TWG)

K. A. Bütefisch, V. Schmidt, H. Vollmers
Institut für Strömungsmechanik, DLR, Bunsenstraße 10, 37073 Göttingen
A. Luczak, V. Beushausen, Laser-Laboratorium-Göttingen e.V., Hans-Adolf-Krebs-Weg 1, 37077 Göttingen
G. Wolfrum, E. Krämer, DASA AG, Postfach 801160, 81663 München

## SUMMARY

In order to show, whether Rayleigh scattering from gas molecules of a wind tunnel flow is applicable for the acquisition of density data in the case of aerodynamically relevant flow problems, first tests in the Ludwieg Tube (RWG) of the DLR Göttingen at M=6.8 have been performed. Based on the experiences made during these first tests, the quantitative determination of the density field above a delta wing has been performed at Mach numbers between 0.5 and 1.4 in the transonic wind tunnel (TWG) of the DLR Göttingen. Three dimensional density fields with a resolution of about 1% have been obtained.

## INTRODUCTION

Earlier investigations in the field of combustion research have shown that it is possible to use Rayleigh scattering of light to determine the density of gas flows. Mainly for burners such applications are well known. Also first considerations on the application in wind tunnel tests have been made [1-5], but several difficulties were found [6]. The fact that the intensity of Rayleigh scattered light increases inversely with the fourth power of the wavelength, suggested to improve the method by applying UV light available from excimer lasers. Hence first tests on the basis of this new light source have been performed in several laboratories.

The rather large density change in front of a blunt body, due to a shock at supersonic Mach numbers made it attractive to test the new method first at high Mach number conditions. As the pulse length of the excimer laser in the range of 20 ns allows for instantaneous exploration of the flow field, the rather short run time of the Ludwieg tube (RWG) [8] of DLR Göttingen seemed not to be a problem when applying the new method. The tests in the RWG were performed at a stagnation pressure of $p_0$ = 25 Bar, stagnation temperatures around $T_0$ = 600 K and at a Mach number of M = 6.8. Due to the high Mach number, the free stream density was merely $\rho = 1/10\ \rho_{ambient}$. As a consequence, care had to be taken to avoid any disturbance by scattering of the light from walls, optical components, and pollution of the gas.

The light of an excimer laser (Lambda Physik, $\lambda$=193 nm, 70 mJ) was converted into a light sheet by a cylindrical lens and fed by a mirror into the test section of the wind tunnel. The light sheet passed a cylinder model close to the stagnation point, perpendicular to the axis of the model and to the free stream direction. By this set up it was possible to obtain the whole flow field in front of the shock, across the shock, and also in front of the stagnation point of the model. The light originating from the gas flow illuminated by the laser light was collected with a video camera together with an UV sensitive intensifier. The camera was attached outside the test section of the wind tunnel. A quartz glass window allowed the optical access to the flow.

The images were observed on a monitor and were recorded using a video recorder for later data evaluation. During the run time of the wind tunnel of about 200 ms about 10 half images with half the resolution of usual video frames could be obtained.

Prior to a wind tunnel run, images were recorded at the lowest density possible in the test section in order to get a reference image for roughly "zero density". The intensity distribution contained the contribution of scattering from the walls and from the surface of the model. Another recording at ambient density provided the reference condition of "homogeneous density". Now the intensity distribution yielded not only the portion of the gas molecules but again the unwanted contributions of the solid surfaces, and in addition the light which was scattered twice: first from the gas molecules and subsequently from the solid surfaces.

Both reference images were used to determine the density information from images which were taken while the flow was switched on. Fig. 1 represents how the images were processed. This was done by digitising the half images and performing digital image processing.

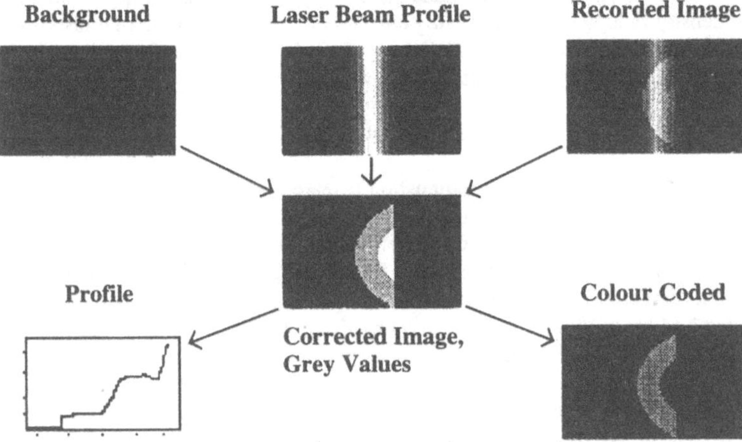


Fig. 1: Scheme of processing the recorded images.

The results gained were very encouraging. It was concluded that it would be feasible to obtain also reliable data in the case of transonic flow conditions in the rather large transonic wind tunnel TWG [9] of DLR Göttingen.

## EXPERIMENTAL SET UP

The experiences of the first tests in the RWG were used to prepare the test in the transonic wind tunnel TWG. As the influence of condensation is not only present in the case of a supersonic wind tunnel, care had to be taken to avoid it. Condensation is possible if the air is not dry enough with respect to the pressure and the Mach number or if, due to further expansion in the flow field around a model, the temperature decreases.

Fig. 2 depicts the set up for the TWG. As excimer lasers could not be operated at lower pressures, it was not possible to bring the laser into the plenum chamber of the wind tunnel. Hence the laser light path was lengthened to about 10 m. An ArF laser with a wavelength of 193 nm was supposed to be used because of the higher Rayleigh scattering intensity. But due to the high absorption strength of the oxygen in the air at 193 nm a KrF laser with a wavelength of 248 nm had to be used.

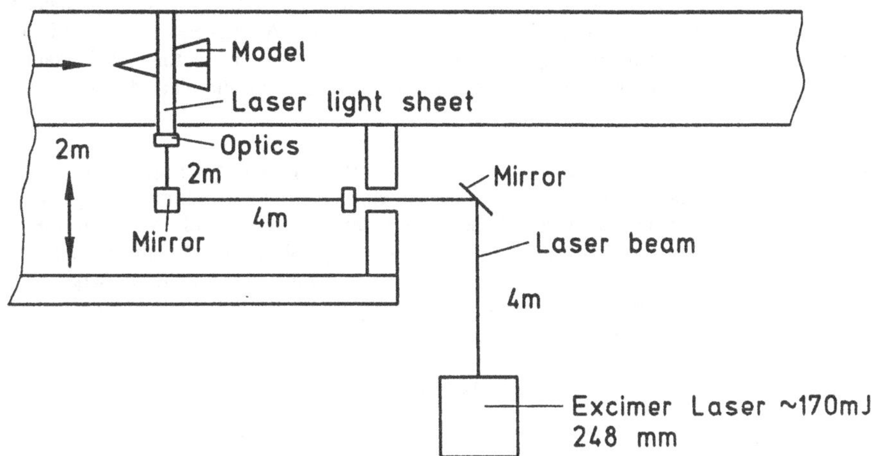


Fig. 2: Experimental set up in the 1m x 1m transonic wind tunnel TWG of the DLR Göttingen. The laser (KrF excimer laser, $\lambda = 248$ nm) had to be placed outside of the plenum chamber.

Fortunately, the wind tunnel is very quiet and does not vibrate too much. So it was very easy to bring all optical parts for generating the laser light sheet into the plenum chamber. Initially the wall of the test section was prepared to be equipped with a quartz window. But the window produced a huge amount of stray light. So it was decided to replace the window by a solid wall with a very small slit as an entrance for the light sheet. As the test section had perforated walls the small slit did not disturb the flow.

The light sheet traversed the test section above a delta wing model. The delta wing was chosen as vortices should be present and density changes could be expected. Reference images without flow at different density levels were taken in order to apply the procedure described above when evaluating the images recorded at the different flow conditions. In order to reduce the amount of stray light originating from the model, the surface of the model was painted black. The reduction of unwanted scattering was considerable.

As the previous investigations showed the quality of images degrades when recorded with a video recorder all images were recorded digitally by a computer. Hence, the image acquisition frequency was only about 2 frames per second, but full images with 384 x 288 pixels were recorded. In the case of the continuous running transonic wind tunnel this mode of operation implied no limitation. Also, the advantage of the method to obtain images of the instantaneous flow field remained valid. The various "realisations" of the flow field were sampled in a statistical manner.

The Rayleigh scattered light was observed by a combination of lenses (f = 30 mm, f-number = 1.0) and a intensified video camera with 12 bit dynamic range rectangularly to the light sheet.

In comparison to the previous tests, this means a major improvement as, due to the unwanted scattered light coming from the solid walls, the dynamic range of the brightness was rather large. The enlarged dynamic range allowed for necessary amplification with respect to the brightness of the light scattered from the molecules and prevented saturation and blooming of the CCD chip in the case of "viewing" bright solid walls. The distance to the light sheet was 500 mm and the field of view had a size of 300 mm x 200 mm. As Rayleigh scattering is sensitive to polarisation a birefringent $MgF_2$ crystal was used to rotate the vector of polarisation of the laser light such that the scattered intensity originating from the illuminated molecules was optimised.

During a run the stagnation pressure was maintained at 0.8 Bar. While the light sheet was held fixed, the model was moved in axial and vertical direction. By this procedure it was possible to obtain several planar images at different locations relative to the model. Assuming the flow conditions remaining constant within the translation range of the model, the planar results could be combined to a 3D density distribution above the wing. Fig. 3 indicates schematically the planes where measurements were made. Note that the position of the model was changed a bit and small deviations may be present.

## EVALUATION

In principal the digital image processing scheme described above in fig. 1 was applied for data evaluation. In order to reduce noise each pixel value was replaced by the average of the directly neighbouring pixels. As a consequence the spatial resolution was about 2 mm or 1 % of the observation field.

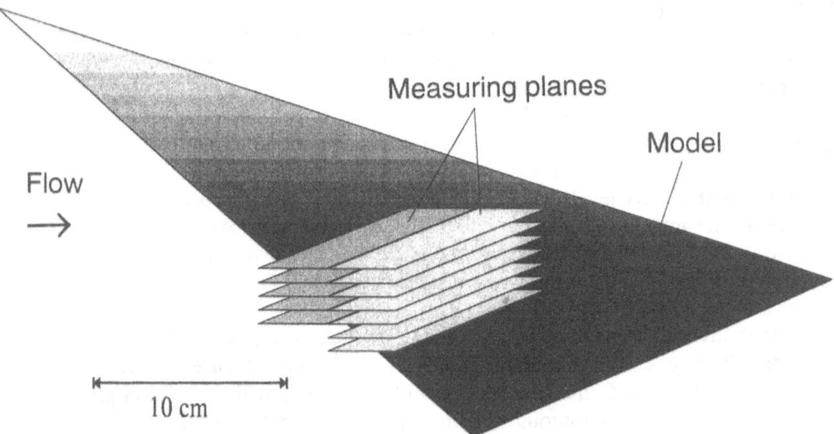


Fig 3: Arrangement of the laser light sheet above the delta wing model.

Another correction was applied for the inhomogeneous laser light brightness. This correction was possible as all video images showed the density distribution above the model as well as in the region where density can roughly be determined from stagnation conditions. From this the variation of the laser light in free stream direction was taken as a reference. The method has the advantage that each individual image can be corrected for the laser beam condition referring to exactly the same image.

A further correction had to be applied because the change of the position of the light sheet relative to the model led to a change of the disturbing contributions to the intensity distribution originating from the model surface. In this first test a simple procedure was applied, namely to adjust the intensity levels of the images of two neighbouring light sheets at their edges.

Again, as in the case of the previous tests in the RWG, the result of these corrections remained incomplete. Due to aerodynamic forces the model was shifted in flow direction with respect to the laser sheet by 3 pixels corresponding to about 3 mm between static and "flow on" conditions. After all corrections from the calibration images obtained at different static density settings, the planar light intensity distributions were converted into planar density distributions.

## RESULTS AND DISCUSSION

In Fig. 4 three original images of a recorded laser light sheet are presented, the first without flow, the second one at M=0.9, and the third one shows the final result after applying the procedures described above. The angle of attack was $\alpha = 10°$. It is obvious that, due to the orientation of the horizontally arranged laser light sheet and the inclination of the model together with the changing shape of the upper model surface, different distances between the light sheet and the model surface cannot be avoided. Any part of the model will contribute a different amount of scattered light generated on the surface of the model. This holds in particular very

Fig. 4: Recorded images at the same position relative to the delta wing model for different conditions: A: constant density, without flow; B: the same with flow at M=0.9; C: final result after processing B.

close to the surface. A more sophisticated correction scheme can compensate for this, but was not tried so far.

Comparing the first two images, the reduced brightness, equivalent to a reduced density, in the case of the M=0.9 flow is visible. In addition, in the vicinity of the leading edge a further reduction of brightness can be observed. Obviously in that part above the wing a vortex is present.

The short pulse length of laser light yielded frozen planar density distributions. Regarding a sequence of several images, Fig. 5, the fluctuation of the density field becomes visible.

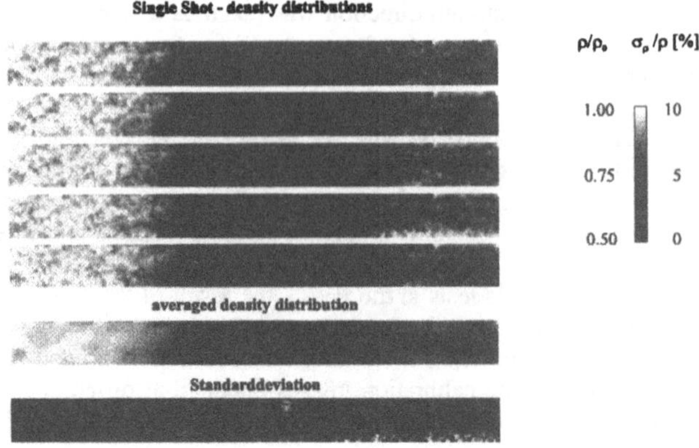


Fig. 5: Fluctuations becoming visible in a sequence of instantaneous planar density distributions.

The images have been taken in time intervals of about a second. The pattern is changing in an uncorrelated manner. Note, that each image is generated by a single shot and represents a "flow realisation" during the tiny time interval of 20 ns. Of course, a certain high frequency scatter in the planar density distribution can be found. On the other hand, the density changes due to the change of the local flow condition i.d. the behaviour of the vortex are obtained. The statistical standard deviation in the region of the free stream is about 3%, whereas in the region of the vortex a significantly higher value of 6% up to 10% is found. Occasionally the model vibrated and as a consequence the solid model surface scattered much more laser light, see arrow in fig. 5.

If a 9 x 9 median filter is used the planar density distribution looks much smoother. The filter does not affect constant gradients and jumps between different constant levels. Hence, contours become more pronounced. The statistical standard deviation in the region of the free stream is now decreased to 1%. The fluctuation level in the region of the vortex remains nearly unaffected.

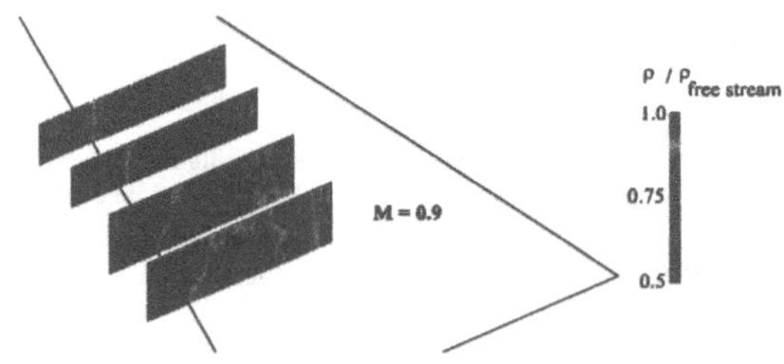


Fig. 6: Density distributions in vertical planes at M=0.9 obtained from results of horizontally arranged laser light sheets.

As images are recorded as a function of the height above the wing, also density distributions in vertical sections can be obtained by interpolation. Fig 6 shows the result for M= 0.9. Two vortices are visible.

In order to obtain the correct position of the vortices relative to the model, not only the exact co-ordinates of the model, but also the precise position of the model relative to the laser light sheet have to be known. For this first test the necessary precision is not well established. The improvement of resolution needs more effort, comparable to generating grids for numerical calculations. The enhancement will be obtained in a later stage of development.

Finally, all images obtained, when traversing in free stream as well as in vertical direction, were used to obtain a 3D representation of the density field above the delta wing, Fig.7. This was possible with the help of the powerful software package COMADI, which allows for interpolations in the case of arbitrarily chosen 3 D grids on the basis of triangles and even if data sets are incomplete. The results for two Mach numbers are shown. Clearly the quite different density levels are visible.

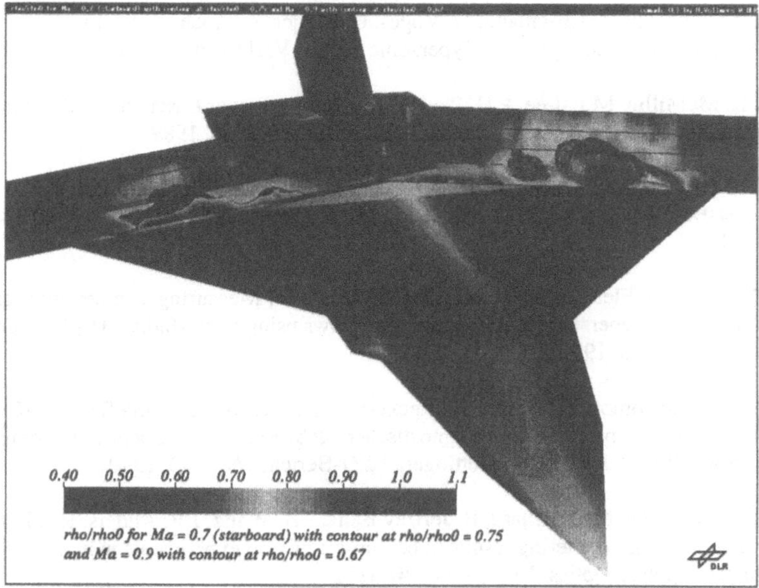


Fig. 7:Representation of the 3 D density distribution above a delta wing at M= 0.7 (left) and M=0.9 (right), experimentally obtained by planar intensity distributions of Rayleigh scattered laser light (PIRS) in the 1m x1m transonic wind tunnel TWG of the DLR Göttingen.

## CONCLUSIONS

For the first time the density distribution on a delta wing has been obtained with the help of the planar intensity distribution of Rayleigh scattered light (PIRS) in an industrial transonic wind tunnel. The Mach numbers ranged from 0.5 up to 1.4. The combination of the 2D density distributions allowed for the quantitative determination of the spatial density distribution of the flow field. Due to the very short pulse time of the UV excimer laser of about 20 ns, single images could be recorded. By averaging several single images the influence of statistical errors could be reduced. Digital image processing was used to correct for the influence of stray light and the inhomogeneous laser light profile. Residual errors remained as the model moved a bit due to the aerodynamic load during the run.

Besides the velocity information, a second flow quantity could be measured. Thus, in the future flow fields may be investigated even if the assumption of constant entropy is no longer valid. It is obvious that the huge amount of acquired data requires a powerful software package to enable analysis, interpretation, and finally the presentation of the results.

## REFERENCES


[1] K.A. Bütefisch, P. Krogmann, H. Voges, G. Meijer, A. Koch, H. Wolfrum; Measurement Techniques for Hypersonic Flows, VKI Lecture Series 1989 - 1990.

[2] B.K. McMillin, M.P. Lee, P.H. Paul, R.K. Hanson; Planar Laser-Induced Fluorescence Imaging of Nitric Oxide in a Shock Tube, AIAA-89-2566, 1989.

[3] R.K. Hanson, A.Y. Chang, M.D. DiRosa, L.C. Philippe, B.K. McMillin, M.P. Lee; Laser-Based Diagnostics for Propulsion and Hypersonics Testing, AIAA-90-1383, 1990.

[4] G. Laufer, D. Fletcher, R. McKenzie; A Method for Measuring Temperatures and Densities in Hypersonic Wind Tunnel Air Flows using Laser-Induced $0_2$ Fluorescence, AIAA-90-0626, 1990.

[5] A. Chryssostomou, H. Voges, W. Reckers, P. Andresen, K.A. Bütefisch, P. Krogmann, H. Wolfrum; Erprobung laserdiagnostischer Meßverfahren an Überschallströmungen im Rohrwindkanal der DLR in Göttingen, LLG-Bericht 1/90/340, 1990.

[6] B. Shirinzadeh, M.E. Hillard, R. Jeffrey Balla, I.A. Waitz, J.B. Anders, R.J. Exton; Planar Rayleigh scattering results in helium-air mixing experiments in a Mach-6 wind tunnel, Applied Optics, Vol. 31 No. 30, 1992.

[7] O. Sieber, K. Fiedler; Quantitative Dichtemessung am ebenen Gitterwindkanal mit Rayleigh-Streulicht, VDI Bericht Nr. 1109, 1994.

[8] H. Ludwieg, Th. Hottner, H. Grauer-Carstensen; Der Rohrwindkanal der Aerodynamischen Versuchsanstalt Göttingen. DGLR Jahrbuch, p. 52, 1969.

[9] Th. Hottner, W. Lorenz-Meyer; Der Transsonische Windkanal der Aerodynamischen Versuchsanstalt Göttingen, Jahrbuch 1968 der WGL, 1968.

[10] H. Vollmers; Diagnostic Visualization Tools for Flow Fields, Proceedings of the 26th International Symposiom on Automotive Technology and Automation, Aachen, 13th-17th Sept. 1993.

# CLOSE COUPLING OF VERY HIGH-BYPASS ENGINES IN COMBINATION WITH TAKE-OFF / LANDING CONFIGURATIONS

W. Burgsmüller, S. Thomé
Daimler-Benz Aerospace Airbus GmbH
D - 28199 Bremen

## SUMMARY

To investigate the installation effects of close coupled engines in combination with high-lift configurations, wind tunnel tests with engine simulators (TPS) were performed in the DA-low-speed windtunnel a Bremen. For these tests on a twin-engine transport aircraft model, the turbofan engine was mounted in the basic and a close coupled position. The results showed, that the negative installation effect for a close coupling can be minimized when the gap between pylon and inboard end of slat is closed. Further investigations are planned with a Very-High Bypass-Ratio engine simulator within the programme "Hochauftriebskonzepte (HAK)", funded by the German ministry of Research and Technology and within the CEC-funded programme ENIFAIR.

## INTRODUCTION

The steady trend to increase engine efficiency resp. reduce fuel consumption led to a steady increase in bypass-ratio. Based on current engines with bypass-ratios of about 5, new concepts with very resp. ultra-high bypass-ratios (VHBR/UHBR) are under design or already in use. Based on todays turbofans, the VHBR-engines promise a reduction in specific fuel consumption (SFC) of 12-14 %, while the UHBR-types could offer up to 19 % reduction (Fig. 1). Since the increased bypass-ratio however is coupled to an increasing engine diameter, solutions must be found for an optimized, close coupled installation under the wing.

## PROBLEMS OF ENGINE INSTALLATION

Beside the aerodynamic problems, engine installation is also heavily influenced by mechanical limitations (Fig. 2). The maximum vertical distance from the wing is limited on one hand by the minimum ground distance, necessary to avoid engine damage in case of flat tires or roll angles during touch down as on the other hand by increasing pylon weight. The forward position is limited in context with flutter boundaries while in case of the aft position limitations exist concerning the consequences of a disintegrating last turbine disc - in this case it must be avoided that fixing points of the pylon, fuel pipes or tanks are destroyed.

In case of VHBR/UHBR engines, these limitations consequently lead to the requirement of a close-coupled installation under the wing, for which aerodynamic solutions must be found (Fig. 3). Potential problems in cruise configuration are mainly coupled with the minimum gully height between engine and wing, where increasing drag due to shocks between engine and wing can occur. Difficulties in the low-speed regime (take-off and landing) can occur due to flow-separations on the wings upper side at high incidence caused by the nacelle body itself, limitations of slat deflection in the nacelle area and negative jet-interferences with the flap-system.

Consequently, the aim of a close coupled VHBR/UHBR engine installation must be, to find a solution where the positive effects due to SFC-reduction and the installation losses reach an optimum (Fig. 4).

## WIND TUNNEL TESTS AT DA

To investigate the aerodynamic phenomena in context with low-speed configurations, some tests were done in the DA-low-speed-windtunnel at Bremen. The first test campaigns were done with company-own fundings, while a continuation is planned within the governmental-funded programme "Hochauftriebskonzepte / HAK".

In a first step, the turbofan engine of a modern twin-engined transport aircraft was mounted under the wing in its standard (or basic) as well as in a close coupled position (Fig. 5). The jet-flow of the engine was simulated by using a so-called TPS (turbo-powered simulator) [1]. For the tests a half-model was used which was mounted to the overhead balance of the DA-LSWT. The close-coupled TPS-installation is shown on Fig. 6.

The upward movement of the engine led to the requirements, either to limit the deflection angle or to cut the length of the inboard slat. The results for the latter case are shown on Fig. 7, combined with those for the engine in basic position. The results show, that the effect of moving the engine upwards has marginal influence on overall drag, the maximum lift however is drastically reduced due to the gap between pylon and inboard slat. To improve this situation, a second test campaign was performed with a special filler plate installed between slat end and pylon. The TPS was run at maximum take-off power (MTO) as well as in a Through-Flow-Nacelle condition (TFN), where net thrust is zero, i.e a TFN is simulated.

A comparison of Fig. 8 and 9 shows, that with the filler plate installed, maximum lift of close coupled and standard engine position are now nearly identical. Also the differences in drag for the engines at TFN-condition are small (Fig. 10, 11). Concerning the jet-interference effects on lift, a positive increment can be seen for the close coupled installation (Fig. 9), coming probably from the jet deviation at the flap.

The jet-effects on drag shown on Fig. 10, 11 are summarized as jet-interference drag (difference between MTO and TFN-setting) on Fig. 12. Here, the close coupling leads to an increase in jet induced drag due to the wing/engine interference effects.

## DESIGN/MANUFACTURE OF ENGINE SIMULATORS AT DA

For the tests, described above, TPS-units manufactured in the USA were used. Due to the long distance and potential time delays during an aircraft programme development caused by overhaul or repair, it was decided some years ago at DA to design and manufacture in the future simulators for own purposes. The first unit produced had a fan diameter of 6.4" and is currently in use within the CEC-funded research programme ENIFAIR (Fig. 13). Lateron , 7.1" turbofan TPS were made for tests on an A 340-model in DNW. Following this, a 7.8" VHBR-simulator was designed for ENIFAIR. A 6" VHBR unit will be used in the follow-up tests at DA-LSWT within the HAK-programme.

## FURTHER ACTIVITIES PLANNED

As a follow-up of the tests with close coupled turbofans, described above, a test with the 6" VHBR engine in the same vertical position on the wing is planned. Different from the turbofan are nacelle size and fan jet flow field. The tests will cover force measurements and wake flow investigations by LDA and pressure probes.

Besides the HAK-programme with governmental funding also the CEC-funded programme ENIFAIR is aiming at the investigation of engine installation phenomena. This 3-year programme with 18 partners from 7 European countries is concentrating on experimental and numerical investigations in the area of pre-competitive research. For the wind tunnel tests in ONERA S1 (high-speed) and DNW (low-speed) the DLR-ALVAST model is used, which represents a twin engined aircraft of an A 320 type in a scale of about 1 : 10.
ENIFAIR is the follow-up programme of the earlier phases DUPRIN I and II [2, 3].

## CONCLUSIONS

- The aerodynamic effects of engine installation are very complex and require detailed investigations
- the tests done so far show potential solutions for close coupled engines in the low-speed regime
- further investigations are necessary and planned in national as well as international programmes
- the testing technique built up by DA includes design and manufacture of TPS-simulators and the results achieved in the DA-LSWT have become a substantial element within the whole Airbus family development.

## REFERENCES


[1] Burgsmüller, W./Akkermann, E.: "Engine Simulation with Turbo-Powered Simulators", Colloquium at the Occasion of Ten Years Operational Phase of DNW, Oct. 16, 1990, paper no. 4.

[2] Burgsmüller, W./Hoheisel, H./Kooi, J.W.: "Engine/Airframe Interference on Transport Aircraft with Ducted Propfans/ The European Research Programme DUPRIN", ICAS-Congress 1994 at Anaheim, USA, paper no. 94-3.7.1.

[3] Burgsmüller, W./Hoheisel, H./De la Puerta, B.: "European Research in Cooperation on Jet Engine Airframe Integration", DLR-Workshop on Aspects of Engine-Airframe Integration on Transport Aircraft, Braunschweig, March 1996, paper no. 6.


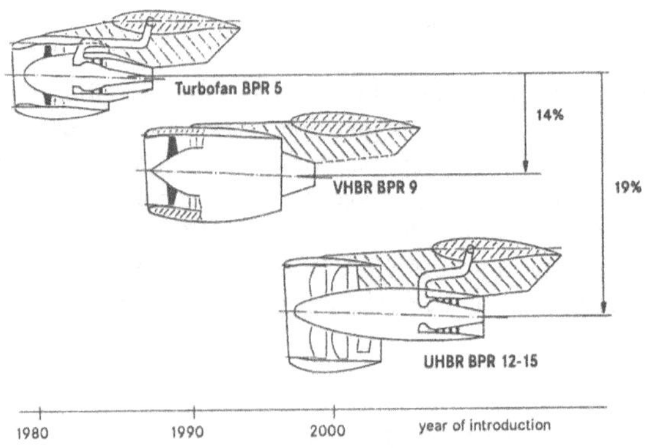


**FIG. 1 Potential of reduced fuel consumption due to new engine types**

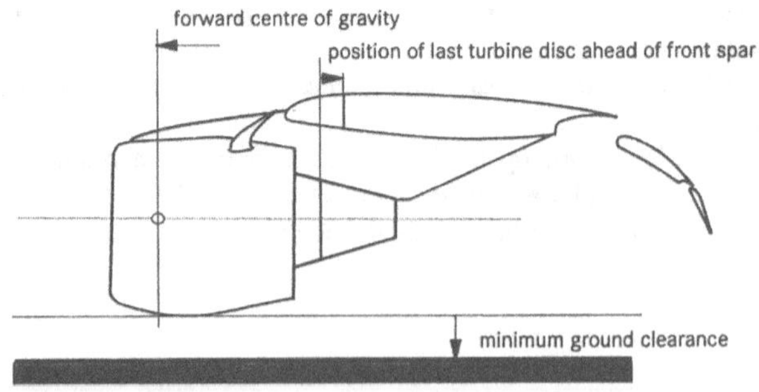


**FIG. 2 Mechanical limitations for engine installation**

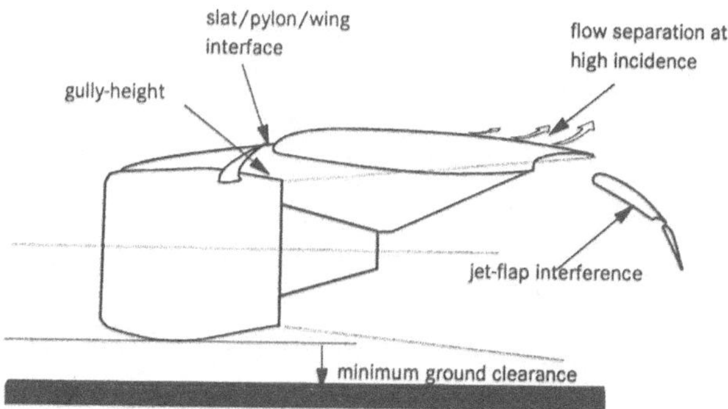


**FIG. 3 Aerodynamic problems of engine installation**

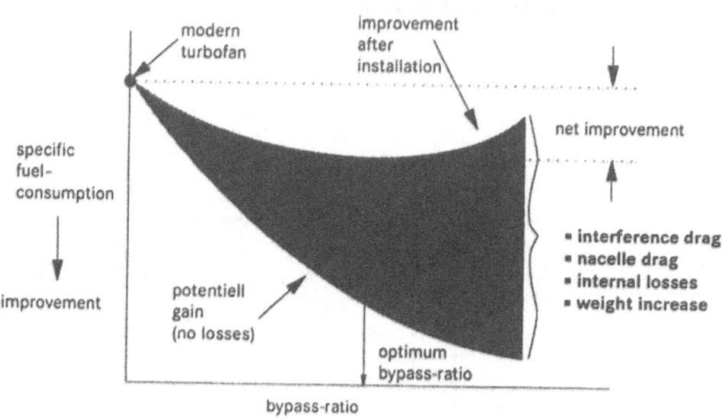


**FIG. 4 Installation effects of modern high-bypass-ratio engines**

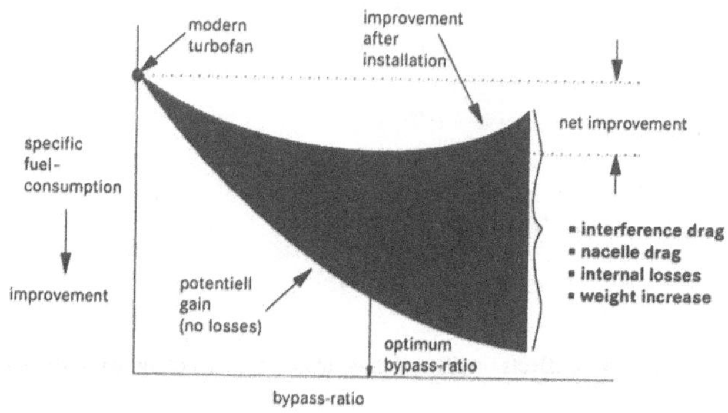


**FIG. 5 Engine in standard resp. close coupled position**

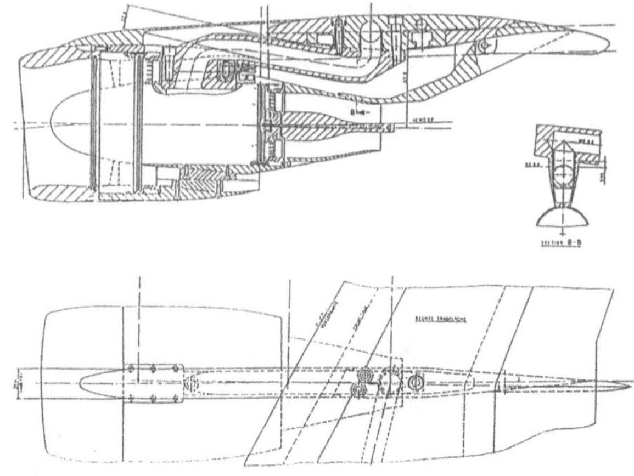

FIG. 6 Close coupled installation of turbofan TPS

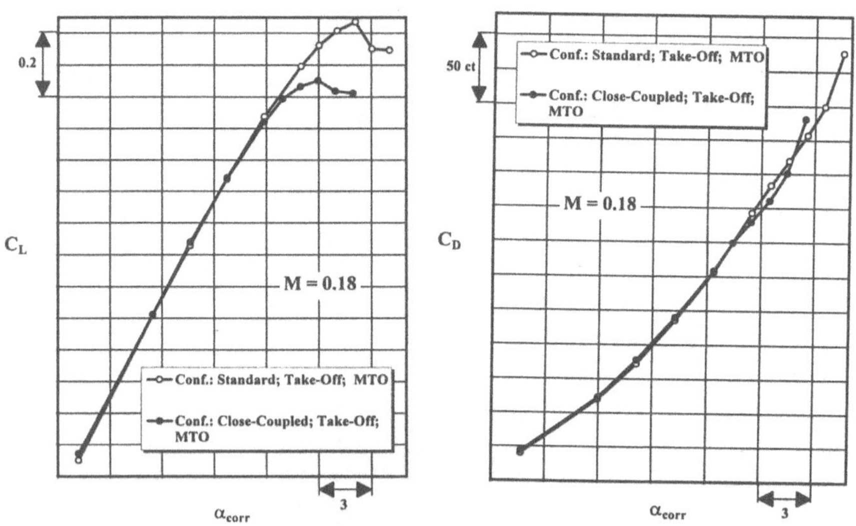


FIG. 7 Installation effects on lift and drag, close coupled config. with I/B slat gap

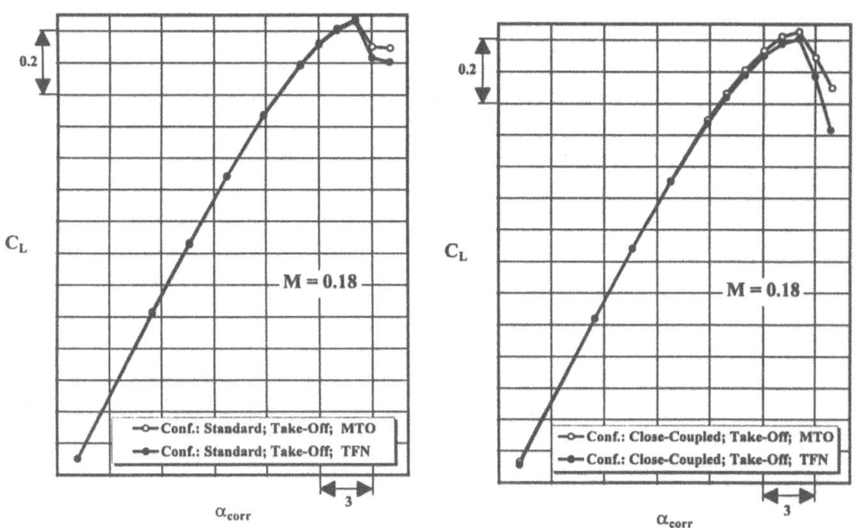


FIG. 8 Power effects on lift, standard engine position

FIG. 9 Power effects on lift, close coupled engine, I/B slat gap closed

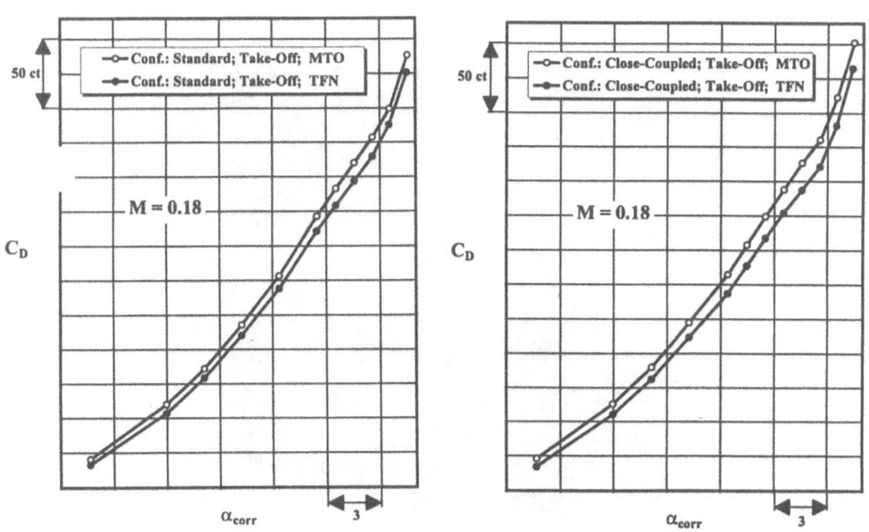


FIG.10 Power effects on drag, standard engine position

FIG.11 Power effects on drag, close coupled engine, I/B slat gap closed

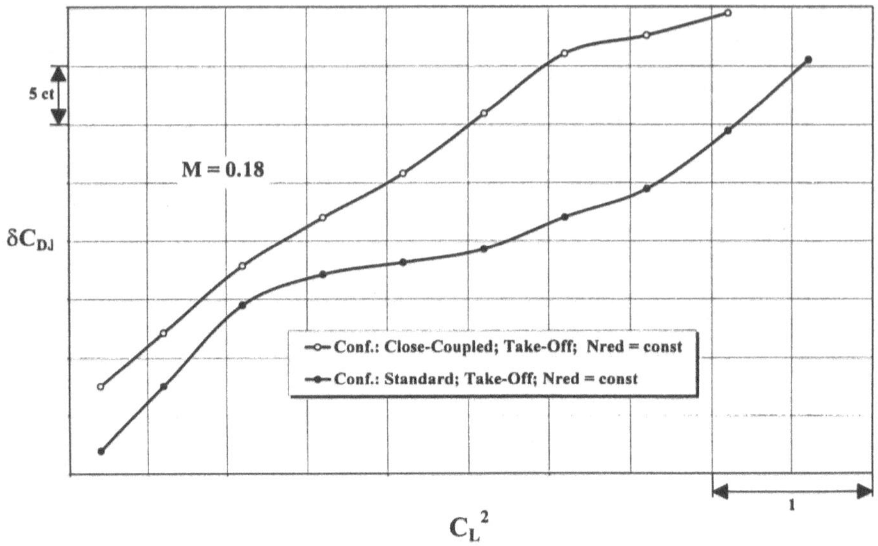


FIG. 12 Jet interference drag for standard resp. close coupled engine position

FIG. 13 Turbofan simulator (TPS), made by DA

# Numerical investigation of boundary-layer instabilities over a blunt flat plate at angle of attack in supersonic flow

Guido Dietz, Alexander Meijering
Aerodynamisches Institut
Rheinisch-Westfälische Technische Hochschule (RWTH) Aachen
Wüllnerstraße zw. 5 u. 7
D-52062 Aachen, Germany

## Summary

The stability characteristics of the supersonic flow over an adiabatic blunt flat plate at angle of attack are investigated using linear stability theory. The Mach numbers range from 2.0 to 2.5 with a unit Reynolds number of $1.26 \cdot 10^7/m$ to $0.99 \cdot 10^7/m$ at incidences of $0^0$ and $-10^0$. The full Navier-Stokes equations for two-dimensional, compressible flow are solved for the laminar basic flow by application of an Advection Upstream Splitting Method (AUSMDV). In the cases presented, first-mode instabilities propagating at wave angles of about $60^0$ have been found. The instability analysis of the boundary-layer flow shows a stabilizing effect of the blunt nose. Decreasing the Mach number as well as decreasing the angle of attack from zero incidence destabilize the boundary layer.

## 1 Introduction

Although transition research has been established for several decades, the laminar-turbulent transition phenomenon still remains an unsolved problem. In order to improve the understanding of the physical mechanism of the transition in supersonic flow, a blunt flat plate at angle of attack is investigated numerically.

The curved bow shock induces an entropy layer, that is swallowed downstream by the boundary layer. Especially, the transition is influenced strongly by the bluntness of the leading-edge, which has been investigated either theoretically on a blunt plate in supersonic flow by Reshotko and Khan [10] and on a blunt cone in hypersonic flow by Kufner [5] or experimentally by Stetson et al. [14]. These experiments have shown an initial stabilization of the boundary layer for small leading-edge blunting, whereas further increase of the bluntness had a destabilizing effect.

## 2 Mathematical model and numerical method

### 2.1 Basic flow model

The governing equations for the flow simulation are the Navier-Stokes equations for two space dimensions and compressible fluids, written dimensionless in generalized coordinates ($\xi$,$\eta$) and in conservation law form:

$$\frac{\partial \hat{Q}}{\partial \tau} + \frac{\partial \hat{E}}{\partial \xi} + \frac{\partial \hat{F}}{\partial \eta} = 0, \tag{2.1}$$

where $\hat{Q} = Sr\frac{\vec{Q}}{J}$ is the vector of conservative variables with $\vec{Q} = (\rho, \rho u, \rho v, \rho E)^T$ and $Sr$ denotes the Strouhal number. In the formulation applied here, the Jacobian $J$ of the coordinate transformation corresponds to the inverse of the cell volume. The equations are non-dimensionalized with the freestream stagnation properties, density $\rho_0$, speed of sound $a_0$ and molecular viscosity $\mu_0$. *Re* denotes the Reynolds number and $\hat{E}, \hat{F}$ denote the transformed flux vectors, e. g. in $\xi$-direction:

$$\hat{E} = \hat{E}_C - \hat{E}_D = \frac{1}{J}\begin{pmatrix} \rho U \\ \rho U u + \xi_x p \\ \rho U v + \xi_y p \\ \rho U H \end{pmatrix} - \frac{1}{JRe}\begin{pmatrix} 0 \\ \xi_x \tau_{xx} + \xi_y \tau_{xy} \\ \xi_x \tau_{xy} + \xi_y \tau_{yy} \\ \xi_x E_{D_4} + \xi_y E_{D_4} \end{pmatrix}, \tag{2.2}$$

$$E_{D_4} = u\tau_{xx} + v\tau_{xy} + q_x, \quad U = \xi_t + \xi_x u + \xi_y v, \quad V = \eta_t + \eta_x u + \eta_y v. \tag{2.3}$$

The rate of strain and the heat fluxes require the computation of the velocity and temperature gradients. For a Newtonian fluid with perfect gas properties, the gradients, e. g. $q_x$, $\tau_{xy}$, and $\tau_{xx}$, are determined as follows:

$$\begin{aligned} q_x &= \frac{k}{(\gamma-1)\,Pr}(\xi_x T_\xi + \eta_x T_\eta), \quad \tau_{xy} = \mu(\xi_y u_\xi + \eta_y u_\eta + \xi_x v_\xi + \eta_x v_\eta), \\ \tau_{xx} &= \tfrac{2}{3}\mu(2(\xi_x u_\xi + \eta_x u_\eta) - (\xi_x u_\xi + \eta_x u_\eta + \xi_y v_\xi + \eta_y v_\eta)). \end{aligned} \tag{2.4}$$

Equation (2.1) is closed with the Stokes hypothesis for the bulk viscosity and Sutherland's law for the molecular viscosity.

*Initial and boundary conditions*
The inflow boundary conditions correspond to the supersonic freestream values, at the outflow boundaries $\rho$, $\rho u$, $\rho v$ and $H$ are extrapolated with first order accuracy. At the solid wall the no-slip condition is applied. The static pressure is determined with the momentum equation, formulated for the direction normal to the wall. The simulations were initialized specifying a parallel flow with freestream properties.

*Method of solution*
The convective and pressure terms are discretized using the Advection Upwind Splitting Method (AUSM) by Liou and Steffen [7] in an improved formulation (AUSMDV) according to Wada and Liou [18] with an entropy-fix. Second order spatial accuracy is achieved with the MUSCL approach according to Van Leer [2]. The flux on the cell interface is constructed from a state variable interpolation and a locally one-dimensional model of wave interactions normal to the cell interface. The interface flux in the $\xi$-direction, for instance, is represented by the Mach-number-weighted average of the left (L) and right (R) values of $\hat{\Phi} = (\rho a, \rho a u, \rho a v, \rho a H)^T$ with $\hat{p}_{\frac{1}{2}} = p_{\frac{1}{2}}(0,\xi_x,\xi_y,0)^T$:

$$\hat{E}_{C\,\frac{1}{2}} = \frac{1}{2}[M_{\xi\,\frac{1}{2}}(\hat{\Phi}_L + \hat{\Phi}_R) - |M_{\xi\,\frac{1}{2}}|(\hat{\Phi}_R - \hat{\Phi}_L)] + \hat{p}_{\frac{1}{2}}. \tag{2.5}$$

The contravariant Mach number in $\xi$-direction $M_{\xi\,\frac{1}{2}} = M^+_{\xi\,L} + M^-_{\xi\,R}$ and the split pressure at the cell interface $p_{\frac{1}{2}} = p^+_L + p^-_R$ are calculated by:

$$M^{\pm}_{\xi\,L/R} = \begin{cases} \alpha_{L/R}\{\pm\frac{(M_{\xi\,L/R}+1)^2}{4} - \frac{M_{\xi\,L/R}\mp|M_{\xi\,L/R}|}{2}\} + \\ \quad + \frac{M_{\xi\,L/R}\pm|M_{\xi\,L/R}|}{2}, & \text{if } |M_{\xi\,L/R}| \lessgtr 1 \\ \frac{M_{\xi\,L/R}\pm|M_{\xi\,L/R}|}{2}, & \text{else} \end{cases}$$

$$\alpha_{L/R} = \frac{2(p/\rho)_{L/R}}{(p/\rho)_{L/R} + (p/\rho)_{R/L/R}}, \quad M_{\xi\,L/R} = \frac{U_{L/R}}{\max(a_L,a_R)}, \quad a_{L/R} = \sqrt{\gamma\frac{p_{L/R}}{\rho_{L/R}}},$$

$$p^{\pm}_{L/R} = \begin{cases} \frac{1}{4} p_{L/R}(M_{\xi\ L/R} \pm 1)^2 (2 \mp M_{\xi\ L/R}), & \text{if } |M_{\xi\ L/R}| \le 1 \\ \frac{1}{2} p_{L/R}(M_{\xi\ L/R} \pm |M_{\xi\ L/R}|)/M_{\xi\ L/R}, & \text{else} \end{cases} . \quad (2.6)$$

The advantages of both the Flux-Vector Splitting (FVS) according to Van Leer [16] or Hänel [12] and the Flux-Difference Splitting (FDS) proposed by Roe [11] are combined in this scheme. The good shock capturing properties of the FVS as well as the low dissipation and the capability to resolve accurately stationary discontinuities of the FDS scheme are achieved with the following formulation of the contravariant momentum flux:

$$\begin{aligned} (\rho U^2)_{\frac{1}{2}\ AUSMV} &= M^+_{\xi\ L} a_L (\rho_L U_L) + M^-_{\xi\ R} a_R (\rho_R U_R), \\ (\rho U^2)_{\frac{1}{2}\ AUSMD} &= \tfrac{1}{2}[(\rho U)_{\frac{1}{2}} (U_L + U_R) - |(\rho U)_{\frac{1}{2}}|(U_R - U_L)], \\ (\rho U^2)_{\frac{1}{2}\ AUSMDV} &= S_{\frac{1}{2}} (\rho U^2)_{\frac{1}{2}\ AUSMV} + (1 - S_{\frac{1}{2}})(\rho U^2)_{\frac{1}{2}\ AUSMD}, \end{aligned} \quad (2.7)$$

with the switching function $S_{\frac{1}{2}}$ at the cell interface

$$S_{\frac{1}{2}} = \frac{1}{2}(\min\{1, k \max_{\iota=i,i+1} \left\{ \frac{p_{\iota-1,j} - 2p_{\iota,j} + p_{\iota+1,j}}{p_{\iota-1,j} + 2p_{\iota,j} + p_{\iota+1,j}} \right\}\} + 1). \quad (2.8)$$

The shear-stress and heat-transfer terms are discretized with central differences.

For the explicit time integration a five-step Runge-Kutta method is used. The coefficients $\alpha_k$ = (0.059,0.14,0.273,0.5,1.) have been chosen to obtain maximum stability with second-order accuracy. Convergence to steady state is accelerated with local time stepping and a multi-grid algorithm [9].

Grid points of the computational grid are distributed in a geometric series in streamwise direction and clustered near the point, where the curvature changes discontinuously. The grids were iteratively adapted to the obtained solutions using grid point distribution functions as proposed by Vinokur [17]. In wall-normal direction, grid points are clustered near the shock as well as in the boundary layer according to the maximum curvature of the streamwise velocity and the temperature. Typically, the number of grid points of the final mesh amounts to about 100,000, with about 80 points in the boundary layer.

### 2.2 Linear nonlocal stability theory

The instability analysis of the compressible boundary-layer flow is performed using the instability solver NOLOT (NOnLOcal Transition analysis) developed at the DLR in Göttingen and KTH/FFA Stockholm. In the present paper only a brief overview of the governing linear stability equations for compressible flows is given, for details we refer to Hein et al. [4]. The flow quantities are decomposed into a steady basic flow part $\bar{q}$ and an unsteady disturbance flow component $\tilde{q}$, which is written according to the normal-mode formulation with the angular frequency $\omega$, the complex wave numbers $\alpha$ in streamwise ($x^1$) and $\beta$ in spanwise ($x^2$) direction:

$$q(x^i,t) = \bar{q}(x^1,x^3) + \tilde{q}(x^i,t), \quad \tilde{q}(x^i,t) = \hat{q}(x^1,x^3)e^{i(\alpha x^1 + \beta x^2 - \omega t)}. \quad (2.9)$$

In the present work only the spatial wave propagation is investigated under the assumption of quasi three-dimensional flow, which corresponds to

$$\alpha = \alpha_r + i\alpha_i, \quad \beta = \beta_r = const. \quad \text{and} \quad \sigma = -\alpha_i + \mathrm{Re}(\frac{1}{\hat{q}}\frac{\partial \hat{q}}{\partial x^1}). \quad (2.10)$$

The indices $r$ and $i$ denote the real and the imaginary part. If the growth rate $\sigma$ is positive, the amplitude is amplified which means, that the flow is unstable, and vice versa. Neutral instability is obtained when $\sigma = 0$. The spatial growth of the wave amplitude $A_1$ along the plate with the initial amplitude $A_0$ is represented in the $N$-factor [13, 15] and the wave angle $\Psi$ is defined as

$$\frac{A_1}{A_0} = e^N \quad \text{with} \quad N = \int_{x_0}^{x_1} \sigma dx, \quad \psi = \tan^{-1}(\beta_r/\alpha_r). \tag{2.11}$$

The linear nonlocal stability equation can be written as a system of eight first order differential equations of the form:

$$\mathcal{A}\hat{\psi} + \mathcal{B}\frac{\partial\hat{\psi}}{\partial x^3} + \mathcal{D}\frac{\partial\hat{\psi}}{\partial x^1} = 0, \text{ with } \hat{\psi} = (\hat{u}^1, \hat{u}^2, \hat{u}^3, \hat{T}, \hat{\rho}, \hat{u}', \hat{v}', \hat{T}')^T, \tag{2.12}$$

where the primes denote the derivative with respect to the wall normal coordinate $x^3$ and $\mathcal{A}, \mathcal{B}, \mathcal{D}$ are $8 \times 8$ matrices. The NOLOT-code allows different subsets of equations to be solved according to different assumptions: For the local parallel flow assumption all streamwise derivatives of amplitude functions are neglected ($\mathcal{D} = 0$) and all terms, which are related to the nonparallel basic flow are removed from $\mathcal{A}$ and $\mathcal{B}$. In the nonlocal parallel case only terms related to the nonparallel flow assumption are removed from $\mathcal{A}$ and $\mathcal{B}$. Analogously one can implement either the local nonparallel or the nonlocal nonparallel flow assumption.

*Boundary conditions*
To solve this eighth-order system of differential equations, eight boundary conditions are necessary, respectively:

$$x^3 = 0 \quad : \quad \hat{u}^1 = \hat{u}^2 = \hat{u}^3 = \hat{T} = 0 \tag{2.13}$$

$$x^3 \to \infty \quad : \quad \hat{u}^1 = \hat{u}^2 = \hat{\rho} = \hat{T} \to 0\,. \tag{2.14}$$

## 3 Boundary-layer instability

The sensitivity of the instability characteristics to the resolution of the basic flow solution has been studied and Navier-Stokes calculations of a flat plate in supersonic flow have been compared to self-similar solutions. Thus, the Illingworth-Stewartson transformation has been applied to the boundary-layer equations for compressible planar flow with zero pressure gradient. The boundary value problem [1] was solved by using a fifth-order Runge-Kutta [3] and a shooting method. The resulting velocity and temperature profiles as well as their first and second derivatives show a sufficient agreement between the Navier-Stokes results and the self-similar solutions for a different number of grid points [8]. Especially the position of the generalized inflection point, where $\frac{\partial}{\partial y}\left(\rho\frac{\partial u}{\partial y}\right)$ vanishes, affects the inviscid stability characteristics of the flow according to Lees and Lin [6]. Its position as well as the positions and values of the maxima in the curvature of temperature and velocity calculated by the Navier-Stokes code fit to the self-similar solutions as shown in Figure 1.a. The comparison of spatial amplification rates in Figure 1.b shows good agreement of the instability characteristics based on the Navier-Stokes solution of the flat-plate flow and the self-similar solutions.

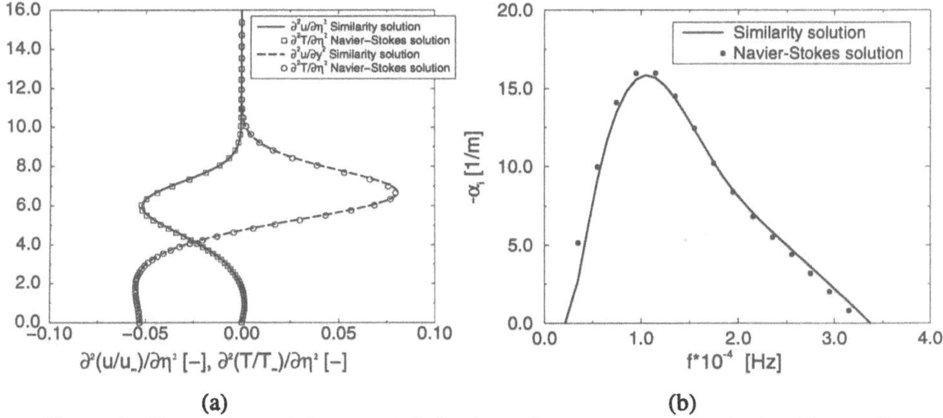


**Figure 1** Comparison of the second derivatives of temperature and velocity (a) as well as of spatial amplification rates (b) for the Navier-Stokes solution of the flat-plate flow and self-similar solutions at $M_\infty = 2.5$, $Re = 3.3 \cdot 10^6$

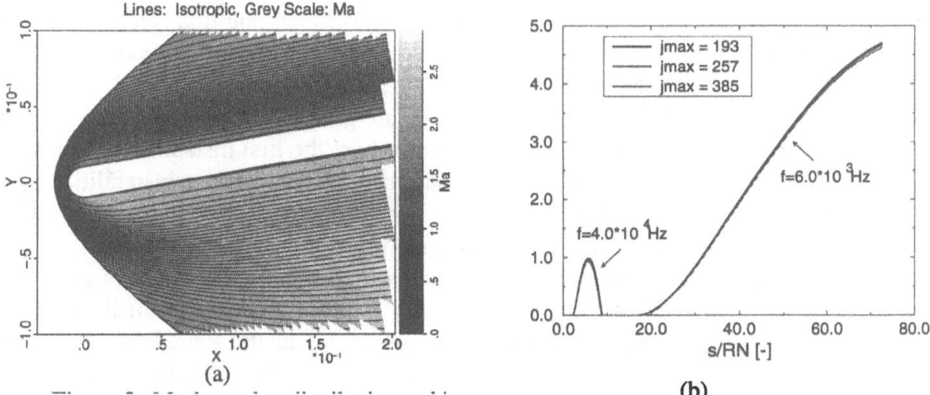


**Figure 2** Mach number distribution and isotropic lines over the blunt flat plate (a). Influence of the number of grid points in wall-normal direction on the $N$-factor for two oblique waves (b). $M_\infty = 2.5$, $Re/l = 0.99 \cdot 10^7/m$, $R_N = 10$ mm, $\alpha = -10^0$

The Mach number distribution and isotropic lines of the flow with a freestream Mach number of $M_\infty = 2.5$ over a blunt plate with nose radius $R_N = 10$ mm at $\alpha = -10^0$ incidence are shown in Figure 2.a. The oblique curved bow shock induces an entropy layer, that is swallowed downstream by the boundary layer. Figure 2.b shows the $N$-factor distribution for two oblique waves over the blunt flat plate obtained on grids with 193, 257 and 385 points in wall-normal direction, applying the local, parallel stability theory. The results are almost independent from the number of grid points normal to the wall.

In Figure 3, the instability characteristics of the Navier-Stokes solution for the blunt plate (a) is compared to those of the self-similar solution (b) with the same flow-properties at the boundary-layer edge $M_e$, $Re_e$. The diagrams show color-encoded amplification rates for different wave frequencies on the ordinate over the local Reynolds numbers calculated with the Blasius length. The distance between the lower and upper

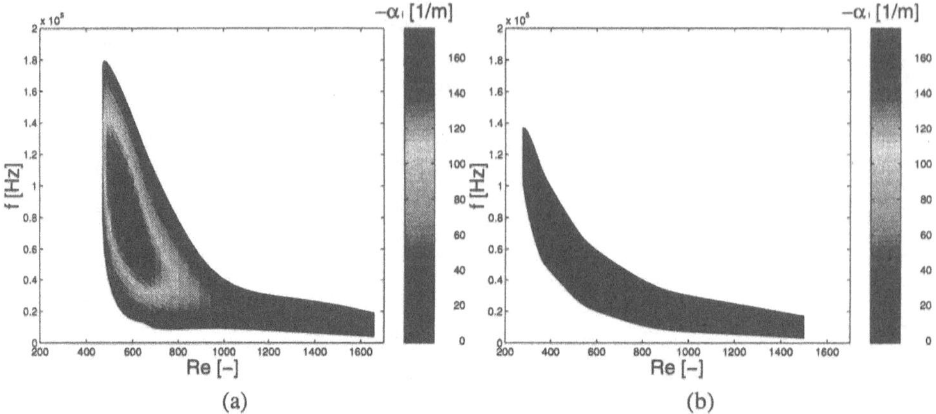


**Figure 3** Comparison of instability diagrams of the Navier-Stokes solution for the blunt plate at $M_\infty = 2.0$, $Re/l = 1.26 \cdot 10^7/m$, $R_N = 10$ mm, $\alpha = -10^0$ (a) to self-similar solutions with the same flow properties at the boundary-layer edge $M_e$, $Re_e$ (b)

branch of the stability curve for the blunt-plate Navier-Stokes solution is larger than for the similarity solution. Additionally, for Reynolds numbers lower than 900, the amplification rate is very high. Here the boundary-layer flow is more unstable, because the outer flow is decelerating due to an adverse pressure gradient. This is in contrast to the observations on blunt cones in hypersonic flows, where the first instabilities are amplified further downstream [5]. In all cases investigated only first-mode instabilities have been found, because the most amplified waves propagate at a wave angle of approximately $\psi \approx 60^0$, irrespective of the frequency. However, the self-similar boundary layer is more unstable to waves with a lower frequency, which are amplified over a longer distance, as can be seen in Figure 4.b, and eventually lead to the onset of transition.

Hence, it is pointed out, that not only the properties at the boundary-layer edge $M_e, Re_e$ affect the instability characteristics considerably, but also the shape of the boundary-layer velocity and temperature profiles. Therefore, in the cases presented here the blunting of the leading edge has stabilizing effects on the boundary-layer flow caused by both, the reduction of the boundary-layer edge Reynolds number and by the vorticity induced by the entropy layer, which affects the boundary-layer profiles.

The increase of the free-stream Mach number results in a bow shock, which is closer to the body and more curved thus producing higher entropy gradients normal to the wall. Figure 4.a shows, that disturbance waves are amplified stronger for the lower Mach number and the corresponding higher unit Reynolds number. The Mach number as well as the Reynolds number behind the shock are higher for the plate at zero incidence compared to the windward side of the plate at $-10^0$ angle of attack. Hence, the presence of the entropy layer causes an additional reduction of both properties. Decreasing the angle of attack from zero incidence destabilizes the boundary-layer flow as shown in Figure 4.b.

The differences shown in Figure 5 result from application of either the local, parallel or the nonlocal, nonparallel stability theory. The influence of the nonlocal, nonparallel theory is small, especially for the most dominant frequencies for the transition. The

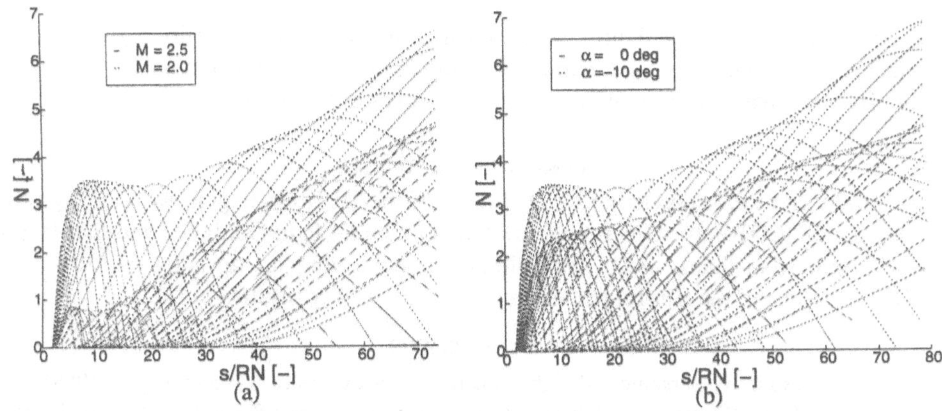


**Figure 4** Influence of the angle of attack and the freestream Mach number on the boundary-layer instability at $Re/l = 1.26 \cdot 10^7/m$, $R_N = 10$ mm; $\alpha = -10^0$ (a), $M_\infty = 2.0$ (b)

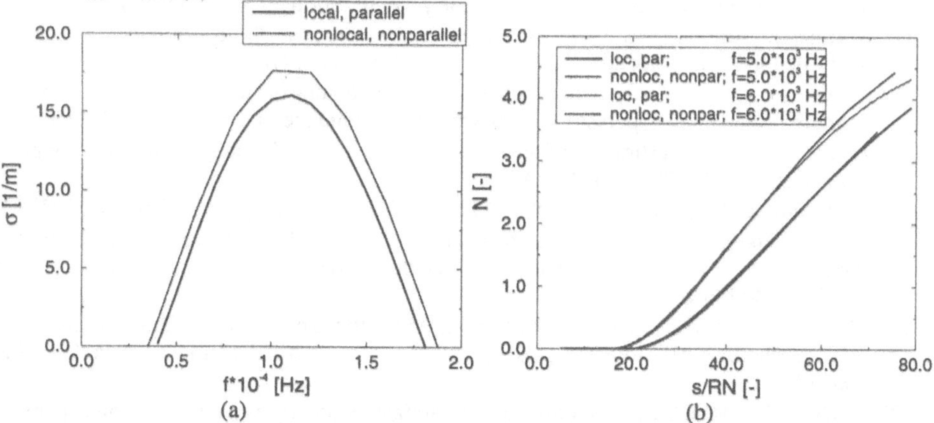


**Figure 5** Spatial amplification rates for different frequencies at $s/R_N = 20$, for $M_\infty = 2.0$ and $R_N = 10$ mm at zero incidence (a). Influence of the nonlocal, nonparallel stability theory on the $N$-factor, for $f = 5.0 \cdot 10^3$ Hz and $\psi \approx 45^o$ and $f = 6.0 \cdot 10^3$ Hz and $\psi \approx 45^o$ (b).

nonlocal, nonparallel effect becomes stronger when the wave angle is increased. However, for these wave angles the wave amplification is not maximal. This is caused by the flat-plate character of the boundary-layer profiles, which fulfill the assumptions of the local, parallel theory well, and no curvature effects are present.

# 4 Conclusion

The presented Navier-Stokes code with a discretization scheme of second-order accuracy in space yields the required base-flow profiles for a subsequent instability analysis. For the investigated Mach numbers of 2.0 and 2.5 at unit Reynolds number of $1.26 \cdot 10^7/m$ and $0.99 \cdot 10^7/m$ it has been shown, that the leading edge-bluntness has a stabilizing effect on the boundary-layer flow. The decrease of the free-stream Mach number as well as the decrease of the angle of attack from zero incidence to $-10^0$ have a destabilizing effect on the boundary layer. Due to the flat-plate character of the boundary-layer profiles and the absence of curvature effects, the stability characteristics obtained by applying the local, parallel stability theory and by the nonlocal, nonparallel stability are considerably small, especially for waves of the most dominant frequencies.

## Acknowledgements

We would like to acknowledge U. Dallmann of the DLR Göttingen, who kindly provided the NOLOT-code, which was developed at the DLR in Göttingen and KTH/FFA Stockholm. Also thanks go to U. Dallmann and S. Hein for advising us to use the NOLOT-code and the fruitful discussion on the instability results.

## References

[1] *J. Anderson*, "Hypersonic and high temperature gas dynamics", McGraw-Hill (1989).

[2] *B. van Leer*, "Towards the Ultimate Conservative Difference Scheme V. A second–order sequel to Godunov's method", J. Comput. Phys. 32, 101–136 (1979).

[3] *E. Hairer, G. Wanner*, "Solving ordinary Differential Equations II. Stiff and Differential-algebraic Problems", Springer Series in Comp. Math. Springer Verlag (1991).

[4] *S. Hein, F. Bertolotti, M. Simen, A. Hanifi, D. Henningson*, "Linear nonlocal instability analysis -the linear nolot code-", DLR IB 223-94 A56 (1994).

[5] *E. Kufner*, "Numerische Untersuchungen der Strömungsinstabilitäten an spitzen und stumpfen Kegeln bei hypersonischen Machzahlen", PhD thesis, DLR FB 95-11 (1995).

[6] *L. Lees, C. Lin*, "Investigation of the stability of the laminar boundary layer in a compressible fluid", NACA TN1115 (1946).

[7] *M.-S. Liou, C. Steffen Jr.*, "A new flux splitting scheme", J. Comp. Phys. 107, 23–39 (1993).

[8] *A. Meijering, G. Dietz, S. Hein, U. Dallmann*, "Numerical investigation of the boundary layer instabilities over a blunt flat plate at angle of attack in supersonic flow", DLR IB 223-96 A44 (1996).

[9] *M. Meinke, D. Hänel*, "Time accurate multigrid solutions of the Navier-Stokes equations", In W. Hackbusch, U. Trottenberg (eds.), "Multigrid methods III", Birkhäuser Verlag, 289-300 (1991).

[10] *E. Reshotko, M. Khan*, "Stability of the laminar boundary layer on a blunted plate in supersonic flow", In "International Union of Theoretical and Applied Mechanics, Laminar-Turbulent Transition", Symposium Stuttgart, Germany (1979).

[11] *P. Roe*, "Approximate riemann solvers, parameter vectors, and difference schemes", J. Comput. Phys. 43, 357–372 (1981).

[12] *R. Schwane, D. Hänel*, "An implicit flux-vector splitting scheme for the computation of viscous hypersonic flow", AIAA Paper 89-0274 (1989).

[13] *A. Smith, N. Gamberoni*, "Transition, pressure gradient and stability theory", Douglas Airkraft Co. TR Rept.ES 26388, El Segundo, California (1956).

[14] *K. Stetson, J. Donaldson, L. Siler*, "Laminar boundary layer stability experiments on a cone at mach 8, part 2: Blunt cone", AIAA Paper 84-0006 (1984).

[15] *J. van Ingen*, "A suggested semi-emperical method for the calculation of the boundary layer transition region", Univ.of Techn. Dept. of Aero.Eng. TR Rept.UTH-74, Delft (1956).

[16] *B. van Leer*, "Flux-vector splitting for the euler equations", Lecture Notes in Physics 170, 507–512 (1982).

[17] *M. Vinokur*, "On one-dimensional streching functions for finite-difference calculations", J. Comp. Phys. 50/2, 215–234 (1983).

[18] *Y. Wada, M.-S. Liou*, "A flux splitting scheme with high-resolution and robustness for discontinuities", NASA TM106452 also AIAA Paper 94-0083 (1994).

# Application of Two-Parameter Turbulence Models to the Prediction of Two-Dimensional Free Turbulent Shear Flows

B. Dreßler
TU Berlin, Hermann-Föttinger-Institut für Strömungsmechanik
Straße des 17.Juni 135, D-10623 Berlin, Germany

## SUMMARY

In the present paper the turbulence models of Chien and of Wilcox are applied to predict three types of two-dimensional free turbulent shear flows. At the end, as an example of a more complex configuration, the flow field behind an axisymmetric body with jet is calculated. The task of the present study was to examine the reliability of two turbulence models in free shear flows, which were implemented especially for use in wall boundary layers. Knowing the shortcomings of the $k$-$\omega$ model in free shear flows expectation is endorsed that the $k$-$\varepsilon$ model of Chien produces better predictions than the model of Wilcox. But the results show that in cases of practical interest the initial profiles, especially those for the turbulence parameters, are more relevant for the prediction of the mean flow parameters in the near field than the applied turbulence models.

## INTRODUCTION

Nowadays turbulence modelling is still one of the major problems for numerical prediction of turbulent flows.
Calculation of turbulent flows, using the Navier-Stokes equations directly (DNS), requires big numerical effort. The discretisation of the flow area has to be made as fine that all turbulent modes up to the smallest eddies can be catched. The size of these eddies decreases with increasing Reynolds number. For that reason DNS is practicable only for simple geometries and low Reynolds numbers. For the majority of practical cases calculations are carried out applying semi-empirical turbulence models.
In complex geometries like airplane configurations the task is to find a compromise between a good modelling of the Reynolds stresses and its required numerical effort. One of possible compromises is assumed in the class of two-parameter turbulence models. Nowadays models of this class are intensively used in all practical applications. In comparison to second order closure one has only two additional equations to solve if scalar turbulence viscosity is sufficient.
In the last decade a lot of papers came up suggesting new two-parameter turbulence models for applications in wall bounded flows (low-Reynolds-number turbulence models). The interest in free turbulent shear flows was much smaller because already the standard $k$-$\varepsilon$ model produces satisfying predictions in those cases.
In engine configurations, a special object of later studies, flow fields with different turbulence structures occur. Here a turbulence model has to work in wall boundary layers as well as in free shear layers. In this paper two-parameter turbulence models implemented in low-Reynolds-number form are applied to calculate some standard test cases of free turbulent shear flows.

## NUMERICAL METHOD AND TURBULENCE MODELS

To solve the Reynolds-averaged Navier-Stokes equations for compressible flows the DLR-CEVCATS Code [1] was applied. This is a cell vertex finite-volume solver working on block-structured grids. The central difference technique is used for the spatial discretization of the convective and viscous fluxes in the mean flow equations. The integration in time is realized by an explicit five-stage Runge-Kutta scheme. Local time stepping, implicit residual smoothing and multigrid technique are applied to accelerate the convergence to a steady-state solution.

In the applied version of the code three different two-equation turbulence models have been implemented, namely the models of Chien [2], Coakley [3] and Wilcox [4]. The implementation was done for working in single grid mode.

In the present study we emphasize on the models of Chien and Wilcox which are known for their ability to predict well turbulent flows in wall boundary layers.

The model of Chien is a low-Reynolds-number model based on the $k$-$\varepsilon$ model. It works with wall damping functions defined by local wall distances and local turbulent Reynolds numbers. Far from the wall and for high local Reynolds numbers it acts like a usual $k$-$\varepsilon$ model. Of course, one of its shortcomings is the dependence on the wall distance, because in complex geometries it is often difficult to define this parameter uniquely.

The advantage of the Wilcox model is its independence of wall functions in the viscous sublayer and its good performance in predicting flows with strong pressure gradients and separations. But in free turbulent shear flows its results are not always satisfying.

The task of the present study is a validation of these two turbulence models in some types of free shear flows. An objective of later studies will be the prediction of the flow field in engine configurations. In that case and in the majority of practical cases free shear flow originates from wall bounded flows which then define the initial conditions. So it is clear that predictions for the characteristic values in the far field as well as predictions of the resulting flow regime in the near field are of special interest.

All calculations were carried out rather from the point of view of an user of the CEVCATS code than from that one of a developper. Some modified procedures prescribing boundary conditions were tested but no changes of the set of turbulence model parameters were made.

## PLANE FREE MIXING LAYER

The simplest case of a free shear flow is a free mixing layer. Here two initially non-turbulent uniform streams separated by a splitting plate become a flow with a turbulent mixing layer. In the classical experiment one of the streams is air at rest. In the far field of this flow regime similar velocity profiles will be established and a spreading rate $\Delta y_c/\Delta x$ can be defined as a characteristic value of the flow.

As sketched in figure 1 the value $\Delta y_c$ is defined as $\Delta y_c = y_{.1} - y_{.9}$.

Considering this flow problem theoretically, it is possible to transform the governing system of partial differential equations into a system of ordinary differential equations by means of similarity functions only depending on the similarity variable $\eta = y/x$. In [5] Wilcox gives the tools to solve such a system of equations. Similarity profiles as well as spreading rates can be computed applying the program MIXER included in [5]. In table 1 a summary of some investigated spreading rates is given.

Now some remarks are made about the fact that the spreading rates corresponding to

the turbulence model of Wilcox are given in form of intervals. Computations based on self-similar equations require boundary values. Especially, in the case of Wilcox model a value $W_\infty$ for $W(\eta)$ with $\eta \rightarrow \infty$ is needed, where $W(\eta)$ represents the vorticity $\omega$ in a transformed form. As pointed out in [5] and [6] the mixing layer predictions using the turbulence model of Wilcox are very sensitive to this quantity. Unfortunately, this is not only a feature of the self-similar equations, the sensitivity of results to the freestream value of $\omega$ has been also found by computations based on the Reynolds equations.
As a conclusion from these facts the results of computations always have to be regarded in connection with the turbulence field in the freestream using the Wilcox turbulence model in free shear flows.

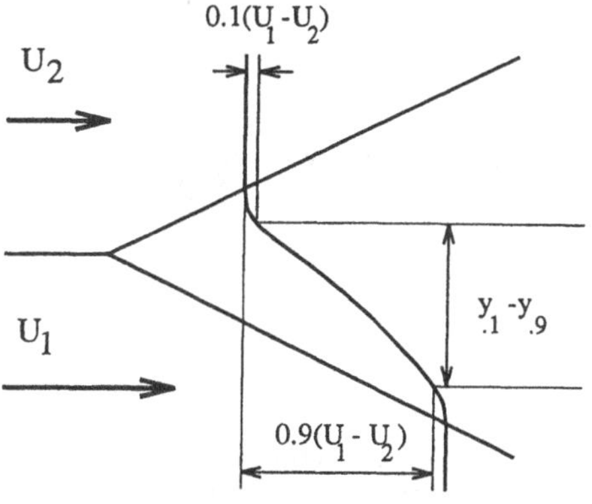


Fig.1 Free mixing layer: spreading rate nomenclature.

Table 1 Spreading rates for free turbulent mixing layers.

| Model | $U_2/U_1$=0 | $U_2/U_1$=0.61 |
|---|---|---|
| MIXER | | |
| Standard- $k$-$\epsilon$ model | 0.152 | 0.0336 |
| $k$-$\omega$ model | 0.15-0.22 | 0.027-0.042 |
| CEVCATS | | |
| Chien | - | 0.033 |
| Wilcox | - | 0.028 |
| Experiment (s.Rodi[8]) | 0.16 | 0.0328 |

Furthermore three remarks have to be made. In an experiment the described flow problem cannot be realized exactly. Every kind of splitting plates produces wall boundary layers and no flow is really non-turbulent.
On the other hand there is a limitation of applications of the CEVCATS code. That flow solver is a code for prediction of compressible flow. A preconditioning for incompressible flow has not been implemented in the applied version. Therefore the local Mach number should not be chosen smaller than 0.1 . For that reason only flows of a non-zero ratio $U_1/U_2$ can be investigated.
The results given in the figures are calculated for the following flow conditions: $M_1$=0.377, ($M_1$-Mach number to $U_1$) , $U_2$= 0.61 $U_1$ , Re=1100000 , Turbulence level: 0.5 % .
The calculation starts at a line defined by the origin of the mixing layer and the computational domain has a length L. As boundary conditions both initial conditions and far field conditions are applied. In the uniform streams a turbulence structure is assumed like those in the far field of grid turbulence. That means a low turbulence level and a small characteristic length scale are chosen.
Figure 2 shows that a similarity profile of mean velocity is established. Because there

no visible difference between the results of both turbulence models was found only the calculated profiles with the model of Chien are presented. The band of experimental data is the one sketched in the validation paper of the BOAT code [7] and based on a review of experimental data by Rodi [8]. Different results are observed in comparing the calculated Reynolds shear stresses. Corresponding to the calculated spreading rate the Wilcox turbulence model produces a too low turbulence level (figure 3).

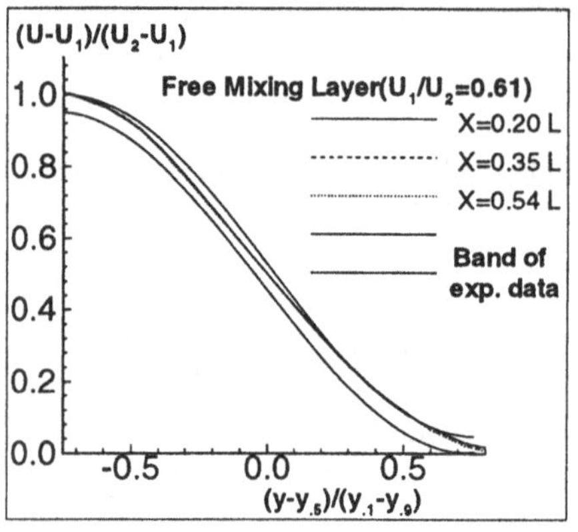


Fig.2 Comparison between predicted profiles of mean velocity and experimental data.

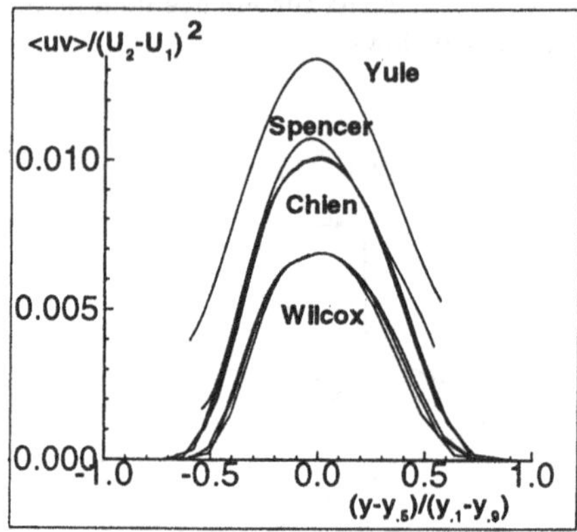


Fig.3 Comparison between predicted and measuered profiles of Reynolds shear stress.

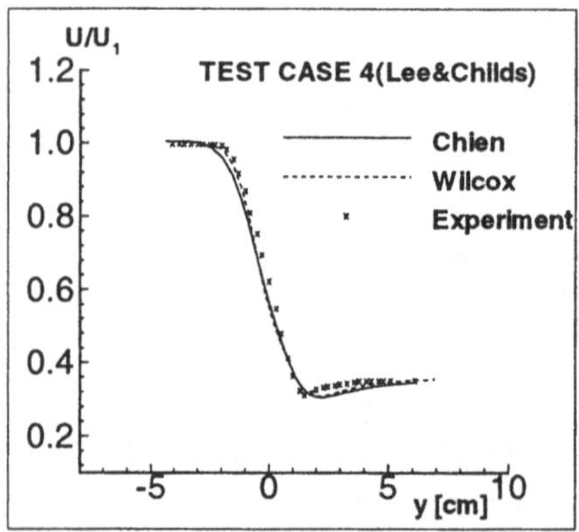


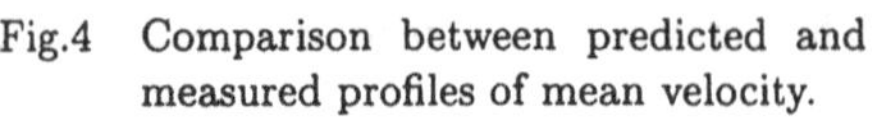
Fig.4 Comparison between predicted and measured profiles of mean velocity.

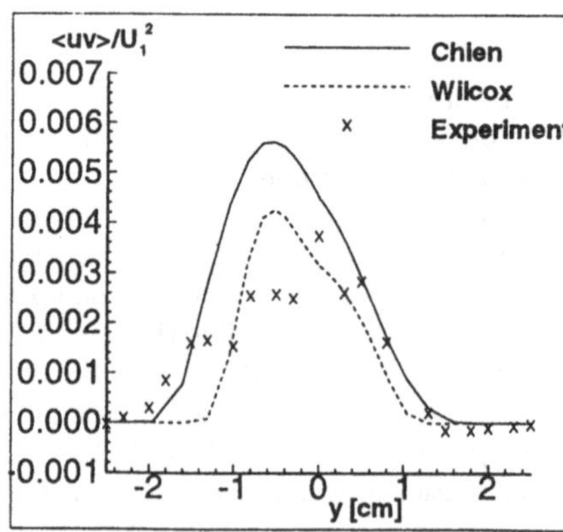


Fig.5 Comparison between predicted and measuered profiles of Reynolds shear stress.

## PLANE FREE MIXING LAYER WITH INITIAL CONDITIONS

Our second example is the test case 4 of the Free Turbulent Shear Flow Conference 1972 [9]. In that case two streams are initially separated by a symmetric airfoil. In the experiment profiles of the mean velocity as well as profiles of the Reynolds shear stresses are measured near the trailing edge and at some stations downstream. Here it is possible to get the initial values for the computations from the experiment directly. The interest is directed towards the near field results emphasizing on the influence of wall boundary layers to the prediction of the flow by means of the turbulence models. In figure 4 results for the axial mean velocity are given at a station in a short distance behind the trailing edge. There is no remarkable difference of the calculated profiles applying the two turbulence models. Predictions of both models agree with the experimental data, but do not reproduce the position of wake correctly. Larger differences in predictions are seen in figure 5. Here the results for the Reynolds shear stresses are given. While the model of Chien overpredicts the level of shear stress, a too slender profile of the Reynolds shear stresses indicates that the spreading rate of the Wilcox model is too small.

## PLANE FREE JET

The flow regime of this flow type is sketched in figure 6. In this case $\Delta y_c$ is defined as $\Delta y_c = y_{.5}$. While on the jet border a mixing layer is evolving the mean centerline velocity decays. The second feature is investigated in this test case. The experimental data are described in test case 13 of the Free Turbulent Shear Flow Conference 1972 [9].

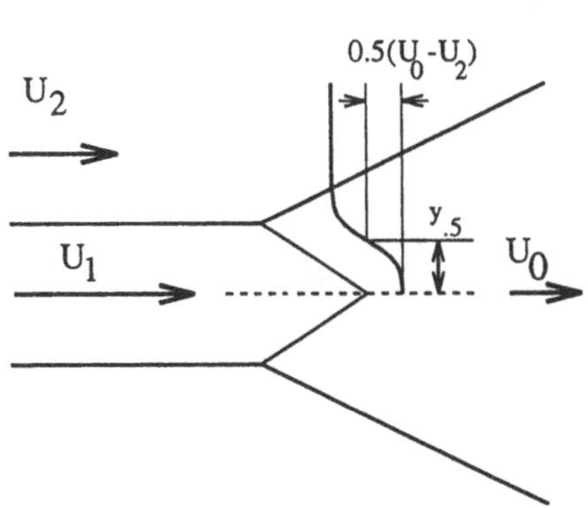


Fig.6 Plane free jet: spreading rate nomenclature.

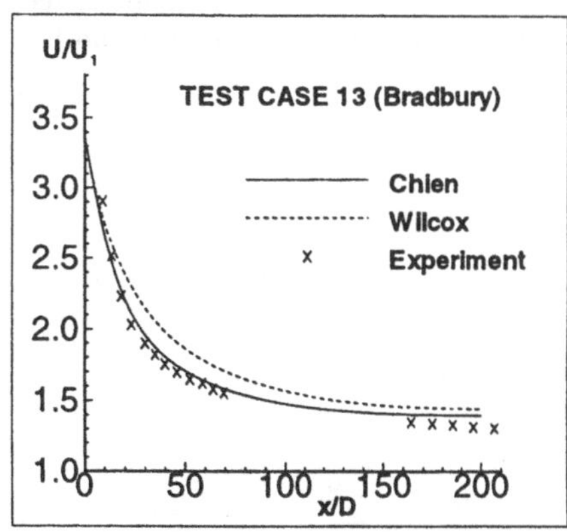


Fig.7 Comparison between centerline velocity decay predictions and data for a plane free jet.

Because the flow is symmetrical the computational domain is reduced to the half of the

flow field. Computations start at the line in the outflow plane of the jet. The turbulence structure of the outer uniform stream is chosen as in the test cases before. Far field, symmetry and initial profiles are the boundary conditions.
Unfortunately, the informations about the experiment are poor, only a velocity ratio and a momentum thickness are given.
For the initial profiles of the jet flow the assumption is used that the flow can be considered approximately as an accelerated tube flow. A variety of computations shows that the decay is extremly influenced by the initial turbulence level. The best agreement with the experimental data is found for a turbulence level of 2-3 % (see figure 7). Because the numerical results do not reflect a potential core there is still a deficiency in the correct form of the initial profiles.
The model of Chien produces a better agreement with the experimental data. This corresponds with the calculated spreading rates. While the model of Chien predicts a spreading rate of 0.026, the Wilcox model gives 0.022, and from the experiment a value of 0.0264 is found.

## AXISYMMETRIC BODY WITH JET

In the AGARD Report 318 [10] a test case is given considering a simplified example of engine configuration.
In the wind tunnel of ONERA Chalais-Meudon an experiment has been carried out, investigating a flow around and behind an axisymmetric body with cold and hot jet. The special interest is directed towards the afterbody behind the cylindrical part of the body with diameter D. The afterbody, partly sketched in figure 8, tapers up to its base downstream. Inside of the afterbody a cylindrical duct ends up in a converging part. Both inner and outer contours are made of rectilinear segments, only the nozzle contour has a radius over a small distance for smooth transition to the outer contour.

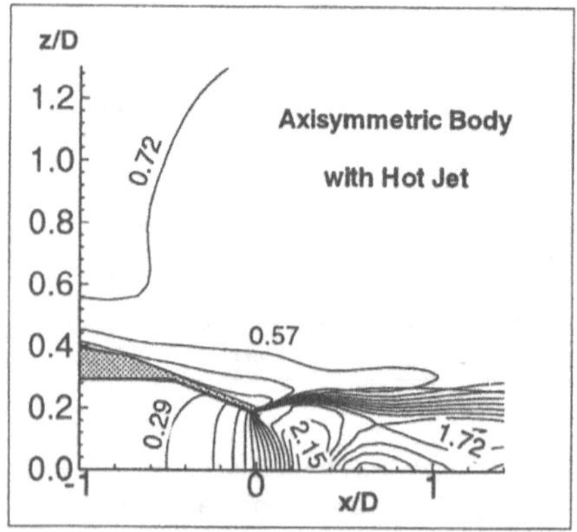


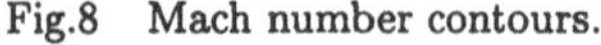
Fig.8 Mach number contours.

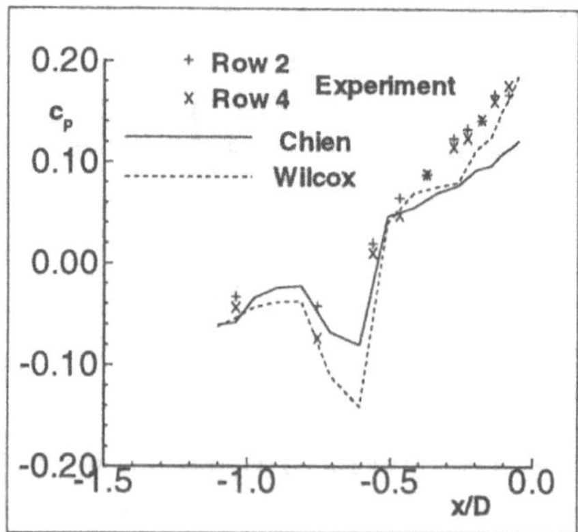


Fig.9 Static pressure coefficient on the afterbody.

In the experiments surface pressures were measured on the outer contour of the afterbody at five rows of taps. At the station x/D=-2.027 pitot pressure measurements at different circumferential positions are given. As a result of these measurements the axiysmmetric hypothesis seems approximately valid. Furthermore, on the basis of pitot pressure profiles a profil of the mean velocity is used as a starting profile for numerical calculations. Unfortunately, at this upstream station there were no informations regarding the turbulence parameters.

Behind the afterbody at different stations measurements were carried out investigating the flow field in detail. Here profiles of the component of mean velocity as well as profiles of the components of Reynolds stresses are given.

For the computations of this test case a computational domain is chosen reaching from x/D= -2.027 up to x/D=5 in x-direction and from z/D= 0 up to z/D= 2.5 in z-direction. The calculation starts with initial profiles outside as well as inside of the afterbody and works with symmetry and far field boundary conditions at the other boundaries of the domain. Taking the measured profiles of the mean velocity the profiles of the turbulent parameters are estimated by an assumed turbulence level and an assumed characteristic length scale. To generate profiles for the flow parameters in the inner flow a modification of the special procedure for engine boundary conditions of the CEVCATS code is used.

The first results of the computations (hot jet case) are shown in the figures 8, 9, 10, and 11. The better prediction of the pressure contribution on the outer contour of the afterbody is given by the turbulence model of Wilcox while the pressure level on the second shoulder of the shape is too low (figure 9). There is no significant difference between the results for the mean axial velocity applying the turbulence model of Chien or the Wilcox model. In both model applications the maximum value is not reproduced (figure 10).

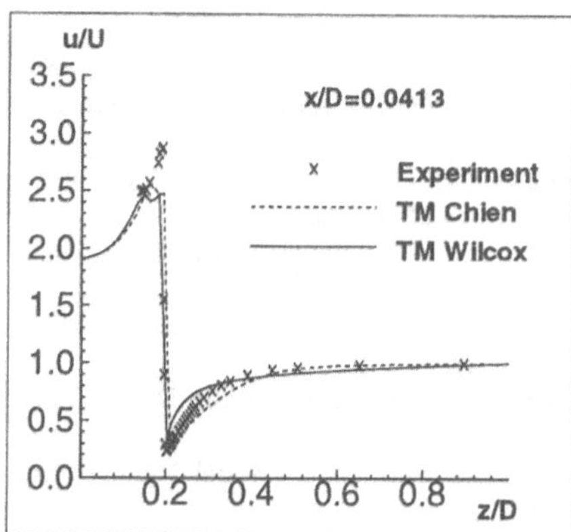


Fig.10 Profiles of mean axial velocity component.

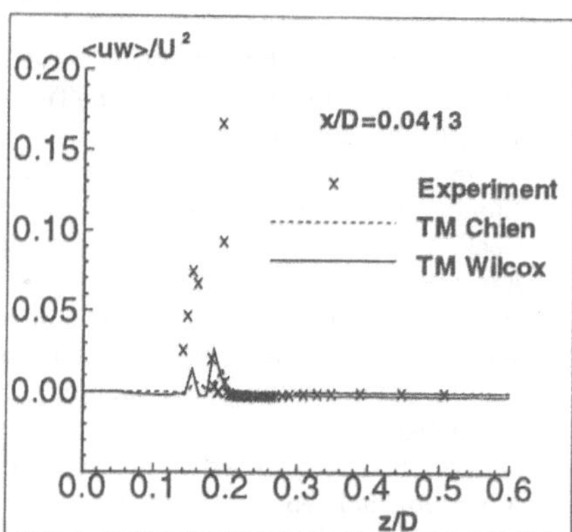


Fig.11 Profiles of Reynolds shear stress.

Especially the level of Reynolds shear stress is much too low (figure 11). For that reason

it is necessary to study the influence of initial profiles of the turbulence parameters extensively. This is the present task and it will be satisfied carrying out systematical variations of the initial data for the turbulence parameters. Only a good adjustment of the initial data, predicting the first completely measured profiles with satisfactory quality, is a basis for turbulence model validation by the flow field behind the afterbody.

## CONCLUSIONS

Free turbulent shear flows are simple and well-determined flow types and therefore good test cases for turbulence models. Future versions of the new FLOWer code will extend the possibilities for investigations of incompressible flow test cases. As a request to the experimentalists additional test cases, especially with informations about all data needed for computations are desirable.
Applying the CEVCATS code and two special two-parameter turbulence models, results of computations for three types of free turbulent shear flows are presented. The better agreement to the experimental data has been observed for the model of Chien. Predictions of the Wilcox model show a lower level of Reynolds shear stresses corresponding to smaller spreading rates.
In ongoing computations a flow about an axisymmetric body with an exhaust jet is considered. Although predicting the mean velocity profiles in the near field behind the body approximately well, the turbulence models predict the turbulence level too low in the same area. A variety of further computations will be carried out to investigate the influence of the upstream turbulence.

## REFERENCES


[1] Rudnik, R.: Berechnung von komplexen zweidimensionalen Profilumströmungen mit algebraischen und Zweigleichungs-Turbulenzmodellen, DGLR-Jahrestagung, Erlangen, 4.-7.10.1994, Bd. III.

[2] Chien, K.-Y.: Predictions of Channel and Boundary-Layer Flows with a Low- Reynolds-Number Turbulence Model, AIAA Journal, Vol.20, pp. 33-38, 1982.

[3] Coakley,T.J.: Turbulence Modeling Methods for Compressible Navier-Stokes-Equations, AIAA Paper 83-1693, 1983.

[4] Wilcox,D.C.: Reassessment of the Scale-Determining Equation for Advanced Turbulence Models, AIAA Journal, Vol.26, pp.1299-1310, 1988.

[5] Wilcox, D.C.: Turbulence Modeling for CFD, DCW Industries, Inc., Palm Drive, California 1994.

[6] Menter, F.R.: Influence of Freestream Values on $k$-$\varepsilon$ Turbulence Model Predictions, AIAA Journal, Vol.30, No. 6, pp. 1657-1659, 1992.

[7] Dash, S.M. and Pergament, H.S.: A Computational Model for the Prediction of Jet Entrainment in the Vicinity of Nozzle Boattails (The BOAT Code), NASA Contractor Report 3075, 1978.

[8] Rodi, W.: A Review of Experimental Data of Uniform Density Free Turbulent Boundary Layers. Launder, B.E., ed.: Studies in Convection, Vol. I, Academic Press (London), pp. 79-166, 1975.

[9] Free Turbulent Shear Flows, Vol. I and II, Conference Proceedings and Summary of Data, NASA SP-321, July 1972.

[10] AGARD-Report No 318: Aerodynamics of 3-D Aircraft Afterbodies 1995.

# Numerical and Experimental Investigations of Flow Induced by Harmonic Motion of a Circular Cylinder

**H. Dütsch, S. Becker, H. Lienhart**

Department of Fluid Mechanics (LSTM),
University of Erlangen-Nuremberg
Cauerstrasse 4, D-91058 Erlangen, Germany

## SUMMARY

A Finite-Volume method to solve the Navier-Stokes equations in an accelerated reference system was applied in order to calculate flow fields around dimensionally stable, moving structures. The two-dimensional velocity field and the force history induced by the harmonic oscillation of a circular cylinder in water at rest at Reynolds number Re=100 and Keulegan-Carpenter number KC=5 was computed accurately in time. Computations have been performed applying different grid levels and time steps. Stable, symmetric and periodic flow patterns occurred showing good agreement with time averaged LDA measurements providing phase resolution of 1 degree.

## INTRODUCTION

For the dynamic analysis of several elastic structures the interaction with fluid flow is of great relevance as it induces additional instantaneous loadings. Especially for large amplitudes the effects of the accelerated motion of solid walls influencing a surrounding viscous fluid and vice versa have to be considered. In this field of dynamic interaction between fluids and structures especially vortex shedding induces non-linear behaviour and loadings. Related investigations are important e.g. for offshore and civil engineering applications as well as for heat exchanger design in particular. In many cases they are concerned with cylindrical structures. Intensive experimental and to an increased extent computational research was performed to derive analytic or semi-empiric models for the fluid behaviour and loading [1, 2, 3]. But still the prediction of the coupled behaviour is not accurate enough in some ranges of application as shown by Bearman et al [4] investigating a circular cylindrical structure.

Numerical computations of these coupled, time-dependent problems may provide a complementary tool for the design and the reliability analysis of elastic structures. The numerical methods for the flow field have to deal with the arbitrary motion of solid boundaries without limitations to amplitudes. For dimensionally stable structures it is one efficient approach to transform the momentum equations of the fluid into an accelerated reference system, which should obviously be the system of the moving structure. The flow description is then changing from the oscillation of the structure towards the oscillation of the fluid.

For the present investigations a Finite-Volume method for the two-dimensional Navier-Stokes equations was applied in order to calculate the laminar flow induced by the translational, harmonic oscillation of a long circular cylinder in a fluid at rest.

The resulting flow field is depending on two major parameters, the Reynolds number Re and the Keulegan-Carpenter number KC [5]:

$$\boldsymbol{Re} = \frac{U_{max} D}{\nu}, \tag{1}$$

$$\boldsymbol{KC} = \frac{U_{max}}{f\, D}. \tag{2}$$

For the Reynolds number the maximum oscillation velocity $U_{max}$, the cylinder diameter D and the kinematic viscosity $\nu$ are representative. The Keulegan-Carpenter number is defined as a function of the maximum oscillation velocity, the cylinder frequency f and the cylinder diameter. It represents the ratio of time scales, which are characteristic for the structure motion $(1/f)$ and for the induced fluid motion $(D/U_{max})$. Therefore it is formally the inverse value of the so-called reduced frequency, which is widely used for oscillating flow around structures. Nevertheless, the authors follow the nomenclature of the recent publications.

In the range of low Reynolds numbers and low Keulegan-Carpenter numbers various flow patterns occur involving asymmetry as well as stability and three-dimensional effects. These patterns were classified into several regimes by Tatsuno and Bearman applying experimental visualisation [6]. Those regimes were reproduced in sample checks by Knörnschild [7] at the LSTM. For the model case the key parameters were set to Re=100 and KC=5 providing periodic and two-dimensional flow behaviour. The pre-determined translational motion $x_i(t)$ of the cylinder was given by

$$x_1(t) = -A\, sin(2\pi f\, t) \qquad \text{and} \tag{3}$$

$$x_2(t) = 0\,, \tag{4}$$

where A denotes the cylinder amplitude and t the time. Here the Keulegan-Carpenter number can be rewritten as $KC = 2\pi\, A\, /\, D$ including the ratio of the amplitude and the diameter.

## NUMERICAL METHOD

For the computational prediction of time-dependent flow, which is induced by moving structures, the deformation of the solid boundaries have to be considered in the momentum equations. In the case of dimensionally stable structures in a fluid environment, which is unbounded in the direction of the structure motion, this can be done without physical restrictions by transforming the momentum equations into an accelerated reference frame. The boundary conditions change from flow induced by moving walls towards unsteady flow around structures at rest. This approach avoids efforts for moving and adjusting the numerical grid as this does not change during the computation. Finally, the location of the grid points has to be calculated by adding the position vector $x_{s,i}(t)$ of the system. Also the resulting vector field $\tilde{u}_i$ has to be retransformed to $u_i$ in the inertial reference frame by adding the system velocity $U_{s,i}$, which derives from

$$x_{s,i}(t) = x_i(t) \quad \text{and} \tag{5}$$
$$U_{s,i}(t) = \dot{x}_{s,i}(t)\,. \tag{6}$$

In order to demonstrate the reliability and efficiency of this approach the harmonic oscillation of a long circular cylinder in water at rest was investigated. For the numerical computations the two-dimensional Navier-Stokes equations for the time-dependent, incompressible and laminar flow of a Newtonian fluid were applied in a non-inertial reference frame:

$$\frac{\partial \tilde{u}_i}{\partial \tilde{x}_i} = 0\,, \tag{7}$$
$$\frac{\partial(\rho\tilde{u}_i)}{\partial t} + \frac{\partial(\rho\tilde{u}_j\tilde{u}_i)}{\partial \tilde{x}_j} = -\frac{\partial P}{\partial \tilde{x}_i} - \frac{\partial \tilde{\tau}_{ji}}{\partial \tilde{x}_j} - \rho\frac{\mathbf{dU_{s,i}}}{\mathbf{dt}}\,. \tag{8}$$

In equation (7) and (8) $\tilde{x}_i$ denotes the Cartesian co-ordinate, $P$ the static pressure, $\tilde{\tau}_{ji}$ the molecularly induced momentum transport and $\rho$ the fluid density. Considering the momentum equations for the fluid in a non-inertial reference frame the acceleration of the system has to be taken into account as given by the term $-\rho\,\mathrm{dU_{s,i}}\,/\,\mathrm{dt}$ in equation (8).

These conservation equations were solved on a O-type structured grid consisting of concentric circles with non-staggered arrangement of the variables. A Finite-Volume approach was applied for a SIMPLE like algorithm [8] with multigrid technique, which is described more detailed in Durst et al [9]. The convection term was approximated by the central difference scheme and the time derivatives by the Crank-Nicolson method providing a complete discretisation of second order. According to the high phase resolution of the experiments 720 time steps were computed per cycle in order to receive time accurate results for comparison. Additionally, another computation with a 12 times higher time step was performed in order to compare the force histories for different time resolution as well.

The flow field far from the oscillating cylinder was assumed to be undisturbed and the fluid there to be fully at rest. In the reference system fixed to the cylinder motion the fluid velocity at the outer boundary $\tilde{u}_{i,b}$ is oscillating with the negative cylinder velocity, while on the cylinder surface no-slip condition was assumed.

$$\text{boundary conditions:} \qquad \tilde{u}_{i,b} = -U_{s,i}(t)\,. \tag{9}$$

The whole outer boundary was treated as inlet of Dirichlet-type providing that no relevant mass error occurs. With respect to the dimensions of the experimental water tank the diameter of the computational domain was set 120 times of the cylinder diameter.

In the undisturbed farfield the additional inertia term fully balances the local acceleration of the fluid. No pressure gradient exists as it would have been expected due to the time-dependent inflow conditions. To avoid any artificial disturbance in the pressure distribution both time derivatives have to be discretised consistently, although the system acceleration can be derived accurately from equation (5) and (6). Figures 1 and 2 qualitatively show a correct and an incorrect pressure field. The difference between the discretised time derivative on the left side of equation (8) and the analytic treatment of the corresponding term on the right side creates an invalid pressure distribution, which is demonstrated in figure 2.

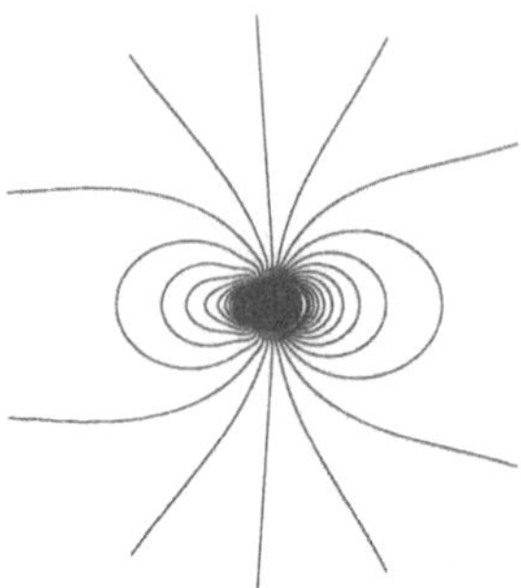

Fig. 1: Correct pressure isolines due to discretised inertia term.

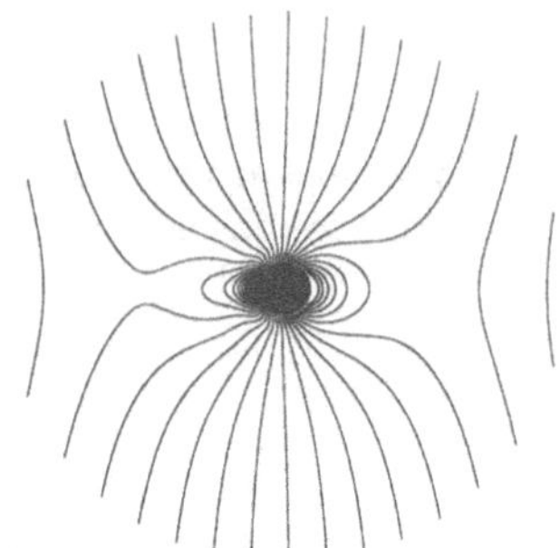

Fig. 2: Incorrect pressure isolines due to analytic inertia term.

As the additional system term in equation (8) influences only the source of the general conservation equation it can easily be implemented in other solution algorithms as well.

The in-line force history $F_1(t)$ was computed by integrating numerically the pressure and shear forces acting on the cylinder surface.

## EXPERIMENTAL SET-UP

In principle the test rig consisted of a fluid tank at rest, in which the cylinder was oscillating sinusoidally actuated by a crank shaft gear drive. Its constructional realisation is sketched in figure 3, which shows its major components: the water tank, the drive system and the LDA probe mounted on top of the two-dimensional traversing unit. In order to achieve the different non-dimensional characteristic numbers Re and KC there were three experimental parameters, that could be adjusted: the rotational speed of the motor drive, the amplitude of oscillation of the gear drive and the diameter of the cylinder. The overall dimensions of the water tank were 1.2 x 0.7 x 0.7 $m^3$, whereas the diameter of the cylinder used for the investigations was 10 mm and its length 0.65 m. The oscillation frequency was set to 0.202 Hz and the amplitude to 0.007958 m. Therefore the influence of the flow field boundaries could be neglected for all the results described furtheron, especially as honeycomb grids were mounted near the tank walls in order to damp out any disturbances. The LDA adopted for the present investigations was a slightly modified LDA system built at the LSTM for wind tunnel applications [10]. A time averaged velocity field in the vicinity of the cylinder was measured with phase resolution of 1 degree. The velocities to be measured were in the order of a few mm/s.

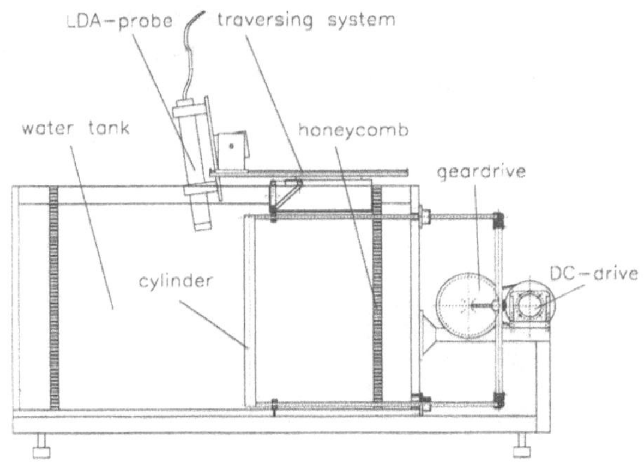


Fig. 3: Test rig in cross section.

## RESULTS

Computations on three different grid levels were performed. In order to receive computational results of at least the same resolution in time as provided by the measurements a very small time step (1/720 of the period length) was chosen. For reasons of comparison the flow field was computed on a medium grid level and for a much larger time step as well. The most important computational parameters are presented in table 1 including the total numbers of control volumes (# of $cv_{total}$) in the flow field, the circumferential grid resolutions (# of $cv_{perimeter}$) and the time resolutions. Considering characteristic values of the in-line force history, such as the phase position and the value of roots and apses, the effects of space and time resolution are presented in table 1 as well.

Table 1: Computational parameters and characteristic results for the in-line force.

| symbol | **A** | **B1** | **C** | **B2** |
|---|---|---|---|---|
| # of $cv_{perimeter}$ | 96 | 192 | 384 | 192 |
| # of $cv_{total}$ | 6144 | 24576 | 98304 | 24576 |
| $f\,\Delta t$ | 1/720 | 1/720 | 1/720 | 1/60 |

| In-line Force $F_1$ (see figures 8 and 9) | | | | |
|---|---|---|---|---|
| phase angle of the 1. root [°] | 35.9 | 33.2 | 32.2 | 33.5 |
| phase angle at minimum [°] | 129 | 125 | 124 | 126 |
| value at minimum [mN/m] | -1.71 | -1.69 | -1.67 | -1.66 |
| phase angle of the 2. root [°] | 216 | 213 | 213 | 214 |
| phase angle at maximum [°] | 309 | 305 | 304 | 306 |
| value at maximum [mN/m] | 1.71 | 1.69 | 1.67 | 1.66 |

In the figures 4 and 5 the computed and retransformed vector field in the vicinity of the cylinder is shown for the numerical case B1. One phase position before and one after the right point of return (phase angle 270°) of the cylinder motion was chosen, see also equation (3).

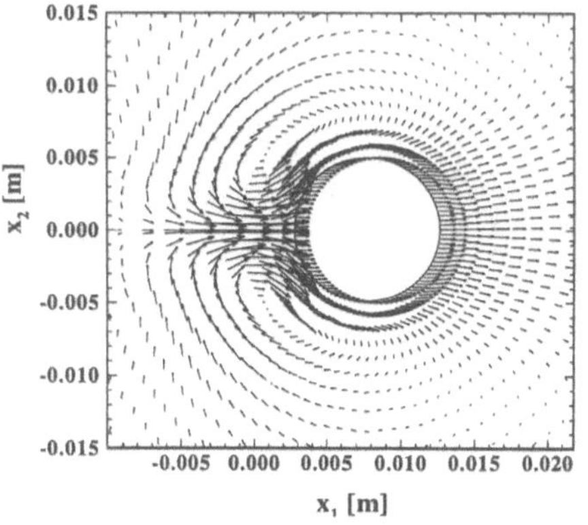


Fig. 4: Computed vector field at phase position 254°.

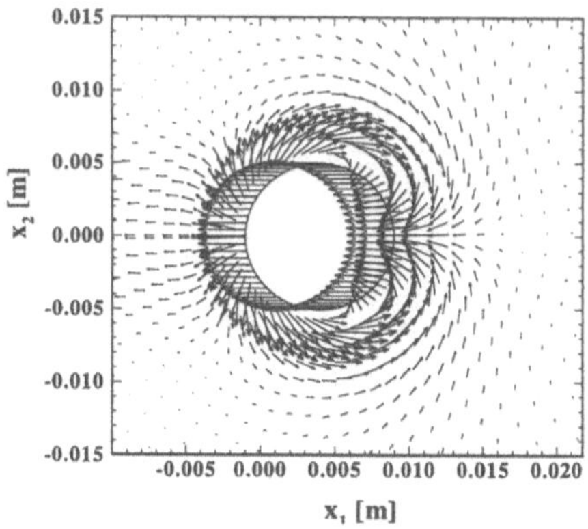

Fig. 5: Computed vector field at phase position 330°.

In figure 4 the cylinder moved from left to right 16 degree before its maximum position. Symmetrically to the plane of oscillation a pair of two large, counterwise rotating vortices was formed due to the separation of the shear layers. In figure 5 the cylinder turned to the left and was 30 degree before its maximum velocity. After changing its direction the cylinder divided the vortex pair. While the vortices were displaced sideward their vorticity dissipated drastically until they were extinguished after being alive for about one cycle. Meanwhile another vortex pair separated into the other wake. The computed and measured velocity components were compared at two cross sections for a constant $x_1$-value in detail for these two phase positions. They are presented in the figures 6a to 7b and show good agreement compared to the measurements.

Adding pressure and shear forces on the cylinder perimeter the history of the in-line and lift force per unit length was recorded during the calculations. As the maximum of the computed lift force was at least three orders of magnitude smaller than of the in-line force an evaluation is not reliable. The time history for the in-line force $F_1$ is presented in figure 8 for the numerical case B1. In figure 9 differences resulting from different grid levels and for the major time step are demonstrated for the second half of a cycle of the periodic regime. Basically, the in-line force had the same frequency as the oscillation, but its maximum lagged about 35 degrees comparing with the $x_1$-position of the cylinder, see equation (3) and table 1. Starting the oscillation from rest the cylinder was exposed to almost maximum forces in the first cycle. Periodic behaviour was received after only a few oscillations.

The shape of the force function was reproduced fairly the same in each computational case. Considering characteristic points like roots or extreme values - see table 1 and figure 9 - differences existed in their magnitude and phase position, which have to be carefully encountered calculating coupled behaviour of fluids and structures.

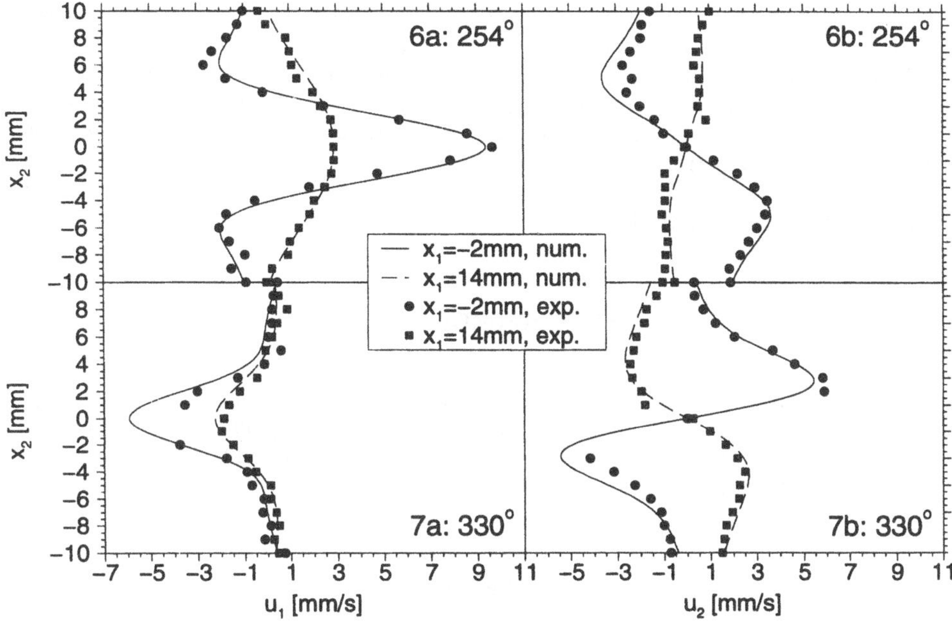


Fig. 6a, 6b, 7a and 7b: Comparison of the numerically and experimentally obtained velocity components.

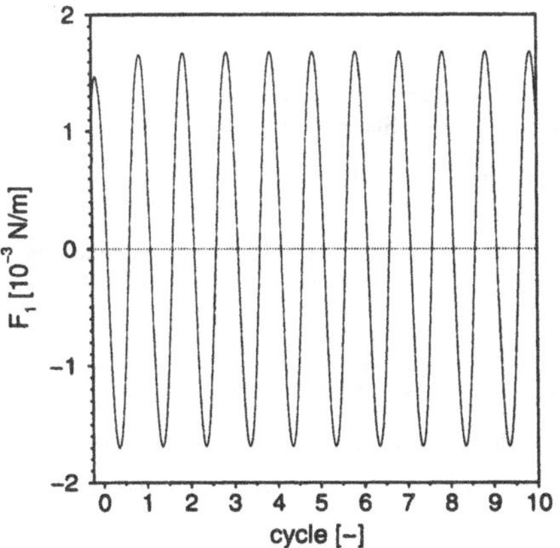


Fig. 8: Time history of the in-line force (case B1).

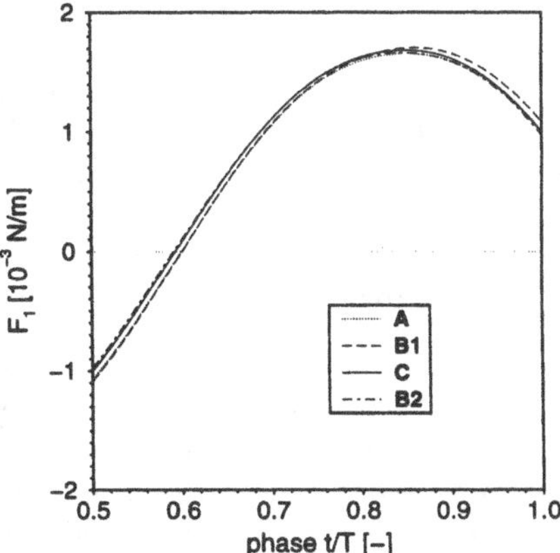


Fig. 9: In-line force for the 2. half cycle in the periodic regime.

## CONCLUSIONS

Investigating harmonic oscillations of a circular cylinder in water at rest a highly resolved, time averaged velocity data base for Re=100 and KC=5 was provided experimentally. Corresponding computations of the two-dimensional Navier-Stokes equations in a non-inertial reference system enabled time accurate results without any effort for grid movement. Good agreement with the measurements was observed comparing velocity components in detail. From the beginning of the motion two counterrotating vortices were separating symmetrically, one pair in each half cycle of the cylinder oscillation. They dissipated rapidly after being divided by the returning cylinder. Considering the recorded in-line force the flow pattern became stable and fully periodic after a few cycles.

## ACKNOWLEDGEMENT

This work was sponsored by the Bavarian Consortium for High Performance Computing, FORTWIHR, and the European Community, the authors gratefully acknowledge. We also greatly appreciate the support by Prof. Dr. Dr. h.c. F. Durst, LSTM, and Prof. Dr. M. Schäfer, TH Darmstadt.

## REFERENCES

[1] J. R. Morison, M. P. O'Brien, J. W. Johnson, and S. A. Schaaf, *The force exerted by surface waves on piles*, Petroleum Transactions, AIME, 189 (1950), pp. 149–154. T.P. 2846.

[2] T. Sarpkaya, *Vortex-induced oscillations*, Journal of Applied Mechanics, 46 (1979), pp. 241–258.

[3] E. D. Obasaju, P. W. Bearman, and J. M. R. Graham, *A study of forces, circulation and vortex patterns around a circular cylinder in oscillating flow*, Journal of Fluid Mechanics, 196 (1988), pp. 467–494.

[4] P. W. Bearman, X. W. Lin, and P. R. Mackwood, *Measurement and predicition of response of circular cylinders in oscillating flow*, in Behaviour of Offshore Structures, vol. 6, 1992, pp. 297–307.

[5] G. H. Keulegan and L. H. Carpenter, *Forces on cylinders and plates in an oscillating fluid*, Journal of Research of the National Bureau of Standards, 60 (1958), pp. 423–440. Research Paper 2857.

[6] M. Tatsuno and P. W. Bearman, *A visual study of the flow around an oscillating circular cylinder at low keulegan-carpenter numbers and low stokes numbers*, Journal of Fluid Mechanics, 211 (1990), pp. 157–182.

[7] U. Knörnschild, *Experimentelle Untersuchungen der Strömungsverhältnisse um einen oszillierenden Kreiszylinder*, Master's thesis, University of Erlangen-Nürnberg, LSTM, Erlangen, October 1994.

[8] S. V. Patankar, *Numerical Heat Transfer and Fluid Flow*, Hemisphere Publishing Corporation, 1980.

[9] F. Durst, M. Perić, M. Schäfer, E. Schreck, *Parallelization of efficient numerical methods for flows in complex geometries*, Notes of Num. Fluid Mech. 38, pp.79-92.

[10] F. Durst, H. Lienhart, R. Müller, *Aerodynamische LDA-Messungen in Windkanälen und im Freiflug*, GALA, Lasermethoden in der Strömungsmeßtechnik, 1992.

# Status and some new results obtained in the HEG

G. Eitelberg, R. Krek, N. Widdecke and W.H. Beck
Institute for Fluid Mechanics DLR,
Bunsenstraße 10, 37073 Göttingen, Germany

## Summary

The HEG (High Enthalpy Tunnel in Göttingen) has recently been equipped with a new conical nozzle, in order to improve its usability for various test conditions. Prior to that, a test series with spherical models was performed. Of particular interest in the course of the presented studies for the nozzle was the spatial and temporal extent of „steady" conditions in the test flow, and for the study with spherical models the detection, if present, of effects of surface catalysis.

## Introduction

There is current interest in providing vehicles that can safely return to Earth from Low Earth Orbit. To achieve this, the vehicle must pass through various flow regimes, Figure 1. The frozen and equilibrium flow regimes are relatively well understood, not so the non-equilibrium flow regime. The HEG facility was built to provide a test bed in which to test various models to help in the understanding of non-equilibrium flow dynamics. Figure 1 shows the operating envelope of the HEG facility.

Here the initial results from the conical nozzle calibration are discussed as the first example of the newest results. Further, results from tests with spheres ranging in diameter from 20 to 160 mm, are reported as the main part of the paper. These tests were performed in order to study the influence of catalycity on heat transfer when the flow chemistry is in non-equilibrium.

## HEG Facility and Test Conditions

A description of the HEG and its mode of operation can be found in earlier references, e.g. Eitelberg (1994). The flow properties at the HEG operating conditions are in the following range: $p_0$ = 40 to 100 MPa, $T_0$ = 6400 to 9085 K, $h_0$ = 11 to 22 MJ/kg, $p_\infty$ = 0.52 to 1.32 kPa, $T_\infty$ = 500 to 1100 K, $\rho_\infty$ = 1.92e-3 to 6.62e-3 kg/m$^3$, $M_\infty$= 9.03; 9.83, $u_\infty$ = 4.7 to 6.2 km/s, for conditions I through VI, see also Figure 1.

These values are based on statistical analysis of the flow conditions and on calculations performed using the codes ESTC (equilibrium shock tube calculation), McIntosh (1968) and STUB, Vardavas (1984). The STUB program includes chemical non-equilibrium and uses an inviscid boundary for the nozzle which is the physical boundary minus the displacement thickness calculated from the NSHYP code for Condition I, Hannemann (1995).

A permanent probe is installed within the test section to measure in all tests the pitot pressure and the stagnation point heat transfer rate on a sphere of radius 10 mm. Also measured is the heat transfer rate at 45° and 90° from the stagnation point on the sphere.

## Calibration of the conical nozzle

A characteristic result from the conical nozzle calibration is given in Figure 2. Here the stagnation pressure measured on the centerline of the test section is shown for both the conical as well as the contoured nozzle. Obvious are the somewhat shorter starting time of the conical nozzle, the lower plateau and the longer time available at constant (normalized) condition. Another effect, not shown in this Figure, is the more uniform stagnation pressure distribution obtained in the conical nozzle than was the case for the contoured one.

## Sphere Models

In our experiments four different diameter spheres were used, d = 20, 40, 80 and 160 mm. Two rakes were used to mount the various spheres in such a way that the use of the test core was maximised, but that the bow shocks produced by the various spheres did not interact and disrupt the flow about the other spheres.

The spheres with diameters 20 and 40 mm were mounted with thermocouples at 0°, 45° and 90° from the stagnation point. For the larger spheres, 15° intervals for mounting the heat transfer gauges were used.

To investigate the effects of catalycity, two coatings were applied to a strip along the thermocouples around the spheres. These coatings, gold, to obtain a non-catalytic surface, and platinum, to obtain a catalytic surface, Hammer and Norskov (1995), were applied by electrolytically depositing them onto the surface. The coatings were approximately 2 µm thick and used a nickel base to help bond the coating to the surface of the spheres.

## Shock Stand-off Distance

The analysis of the shock stand-off distance is important in order to understand and characterise the free stream conditions. Whereas the stagnation point heat transfer $q_{t2}$ and the stagnation pressure $p_{t2}$ can be utilised to obtain a measure of the total enthalpy of the flow, these parameters are not a measure of the free stream species' concentrations. With the shock stand-off distance the capability to produce relaxation effects can be analysed, Wen (1994) and Wen and Hornung (1995).

The shock stand-off distances from our experiments are displayed in two non-dimensional presentations. In the first one the shock stand-off distance $\Delta$ is normalised with the sphere radius R and plotted as a function of the size of the sphere. Here also known correlations from Billig (1967) and Hornung (1972) are given. Whereas the Billig correlation is only a function of the free stream Mach number and applicable to perfect (frozen) gases, the Hornung correlation can be adapted to any state behind the shock.

The second display, also in Figure 3, right, shows the non-dimensional shock stand-off distance, this time normalised with the density ratio across the shock against a non-dimensional reaction rate parameter $\Omega^*$ as defined by Wen and Hornung (1995). This non-dimensional parameter describes the rate of relaxation, $d\alpha/dt$ (defined for air as a weighted average over all species), directly at the frozen shock normalised with the flow rate for the free stream $u_\infty/R$ : $\Omega^* = d\alpha/dt \times R/u_\infty \times K$. In this expression K is a kinetic energy parameter accounting for the free stream conditions.

As our results fall between the frozen and equilibrium limits given on the left and right edges of the presentations in Figure 3, it is concluded that the presented experiments were performed in the flow regime that produced a non-equilibrium shock layer. The exact interpretation of the curves in Figure 3, right, however, suffers somewhat from the lack of reliable reaction rate data required for the term $d\alpha/dt$. The parameter L used in Figure 3 has the same meaning as the factor 0.852 from Hornung's correlation, is empirical and correlates the average density jump to the shock stand-off distance. The literature values range from L = 0.41 (Wen (1993)) and 0.5 (Hayes and Probstein (1966)). This range is also given in Figure 3, right.

## Heat Flux Results

### Literature survey

One solution for the stagnation point heat flux on a spherical nose that is widely used in the aerothermodynamic community is the Fay and Riddell (1958) formulation

$$\dot{q}_s = 0.76 Pr^{-0.6} (\rho_e \mu_e)^{0.4} (\rho_w \mu_w)^{0.1} K \sqrt{\left(\frac{du_e}{dx}\right)_s} (h_0 - h_w), \tag{1}$$

where the index w is used to denote wall values and the index e for values at the boundary layer edge. K is a constant depending on the state of the boundary layer and wall catalycity:

$$K = 1 + (Le^{\lambda} - 1)\left(\frac{h_D}{h_0}\right), \tag{2}$$

where $\lambda$ = 0, 0.63 and 0.52 for frozen boundary layer and non-catalytic wall, for frozen boundary layer and catalytic wall and for equilibrium boundary layer respectively.

The stagnation point velocity gradient is obtained from Newtonian theory:

$$\left(\frac{du_e}{dx}\right)_s = \frac{1}{R}\sqrt{\frac{2(p_e - p_\infty)}{\rho_e}} \tag{3}$$

and $h_D$ is defined as

$$h_D = \sum_s c_s h_s^0 . \tag{4}$$

Other, simpler, correlations have been proposed by Sutton and Graves (1971), Rose and Stankevics (1963), Detra, Kemp and Riddell (1957), Scott et al (1985) and Verant (1995). The last author refers to all the above references and reformulates their correlations to a form

$$\dot{q}\sqrt{\frac{R}{p_{t2}}} \approx A\left(\frac{h_0 - h_w}{R_g T_{ref}}\right)^n \tag{5}$$

where A is a constant 23.7<A<31.7 and the exponent 1.025<n<1.16 for the various correlations. The results obtained with the above methodology for the HEG conditions are listed in the following Table 1 along with the experimental results from the uncoated permanent spherical probe.

**Table 1** Summary of stagnation point heat flux calculations for the radius of the sphere,, R = 0.01 m and the test gas air.

| Correlation | | $q_{t2}$ (MW/m$^2$) Cond. I | $q_{t2}$ (MW/m$^2$) Cond. III |
|---|---|---|---|
| Detra, Kemp & Riddell | | 32.56 | 20.67 |
| Fay & Riddell | Froz. bl., Non-Cat. | 14.93 | 9.87 |
| | Froz. bl., Cat. | 16.99 | 10.87 |
| | Equil. bl. | 16.60 | 10.68 |
| Sutton & Graves | | 18.24 | 11.91 |
| Scott et al | | 19.03 | 12.38 |
| Rose & Stankevics | | 15.04 | 9.05 |
| Verant | | 23.25 | 14.66 |
| **Experiment** | | **16.24** | **11.15** |

**Effect of coating on the heat flux measurements**

In the following diagrams, Figures 4 and 5, the results of compiling a large amount of information into a somewhat reduced number of figures is attempted. Here similar spheres are grouped together, and compared according to their coating and flow condition. As a reference an uncoated first shot is also given. The uncoated further HEG shots, performed between the coated shots, are thereafter grouped together. The averaging approach, of course, smears over singular events or step function types of developments. All results have been normalised with respect to the uncoated permanent probe stagnation point heat flux $q_{t2}$.

Some spheres which were coated with gold first, then coated with platinum, failed to reproduce the high heat flux measurements of the spheres which were coated only with platinum prior to any other tests. This behaviour can, among other things, be not just due to stochastic behaviour, but due to the sensitivity of catalytic surfaces to contamination. It is conceivable that the thin Pt coating, when new and clean, acts as a catalyst, and loses its catalytic behaviour completely, when traces of contaminants appear. This suggests that any measurements conducted on coated surfaces are highly dependent on the initial coating deposited.

For the small spheres, D = 20mm, Figure 4, left, the first uncoated shot at Condition I gave anomalously high heat transfer, when normalised with the permanent probe of the same size. Later another five shots with uncoated small spheres were performed at the same condition. The average of all these is presented by the dashed line in the Figure 4. In the condition I, a weak catalycity/coating effect can be observed. The Au coating produces an approximately 15% lower heat transfer than Pt coating or the averaged uncoated spheres.

For the larger sphere, D = 40mm, Figure 4, right, the uncoated reference shot agrees well with the $\sqrt{R}$ scaling from the reference permanent probe. The same is true for the Au coated shot average, obtained from four tests with two gold coated spheres each. The Pt coating, however produces a different behaviour altogether. It shows an unreasonably high heat transfer in the stagnation point, especially at the start of the flow. This trend is even more obvious, when single signals are studied.

For Condition III and the medium size spheres, D = 40mm, Figure 5, the evolution from the uncoated reference shot over the following Pt and Au tests, all scaled with the permanent probe is shown for four spheres. The average value of the Pt coated tests yields a much higher heat transfer result than could be expected from the standard $\sqrt{R}$ scaling. When this single shot data is evaluated, again the fact that the earlier shots produce higher values of q becomes discernible.

## Conclusions

Tests have been conducted on various spheres, with diameters d = 20, 40, 80 and 160 mm, at two of the HEG operating conditions, I and III, with air as the test gas. Heat flux distributions were measured on all the spheres tested and Schlieren and holographic interferometery images were produced of the flow about the spheres. The spheres were coated with gold, to provide a non-catalytic surface, and platinum, to provide a catalytic surface, in order to investigate the effects of surface catalysis.

Comparisons of stagnation point heat flux empirical relationships and experiment show that the Fay and Riddell correlations are in best agreement with our experiments. One way to interpret the results is to assume that all presented measurements, except those with a clean Pt coating, are for frozen boundary layer conditions and no catalytic effects occur on the sphere surfaces. This is possible, since the residence times of fluid particles in the stagnation point boundary layer are still finite, despite a zero velocity in the singular point, e.g. Anderson (1989).

The stagnation point heat flux levels indicate that there is no measurable effect of variation of the radius of the spheres when the $\sqrt{R}$ scaling is applied to standard sphere conditions with no coating or coating with Gold. This indicates that non-equilibrium chemistry has negligible effect on the stagnation point heat flux levels or that the measurement technique cannot resolve these effects. When Platinum coating is applied to the spheres, significant deviations from the standard configuration can occur.

A possible cause for the high heat flux measured at Pt coated stagnation points could be due to gas-surface interaction effects, not accounted for in our standard modelling, which only accounts for the air chemistry. The conceivable gas surface mechanisms that could be responsible for the noted behaviour are a dissociative chemisorption process (Lennard-Jones (1932) in Somorjai (1994)) or adsorption and desorption mechanisms. In those cases the numerical simulation methods considering only air chemistry without the detailed modelling of the gas/surface interaction would have to be further completed. Unfortunately no clear cut suggestions for modelling could be drawn from the reported study.

## References


Anderson, J. D., 1989, *Inviscid and Viscous Hypersonic Aerodynamics - A Review of the Old and New.* AGARD-FDP/VKI *Special Course on Aerothermodynamics of Hypersonic Vehicles.*, AGARD Rep. 761.

Detra, R. W., Kemp, N. H. and Riddell, F. R., 1956, *Addendum to 'Heat Transfer to Satellite Vehicles Re-entering the Atmosphere.*, Jet Propulsion, Vol. 27, No. 12, pp. 1256 - 1257.

Eitelberg, G., 1994, *First Results of Calibration and Use of the HEG.*, AIAA 94-2525.

Fay, J. A and Riddell, F. R., 1958, *Theory of Stagnation Point Heat Transfer in Dissociated Air.*, Journal of the Aeronautical Sciences, Vol. 25, No. 2, pp. 73-85.

Hammer B. and Norskov, J.K., 1995, *Why Gold is the noblest of all the metals.*, Preprint, Personal communication.

Hayes W.D. and Probstein, R.F., 1966, *Hypersonic flow theory,* Academic Press, NY.

Kastell, D. and Eitelberg, G., 1995, *A Combined Holographic Interferometer and Laser-Schlieren System Applied to High Temperature, High Velocity Flows.*, ICIASF Records, Dayton, Ohio.

Krek, R. M., Kastell, D., Beck, W. H., Ueda, S., Eitelberg, G. and Le Bozec, A., 1994, *HERMES Model 3036-2 and 3041 Experiments in the HEG.*, DLR-IB 223-94 C 02.

Krek, R. M., Eitelberg, G. and Kastell, D., 1995, *Coating Investigations on the Hyperboloid Flare.*, DLR-IB 223-95 A 42.

Krek, R. M., Ueda, S., Kastell, D. and Eitelberg, G., 1995, *ELECTRE Experiments in the HEG Facility.*, DLR-IB 223-95 A 65.

Krek, R. M. and Eitelberg, G., 1994, *Classical Characterisation of HEG (Nozzle, Freestream Flow Field).*, MSTP Phase 1, Report No. HT-TR-E-1-201-DLRG, WP 2110.

Lees, L., 1956, *Laminar Heat Transfer Over Blunt-Nosed Bodies at Hypersonic Flight Speeds.*, Jet Propulsion, Vol. 26, No. 4, pp. 259 - 269.

McIntosh, M. K., 1968, *Computer Program for the Numerical Calculation of Frozen and Equilibrium Conditions in Shock Tunnels.*, Department of Physics, School of General Studies, Australian National University, Canberra.

Olivier, H., 1993, *An Improved Method to Determine Free Stream Conditions in Hypersonic Facilities.*, Journal of Shock Waves, No. 3, pp. 129-139.

Rose, P. H. and Stankevics, J. O., 1963, *Stagnation-Point Heat-Transfer Measurements in Partially Ionized Air.*, AIAA Journal, Vol. 1, No 12, pp.2752 - 2763.

Scott, C. D, Ried, R. C., Maraia, R. J., Li, C. P. and Derry, S. M., 1985, *An AOTV Aeroheating and Thermal Protection Study.*, Progress in Astronautics and Aeronautics, Vol 96, Thermal Design of Aeroassisted Orbital Transfer Vehicles, Ed. H. F. Nelson, pp. 198 - 229.

Somorjai, G.A. 1995, *Introduction to Surface Chemistry and Catalysis*, Wiley, NY.

Sutton, K. S. and Graves, R. A., 1971, *A General Stagnation-Point Convective-Heating Equation for Arbitrary Gas Mixtures.* NASA Report TR R-376.

Vardavas, I. M., 1984, *Modelling Reactive Gas Flows within Shock Tunnels.*, Australian J. Phys., Vol. 37, pp. 157-177.

Verant, J. L, 1995, *Numerical Enthalpies Rebuilding for Perfect Gas and Nonequilibrium Flows. Application to High Enthalpy Wind-Tunnels.*, ONERA Report RT No. 69/6121 SY, HT-TN-E-1-201-ONER.

Wen, C., 1994, *Hypervelocity Flow over Spheres.*, PhD thesis GALCIT, Caltech.

Wen, C-Y. and Hornung, H., 1995, *Non-equilibrium dissociating flow over spheres.*, J. Fluid Mech. Vol. 299, pp. 389-405.

## Acknowledgements

The authors wish to thank ESA for financial support of these tests (Contract supervisors G. Durand and D. Vanmol). Thanks must also go to M. Carl for many fruitful discussions, J. Lenz for the tunnel operation and D. Garbe for the model instrumenting, interfacing and maintenance.

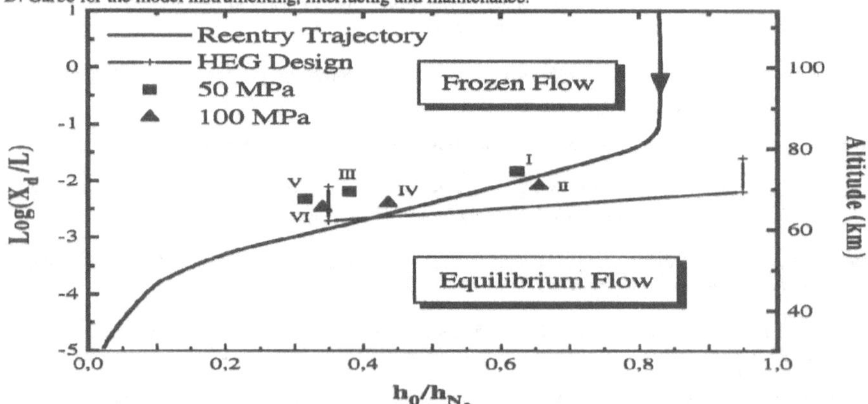


**Figure 1.** Typical re-entry trajectory plotted as non-dimensional dissociation length versus non-dimensional total specific enthalpy. Also shown is the operating envelope of the HEG. The non-equilibrium regime is between the frozen and equilibrium domains.

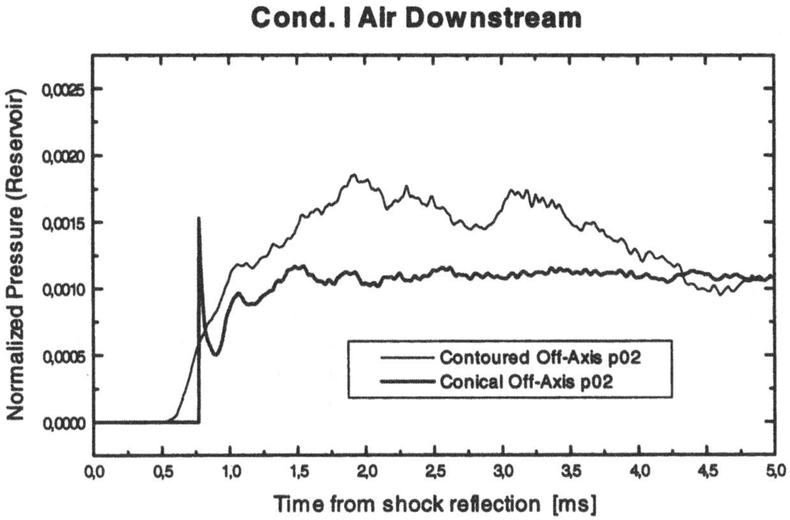

Cond. I Air Downstream

Normalized Pressure (Reservoir)

Time from shock reflection [ms]

Contoured Off-Axis p02
Conical Off-Axis p02

**Figure 2.** Comparison of stagnation pressures just off the centreline, normalized with the reservoir pressure for the contoured and conical nozzles, condition I. air.

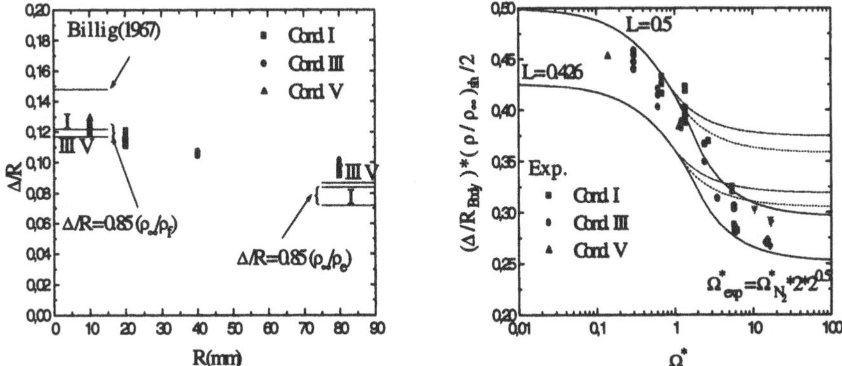


**Figure 3.** Dimensionless shock stand off distance $\Delta/R$, versus sphere radius and dimensionless shock wave stand-off distance versus reaction rate parameter for air as test gas. The Billig (1967) correlation in the left is given as $\Delta/R = 0.143 \exp(3.24 M_\infty^{-2})$. The theoretical curves in the right are: —— for Cond. I, ······ for Cond. II, and –·–·–·for Cond. V.

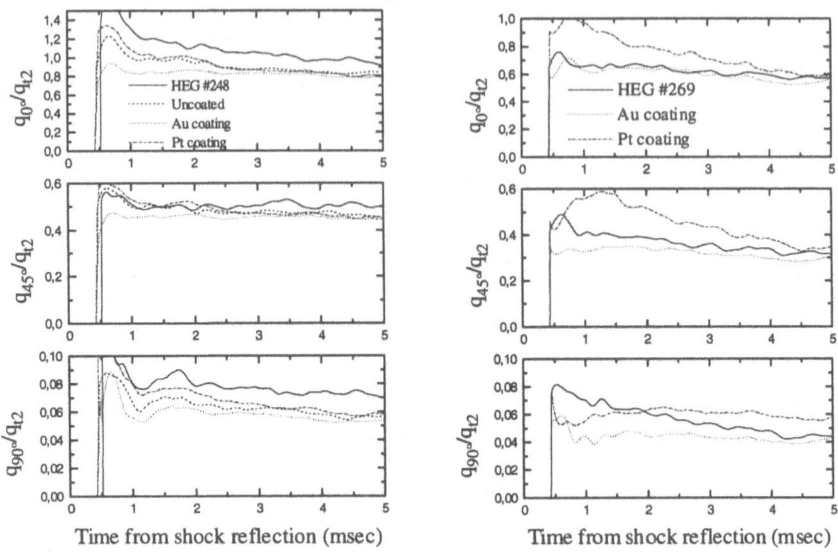


**Figure 4** q/$q_{t2}$ versus time for the data averaged over the four spheres and coating condition. Flow Condition I. Left figure: sphere diameter, d = 20 mm, right figure: sphere diameter, d = 40 mm,.

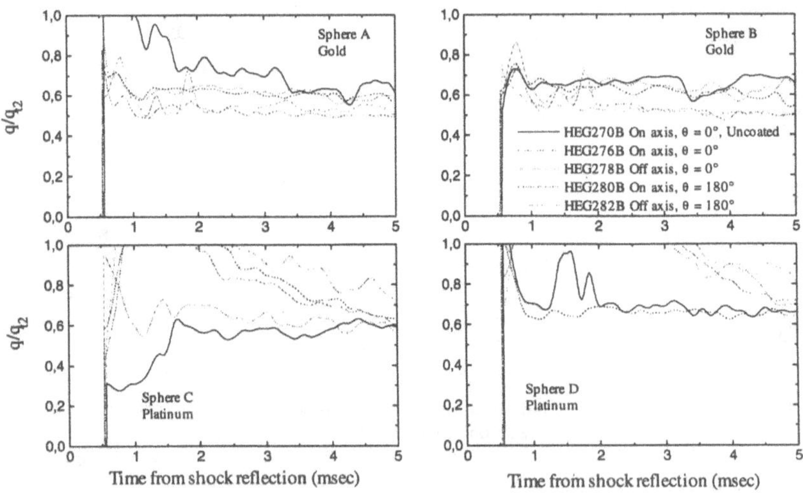


**Figure 5.** Evolution of the measured heat transfer on the stagnation point of the R = 40mm sphere in time and from shot to shot for condition III. Note that a number of the evaluated curves for the Pt coated spheres go off

# MESH MODIFICATION TECHNIQUES FOR NAVIER-STOKES MESHES

## Demonstration of Viscous Adaptation, Geometry Change and Mesh Enrichment

E.Elsholz, H.Steinmeyer

Daimler-Benz Aerospace Airbus GmbH, Dept. EFV

Hünefeldstraße 1-5, D- 28199 Bremen

## SUMMARY

A mesh modification system is described that is suitable for re-designing a 3D mesh without depending on the mesh generation tool originally used. All modifications are performed on the basis of unidirectional interpolation with the ability to detect and preserve geometric discontinuities. The concept also includes the modification of the solution to minimize computational turn-around times. The mesh-modifier runs in a workstation environment and is closely coupled to a commercial graphical tool for instant re-check of subsequent modifying steps.

A demonstration of applicability is presented for a number of viscous mesh problems. These are the adaptation of the mesh to the boundary layer and wake surfaces of a wing-alone configuration, massive mesh enrichment in the rear-fuselage/tail area of a complex wing/-body/tail transport aircraft configuration and the deformation of the outer wing due to simulated elasticity.

## MOTIVATION

When computing 3D Navier-Stokes solutions around complex aircraft configurations, the design engineer often encounters numerical difficulties due to unsatisfactory mesh resolution or solutions that suffer from mesh deficiencies. In general, these situations require tedious re-engineering of the meshes, which may be difficult or impossible when the mesh generation system that had been used originally, should not be at hand. So an easy-to-use mesh-modification system that is independent of any mesh generator appears to be a valuable tool. This is of utmost importance in those cases, where only smaller changes are necessary, e.g. local refinements in shear layer regions.

In contrast to 2D Navier-Stokes resolution, where meshes of 'gold standard' (i.e. fine meshes that produce mesh-independent solutions) are possible today, the situation for 3D simulations is much more difficult. Here, the computer power available does not allow the construction of fine meshes throughout the field. So any 3D solution necessarily depends on the local resolution of the mesh - and unfortunately, there is only little knowledge up to now about the influences of coarser regions on the overall solution. Again, a specific tool to enable local mesh modifications seems to be a good basis to experience these unknown influences and learn optimizing the 3D mesh generation process in general.

Besides this, qualified mesh adaptation to the computed flow field is a great challenge. The same is true for easy-to-do mesh adaptation to local geometry changes, which is required in the field of design-aerodynamics, but also when dealing with icing effects, effects of tripping devices and aeroelasticity problems.

## THE MESH-MODIFICATION SYSTEM

The mesh modification techniques discussed herein have to be applicable within highly complex, massively block-structured meshes. Modifications should become possible inside the blocks or part of blocks only, but at the same time across the block boundaries as well. The modifying procedure described here is designed as a quasi-interactive procedure, i.e. the program runs under the control of a data file in batch mode producing also the necessary input files for a subsequently applied graphical tool (CFView [1]). However, due to the limitations of a workstation, medium to coarse meshes have to be used. In a final re-run, possibly on a main-frame computer, the current solution belonging to the original mesh may be modified (interpolated) also in order to decrease the overall computation and turn-around time by using the code's restart option.

The system performs mesh modifications by means of interpolation. The interpolative method is a C2-continuous, uni-directional technique that applies along a single mesh line direction. By this, the method may detect directional geometric discontinuities, e.g. the wing/body intersection line, which is a discontinuity with respect to two mesh directions but is smooth along the intersection line itself. This detection is triggered by the 3D angle between the adjacent vectors representing the mesh. Once a discontinuity is detected, the interpolation switches back to C1-continuous formulations to avoid overshoot but accurately preserving the discontinuity. This technique is demonstrated in FIG. 1 where the surface and wake mesh portion over an aircraft fin is enriched by interpolation. In the left hand figure the discontinuity detection is switched off, producing massive overshoot that is caused by the very closely spaced tip definition, while on the right hand side, the detection prevents the interpolated mesh lines from overshooting, now properly representing the fin.

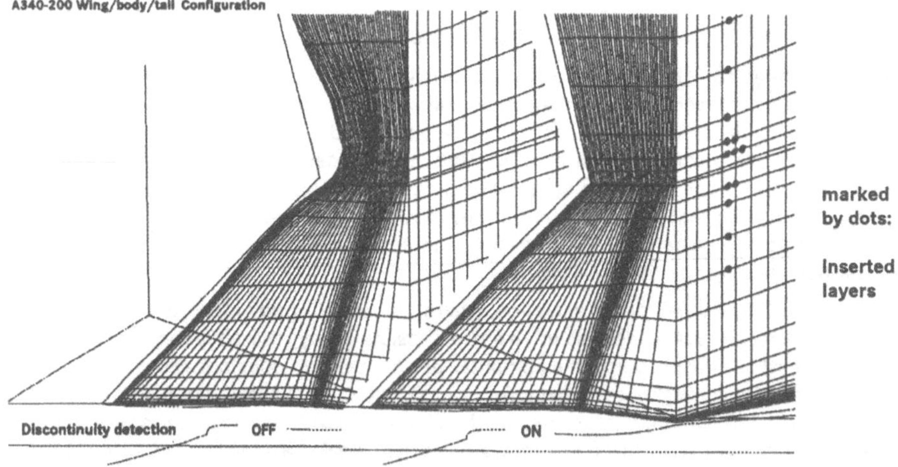


Fig. 1 Interpolated surface mesh on the fin (medium resolution)

Besides the uni-directional interpolation principle, multi-dimensional smoothing techniques are incorporated, applied either to the mesh nodes themselves or to the spacing between the nodes (progression rates).

A series of modifications may be done sequentially:

- After reading the complete original mesh to an internal temp-file system, adjacent blocks or parts of blocks are loaded/concatenated into memory from the temp-file system. On request, the geometric vector may be extended to the geometry plus solution vector, to enable the solution to be modified precisely the same way as the mesh.
- By calling several pre-defined actions, deformations such as warping or smoothing may be applied on the memory area. Since the memory area may contain parts of adjacent blocks, the original block boundaries may be modified in an easy way.
- After a sequence of actions has been applied, the memory area may be re-split into the original or even newly definable parts of blocks and the modification process may be repeated for another region of the mesh.
- Finally, the modified mesh (and solution) is written back to the original disk format and/or an input file may be generated for the CFView graphics tool.

The pre-defined actions are split into basic and more complex categories. The basic tools are:

- Check for negative volumes. This option has been taken from the FLOWer code [2] to ensure valid pre-checking.
- Warp the mesh according to given changes of (memory area) faces, either along the face-normal mesh lines or by a 3D vector shift. This option may be used for geometry changes and flow adaptation.
- Smooth either the grid coordinates in a single- or multi-dimensional fashion or apply this function to the progression rates, i.e. modify the spacing of the mesh.
- Orthogonalize the mesh lines with respect to a specified face and blend over to the original lines
- Coarsen or refine a block with respect to pre-defined direction(s).

More complex actions available are:

- Redistribute the layers with respect to one direction. Three regions are established: for the inner and outer region, the number of layers and progression rates are specified, while in the middle region, equidistant layers are interpolated.
- Adapt a specified layer or face to the boundary layer edge or wake surface, both scanned from the solution. This option may also be used to adapt a block face to a modified neighbouring face. However, this technique is not yet fully operational.
- Blending: most of the aforementioned actions may be processed under the control of a 3D linear blending frame to avoid sudden changes inside the mesh structure.

Note that each call to one of these actions requires in most cases a single data input record defining the type of action, direction and domain of application plus one or two additional parameters related to the specific action.

## DEMONSTRATION OF APPLICABILITY

### Viscous adaptation over a wing

Based on the solution of the transonic M6-wing, the adaptation of the mesh to the boundary layer edge and wake center surfaces will be demonstrated. The mesh used has 112 x 24 x 16 cells only (2 blocks) with approx. 10 out of the 24 cells covering the boundary layer over the wing. The computed pressure distribution is given in FIG.2, clearly showing the lambda shock system, resulting even from this coarse mesh. By calling the boundary layer and wake adaptation options, the boundary layer thickness and the locus of minimum velocity in the wake are scanned from the solution. After some smoothing of these surfaces, the layers J=1 and J=12 of the original mesh are adapted to these surfaces (wake and b.l.edge, respectively). FIG. 3 shows the adapted inner mesh area at an outbord wing station, where the two separate shocks (indicated by condensed iso-pressure lines in FIG.2) have already merged at a forward position x/c<0.3. At this forward position (upper side) and at the trailing edge (both upper and lower), the increase of the boundary layer thickness now is visible in the mesh, resulting in a constant number of grid points throughout the boundary layer. In FIG 4 the adapted surfaces are presented. The wake center surface (FIG. 4a) obviously shows the influence of the inward moving wing tip vortex. Consequently, this effect overlays the boundary layer edge surface given in FIG. 4b.

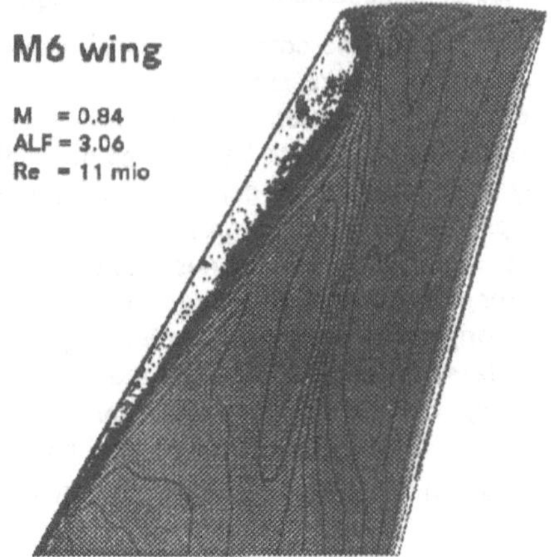


Fig. 2 Computed pressure distribution on upper wing (Demo mesh)

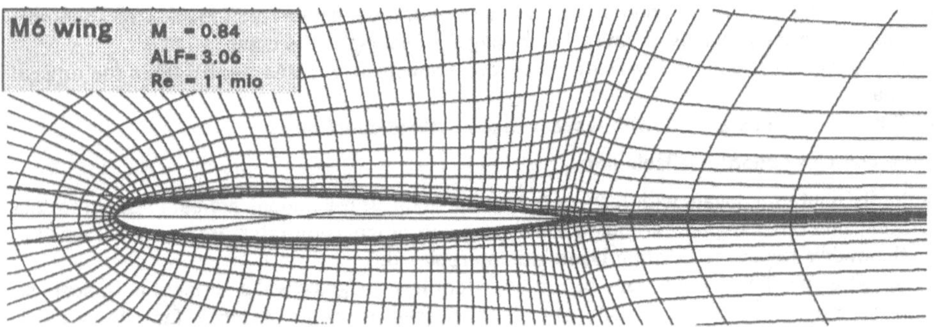


Fig. 3 Adapted mesh at outbord station (Demo mesh)

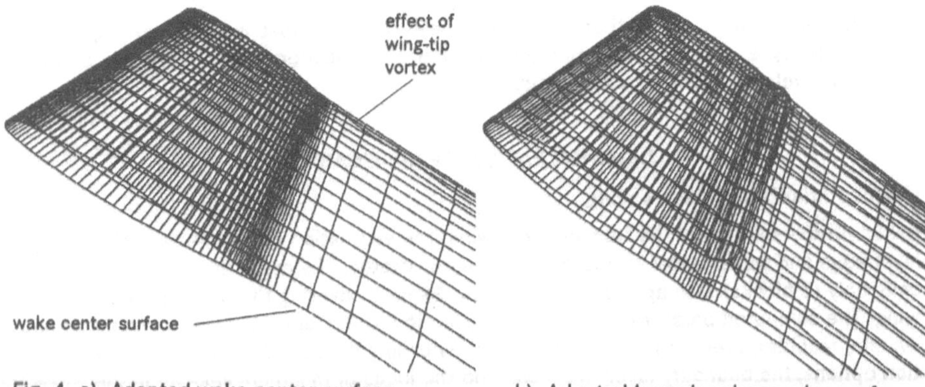


Fig. 4 a) Adapted wake-center surface b) Adapted boundary layer edge surface

**Wing/body/tail configuration: mesh enrichment**

A more complex application is demonstrated on the grid of an A340-200 transport wing/body/tail configuration, the surface mesh of which given in FIG. 5 (medium grid level). The mesh is an 8-block configuration of over 2.8 million cells in total. However, due to time constraints when generating this mesh [3], using the IGG mesh generation system [4], the resulting mesh turns out to be rather crude and therefore does not yet allow a really convergent solution. In FIG. 6, a view of the most critical area in the tail region is given, showing the surface mesh over the wing/body part and the outer boundaries of the embedded C-H and C-O meshes surrounding the fin and horizontal tail, respectively. In addition, the layer I=36 of the main block 3 indicates the very coarse representations of the tip regions of both, fin and horizontal tail. So here, some refinements and redistributions become necessary that in consequence will affect most of the other mesh areas, too. FIGs 7,8 show the refined and redistributed surface meshes of the fin and horizontal tail in medium resolution. The mesh area in between (block 3) is given in FIG. 9 after being adapted to the embedded meshes of FIGs 7,8. On the left of FIG. 9, the instant outcome of the adaptation process is shown (close-up and inner mesh area), exhibiting massive difficulties arising from the original mesh structure. However, after subsequent smoothing, the mesh appears more reasonable but some of its properties such as orthogonality on solid surfaces, may have been affected. A re-orthogonalization of the mesh near solid surfaces has not yet been tried but should be possible in general. Note that the smoothing mentioned above had been performed throughout the concatenated memory area containing the 'main block 3' plus the embedded mesh around the fin. As a consequence, the block boundaries appear changed, i.e. smoothed after re-splitting into the original block sizes.

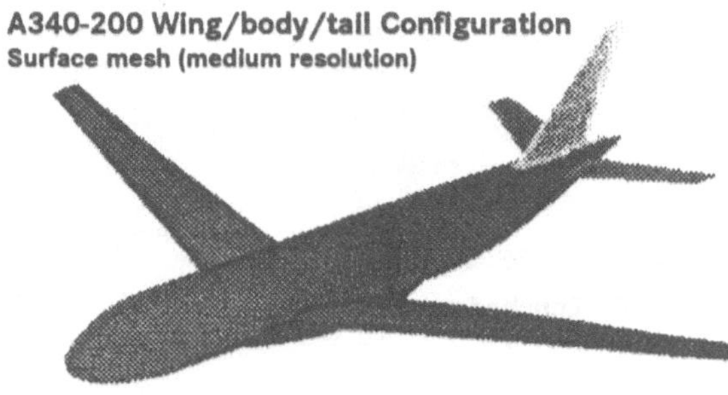


Fig. 5 Wing/body/tail Configuration

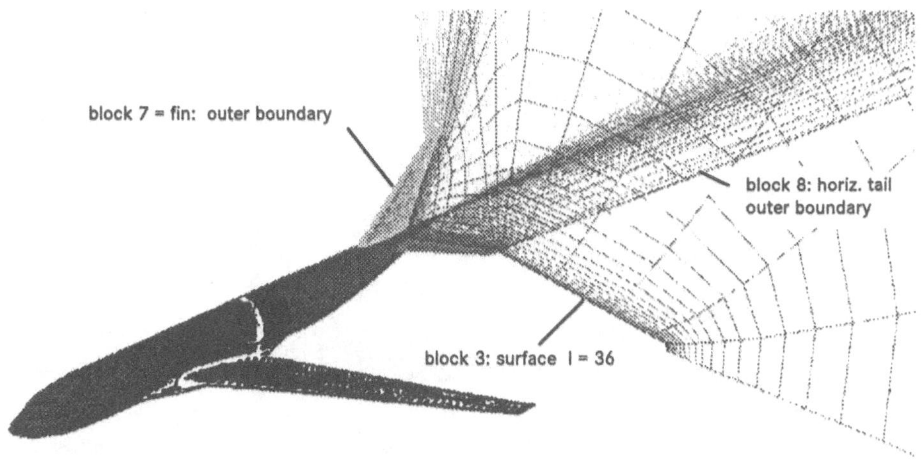

## A340-200 Wing/body/tail Configuration

(medium resolution)

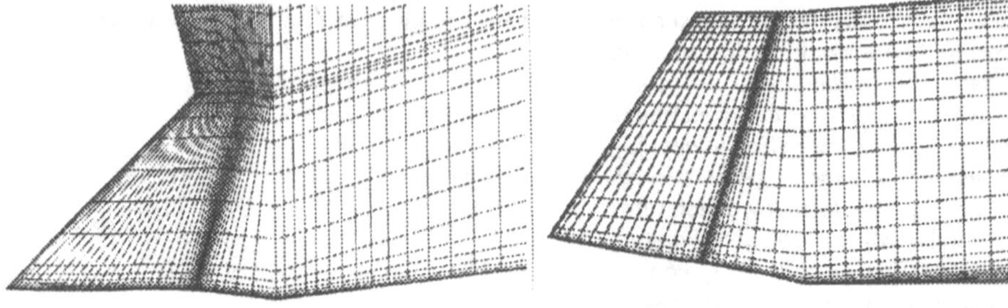

Fig. 7 Enriched mesh on fin

Fig. 8 Enriched mesh on horizontal tail

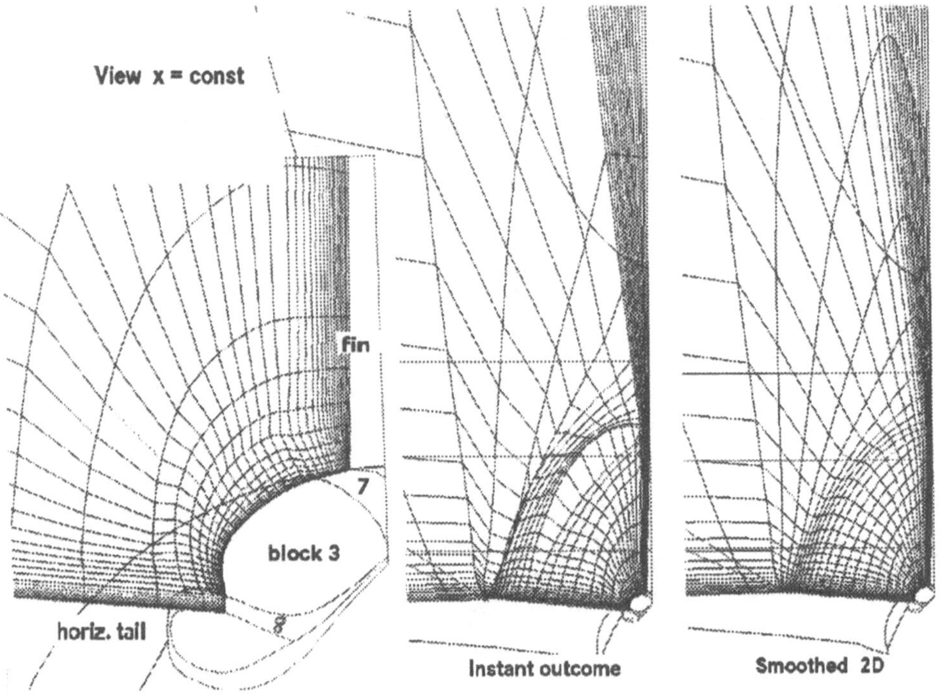


Fig. 9 Main block 3, connecting fin (block 7) and horizontal tail (block 8)

FIG. 10 gives an idea on how the main block 3 may be modified further. The figure shows the face-aligned grid of this block, fitting here to the outer boundary of the embedded mesh around the fin. Note that upstream and downstream of the fin area, application of massive cascaded smoothing is able to resolve the very thin cells towards the block boundaries. This technique is used especially to allow large step-sizes downstream of the configuration but not to encounter the drawbacks of extremely thin cells from the necessary upstream boundary layer resolution.

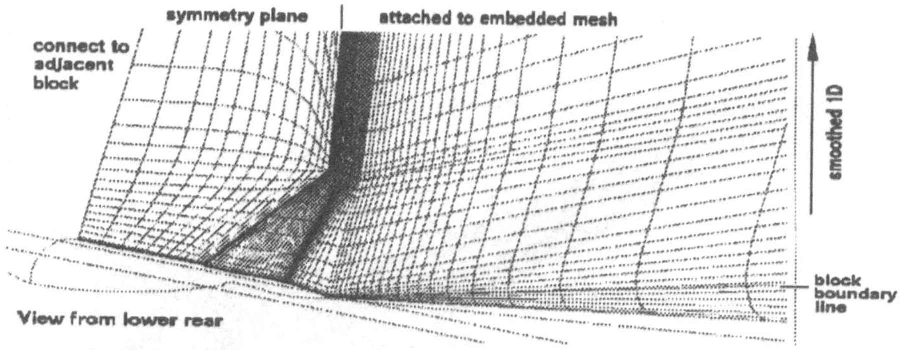


Fig. 10 Possible modifications of main block

### Changes in geometry

In the third example, the solid wing surface of the previous A340-200 configuration has been mildly bent upwards, simulating by this common elasticity effects. FIG. 11 shows the original wing (top) and the wing under simulated aeroelastic deformation (bottom). Simultaneously, the mesh near the wing surface follows this deformation without any problems (FIG. 12). Again, the mesh appears crude at the inner and outer wing block boundary (top) which has been smoothed across the block boundary (bottom). Finally, FIG. 12 shows the surface grid on wing/near-wake and forebody, including the block boundary of the innermost block over the wing under simulated aeroelastic deformation. For downstream continuation, the deformation of the mesh structure should decay before reaching the tail structures.

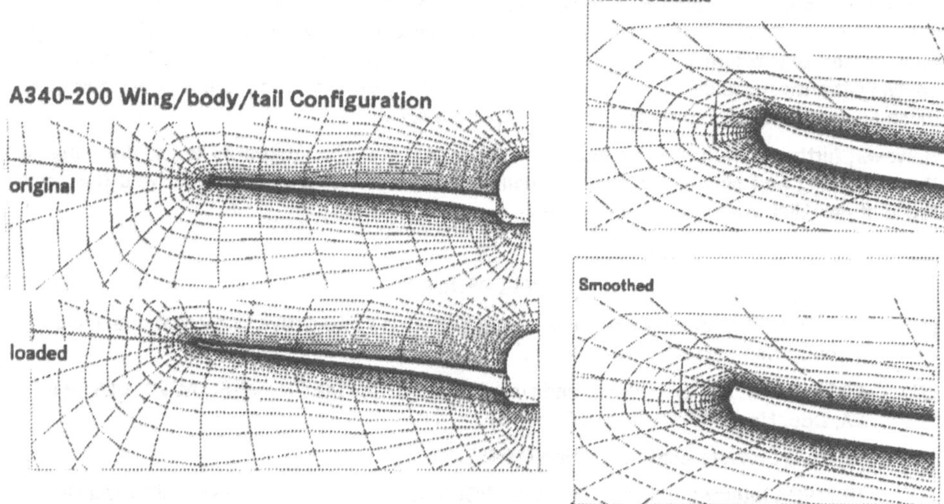


Fig. 11 Simulated aeroelastic deformation of aircraft wing
Left: Comparison of original and loaded wing
Right: Close-up of instant outcome and smoothed mesh at wing tip

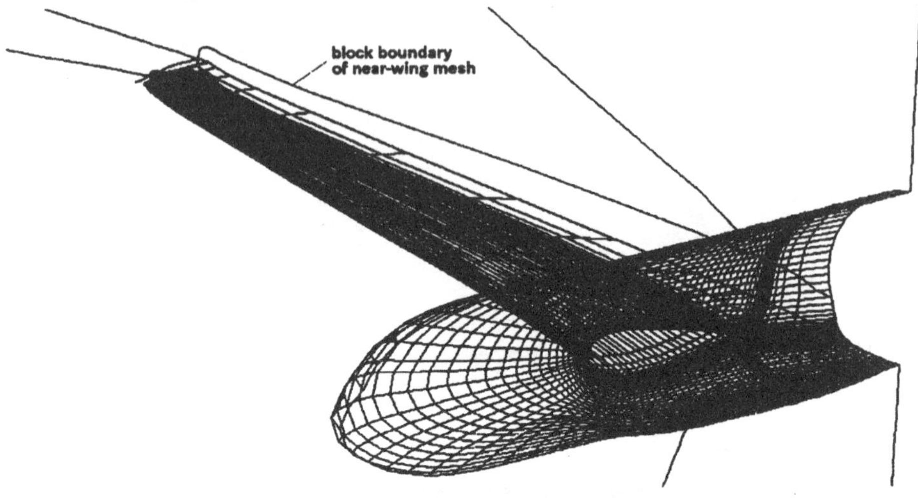


Fig. 12 Wing surface and wake after simulated aeroelastic deformation

## CONCLUSION

The mesh modification system described has been applied on a number of common or near-future viscous mesh problems to demonstrate its main features, including the built-in discontinuity detection capabilities. By this, the system had been shown to be a useful tool. It may be expected that future problems, similar to those shown here, might be easily resolved by applying the mesh modification principle instead of partially or completely re-designing the mesh by using the original mesh generating system again.

However, further work has to be done concerning some of the integrated tools. Here, mainly the smoothing and blending frame techniques need further improvement to avoid the occurence of new disturbances inside the mesh.

## REFERENCES

[1] NUMECA International s.a., Brussels: "CFView - An Interactive Computational Field Visualization System". User Manual V 3.6 (1994).

[2] KROLL, N. et al: "Installation and User Handbook for the Project FLOWer". DLR SM-EA, Doc. No. QS-FLOWer-3008 (1996).

[3] JOHN, D.: private communication (1995).

[4] NUMECA International s.a., Brussels: "IGG - Interactive Geometry Modeller and Grid Generator System". User Manual V 3.2 (1993), V 3.4 (1995).

**Flight experiment guidance technique for research on transition with Grob G109 aircraft of the Technische Hochschule Darmstadt**

P. Erb, B. Ewald, M. Roth
Technische Hochschule Darmstadt, Fachgebiet Aerodynamik und Meßtechnik,
Flughafenstr. 19, 64347 Griesheim, Deutschland

**Summary**

The inflight research on boundary layer transition is subject of the DFG research project „Transition", which is carried out by the universities of Aachen, Berlin, Darmstadt, Erlangen and Stuttgart. To achieve flight test settings, predefined in terms of Reynolds number and angle of attack, a flight guidance system is developed by the TH Darmstadt. This system enables the pilot to reproduce a flight setting independent of variable flight parameters such as aircraft weight, center of gravity and weather data. The current flow characteristic along the laminar glove is investigated by inflight experiments and by a 3-D panel method. The knowledge about these local flow characteristics gives a database for direct numerical simulation performed by the University of Stuttgart.

# 1 Introduction

Within the DFG research project „Transition" the group IV „free transition and measurement technique" investigates the laminar turbulent transition by inflight experiments on a laminar glove on a Grob G 109b airplane [1].

The main objective is to observe predefined particular flight test setting, and to compare experimental results with those from equivalent numerical simulations. Each experimental and numerical flow analysis is taken under controlled exitation of the laminar boundary layer.

The individual members of this research project and their measurement methods are listed below:

- RWTH Aachen: Investigation by CCA-multisensor hot-film arrays
- TU Berlin: Piezo-foil arrays, CPM-3 measurement
- TH Darmstadt: CTA-hot-film arrays, flight test guidance technique
- UNI Erlangen: Laser-Doppler-anemometry
- UNI Stuttgart: Direct numerical simulation

The flight test guidance technique developed by the TH Darmstadt has to guarantee exactly defined flow conditions along the laminar glove during flight experiments. The flow characteristics are defined by Reynolds number and pressure distribution (2-dimensional in parallel flow). Following requirements must be fulfilled by the flight test guidance technique:

1. Measurement of the flight data: angle of attack, yaw angle, ambient temperature and humidity, static and dynamic pressure

2. Calculation of flight setting parameters as input for flight guidance instruments for predefined test cases

3. Examination of the upcoming local flow parameters along the laminar glove: pressure distribution, crossflow effects

## 2 Research aircraft G 109b of the Technische Hochschule Darmstadt

Fig. 1 Research airplane G 109 B with measurement equipment

The G109b manufactured by Grob is a two place powered glider. This airplane enables an extensive and economical flight testing in the scope of the joint research project at comparatively low cost. On the other hand the test flights can be carried out avoiding most disturbances when the experiment is taken with the engine shut down.

Figure 1 shows the research aircraft in measurement flight. The removable laminar glove can be seen on the starboard wing. Every measurement group from Aachen, Berlin, Darmstadt and Erlangen has its own glove, those being identical in their geometry. Each glove can be individually equipped with different measurement technique. The port side wing carries a probe mast, which provides various sensors for measurement of flight data. The complete instrumentation can be installed in two pods attached to the lower surface of each wing and on a shock and vibration absorbing platform behind the crew seats. An additional 24V /16Ah power supply for the flight measurement equipment was established. This supply is operating independent of the onboard electrical system.

## 3 Flight data measurement

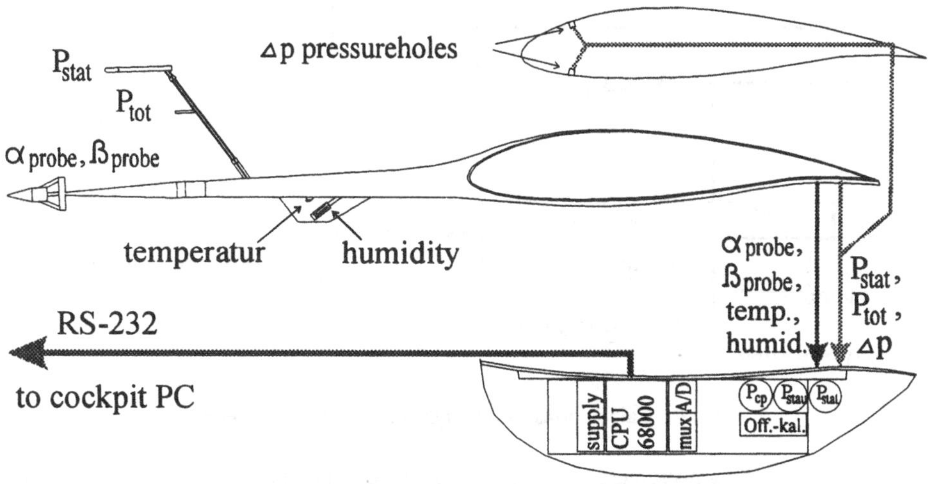


Fig.2 Inflight data measurement system

Figure 2 shows the inflight data measurement system installed on the port side wing. The head of the probe mast carries a probe for angle of attack and yaw angle. The probes for static and dynamic pressure are mounted above. Two sensors for ambient temperature and humidity are located at the bottom of the mast . These sensors are installed behind a metal shield for protection of direct sunlight. For reproduction of a defined flight setting, it is essential to know the exact angle of attack. Therefore, the angle of attack is determined by two independent means: one is the probe mast's angle of attack sensor, the other data source consists of two pressure taps located in the wing's nose section. The pressure difference between the two holes is divided by the dynamic pressure. The resulting value Δcp is a linear function of the angle of attack in the range of -4 to 6°, which is established by flight testing (see fig. 6). The electrical signals, produced by the probes for angle of attack, yaw angle, ambient temperature and humidity are analogically transmitted to the electronic pod. All pressure signals are carried by hoses to the pressure transducers, installed in the pod. The pressure transducers are equipped with a pneumatical offset calibration feature. The analogical signals are A/D-converted by a computer in the pod which sends them via RS232 connection to the boardcomputer (flight guidance PC).

## 4 Flight guidance

The incoming flight data recorded by the computer in the pod are processed by the flight guidance PC which is installed inside the instrument panel. First the parameters essential for flight guidance are calculated as it is shown in the flowchart figure 3.

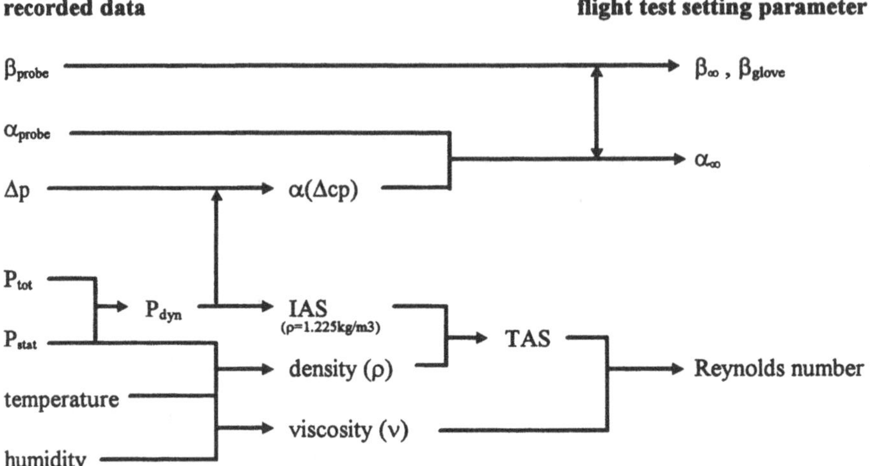


Fig. 3 Calculations of the flight test setting parameter

Two additional analog instruments were installed in the cockpit, enabling the pilot to maintain a predefined fligt test setting. The instruments are used to display any flight test setting parameter. The test engineer can select the parameter, its offset and its amplification from a menue of the flight guidance programm. The flight PC is operated via TFT-VGA display, keyboard and trackball. The operating system is IBM OS/2 with full multitasking ability, allowing to use the flight PC for multiple tasks such as running several processes of the flight guidance programm (saving of ambient data, calculation of flight test setting parameters, providing commands to the flight guidance instruments) or for use as a control unit during measurement campaigns of participating universities. Figure 4 gives an overview of the whole flight guidance system.

As already mentioned in the introduction, the central point of the inflight experiments is the examination of a particular flight test setting and to compare its results with those of the direct numerical simulation. The flight test setting, which must be exactly reproduced, is defined by Reynolds number and pressure distribution along the glove. The pressure distribution of the fixed mounted glove can only be influenced by alingment of the airplane in the flow via angle of attack and yaw angle. In wind tunnel tests Reynoldsnumber can be changed independently from angle of attack or lift coefficient ($Re \neq f(\alpha)$). This is not true for flight testing, where the Reynolds number is dependent on the angle of attack ($Re = f(\alpha)$ !!!). By variation of aircraft weight and by adequate selection of altitude (depending on actual atmospherical data), the pilot may influence the Reynolds number. These relationships are shown in Figure 5.

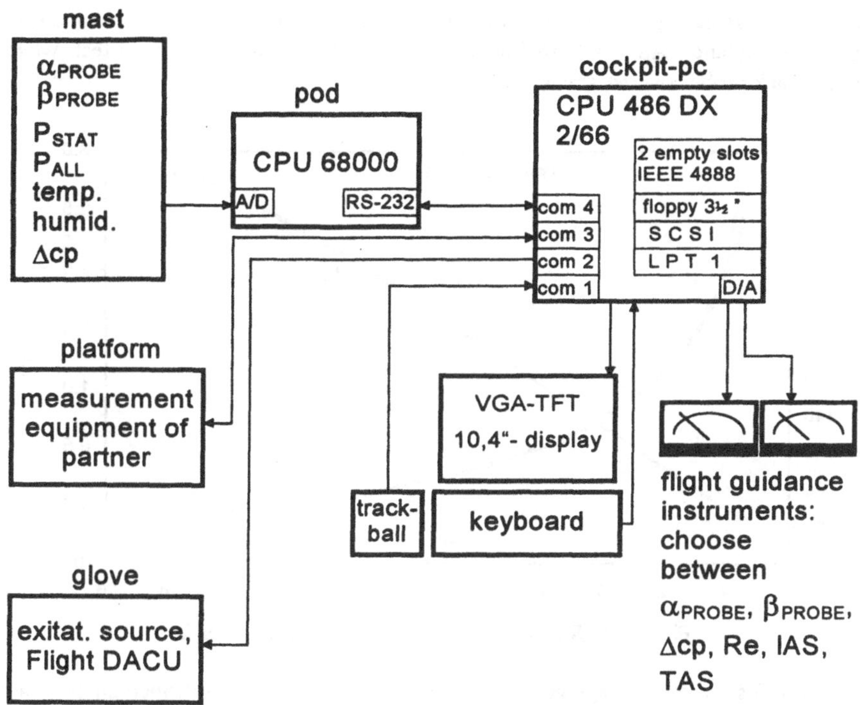


Fig. 4 The flight guidance system of the research aircraft G-109b

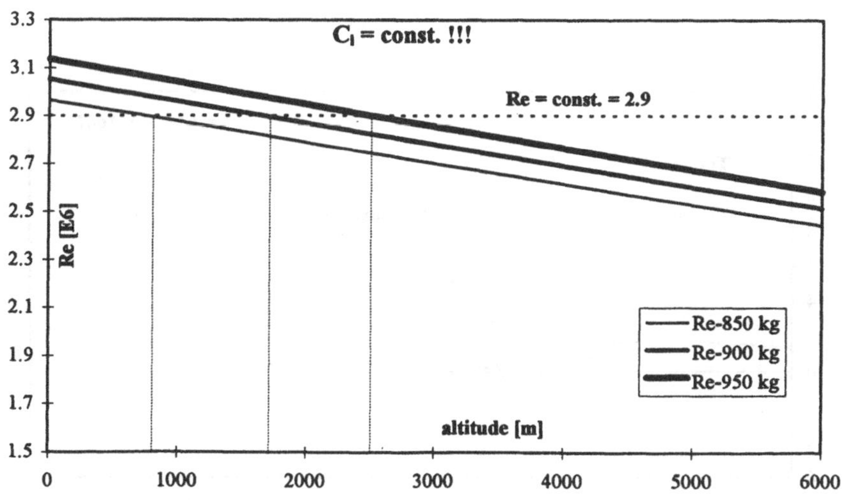


Fig. 5 Reynolds number as a function of altitude with constant lift coefficient

**Repeatability of gating:** Figure 6 shows a recording of a flight test. The complete speed range of the airplane (IAS = indicated airspeed) was covered during this flight test, while the angle of attack probe covered a range of 18°.

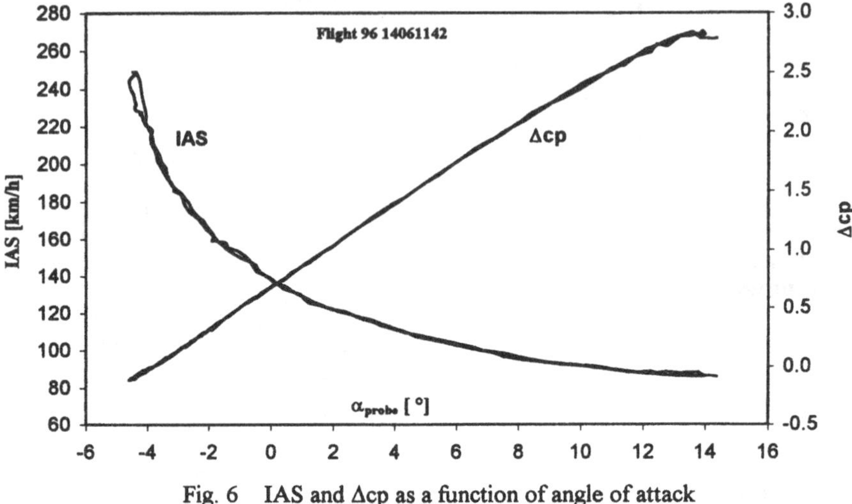


Fig. 6 IAS and Δcp as a function of angle of attack

Figure 7 and 8 show the average grade of precision when aiming for a defined angle of attack during test flight. In acceptable meteorological condition (light turbulence and low thermal activty) it has a deviation of ± 0.1°. Compared with the overall range of the angle of attack this is a relative precision of 0.6 %. This is equivalent to a speed deviation of ± 1 km/h.

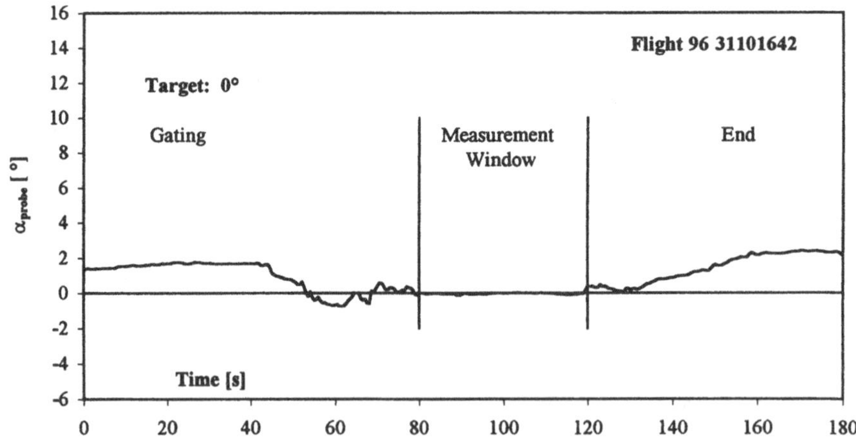


Fig. 7 Recording of angle of attack during flight test

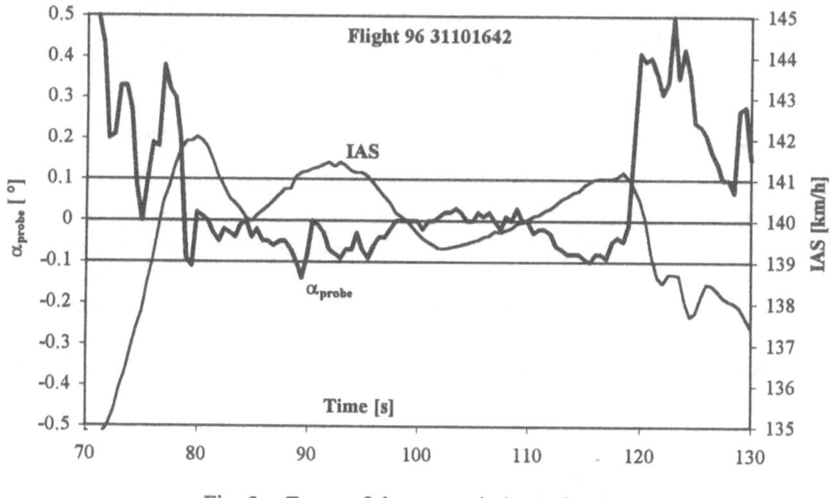


Fig. 8 Zoom of the „test window" (in Fig. 7)

## 5 Aerodynamic conditions on the glove

Using the flight guidance system, the option exists to reproduce specially defined flight conditions. These flight conditions are defined by angle of attack (probe) and Reynolds number. For a useful interpretation of the glove measurements, especially for carrying out the direct numerical simulations of the transition, the local aerodynamical conditions along the glove must be known thoroughly. At first theoretical investigations were done, using a 3-D panel method (VS-AERO). Fig. 9 displays the panel-model used for that analysis.

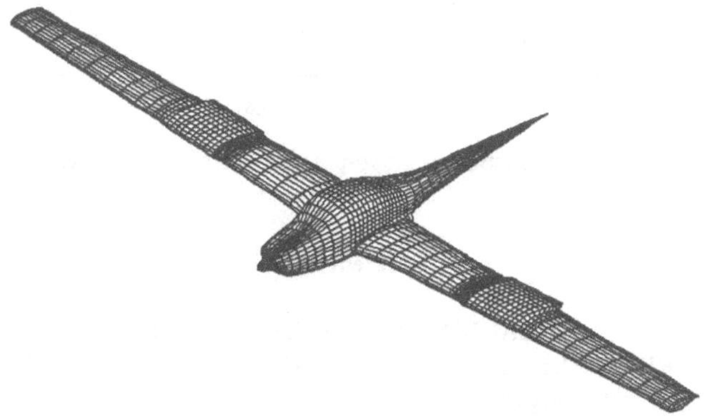

*** G109b FRBH V3° AF2.3° AG-0.66° BG-0.50° AS1.00° BS-0.32°, BH0.00°

Fig. 9 Panelmodel of the configuration fuselage wing with pod and glove (sym.)

By altering the aircrafts angle of attack and yaw angle, the calculation of the local angle of attack and yaw angle of the probe and yaw angle of the glove (middle of glove, on 25% chord) was enabled. With these results, calibration functions in the form

$$\alpha_{\infty},\ \beta_{\infty},\ \beta_{glove} \ =\ f(\alpha_{probe},\ \beta_{probe},\ \alpha_{probe}^{\ 2},\ \beta_{probe}^{\ 2},\ \alpha_{probe} * \beta_{probe})$$

were established. The calibration function of the local yaw angle at the glove enables the calculation of the gating parameter $\beta_{probe} \mid \beta_{glove} = 0 \mid = f(\alpha_{probe})$ for measuring the pressure distribution of the glove under side slip free conditions. This function was confirmed by flight tests (using tufts as visual devices). Figure 10 shows a typical pressure distribution as measured in flight. Presented are the unsmoothed raw data of 10 single shots within a two minute measuring window. Each second measurement, an offset-calibration of the pressure modules was done. The target angle of attack was 0.5°. The mean value of the recorded angles of attack associated to the single shots is 0.45°.

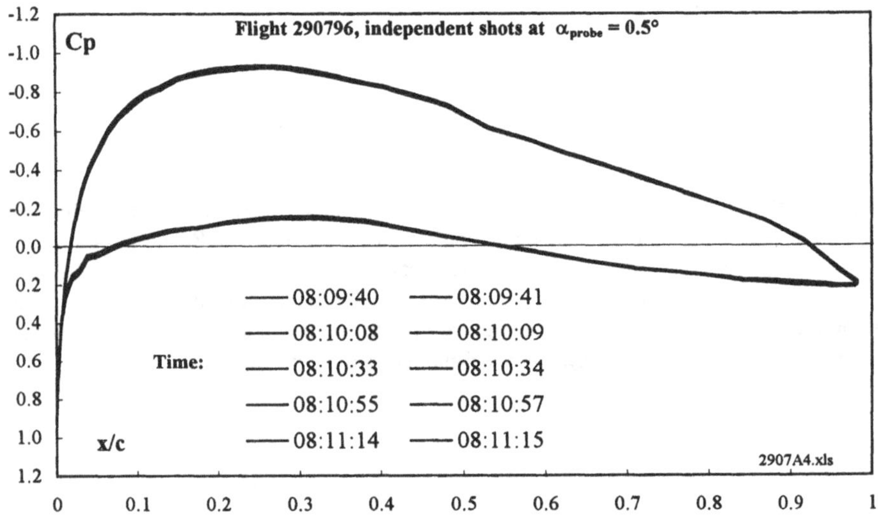


Fig. 10 Successful reproduction of glove-pressure distributions measured in flight

## Reference

[1] J. SUTTAN, M. BAUMANN; TU Berlin; S. FÜHLING; RWTH Aachen; P. ERB; TH Darmstadt; S. BECKER; Universität Erlangen; C. STEMMER; Universität Stuttgart, „Inflight research on boundary layer transition - works of the DFG- university research group“, Tagungsband der AGSTAB 1996.

# Alternative drag calculations from off body flow quantities using the FLOWer code

B. Ewald, P. Kreuzer
Technische Hochschule Darmstadt, Fachgebiet Aerodynamik
und Meßtechnik, Petersenstr. 30, D-64287 Darmstadt, Germany

## Summary

A technique for the evaluation of the thermodynamic drag from flow field solutions based on Oswatitsch's theorem is discussed. The technique allows the decomposition of thermodynamic drag into wave drag and viscous drag. Consequently, it provides more physical insight into the drag sources than conventional surface force integration techniques. The method is applied to two-dimensional flow field solutions obtained from the FLOWer code. Qualitative comparisons of the two techniques are presented for a NACA 0012 airfoil at different Mach numbers.

## Introduction

One of the main tasks in the aerodynamic design process of an aircraft is the accurate drag prediction for all flying conditions. Computational methods based on the Euler and Navier-Stokes equations like the FLOWer code [1] calculate the aerodynamic drag by body surface integration of the normal- and tangential forces. Reliable integration of the normal forces can be done only with a very fine grid, because there exists a near cancellation of a large force component in thrust direction and a slightly larger force in the drag direction [2]. Also, this technique does not allow the separation of drag in its physical components (Fig. 1). But this is often needed during an aerodynamic design process in order to detect possible drag sources. Therefore an alternative way is proposed to evaluate the wave drag and the viscous drag using Oswatitsch's theorem [3].

## Wave drag calculation

Oswatitsch has derived a formula, which combines the mass, momentum and energy conservation and which makes it possible to extract thermodynamic drag (wave-, form- and friction drag) from total drag:

$$D_S = \frac{T_\infty}{U_\infty} \iint_S \rho \vec{q} \cdot \vec{n} (s - s_\infty) dS \qquad (1)$$

This expression is valid for a closed surface and calculates the drag coming from the entropy production in the enclosed volume (Fig. 2). $\rho \vec{q} \cdot \vec{n} dS$ denotes the local mass flux across the surface, $s$-$s_\infty$ the local entropy deviation from the free stream entropy and $U_\infty$, $T_\infty$ the velocity and static temperature of the undisturbed flow. In order to separate wave drag from thermodynamic drag, the surface of the control volume has to be contracted to the shock discontinuity. The resulting entropy jump can now be calculated from the exact Rankine-Hugoniot relation. A cell vertex based computational method like the FLOWer code cannot resolve a shock discontinuity. Therefore this shock discontinuity has to be constructed on the cell edges by using the local subsonic and supersonic machnumbers in the cell vertices [4]. Furthermore, in order to evaluate equation (1), the velocity and density is needed upstream of the shock. They can be obtained by applying an upstream streamline calculation which starts in the corner points of the constructed shock surface (Fig. 3).

## Viscous drag calculation

As mentioned above, Oswatitsch's theorem makes it possible to calculate the total thermodynamic drag of the flow field by extending the control surface to the far-field boundaries. Initial computational experiments with Euler codes revealed, that a direct application of equation (1) gives unacceptable errors for only moderately converged solutions [5]. The reason is due to the fact that the residuals in a numerical solution act as small point sources or sinks throughout the flow field. Therefore a correction term is introduced as follows [5]:

$$D_{scorr} = D_s - U_\infty \iint_S \vec{q} \cdot \vec{n} dS . \tag{2}$$

The second term in equation (2) is directly proportional to the net amount of mass generated within the control surface. It takes into account that a source or a sink immersed in a flow field experiences a thrust force or a drag. The viscous drag now can be calculated very simply:

$$D_{vis} = D_{scorr} - D_{wav} . \tag{3}$$

## Numerical results

FLOWer, the code used to generate the flow field solutions, has been developed by the DLR Braunschweig in Germany [1]. FLOWer is a finite volume method based on a cell vertex scheme. It uses an explicit multi-stage Runge-Kutta time-marching algorithm with central differences for the fluxes. Both blended second-order and fourth-order dissipation terms are added for numerical stability.Turbulent calculations can be performed by using the Baldwin-Lomax turbulence model. Multigrid technique and implicit residual smoothing as well as local time stepping and enthalpy damping accelerate the convergence.

In Fig. 4 the drag coefficient obtained from a 2-D Euler computation for a NACA 0012 airfoil is plotted versus the free stream Mach number. The surface-pressure integration predicts incorrect values for the wave drag below $M_\infty$ =0.75, because no supersonic point can be detected in the whole flow field. More acurracy, i. e. no wave drag below $M_\infty$ = 0.75, is obtained by the proposed shock integration technique. This trend is confirmed at higher Mach numbers. Simliar results can be seen for $\alpha$ = 1.25° in Fig. 5, where shocks with different strenghts (and on different locations) can be observed for this flow case. Fig. 6 illustrates the influence of the source correction term. It can be shown that a direct application of equation

(1) overestimates the wave drag and more realistic results are obtained by using equation (2). The next case deals with a 2-D Navier-Stokes compution for the same airfoil at $\alpha = 0^\circ$ [Fig.7]. As expected large differences can be observed between the surface force integration ($C_{Dp}$ + $C_{Df}$) and the proposed thermodynamic drag calculation ($C_{D\bullet corr}$). In the beginning shock region, where form drag is negligible, there is a good agreement between the sum of wave and friction drag and the thermodynamic drag. Note that all three drag components have been calculated at three different locations in the flow field.The development of the viscous drag near $M_\infty = 0.8$ can be explained with the decrease of the local friction behind the shock. This is compensated by the form drag at higher Mach numbers because separation occurs and the movement of the shock to the trailing edge produces an additional drag increment.

## Further investigations planned

The performance of the method has to be tested for 3-D flow fields. Furthermore, it is planned to evaluate the induced drag from a wake integral method which has been successfully applied in the past by several authors in computations and experiments [4, 6, 7, 8].

## Acknowledgments

This work has been carried with the support of the German Ministry of Education and Research. The authors would also like to thank Mr. H. Jakob from Daimler-Benz Aerospace Airbus for the supply of the EULWID code.

## References

[1] Kroll, N., Radespiel, R. and Rossow, Cord-C: "Accurate and Efficient Flow Solvers for 3D Applications on Structured Meshes ", Special Course on Parallel Computing in CFD, AGARD-Report R-807, 1995.

[2] Lock, R. C.: " The Prediction of the Drag of Airfoils and Wings at High at High Subsonic Speeds ", Aeronautical Journal, Vol 90, No. 896, 1986, pp. 207-226.

[3] Oswatitsch, K.: " Der Luftwiderstand als Integral des Entropiestromes ", Nachrichten der Akademie der Wissenschaften Göttingen, Mathematisch-physikalische Klasse, Heft 1, 1945, pp. 80-90.

[4] Jakob, H.: " Folgeprogramm EULWID zum Euler-N/S-Löser Melina für die Berechnung von Einzelanteilen des Widerstandes ", Technischer Bericht DA-EF11/96, Daimler-Benz Aerospace Airbus, 1996.

[5] Yu, N. J., Samant, S. S. and Rubbert, P. E.: " Inviscid Drag Calculations for Transonic Flows ", AIAA Paper 83-1928, July 1983.

[6] van Dam, C. P., Nikfetrat, K., Wong, K. and Vijgen, P. M. H. W.: " Drag Prediction at Subsonic and Transonic Speeds Using Euler Methods ", Journal of Aircraft, Vol. 32, No. 4,1995, pp.839-845.

[7] Brune, G. W. and Bogataj, P. W .: " Induced Drag of a Simple Wing From Wake Measurments ", Society of Automotive Engineers, Paper 901934, October 1990.

[8] Kreuzer, P.: " Weiterentwicklung und experimentelle Überprüfung eines 3-D Panelverfahrens im Falle einer Tragflügelanordnung mit Winglets ", Dissertation, Technische Hochschule Darmstadt, 1993.

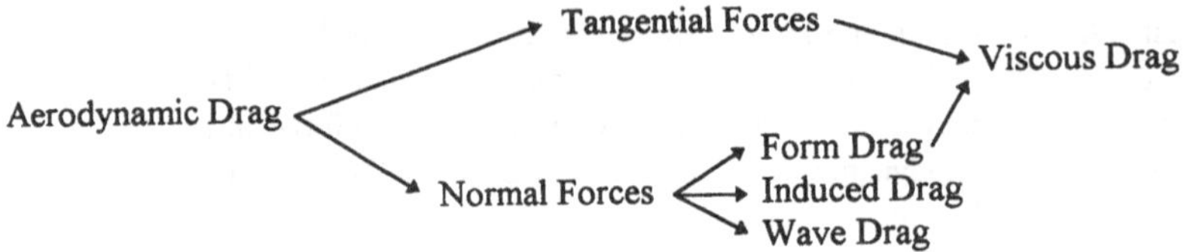


Fig. 1: Drag decomposition by forces and physical mechanism

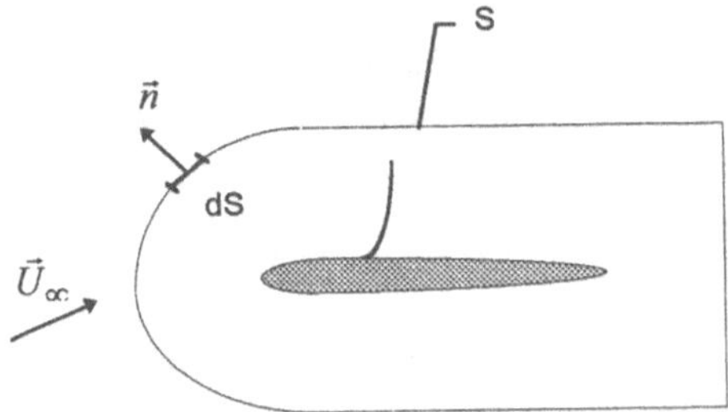


Fig. 2: Possible integration surface for evaluating Oswatitsch's theorem

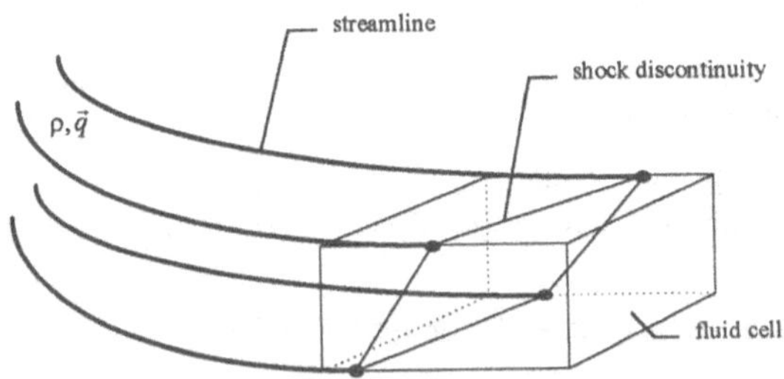


Fig. 3: Interpolated shock discontinuity in fluid cell

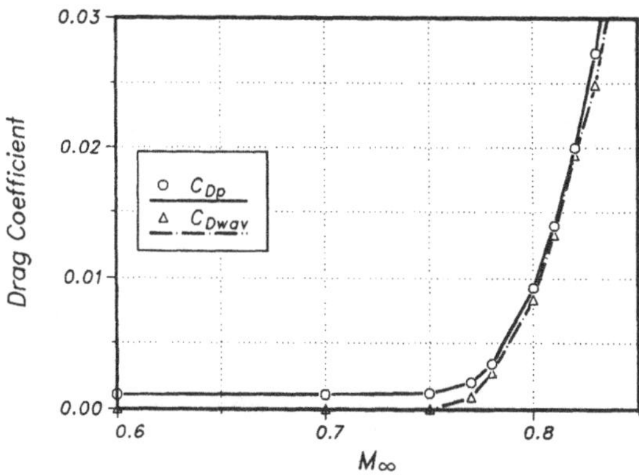


Fig. 4: 2-D Euler computation for NACA 0012 airfoil at
$\alpha = 0^\circ$, O-type grid (164x33 points),
drag coefficients $C_{Dp}$ obtained by surface pressure integration and
$C_{Dwav}$ by Oswatitsch's theorem applied to the shock

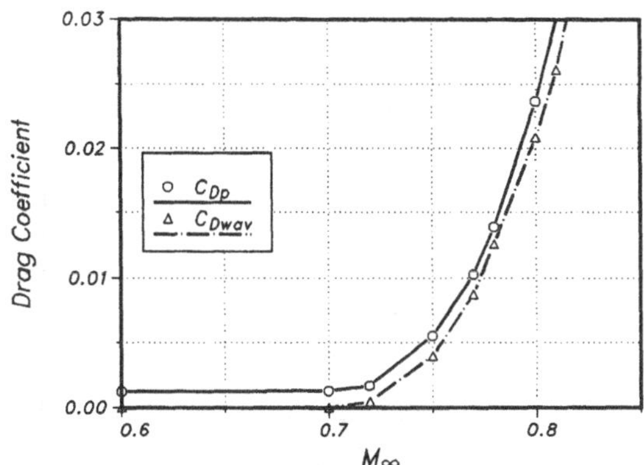


Fig. 5: 2-D Euler computation for NACA 0012 airfoil at
$\alpha = 1.25^\circ$, O-type grid (164x33 points),
drag coefficients $C_{Dp}$ obtained by surface pressure integration and
$C_{Dwav}$ by Oswatitsch's theorem applied to the shock

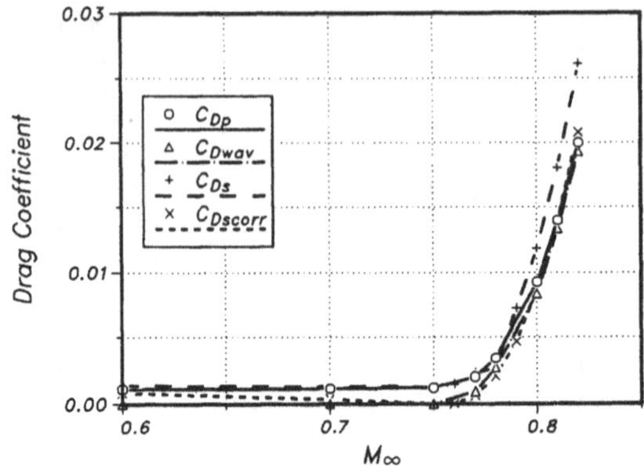


Fig. 6: 2-D Euler computation for NACA 0012 airfoil at $\alpha = 0^\circ$, O-type grid (164x33 points), drag coefficients $C_{Ds}$ and $C_{Dscorr}$ obtained by Oswatitsch's theorem applied to the far-field boundaries; $C_{Dp}$ and $C_{Dwav}$ as in Fig. 5

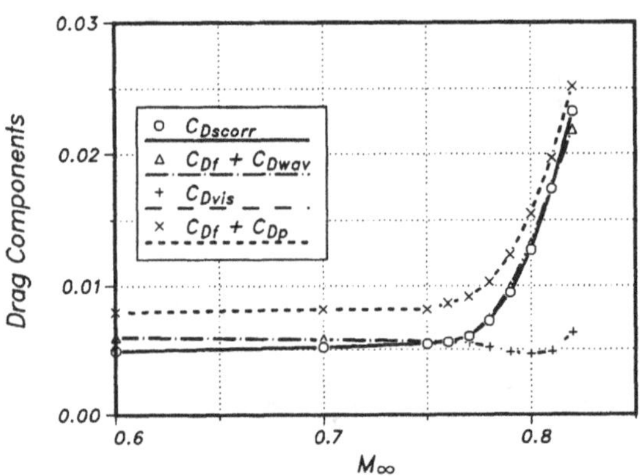


Fig. 7: 2-D Navier-Stokes computation for NACA 0012 airfoil at $\alpha = 0^\circ$, Re = 9 Mio., C-type grid (257x62 points, drag components obtained by Oswatitsch's Theorem applied to the shock and far field-boundaries; $C_{Df}$ is obtained by integrating the local friction force over the airfoil surface

# Numerical Simulation of the X-31A Flow Characteristics

M. Franke
DLR, Institute of Design Aerodynamics,
Lilienthalplatz 7, D-38108 Braunschweig, Germany [1]

## Summary

In this study, Euler simulations of the X-31A experimental aircraft are performed in order to analyse the flow characteristics of the configuration as well as their influence on the aerodynamic coefficients. The discretized geometry is based on the full configuration with the exception of the inlet, which has been faired for reasons of grid generation simplicity, and the assumption of sharp leading edges for numerical reasons. Solutions are obtained with the DLR flow solver CEVCATS, which is based on a cell-vertex finite volume approach on block structured grids. Comparison with available wind tunnel data shows good agreement of the aerodynamic coefficients up to maximum lift. The vortex formation and vortex burst phenomena on the canard and the wing are investigated and their influence on the pitching moment coefficient characteristics are identified.

## 1 Introduction

Steady flows over delta wings have been subjected to extensive numerical research, see e.g. [1]. Furthermore, also studies on complex fighter aircraft configurations have been performed, see e.g. [2, 3]. This motivated further studies of the complex flow around this type of configuration. The X-31A (see Fig. 1) is chosen as experimental data of the aerodynamic characteristics are readily available [4]. The goal of this work is the numerical simulation of the flow field around the X-31A with emphasis on the vortex formation and vortex burst phenomena observed on its canard and wing and their influence on the aerodynamic characteristics of the aircraft.
The X-31A experimental aircraft was developed under a joint program by MBB (now DASA-LM) and Rockwell to evaluate possible requirements for future fighter aircraft. Its special feature is the ability of controlled flight in the post-stall regime achieved by the utilisation of Enhanced Fighter Manoeuverability (EFM) concepts including thrust vectoring and an advanced digital flight control system. The aircraft completed an extremely successful flight test program, see e.g. [5].
Considering its aerodynamic characteristics, it should be mentioned that the X-31A is not an optimised wing-canard configuration. The wing-canard coupling is extremely long - the distance between the trailing edge of the canard and the leading edge of the wing at the respective root location is more than 20 % of the aircraft length - whereas the area of the canard is small

[1] New address: TU Berlin, Hermann-Föttinger-Institut für Strömungsmechanik, Müller-Breslau-Str. 8, D-10623 Berlin, Germany

compared to typical wing-canard configurations. This is due to the fact that the canard chiefly serves stability purposes during recovery from high angle-of-attack manoeuvers. Furthermore, it should be noted that, compared to typical delta wings, the aspect ratio of the wing is relatively high ($AR = 2.30$), whereas the sweep angle of the inner as well as the outer wing is comparably low ($\Lambda_i = 56.6°, \Lambda_o = 45°$).

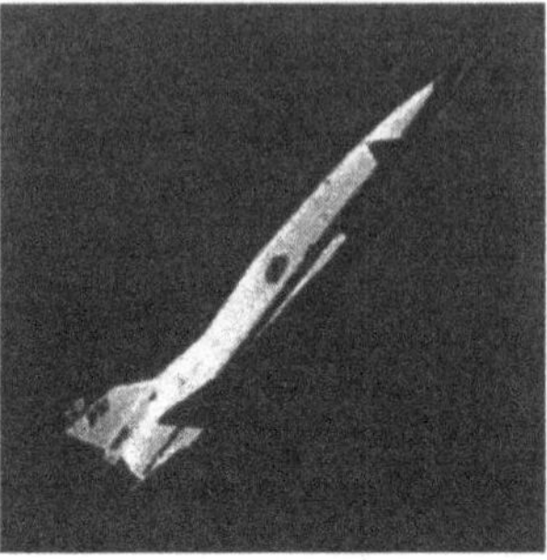

**Figure 1** The X-31A in flight (Photo by NASA Dryden)

## 2 Computational Approach

Compared to numerical Euler solutions, the numerical solution of the Navier-Stokes equations is extremely expensive with respect to computer time and memory requirements. Furthermore, adequate modelling of turbulence and an increased complexity in grid generation pose additional difficulties. Therefore, in this study a numerical Euler method was chosen. In the following, remarks are made about the capability of Euler computations to model the vortical flow field structure of delta wings as well as the solver and the grid used. For a general description of the flow characteristics of delta wings see [6, 7].

It is a well-known fact that Euler calculations can capture the separating shear layer generated at sharp leading edges of delta wings and the subsequent development of primary vortices, see e.g. [8]. However, in case of rounded leading edges, which are present on the X-31A, the separation is no longer fixed by the geometry but depending on the numerical dissipation of the scheme employed. As this is not tolerable, the separation has to be fixed at the leading edge or, if known, at the physical separation line. In this study the geometry is locally modified to create a sharp wedged leading edge. Furthermore, as Strohmeyer et al. [9] have shown, a steady state, time-marching Euler procedure can produce effects which simulate vortex burst and oscillate periodically with the proceeding iterations. This in turn leads to a periodic variation of the aerodynamic coefficients. Secondary vortices, however, cannot be simulated with an Euler procedure, as they are an inherently viscous effect.

The study is performed using the DLR solver CEVCATS [10], which is characterised by a cell-vertex finite volume procedure. Spatial discretization is based on central differences with added artificial viscosity. An explicit five-stage Runge-Kutta scheme is used to integrate in time. Convergence accelerators employed are local time stepping, implicit residual smoothing and a multigrid algorithm. The code is optimised for vector performance.

The discretized geometry of the aircraft includes wing, canard, fuselage, tail and canopy. To facilitate grid generation and to save grid points, the inlet is faired over. The canard is fixed in zero position, which is consistent with the selected experimental data. As only cases without yaw angle are investigated in this study, a half model of the configuration is used. The block-structured grid is arranged in a modified H-O topology. In order to ensure the cell spacing at critical points, initially 31 blocks had to be generated. To enhance vector performance, the blocks are concatenated to 7 resulting ones. As two multigrid levels prove sufficient for maximum convergence acceleration, the mesh is restricted to these two levels, which allows for greater flexibility in mesh point distribution. Still, the grid consists of approximately 2.3 million points on the fine level.

# 3 Results and Discussion

Fig. 2 shows the computed aerodynamic lift, drag and pitching moment coefficient plotted versus angle-of-attack for an angle-of-attack range of $-5° < \alpha < 28°$ at a Mach number of $M = 0.2$ and compares them to experimental data from [4].

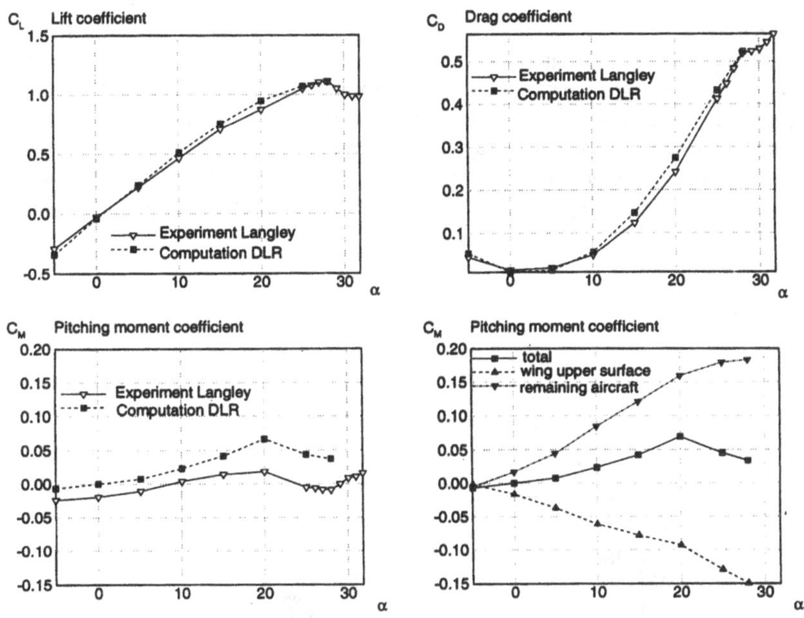


**Figure 2** Aerodynamic coefficients versus angle-of-attack for $M = 0.2$

For $\alpha \leq 10°$, the wing vortex does not burst above the wing surface, therefore the coefficients converge to a steady state. However, for $\alpha > 10°$, the vortex breakdown location is above the

wing surface. In this case, as generally shown in [9], the aerodynamic coefficients vary periodically (for the X-31A, $C_L$ typically varies about 1%). In order to be able to compare with the experimental data, the coefficients have been averaged.
It can be seen that the lift and the drag coefficient compare favourably with the wind tunnel data up to maximum lift. The fact that lift as well as drag are slightly overpredicted in the region between $5^\circ < \alpha < 28^\circ$ is consistent with the fact that Euler solutions cannot model secondary vortices. In their absence, the location of the primary vortex shifts outboard and closer to the wing surface, thereby leading to higher suction peak on the surface (see also [1]). For small angles-of-attack, the computed drag is smaller than the measured one. This is due to the fact that Euler simulations cannot capture friction drag, which is dominant in this region.
The computed pitching moment coefficient is in qualitatively good agreement with the experimental data, although an offset can be observed. In this plot a positive pitching moment corresponds to a nose-up moment. It is interesting to note that up to an angle-of-attack of $\alpha = 20^\circ$ the X-31A shows a behaviour typical of a delta wing configuration, i.e. with increasing angle-of-attack the pitching moment is increasing too. However, for $\alpha > 20^\circ$ the pitching moment is decreasing again, which can be considered as rather untypical for a classic slender delta wing configuration.
A detailed analysis (see Fig.2) has shown that mainly the upper surface of the wing is responsible for this behaviour. The pitching moment contribution on top of the wing is decreasing with increasing angle-of attack, whereas the contribution of the remaining aircraft increases as expected. Up to $\alpha = 20^\circ$, the top side of the wing is not able to balance the increase generated by the remaining aircraft. This situation changes above $\alpha = 20^\circ$, where the rate of pitching moment decrease of the upper surface of the wing grows significantly while the rate of pitching moment increase of the remaining aircraft slightly decreases, so that the pitching moment of the total configuration starts decreasing again. An explanation of this behaviour will be given after a brief overview of the pressure distribution on the aircraft surface induced by the vortical flow field for three different angles-of-attack, where special attention will be paid to the wing surface.

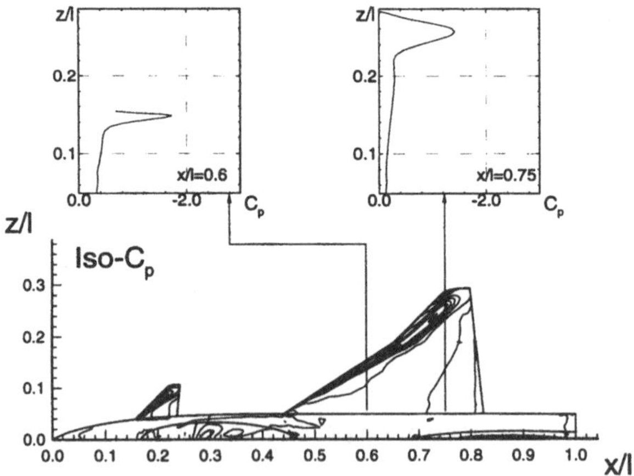


**Figure 3** Pressure contours on the suction side and the pressure distribution in two selected cuts on the wing surface for $\alpha = 10^\circ$ and $M = 0.2$

Fig. 3 displays surface pressure conturs on the suction side as well as the pressure distribution in two selected cuts on the upper surface of the wing at $\alpha = 10°$ and $M = 0.2$. The coordinate system is chosen such that the origin is in the nose apex of the aircraft and the distance from the plane apex to the plane base is unity. The suction peaks induced by the primary vortices on the wing and the canard are clearly visible. Neither shows breakdown above the respective surface. For the wing vortex, this is also verified in the two cuts in spanwise direction, where the pressure distribution shows the high, narrow suction peak typical of a vortex that has not burst. Due to the long wing-canard coupling and the relatively high angle-of-attack, interaction of the canard and the wing vortices is negligible.

A similar analysis for $\alpha = 20°$ and $M = 0.2$ can be seen in Fig. 4. As remarked before, this case includes vortex burst over the lifting surfaces which leads to periodic oscillations of the solution, which in turn leads to a variation of $C_L \approx 1\%$. The surface pressure distribution is displayed for the maximum $C_L$ for this case, whereas the plots showing the pressure distribution in the cuts include curves for both maximum and minimum lift. Clearly, the wing vortex is much stronger in the front part of the wing than at $\alpha = 10°$. It is noteworthy that in the cut at $x/l = 0.6$, where the vortex has just started to break down, the solution shows almost steady behaviour. The situation is very different at $x/l = 0.75$, where the vortex is fully burst. It can be seen that the suction peak oscillates between two different spanwise locations while being much lower and broader than at $x/l = 0.6$. This clearly indicates that the vortex breakdown causes the unsteady behaviour of the aerodynamic characteristics. The magnitude of the peak, however, remains approximately constant.

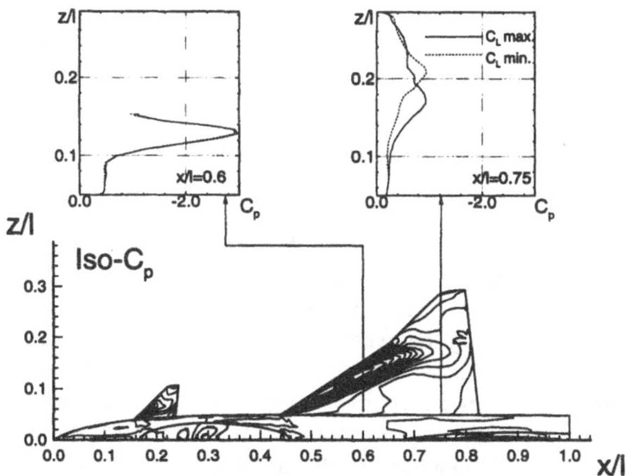


**Figure 4** Pressure contours on the suction side and the pressure distribution in two selected cuts on the wing surface for $\alpha = 20°$ and $M = 0.2$

Fig. 5 shows the same plots for the maximum lift case at $M = 0.2$, where $\alpha = 28°$. It can be seen that on the canard the vortex burst location has reached the leading edge so that no vortical flow region is generated, which contributes to the reduction in the pitching moment gradient of the remaining aircraft described above. The wing vortex still exists, but breakdown already occurs at $x/l \approx 0.47$. The cut at $x/l = 0.6$ shows a typical pressure profile of a burst vortex. A very interesting flow situation can be found at $x/l = 0.75$. The burst vortex has spread spanwise to such an extent that it basically covers the whole wing span. This leads to an almost constant suction on a moderate level in spanwise direction (compare to Fig. 4), where the unsteadiness of the breakdown region does not alter the pressure profile to a great extent.

In order to determine the pitching moment characteristics of the top side of the wing, it is necessary to analyse the chordwise lift distribution with respect to the moment reference center. To achieve this, a spanwise integration of the pressure coefficient $C_p$ is performed at different $x/l$-stations. The results are then plotted over $x/l$. The coordinate system is not changed to a wing system for reasons of consistency.

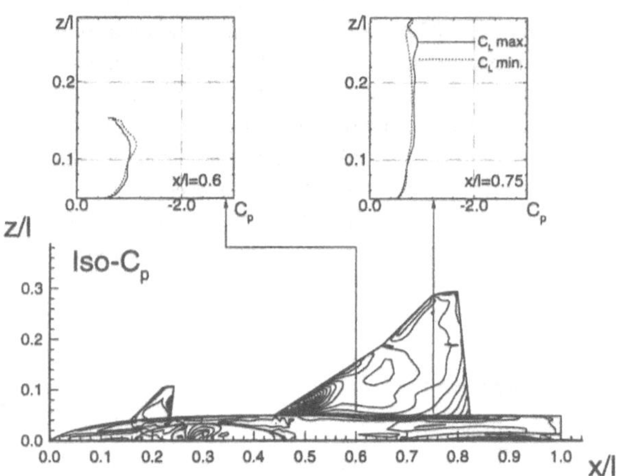


**Figure 5** Pressure contours on the suction side and the pressure distribution in two selected cuts on the wing surface for $\alpha = 28°$ and $M = 0.2$

Fig. 6 shows the chordwise lift distribution for three different cases, viz. $\alpha = 10°$, $\alpha = 20°$ and the maximum lift case $\alpha = 28°$. Here, the local contribution to lift was not averaged in the vortex breakdown locations but plotted separately in order to demonstrate that the unsteadiness of the breakdown region is not altering the general characteristics of the chordwise lift distribution. Taking a look at the curve for $\alpha = 10°$, where the vortex does not burst, it can be seen that the lift distribution is increasing in chordwise direction. This is due to two reasons: First, the integration path increases when moving downstream, and second, the vortex induced suction peak increases too. The sudden increase of the gradient at $x/l = 0.6$ is due to the fact that the rate of increase of the integration path rises as the sweep angle decreases at the outer wing. The curve reaches its maximum at $x/l \approx 0.72$, where the vortex leaves the vicinity of the wing surface. It should be pointed out that the local contribution to lift is zero at the front and the end of the wing simply because the integration path becomes zero. Regarding the curve for $\alpha = 20°$, it is evident that the increase in the local contribution to lift is much higher than at $\alpha = 10°$, which is due to the higher suction peak induced by the vortex. The vortex breaks down at $x/l \approx 0.57$, which leads to an unsteady and decreasing contribution downstream of this location. However, at $x/l = 0.75$, the burst vortex has spread so widely in spanwise direction, that its rather low induced suction integrated over the relatively wide span at this location is sufficient to lead to a local maximum in the local contribution to lift in the very rear part of the wing. The difference between the two different solutions for maximum and minimum lift for this angle-of-attack is evident, still, the general behaviour is not altered to a great extent. Finally, regarding the curve for $\alpha = 28°$, it can be seen that downstream of the breakdown location first a local maximum (due to the balancing of the loss of suction caused by the breakdown and the increasing integration path) and a subsequent local minimum (due to the loss of this balancing) in the local contribution to lift is visible. Starting at $x/l \approx 0.57$ however, the local lift increases steeply to reach a total maximum in the very rear part of the wing. The reason for this can be found in the spanwise widening of the burst vortex, which covers the whole span in the rear part of the wing. Here too, the unsteadiness of the burst region does not alter the general behaviour.

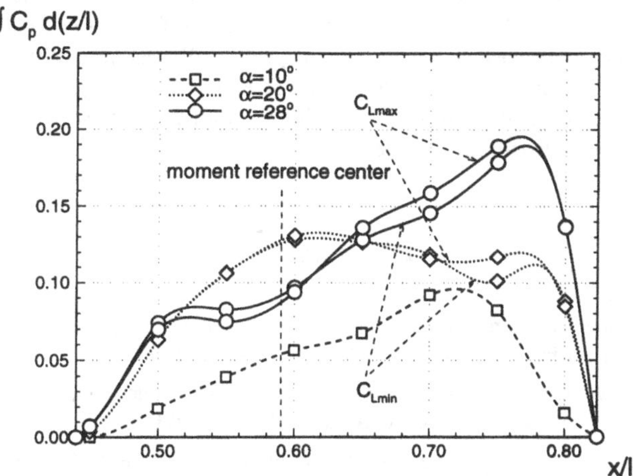


**Figure 6** Chordwise lift distribution on the upper surface of the wing for $\alpha = 10°$, $\alpha = 20°$ and the $\alpha = 28°$ at $M = 0.2$

Taking the location of the moment reference center into account, it is evident that the major part of the lift created on the upper surface of the wing is indeed created behind the reference point. It is also obvious that the higher the angle-of-attack becomes, the bigger the fraction of lift created behind the reference center and the more negative the pitching moment gets. Finally, it should be pointed out that the maximum of the lift contribution at $x/l = 0.72$ for $\alpha = 10°$ and the one at $x/l = 0.77$ for $\alpha = 28°$ have very different physical reasons, although the comparably long integration path is important in both cases: Whereas the former is reached at the point where the intact vortex is still close to the surface, so that the sharp suction peak is still present while the integration path has constantly increased in downstream direction, the latter is due to the fact that the burst vortex spreads over the whole wing span and imposes a moderate but still significant suction along the whole span.

The question remains what causes the offset to the wind tunnel data in the pitching moment shown in Fig. 2. Preliminary computations with a generic inlet modelling seem to indicate that the offset is at least to some extent caused by the fairing. However, the insufficient resolution of the inlet created numerical problems so that this could not be investigated any further without a complete redefinition and new generation of the mesh.

## 4 Conclusion

In this study, an Euler simulation of a full X-31A aircraft configuration is performed. A qualitatively good agreement of the aerodynamic coefficients with wind tunnel data up to maximum lift can be achieved. The structure of the primary vortices is described and the unsteady vortex burst effects generically described in [9] are investigated. The unusual pitching moment development with increasing angle-of-attack is explained by determining the chordwise lift distribution on the upper surface of the wing, while the inlet fairing is at least partly responsible for the offset found.

Future investigations should include Euler computations with a fully discretized inlet as well as Navier-Stokes simulations. The advantage of the latter is the fact that no leading edge fix is necessary and secondary vortices can be captured as well. Still, increased difficulties concerning grid generation and turbulence modelling should be expected. Finally, it seems desirable to perform a generic study on delta wings to determine the influence of sweep angle and aspect ratio on the chordwise lift distribution of the suction side.

## References


[1] *J.M.A. Longo*: Numerische Analyse der Strömungsfelder von Deltaflügelkonfigurationen ohne und mit Canard. ZLR Forschungsbericht 93-09, TU Braunschweig, 1993.

[2] *D.T. Yeh, M.W. George, W.C. Clever, C.K. Tam* and *C.J. Woan*: Numerical Study of the X-31 High Angle-of-Attack Flow Characteristics. AIAA Paper 91-1630, 1991.

[3] *F. Ghaffari*: Navier-Stokes, Flight and Wind Tunnel Flow Analysis for the F/A-18 Aircraft, NASA Technical Paper 3478, 1994.

[4] *D.W. Banks, G.M. Gatlin* and *J.W. Paulson*: Low-Speed Longitudinal and Lateral-Directional Aerodynamic Characteristics of the X-31 Configuration. NASA Technical Memorandum 4351, 1992.

[5] *M.S. Francis*: X-31 Enhanced Fighter Maneuverability Demonstrator: Flight Test Achievements. ICAS Paper 94-7.2.2, 1994.

[6] *D. Hummel*: Untersuchungen über das Aufplatzen der Wirbel an schlanken Deltaflügeln. Z.Flugwiss. 13(1965), pp. 158-168.

[7] *D. Hummel*: Experimentelle Untersuchung der Strömung auf der Saugseite eines schlanken Deltaflügels. Z.Flugwiss. 13(1965), pp. 247-252.

[8] *J. Rom*: High Angle of Attack Aerodynamics. Springer, Berlin, 1992.

[9] *D. Strohmeyer, M. Orlowski, J.M.A. Longo, D. Hummel* and *A. Bergmann*: An Analysis of Vortex Breakdown Predicted by the Euler Equations. ICAS Paper 96-1.6.3, 1996.

[10] *N. Kroll, R. Radespiel* and *C.-C. Rossow*: Accurate and Efficient Flow Solvers for 3D Applications on Structured Meshes. VKI Lecture Series *Computational Fluid Dynamics* 1994-05, 1994.

# Experimental investigation of active (dynamic) control of hydrodynamic instability in a two-dimensional boundary layer

B. Gölling, F.-R. Grosche
DLR, Institute of Fluid Mechanics
Bunsenstraße 10, D - 37073 Göttingen, Germany

## SUMMARY

An experimental study has been made of the concept of controlling boundary layer transition by superposition of anti-phase disturbances on the Tollmien-Schlichting (TS) waves developing in the Blasius boundary layer over a flat plate. The Tollmien-Schlichting waves were excited artificially by periodic suction and blowing through a spanwise slot in the surface of the plate. The anti-phase waves were generated through a second slot downstream of the first one. Phase and amplitude of these waves were adjusted by hand to obtain minimum amplitude of the resulting flow disturbances downstream of the excitation region. TS waves with initial amplitudes of up to 1.7 % of free-stream velocity were investigated. The parameters of the flat plate and the other experimental conditions allowed to investigate the complete development of the TS waves. For comparison, amplitude and phase of the TS waves were measured by using a single hot-wire sensor – the primary TS wave, the anti-phase TS wave and the wave resulting from superposition of both. The aim of this work is to discuss problems in optimum abatement of TS waves with dependence on amplitude and Reynolds number.

## INTRODUCTION

Control of laminar-turbulent transition is of considerable interest in fluid dynamics. Delay or prevention of transition can significantly reduce the viscous drag. Besides passive control methods which influence the basic flow by special geometry, stationary suction etc., dynamic control methods work in a specific manner selecting the most amplified instability mode and attenuating it.

The principle of the anti-phase control method is based on the linear character of the TS wave in the linear instability regime. The natural oncoming TS wave will be reduced or extinguished by interference with a second artificially excited TS wave with the same amplitude but with 180° phase shift.

Artificial excitation techniques were developed to test the anti-phase control method. The pioneering investigations of LIEPMANN and NOSENCHUCK [1] and of MILLING [2] are important milestones in this field. Furthermore, the work of THOMAS [3] illustrated the effectiveness of this method , a TS wave with an amplitude of roughly 1 % of the free-stream velocity was attenuated to 0.2 %. An effective delay of transition by about a few wavelengths was shown by GILEV and KOZLOV [4, 5].

One has to distinguish between the linear and the non-linear instability regimes of the TS wave. They depend on the local amplitude. Although there is no general critical value of the TS-wave-amplitude, we assume a critical value of roughly 1 % of the free-stream velocity. Above this value the boundary layer will become unstable against secondary instability structures [6, 7]. We will demonstrate that in the linear instability regime the effectiveness of the anti-phase control method does not depend on the Reynolds number of the excitation position of the anti-phase wave. Also we will show how optimal adjustment of the anti-phase excitation can be obtained. Especially, we want to use computed local linear superposition to answer this question (see also LAURIEN and KLEISER [8]). In the last part of this paper we will compare the behaviour of the boundary layer with and without application of the anti-phase control.

## EXPERIMENTAL CONSIDERATIONS

The experiments were conducted in a low-turbulence wind tunnel at DLR Göttingen which has an open square test section with a size of $400 \times 400$ mm$^2$, a maximum free-stream velocity of 28 m/s, and an average turbulence intensity in streamwise direction of 0.16%.

The model investigated was a flat plate that had been used already by other authors in previous experiments [9, 10]. It is shown schematically with some details of the experimental set-up in Fig. 1. The flat plate is 800 *mm* long, 300 *mm* wide and 30 *mm* thick. The leading edge of the plate has an elliptic half-profile with a ratio of major to minor axes of 6:1 to avoid leading edge separation. The plate was mounted horizontally and was enclosed by three Plexiglas plates above and at both sides of the model plate. This arrangement allowed to separate the flow field under investigation from the mixing layer of the free jet of the wind tunnel flow and to set the pressure gradient to zero along the centreline of the model by adjusting the angle of the top plate.

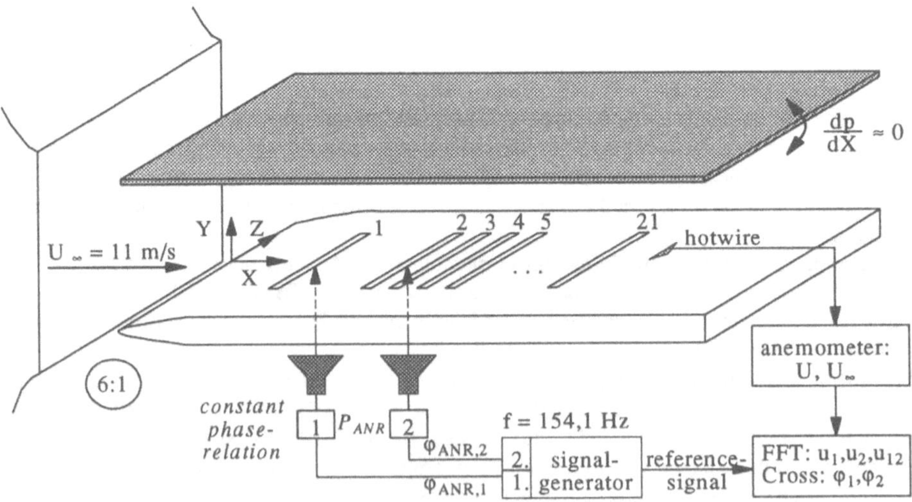


Figure 1 Schematic description of the flat plate model with some details of the experimental set-up.

There are 21 spanwise pressure chambers inside the plate with narrow slots in the surface of the plate. Periodic suction and blowing through these slots is obtained if pressure fluctuations are introduced into the chambers from the horn driver loudspeakers indicated in Fig. 1, and thus TS waves can be excited.The first chamber is only 180.5 mm downstream of the leading edge of the plate and is used to excite the initial TS wave. Further downstream the remaining 20 chambers are located. The streamwise positions of the slots of the chambers 2 to 21 are given by $X_{No} = 282.5\,mm + 6.0\,mm \cdot (No - 2)$. One of these chambers can be chosen to supply the anti-phase excitation to the boundary layer. If different chambers are used, different positions (or local Reynolds numbers of the boundary layer) of the anti-phase excitation can be investigated.

Sinusoidal pressure fluctuations were generated by means of a computer-based multi-signal generator connected to loudspeakers (horn drivers) through power amplifiers. so that the pressure fluctuation intensity and the phase of the signals could be adjusted independently. The sound pressure fluctuations of the loudspeaker were introduced through tubes into the chamber inside the plate underneath a narrow, 1 mm wide slot, and then emitted into the flow field through the slot so that a periodic suction and blowing of air is realised. The average pressure fluctuation intensity above one slot was approximately constant over the span and the signal-to-noise ratio was about 95 dB for the frequency which was used for the case of strongest excitation. To avoid disturbances caused by the edges of the slot, fine metal mesh was glued tightly onto the upper surface of all slots.

The free-stream velocity of $U_\infty$ = 11 m/s and the excitation frequency of $f = 154{,}1\,Hz$ were chosen according to the linear stability theory so that the initial TS wave would be excited some wavelengths upstream of the unstable region, but the excitation position of the anti-phase TS wave would be in the unstable region. It should be noted that the Reynolds number is made non-dimensional by the Blasius-length, because the flow is accelerated in the region of the leading edge [11] (see Fig. 2).

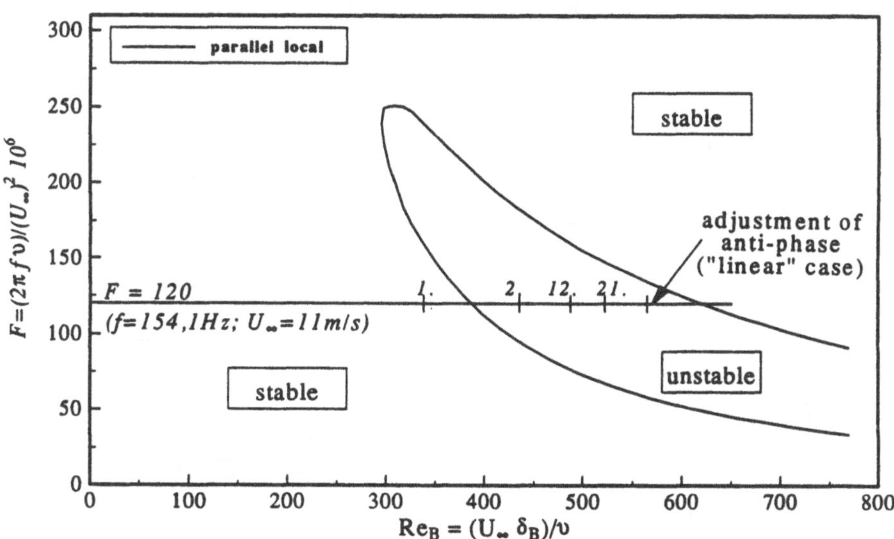


Figure 2 Neutral stability curve according to the linear stability theory with some experimental parameters.

## REALISATION OF THE ANTI-PHASE CONTROL

In this experimental investigation the anti-phase control was realised in the following way. The hot-wire was situated in the amplitude-maximum of the initial TS wave at a defined position (Reynolds number) sufficiently far downstream of the excitation positions for the initial and the anti-phase TS waves. The root-mean-square(RMS)-amplitude and phase of the initial TS wave at this position was measured and stored. Then the initial TS wave was switched off and the anti-phase signal was switched on. The amplitude of the anti-phase TS wave was adjusted so that it was approximately equal to the amplitude of the initial TS wave at the hot-wire position. Now, the initial TS wave was again excited while the anti-phase TS wave was still on. The excitation-phase of the loudspeaker for the anti-phase TS wave (and therefore the phase of the this TS wave) was adjusted by hand so that the amplitude of the resulting TS wave at the hot-wire position decreased. This procedure was continued until a minimum of the RMS-amplitude of the attenuated TS wave at this streamwise position, and at another streamwise position further downstream in the stable region according to the linear stability theory, was reached. All these adjustments of the anti-phase control were carried out with the hot-wire at the symmetry line of the flat plate at Z = 0.0 mm, although the effectiveness of anti-phase control was checked at different spanwise positions too.

## RESULTS

The anti-phase control is quite effective in the linear instability regime with amplitudes of the TS waves smaller than 1 % of the free-stream velocity. But the question remains, if the effectiveness of this method is independent of the Reynolds number of the excitation position for the anti-of-phase TS wave in the regime of linear instability development.

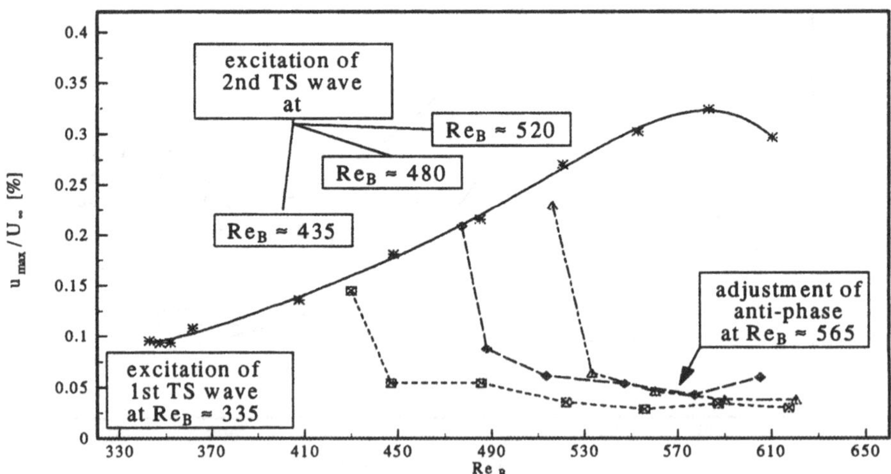


Figure 3 Anti-phase method in the linear instability regime with TS wave amplitudes smaller than 1 % of the free-stream velocity.

In Fig. 3, three cases are shown where the Reynolds number of the excitation position for the anti-phase TS wave was varied, while the excitation position of the initial TS wave remained unchanged. In all three cases the hot-wire was at the same streamwise position $Re_B \approx 565$ where the signal-to-noise ratio had the largest value. In all three cases the attenuated TS waves reached about the same amplitude.

Proportional increase of the amplitudes of the TS waves with increasing pressure level of the excitation was observed and amplitudes of the TS waves above 1 % of the free-stream velocity were obtained so that the development of the TS waves was no longer linear. Higher harmonics were found in the frequency spectrum although the corresponding excitation spectrum did not show significant higher harmonics.

Now, it is important to know at which Reynolds number the adjustment of the anti-phase control will be realised. If the anti-phase adjustment is performed too close to the excitation position of the anti-phase TS wave, the TS waves cannot interact well. On the other hand, if the anti-phase TS wave is adjusted at such Reynolds numbers where the boundary layer is nearly turbulent, a stable anti-phase adjustment is no longer possible.

With the linear local superposition of the initial and anti-phase TS waves it was possible to find a streamwise position where optimal anti-phase adjustment was achieved. The amplitude- and phase-profiles of the initial and anti-phase TS waves were measured. The RMS-amplitude of the computed local linear superposition of these TS waves $u_{12,simu}$ was compared with the measured RMS-amplitude $u_{12}$ of the attenuated TS wave. The computed amplitude is given by

$$\left| u_{12,simu}(Y) \right| = \left| u_1(Y)\; e^{i\varphi_1} + u_2(Y)\; e^{i\varphi_2} \right|$$

with $u_1$, as RMS-amplitude and $\varphi_1$ as averaged phase of the initial TS wave, and with index 2 marking the anti-phase TS wave. An optimal adjustment of the anti-phase control is achieved, if the amplitude profiles of the attenuated and the simulated TS wave are very similar up to the streamwise position where the anti-phase adjustment was performed, and if the maximum amplitude of the attenuated TS wave becomes minimal downstream of the upper branch.

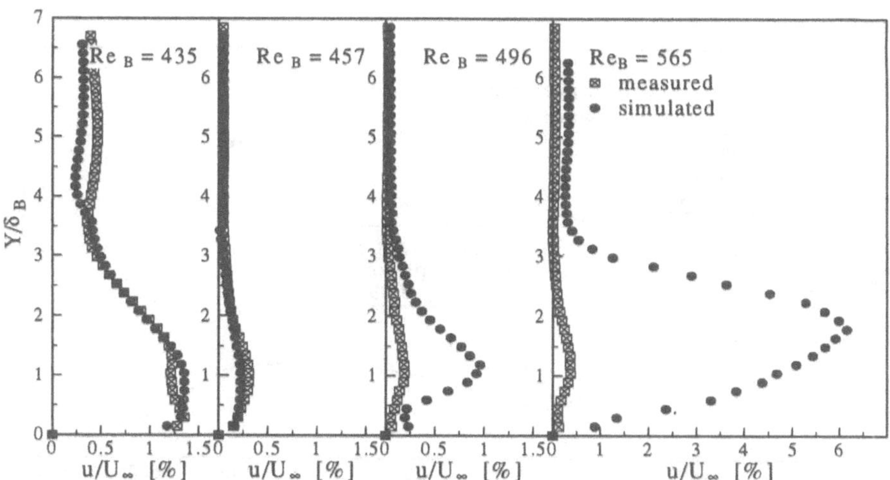


Figure 4 Comparison of the amplitude profiles of the measured (attenuated) TS wave with the TS wave computed by linear superposition.

One case in the regime of large TS wave amplitudes is illustrated in Fig. 4. The amplitude profiles of the measured attenuated TS wave and the computed TS wave are shown for different Reynolds numbers (streamwise positions). Up to a Reynolds number of $Re_B \approx 490$, the amplitude profiles are similar. The anti-phase adjustment was realised at this Reynolds number, and we got the development of the attenuated TS wave as seen in Fig. 5. By using this technique, the initial TS wave with a maximum amplitude of 1.7 % of the free-stream velocity is reduced by about 90 %. Furthermore, the attenuated TS wave behaves according to the linear stability theory. Downstream of the adjustment Reynolds number the attenuated TS wave undergoes a slow amplification (according to its instability in this regime). But downstream of the upper branch of the instability diagram the amplitude decreases again, see Fig. 5. In contrast, the artificially excited boundary layer without anti-phase control is already turbulent at this Reynolds number.

Another interesting result was found comparing the frequency spectra of the boundary layer – excited only by the initial or only by the anti-phase TS wave, or by superposition of both TS waves. Here, we want to discuss the frequency spectra at two streamwise positions – at a Reynolds number $Re_B \approx 565$ (Fig. 6, top) which corresponds to the upper branch of the stability diagram, and at $Re_B \approx 620$ (Fig. 6, bottom) further downstream.

With increasing amplitude of the TS wave higher harmonics appear, and non-linear interactions of these modes with inherent most powerful low-frequency disturbances take place [12]. The low-frequency component, called $f_{low}$, had a value of roughly 20 Hz. The resulting frequencies, e.g. $f_0 \pm f_{low}$, $f_1 \pm f_{low}$, $f_2 \pm f_{low}$ appear in the spectra of the initial and of the anti-phase TS wave, respectively. The low-frequency component of 20 Hz is not understood and more investigations are needed. The frequency spectra show that the turbulent energy in the boundary layer rises with increasing Reynolds number. The attenuated boundary layer shows a more or less linear behaviour. The amplitude of the harmonic decreases to an amplitude far smaller than 1 %, and all higher harmonics were reduced and had no significant amplitudes. Only low-frequency disturbances remained but decreased also further downstream of the adjustment-Reynolds number.

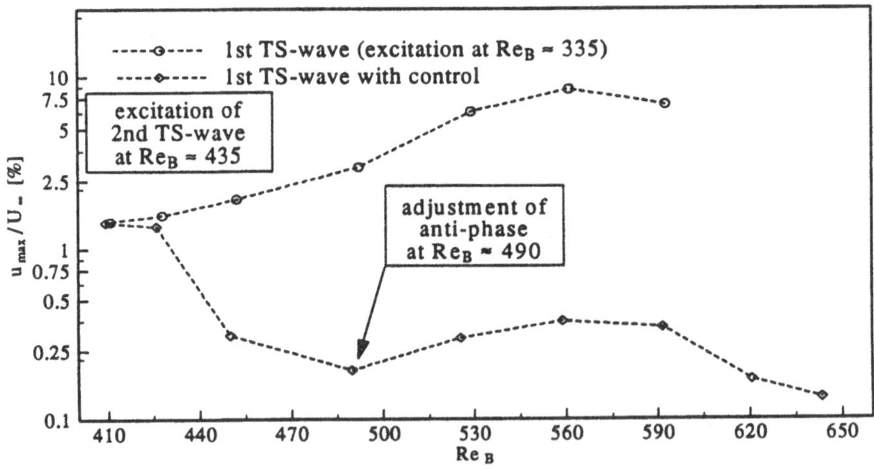


**Figure 5** Anti-phase method in the non-linear instability regime with TS wave amplitudes larger than 1 % of the free-stream velocity.

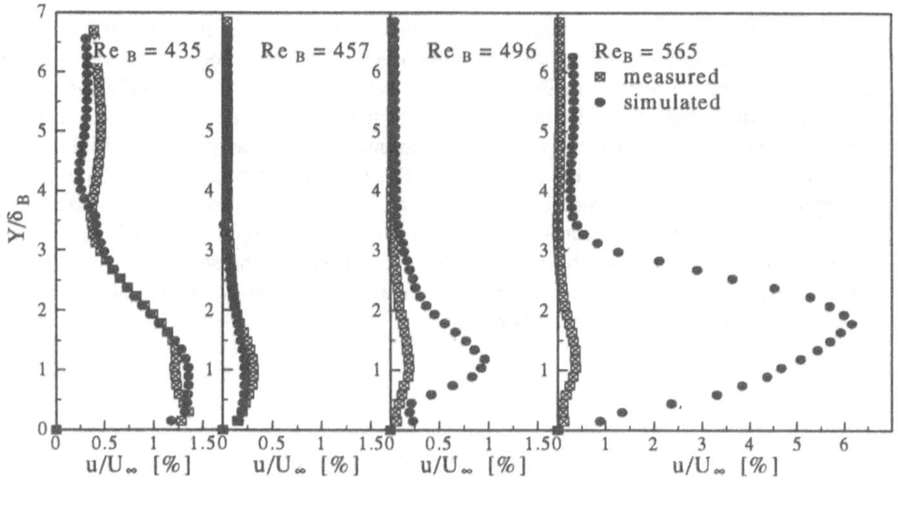

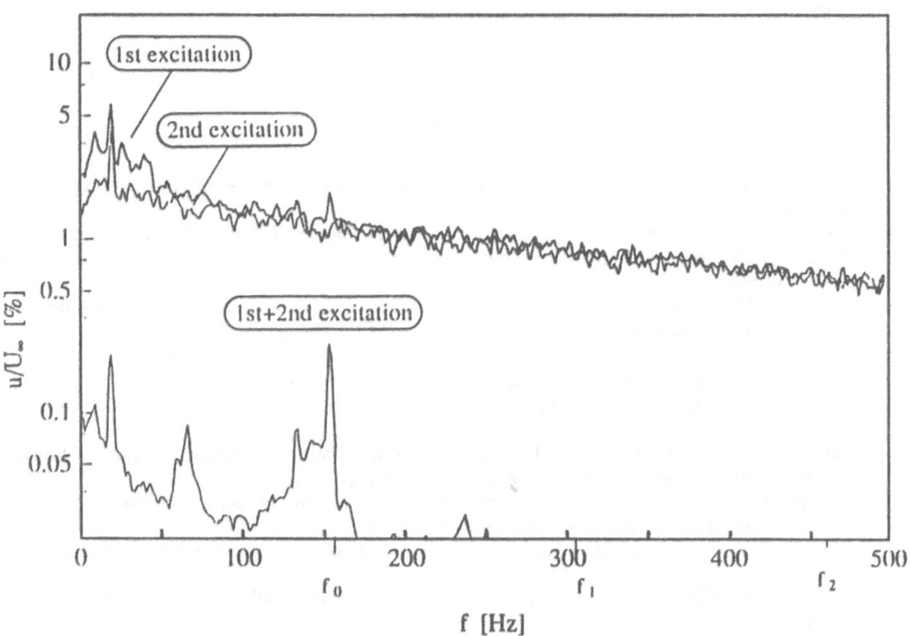


Figure 6 Frequency spectra of the boundary layer excited by the initial and by the second TS waves, and by both TS waves. Measured at a Reynolds number of $Re_B \approx 565$ (top) and $Re_B \approx 620$ (bottom).

## CONCLUSIONS

The anti-phase control technique is very useful for TS waves not only of small, but also large amplitudes up to 1,7 % of the free-stream velocity. In the linear instability regime with amplitudes of the TS waves smaller than 1 % it has been shown that the effectiveness of this method is independent of the Reynolds number of the excitation position of the anti-phase TS wave. Furthermore, it is possible to find an optimum anti-phase adjustment for TS waves with amplitudes larger than 1% of the free-stream velocity by comparing the computed linear local superposition of the oncoming and the controlling TS wave with the measured attenuated TS wave. Especially, an oncoming TS wave with a maximum-amplitude of 1.7 % of the free-stream velocity was reduced by about 90 %. Of course, the present computed linear local superposition can only be one first step to increase the efficiency of the anti-phase method, especially, in the non-linear instability regime.

## ACKNOWLEDGEMENTS

The authors would like to express their gratitude to Dipl.-Ing. H. Denecke for his support in designing many electronic components in the experimental set-up. We would also like to thank Mr. D. Baumgarten for his help in the technical set-up.

## REFERENCES

[1] LIEPMANN, H.W. and NOSENCHUCK, D.M.; Active control of laminar-turbulent transition. Journal of Fluid Mechanics, vol. 118, 1982, pp. 201-204.

[2] MILLING, R.W.; Tollmien-Schlichting wave cancellation. Physics of Fluids, vol. 24, 1981, pp. 979-981.

[3] THOMAS, A.S.W.; The control of boundary-layer transition using a wave-superposition principle. Journal of Fluid Mechanics, vol. 137, 1983, pp. 233-250.

[4] GILEV, V.M. and KOZLOV. V.V.; Use of small localized wall vibrations for control of transition in the boundary layer. Fluid Mechanics – Soviet Research, vol. 14, 1985, pp. 50-54.

[5] GILEV, V.M. and KOZLOV. V.V.; Effect of alternating injection and suction on transition in the boundary layer. Fluid Mechanics – Soviet Research, vol. 16, 1987, pp. 61-67.

[6] HERBERT, Th.; Secondary instability of boundary layers. Annual Review of Fluid Mechanics, vol. 20, 1988, pp. 487-521.

[7] KLEBANOV, P.S.; TIDSTROM, K.D.; SARGENT, L.M.; The three-dimensional nature of boundary-layer instability. Journal of Fluid Mechanics, vol. 12, 1962, pp. 1-34.

[8] LAURIEN, E. and KLEISER, L.; Numerical simulation of boundary-layer transition and transition control. Journal of Fluid Mechanics, vol. 199, 1989, pp. 403-440.

[9] TENG , Y.-G. and GROSCHE, F.R.; A preliminary study of control of boundary layer transition by active wall motion. DLR-IB 222-90 A 27, 1990.

[10] ZHOU, A. and GROSCHE, F.-R.; Experimental study on active control of boundary layer by a multiple slot blowing/suction system. DLR-IB 223-94 A 33, 1994.

[11] SARIC, W. S.; Low-speed experiments: requirements for stability and measurements. In Hussaini, M.Y. and Voigt, R.G.: Instability and transition. Springer-Verlag, N.Y., 1990.

[12] NISHIOKA, M.; Some fundamental problems on transition to wall turbulence. IUTAM Symposium, Sendai/Japan, pp. 15-26, 1994.

# Simulation of Unsteady, Three-Dimensional, Viscous Flows Using a Dual-Time Stepping Method

Ralf Heinrich, Hans Bleecke
DLR, Institute of Design Aerodynamics, Lilienthalplatz 7
D-38108 Braunschweig, Germany

## Summary

A time-accurate flow solver for the three-dimensional Reynolds-averaged Navier-Stokes equations has been developed. An implicit time discretization is used, and the resulting set of coupled non-linear equations is solved iteratively. This is accomplished using well proven convergence acceleration techniques for explicit schemes such as multigrid, residual averaging, and local time stepping, in order to achieve large computational efficiency in the calculation. The approach, known as dual-time stepping in literature, allows the physical time step to be chosen on the basis of accuracy rather than stability. The method is parallelized using the domain decomposition technique. Results are presented for three-dimensional transonic flows around the oscillating LANN-wing. The large speed-up made possible through dual-time stepping and parallelization is pointed out. Using the Baldwin-Lomax turbulence model, good agreement of the pressure coefficient with experimental data is shown for attached flows. The shock induced separation for the LANN-CT9 test case is predicted too far downstream.

## Introduction

The prediction of unsteady airloads on airfoils and wings plays an increasing role in the aircraft design process. So it is essential to develop numerical schemes which provide time accurate solutions at reasonable costs.

The algorithms for computing unsteady flows are of two main types, explicit and implicit. Explicit methods are subject to severe stability restrictions. This often necessitates the use of allowable time steps, which are much smaller than those required to obtain accuracy, hence leading to a requirement for a large number of time steps. In addition, many of the acceleration techniques used for steady flows cannot be used, because they destroy time accuracy.

An implicit discretization of the problem helps to bypass the time step limitation. Jameson [4] has developed a very efficient multigrid-driven implicit approach for the solution of the unsteady Euler equations. Using central differences in space and an implicit multistep discretization in time, a large set of non-linear equations is formed and marched to steady-state in *pseudo* time through a multigrid algorithm within each physical time step. This approach has been recently applied to the solution of the unsteady Euler and Navier-Stokes equations with great promise, see for example [5], [6], [8].

Following Jameson [4], the DLR multi-block code FLOWer [1] has been extended to the solution of unsteady viscous flows.

Another powerful tool to reduce the computational costs is parallelization. In the FLOWer code it is realized through domain decomposition. Every processor of a computer or workstation cluster works on one or more blocks of the whole block structure. The communication between them is done by a special high level communication library based on message passing [7].

## Governing equations

The integral form of the three-dimensional Reynolds-averaged Navier-Stokes equations in a moving Cartesian coordinate system can be written as

$$\frac{d}{dt}\int_{V(t)} \vec{W}\,dV + \int_{\partial V(t)} (\bar{\bar{F}}^c - \bar{\bar{F}}^v)\cdot\vec{n}\,dS + \int_{V(t)} \vec{G}\,dV = 0 \qquad (1)$$

with

$$\vec{W} = \begin{bmatrix}\rho\\ \rho\vec{q}\\ \rho E\end{bmatrix},\quad \bar{\bar{F}}^c = \begin{bmatrix}\rho(\vec{q}-\vec{q}_b)\\ \rho\vec{q}\bullet(\vec{q}-\vec{q}_b)+p\bar{\bar{I}}\\ \rho E(\vec{q}-\vec{q}_b)+p\vec{q}\end{bmatrix},\quad \bar{\bar{F}}^v = \begin{bmatrix}0\\ \bar{\bar{\sigma}}\\ \vec{q}\bar{\bar{\sigma}}-k\vec{\nabla}T\end{bmatrix},\quad \vec{G} = \begin{bmatrix}0\\ \rho(\vec{\omega}\times\vec{q})\\ 0\end{bmatrix}.$$

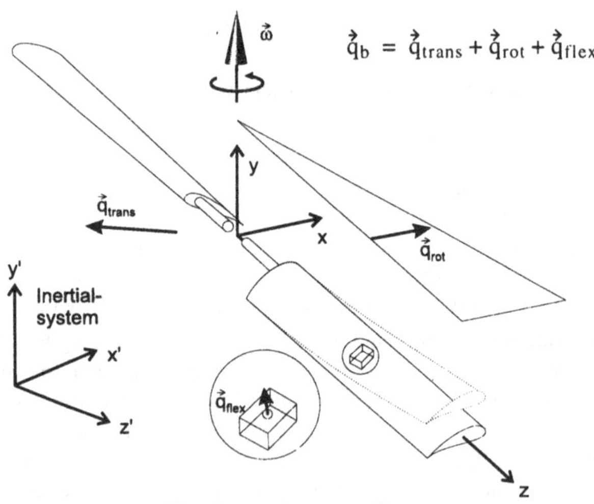


**Fig. 1** moving coordinate system

$\vec{W}$ is the vector of conserved quantities with $\rho$, $\vec{q}$ and E denoting the density, velocity and specific total energy, respectively. $\vec{q}$ is the absolute velocity vector transformed to the moving coordinate system and not the relative speed. V denotes an arbitrary deforming control volume with boundary $\partial V$ and the outer normal $\vec{n}$. Note, that the total derivative with respect to time in (1) is not the substantial time derivative. A source term $\vec{G}$ has to be added, including the time derivatives of the unit base vectors of the coordinate system rotating with the angular velocity $\vec{\omega}$. The gas is assumed to behave like a calorically perfect gas with constant specific heats.

The fluxes $\bar{\bar{F}}$ may be divided into its inviscid (convective) part $\bar{\bar{F}}^c$ and its viscous part $\bar{\bar{F}}^v$. For the calculation of the convective fluxes the boundary velocity $\vec{q}_b$ has to be introduced. It contains three portions induced through translation ($\vec{q}_{trans}$) and rotation ($\vec{q}_{rot}$) of the coordinate system and the deforming of the mesh ($\vec{q}_{flex}$), so that

$$\vec{q}_b = \vec{q}_{trans} + \vec{q}_{rot} + \vec{q}_{flex}\,. \tag{2}$$

## Description of the Algorithm

The approximation of the governing equations follows the method of lines, which decouples the discretization of space and time. The spatial discretization is based on a finite volume method which subdivides the flow field into a set of non-overlapping hexahedral cells. In FLOWer [1] the cell-vertex approach is realized, in which the flow variables are associated with the cell vertices of the cell. The spatial discretization leads to an ordinary differential equation for the rate of change of the conservative flow variables in each grid point

$$\frac{d}{dt}(\vec{W}_{ijk}V_{ijk}) + \vec{R}(\vec{W}_{ijk}) = 0 \tag{3}$$

with the residual

$$\vec{R}_{ijk} = \vec{R}^C_{ijk} + \vec{R}^V_{ijk} + \vec{R}^S_{ijk} + \vec{R}^D_{ijk} \tag{4}$$

where $\vec{R}^C_{ijk}$, $\vec{R}^V_{ijk}$ and $\vec{R}^S_{ijk}$ represent the approximation of the inviscid and viscous net fluxes and the source term for a particular control volume arrangement with volume $V_{ijk}$ surrounding the grid node (i,j,k). The fluxes are approximated using a central discretization operator. Therefore dissipative terms have to be explicitly introduced, to damp high frequency oscillations in the solution. To avoid these spurious oscillations, in FLOWer the well known scalar dissipation model of Jameson [3] is implemented. It uses a blend of second and fourth differences of the flow variables. In order to preserve the conservation form of the numerical scheme, the artificial dissipative terms are introduced by adding dissipative fluxes ($\vec{R}^D_{ijk}$) to the semi-discrete system.

If we discretize the time derivative term with a backwards difference second order accurate operator, we obtain, dropping the i,j,k subscripts and the vector signs for clarity (in FLOWer first, second and third order operators can be chosen)

$$\frac{3W^{n+1}V^{n+1}}{2\Delta t} - \frac{2W^nV^n}{\Delta t} + \frac{W^{n-1}V^{n-1}}{2\Delta t} + R(W^{n+1}) = 0\,. \tag{5}$$

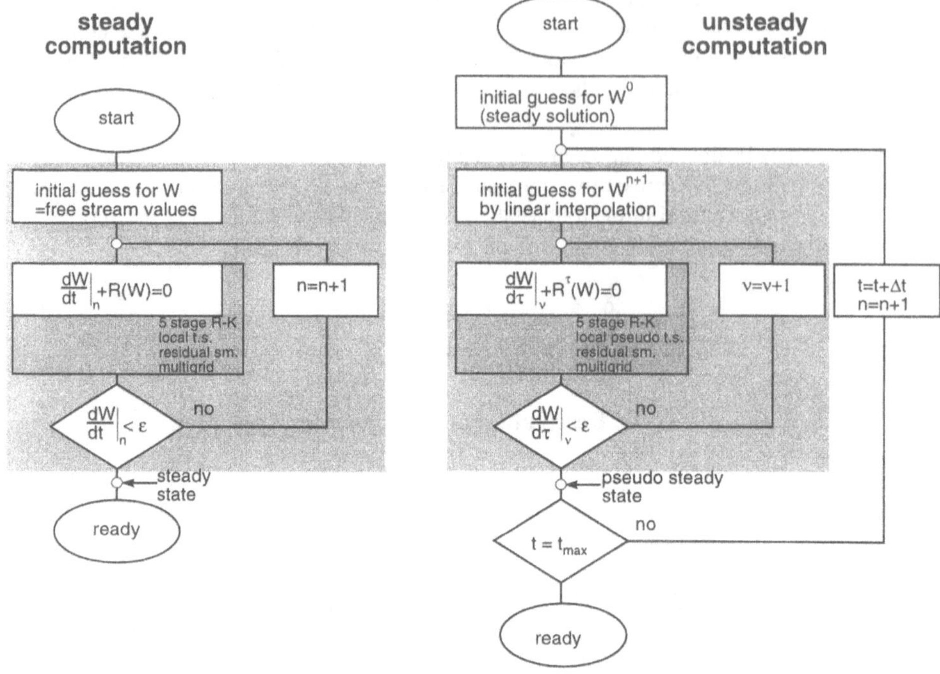


**Fig. 2** Flowchart for steady and unsteady calculations

This equation for the unknown $W^{n+1}$ is non-linear due to the presence of the term $R(W^{n+1})$ and cannot be solved directly. One must therefore resort to iterative methods. The solution of equation (5) at each physical time step can then be seen as a modified pseudo time steady-state problem with a slightly altered residual [4]

$$R^\tau(W) = \frac{3WV^{n+1}}{2\Delta t} - \frac{2W^nV^n}{\Delta t} + \frac{W^{n-1}V^{n-1}}{2\Delta t} + R(W)\ . \tag{6}$$

In this case, the vector of flow variables W which satisfies the equation $R^\tau(W) = 0$ is the $W^{n+1}$ vector we are looking for. In order to obtain the solution vector, we can reformulate the problem at each time step as the following modified steady-state problem in a fictitious pseudo time $\tau$

$$\frac{d}{d\tau}W + R^\tau(W) = 0\ , \tag{7}$$

to which one can apply the acceleration techniques used for steady-state calculations. Applying this process repeatedly, one can advance the flow field solution forward in time in a very efficient fashion. For each physical time step the system of equations (7) is integrated in pseudo time using an explicit five-stage Runge-Kutta method. The acceleration techniques used to march to a pseudo-steady state are local pseudo time stepping, implicit residual smoothing and multigrid. To provide a good initialisation of the solution at the new physical time step n+1, W is estimated by linear extrapolation with the known values [5][6]. In fig. 2 the solution process for steady and unsteady computations is compared. The *kernel* algorithm almost remains unchanged. Only a new loop has to be constructed around the *kernel* to progress in the physical time direction.

## Results

In this paper results are presented for three-dimensional periodic flows over the oscillating LANN-wing. The harmonic angle of attack oscillation is described by $\alpha = \alpha_m + \alpha_0 \sin(\Omega t)$ .

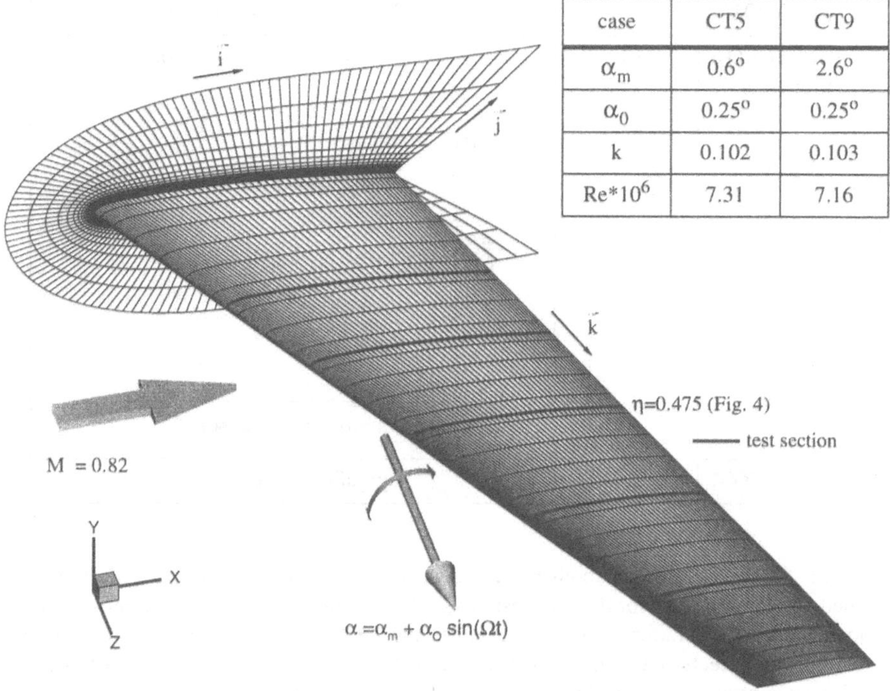

| case | CT5 | CT9 |
|---|---|---|
| $\alpha_m$ | 0.6° | 2.6° |
| $\alpha_0$ | 0.25° | 0.25° |
| k | 0.102 | 0.103 |
| Re*10$^6$ | 7.31 | 7.16 |



**Fig. 3** Surface grid of the LANN-wing

The angular velocity $\Omega$ is given through the reduced frequency $k = \Omega c/(2q_\infty)$ with the root chord length *c* and the free stream velocity $q_\infty$. No elasticity effects are taken into account. So the movement of the wing and the mesh is described by a rigid body movement. For each case the steady solution for $\alpha = \alpha_m$ was calculated first and used as a starting solution for the time-accurate calculations. The time-accurate computations were stopped, when two successive periods showed the same pitching moment as a function of $\alpha$. This is typically the case after three periods. The surface of the LANN-wing is defined by a linear interpolation between a first supercritical airfoil for the root and a second for the tip section (Zwaan, 1982 [9]). The model has a span of one meter, a quarter-chord sweep angle of 25°, an aspect ratio of 7.92 and a taper ratio of 0.4. The airfoil thickness is about 12%. The model was excited in a pitch motion around an unswept axis at 62.1% of root chord. Experimental results for the 0-th and the 1-st harmonics are given in 6 test sections. Two test-cases, CT5 and CT9, were selected for the comparison of experimental data with numerical results.

**Table 1:** computational time for one period (CT9, Navier-Stokes, 420000 cells, 50 physical time steps per period)

| computer | time[h] | comments |
|---|---|---|
| CRAY-J916 | 8.5 | 200 megaflops performance |
| IBM-SP2 (16 processors) | 3.13 | speed-up through parallelization 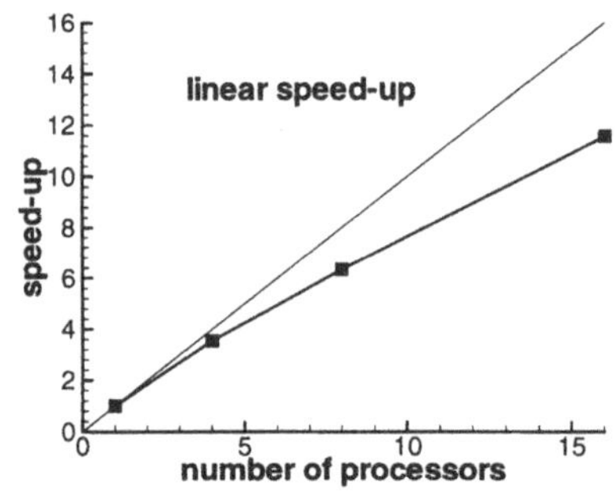<br> |
| NEC-SX4 | 1.7 | 1 gigaflop performance |

The main difference between both test-cases is the mean angle of attack. The flowfield is attached all over the wing for the CT5 test-case, whereas the CT9 case is characterized through a shock induced separation. For Euler calculations a CH grid was generated with the C in wraparound (i-)direction and the H in spanwise (k-)direction, containing 160x32x40 cells [2]. For viscous computations a grid was constructed on the basis of the coarse CH-grid. 32 points were added in the wall normal direction, with a minimum non-dimensional distance to the wall of $2 \times 10^{-5}$ chord length.

To test the feasibility of parallelization for unsteady flows, the Navier-Stokes grid was subdivided in 1, 4, 8 and 16 blocks of equal size. The grids containing 1, 4 and 8 blocks were splitted in i-direction only, whereas the 16 block grid was split additionally in spanwise k-direction.

For both test cases three oscillations of the LANN-wing were performed with 50 physical time steps per cycle (Euler and Navier-Stokes). In table 1 the wall clock times for the third oscillation cycle for the test-case CT9 (Navier-Stokes) are plotted for different architectures. A Cray J-916 needs 8.5 hours, a NEC-SX4 1.7 hours for one cycle. Using 16 processors of an IBM-SP2 computer the time is reduced to 3.3 hours. The figure in the table shows the resulting speed-up over the number of processors. If 16 processors are used, a speed-up of 11.2 is realized. Increasing the number of blocks the parallelization efficiency decreases. For moderate processor numbers, as shown in [7], it follows that the deviation from a linear speed-up is largely due to the double flux computations at the block interfaces and not to the limitations of the communication system. For a NEC-SX4 with 16 processors the computational time maybe estimated as reducible to about 10 minutes.

Very impressive is 'the speed-up of the dual-time stepping method compared to the basic explicit global time stepping scheme. The maximum dimensionless time step due to stability reasons is about $9.8 \times 10^{-6}$. So more than 3 million iterations per cycle have to be carried out. With 15 seconds for a time step on a CRAY-J916 computer, one cycle *costs* more than 500 days. With about 40 inner iterations and 50 physical time steps, only about 2000 multigrid cycles have to be carried out for the dual time stepping method. Only about half a day is needed for one cycle, resulting in a speed-up of more than 1000.

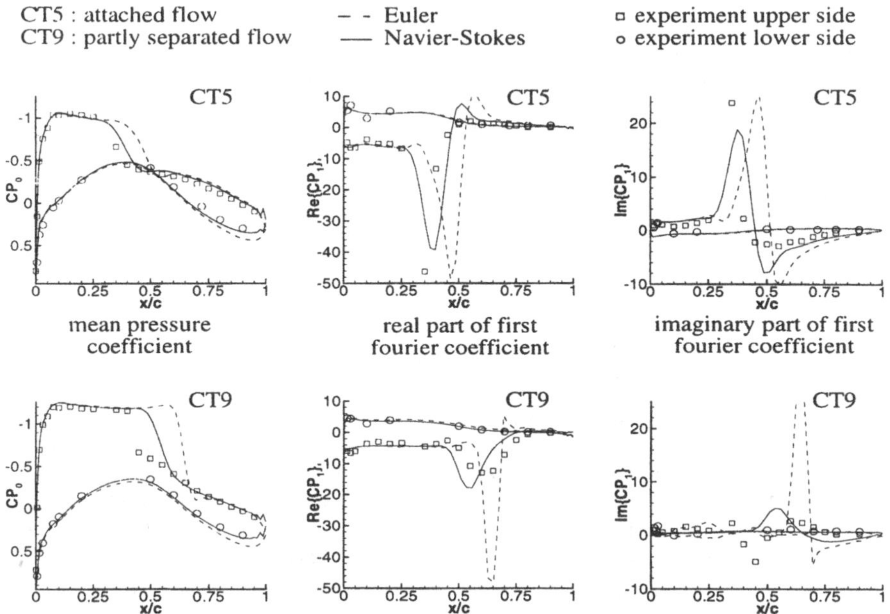


**Fig. 4** Mean pressure distribution and first Fourier components for the CT5 and the CT9 test cases

The results of the Fourier analysed pressure distributions are shown in Fig. 4 for one typical section ($\eta$=0.475) in comparison to experimental data. For both test cases, the shock position is predicted too far downstream for the Euler calculations. The agreement for all Fourier components of the pressure becomes much better applying the Navier-Stokes equations to the CT5 test case, where the flow is attached all over the wing surface. The difference between Euler and Navier-Stokes calculations for the lift is about 16%, which shows the importance of taking viscous effects into account. In the case of separated flow, the shock position is predicted too far downstream. Compared to the Euler computations, the results are improved, but there are still remarkable differences between simulation and experiment. The use of a k-ω turbulence model permits a better description of the flow phenomena in the regions of separated flow. So our next step will be the implementation of the k-ω model for unsteady computations.

## Conclusions

A solver for the unsteady Navier-Stokes equations has been presented. The comparison of experimental data for the transonic flow about the oscillating LANN-wing shows good agreement of the results achieved by viscous calculations using the Baldwin-Lomax turbulence model for attached flows. For partly separated flows, Navier-Stokes calculations improve the results compared to Euler computations. But there are still remarkable differences in comparison to the experiment. Therefore we plan to investigate the influence of an improved turbulence model in future.

The speed-up due to dual-time stepping versus explicit methods is about three orders of magnitude for viscous calculations. An additional speed-up is realized through parallelization of the code. Using 16 processors a speed-up of more than one order of magnitude is obtained. The combination of the implicit dual-time stepping method and the parallelization has been shown to be a very promising tool for predicting unsteady airloads.

This work was supported by DASA-DA (Bremen).

The authors thank R. Voß of the DLR-Institute of Aeroelasticity for providing the 3D Euler grid.

## Bibliography

[1] Kroll, N.; Radespiel, R.; Rossow, C-C. *Accurate and Efficient Flow Solvers for 3D Applications on Structured Meshes.* Lecture Series 1994-05 of the von Karman Institute for Fluid Dynamics, March 1994.

[2] Müller, U. R.; Schulze, B.; Voß, R. *Computation of Transonic Steady and Unsteady Flow about the LANN Wing.* in 'Notes on Numerical Fluid Mechanics'. Vieweg Verlag Braunschweig, to appear 1996.

[3] Jameson, A.; Schmidt, W.; Turkel, E. *Numerical Solutions of the Euler Equations by Finite Volume Methods Using Runge-Kutta Time Stepping Schemes.* AIAA-Paper 81-1259, 1981.

[4] Jameson, A. J. *Time dependent calculation using multigrid, with applications to unsteady flows past airfoils and wings.* AIAA-Paper 91-1596, 1991.

[5] Yang, Y.; Pahlke, K. *Implementation of a Dual-Time Stepping Method for the Numerical Solution of N-S Equations for Rigid Airfoils in Arbitrary Unsteady Motion.* DLR, Institut für Entwurfsaerodynamik, IB 129-95/19.

[6] Arnone, A. *Multigrid time accurate integration of Navier-Stokes equations.* AIAA-93-3361-CP, 1993.

[7] Eisfeld, B.; Bleecke, H.-M.; Kroll, N.; Ritzdorf, H. *Parallelization ob Block Structured Flow Solvers.* AGARD R-807, pp 5.1-5.20, 1995.

[8] Gaitonde, A. L. *A Dual-Time Stepping Method for the Solution of the Unsteady Euler Equations.* Aeronautical J., October 1994, 98, pp 283-291.

[9] Zwaan, I. R. J. *LANN-Wing Pitching Oscillations.* Compendium of Unsteady Aerodynamic Measurements, Addendum No. 1, AGARD R-702, 1982.

# Unsteady boundary layer separation on swept and unswept wings in sinusoidal dynamic stall motion

J. Henkner, H. Ranke*

Lehrstuhl für Fluidmechanik, Technische Universität München, Boltzmannstr. 15, 85747 Garching, Germany

## Summary

Unsteady or turbulent flowfields require detailed measurements to improve the basic knowledge about essential fluid mechanics phenomena e.g. about unsteady boundary layer separation which are not fully understood until now. Flowfield measurements on sinusoidal pitching swept and unswept wings contribute to a better understanding of the unsteady boundary layer separation at fast changing angles of attack called dynamic stall.

In order to fulfill this task two new types of hotwire probes were built. The operationality of the new probes is proved by a good agreement of flowfield visualisation and hotwire measurements as well with similar experiments [1, 2, 3] and boundary layer computation [4, 5].

The experimental data gives a step by step insight to the processes leading to the dynamic stall and the reattachment.

## Nomenclature

| | |
|---|---|
| $b$ | wing span [m] |
| $c$ | airfoil chord [m] |
| $E_i$, | anemometer output voltage of hot-wire number i [V] |
| $f$ | frequency [Hz] |
| $k$ | reduced frequency, $2\pi fc/U_\infty$ |
| $Re$ | Reynolds number, $U_\infty c/\nu$ |
| $t$ | time [s] |
| $T$ | time of one oscillatory cycle [s] |
| $U$ | streamwise calibration velocity [m/s] |
| $U_c$ | total calibration velocity [m/s] |
| $U_\infty$ | freestream velocity [m/s] |
| $V$ | lateral calibration velocity [m/s] |
| $u, v, w$ | streamwise, lateral and vertical velocity components [m/s] (wind tunnel–axis system) |
| $x, y, z$ | streamwise, lateral and vertical coordinates of the wind tunnel–axis system or model coordinate system measured from the nose [m] |
| $\Delta x$,$\Delta y$ | measurement grid distance in streamwise and vertical direction |
| $\alpha$ | angle of attack [deg] |
| $\beta$ | yaw-angle [deg] |
| $\alpha_c$ | calibration pitch angle [deg] |
| $\beta_c$ | calibration yaw angle [deg] |
| $\nu$ | kinematic viscosity [$m^2$/s] |

*now: Process Engineering and Contracting Division, Linde AG, Dr.-Carl-von-Linde-Str. 6-14, 82089 Höllriegelskreuth, Germany

# 1 Introduction

It is well known that under unsteady flow conditions the boundary layer velocity profiles, in which flow reversal can occur, and the separation process is different from that under steady conditions [6].

The experimental analysis of the unsteady flowfields has always been a difficult task. Until the late 70's contemporary use of flow visualisation by smoke flow and yarn tufts as well as boundary layer mesurements by hotwire-anemometry provided to be a good but difficult way to properly analyze unsteady flowfields [7].

New measurement techniques such as particle image velocimetry have some advantages above hot-wire anemometry i.e in dectecting flow reversal regions with the help of velocity shifting [8] but as well some disadvantages such as i.e. the much more complex and expensive instrumentation and the disability to measure the flow velocities in a frame fixed to the oscillating wing. Therefore near wall data cannot be obtained with such techniques.

Neverthess traditional anemometers are not able to distinguish directly between i.e. forward or reverse flow because the responses for both directions are the same. Therefore, at the Lehrstuhl für Fluidmechanik of the Technischen Universität München a hot-wire system for two-dimensional flow has been developed which circumvent the problem of equivocation [9, 5]. For the use in three-dimensional flows a new hot-wire probe will be presented which produces a unique output signal for all possible flowdirections.

The results of this investigation show apart from other observations that the onset of the dynamic stall vortex and the reattachment are boundary layer driven effects. This agrees well with the proposed description of the phenomenon of unsteady separation [5]. The experimental results obtained from a NACA 0012 airfoil in dynamic stall motion show a good agreement with similiar PIV experiments from Raffel [8] and surface pressure data from Galbraith et al. [3].

# 2 Hot-wire Experimental Techniques and Test Programs

## 2.1 Facility

All experiments were carried out in the low-speed wind tunnel of the Technische Universität München. The nozzle diameter of this closed-return facility is $1.5\,m$. Maximum usable velocity is $55\,m/s$. Turbulence intensity is known to be under 0.3% - 0.4% over the speed range of interest.

The tunnel is equipped with an automated data acquisiton and control system. A personal computer with a high-speed A/D-board is used for data acquisition of the hot-wire, pressure, temperature and further additional signals as well as for the control of the wind tunnel and the 3-axis model support. A second computer system is required for data storing, reduction and processing by statistical means of the huge amount of data during instantaneous measurement cycles. This is done by online data transfer and communication with several work stations which realizes then a completely automatic process of flowfield measurements.

## 2.2 Hot-wire Anemometry

The probe used for the two-dimensional dynamic stall study is a cross-wire probe with two additional wires to measure both streamwise and reverse flow, Fig 1, developed by H. Ranke et al [10, 11, 12]. The probe axis is mounted normal to the mean flow direction and has, therefore, a very thin probe body and strong outward curved prongs. The measuring volume is about $2\,mm$ in diameter. One of the two additional wires is operated as a third hot-wire (direction-wire). Very close to it, a strongly bended so called heating-wire is mounted which causes due to the forced convection a significant change in the output signal of the direction-wire to decide whether streamwise or reverse flow is present. The probe operates at all calibration velocities ($-40\,m/s < U_c < 40\,m/s$) in a pitch angle range from $\alpha_c = -40\,deg$ to $\alpha_c = +40\,deg$.

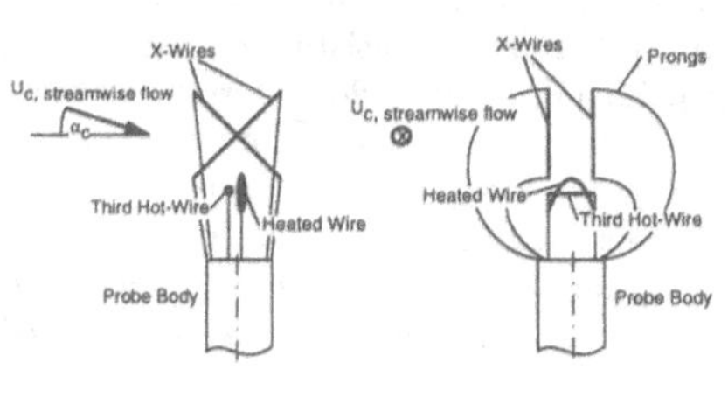


Figure 1: Probe for two-dimensional flow

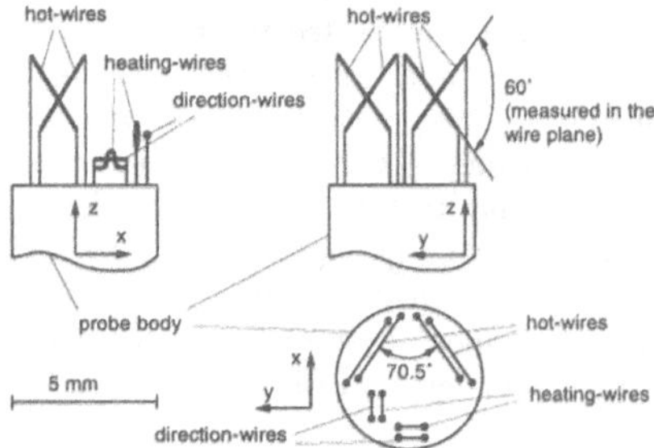


Figure 2: Probe for three-dimensional flow

The eight-wire probe, Fig. 2 consists of a quadruple-wire probe with two perpendicular pairs of additional wires. The use of a quadruple-wire probe allows with respect to a triple-wire probe mesurements within an enlarged range of pitch- and yaw- angles but due to probe symmetry an equivocation between the four quadrants of the yaw-angle remain. This equivocation is removed by the additional wires in which one of the two in each pair is operated as a hot-wire (direction-wire). A unique output signal therefore can be archieved at pitch angles in the range from $\alpha_c = -40\,deg$ to $\alpha_c = +40\,deg$ and yaw angles from $\beta_c = -180\,deg$ to $\beta_c = 180\,deg$ within the velocity range from $2\,m/s$ up to $40\,m/s$. Even with the need to place 16 prongs on the probe the measuring volume stays within a cube of $5\,mm$ sidelength.

The self-manufactured sensors consist of $5-\mu m$ platinum-plated tungsten wires giving a length to diameter ratio of 250. An additional temperature probe is used to correct the anemometer output-voltages if ambient flow temperature varies. The hot-wire probes are operated by a AA LAB eight-channel constant-temperature anemometer system. The total sampling frequency is limited to 160 kHz by a eight channel 16-bit A/D-board.

## 2.3 Calibration and Evaluation Procedure

To calibrate the hot-wire probes a computer-aided fully automated procedure is developed based on a velocity and flow-angle dependent, temperature corrected method [13]. A cross-wire calibration surface is achieved by pitching the probe in the freestream of the flow at different velocities. The pitch angle range was ±45 deg, the velocity magnitude ranges from $2m/s$ to $40m/s$. In order to measure the considered velocity magnitudes below $2m/s$ with high accuracy it is necessary to calibrate the probe in addition to the described calibration method with an in situ swinging arm calibration method, similar to that of Zabat et al. [14]. For each phase angle of the swinging arm the probe velocity can be calculated, see Ref. [5, 12]. For each pitch angle all velocities between −2 m/s and +2 m/s are calibrated with only one record of a timeseries. Thus a streamwise and a reverse flow calibration grid, each composed of a wind tunnel and an in situ swinging arm calibration grid, are obtained, Fig 3a,b. In order to select the streamwise or reverse flow correspondig look-up table the two additional wires produce a decision signal quantified as a difference between the output voltages, Fig 3c. The resolution of each grid is very fine in the low velocity range (in situ swinging arm calibration) and more coarse for the higher velocities (wind tunnel calibration). In order to select the streamwise or reverse flow corresponding look-up table the two additional wires produce a decision signal.

The calibration of eight-wire probes requires additional yawing of the probe. The probe was calibrated in a pitch angle range from $-45\,deg$ to $+45\,deg$ and yaw angle range from −180 deg to +180 deg at a velocity range from $2m/s$ to $40m/s$. This is an execptionally difficult task, as even a coarse calibration grid with a stepsize of 5 *deg* in pitch- and yaw-angle for eight different velocity magnitudes consists of approx 10,000 calibration points. This calibration procedure takes at the moment about four days alone for the data acquisition. A further reduction, not yet implemented, is expected to bring the total calibration time down to about two days.

The look-up tables for the 8-wire probe, a total of 4 tables, each consisting of 4 subtables, cannnot be easily be presented as every measurement point (V, $\alpha$, $\beta$) is characterized by three planes intersecting in one point, but the overall scheme remains the same as for the crosswire probe. According to the two direction signals the appropriate look-up table can be found. In each table representing a quadrant, up to four subtables may contain a valid result for the flow direction and magnitude which can only differ by the accumulated error by measurement and calibration. If more than one subtable with a valid entry is found the result is interpolated between all valid entry of the subtables.

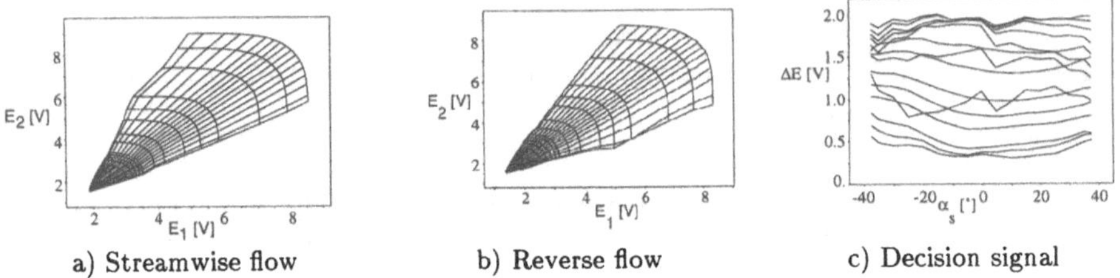

a) Streamwise flow b) Reverse flow c) Decision signal

Figure 3: Look-up tables for the crosswire probe

Using these calibration techniques actual experimental flow conditions, i.e. varying flow angle and velocity magnitude are considered during the calibration procedure. Consequently the interference between prongs and wires and the wires themselves is also covered as well as manufacturing inaccuracy and divergence from the ideal probe geometry. For this direct calibration method there is no need of simplifying assumptions concerning the sensor characteristics or about physical laws governing the sensor cooling and no yaw or pitch angle factors have to be introduced.

The calibration grid itself is too rough to determine the magnitudes and the associated directions of the measured velocity vector exactly. Therefore, the calibration grid is processed to a highly refined grid called look-up table. For the generation of look-up tables a by Lueptow et al. [15] proposed method is being developed for the eight-wire probe, using weighted derivatives of first and/or second order of a two-dimensional Taylor serial expansion, bicubic spline interpolation, or polynomial approximation.

## 2.4 Wind-tunnel models

In the dynamic stall tests a carbon fibre wing with a NACA 0012 airfoil, Fig. 4, is used to investigate unsteady separation. The model has a wing span of $b = 1\,m$ between the endplates and a chord of $0.3\,m$. In order to enable a sinusoidal dynamic stall motion an eccentric disk drives a parallel guided rod. The maximum frequency is 6 Hz. Amplitudes up to 15 deg with a mean angle of attack of $-10$ to $10\,deg$ pitching around the quarter-chord axis are possible. Large end plates simulate a large aspect ratio for quasi two-dimensional flow in the mid section. The actual position of the wing is determined using a potentiometer driven by a gearrack on the rod.

## 2.5 Description of the Tests

For the dynamic stall experiments a mean angle of attack of 10 deg and an amplitude of $\pm 10$ deg is chosen. The pitching frequency of the airfoil is $2\,Hz$ at a freestream velocity of $12.5\,m/s$

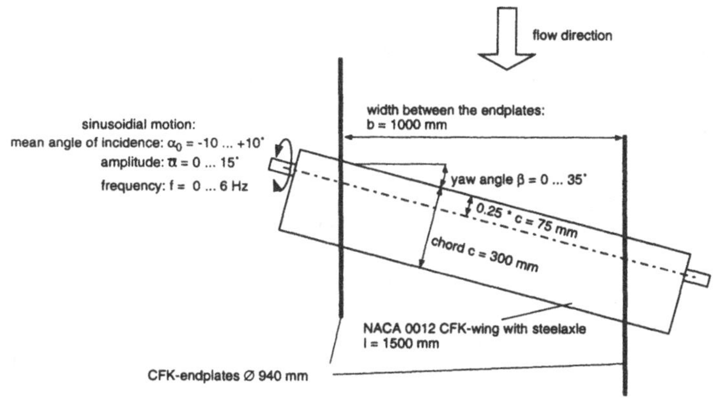


Figure 4: Schematic view of the testbed for the dynamic stall experiments

resulting in a reduced frequency of $k = 0.3$. The Reynolds number is $2.5 \cdot 10^5$ based on the wing chord. For the two-dimensional case ($\beta = 0\,deg$) the measurement plane contains 15 points in the chord direction and 12 points vertically giving 180 measurement points for the coarse grid respectivly 21 by 12 points for a finer grid. The sampling frequency for each channel is set to 600 Hz. At each grid location 100 amplitudes of the phase angle are measured. This leads to a sampling time of 50.2 s and for 6 channels (4 wires, trigger and temperature) to 180000 samples at each measurement point. The velocity data is aquired in a frame fixed to the oscilating wing and the results are therefore expressed as velocities relative to the wing surface. In Ref. [5] a more detailed description of the dynamic stall investigations results is presented. The smoke visualisation for swept wing is carried out with a yaw angle of $\beta = 15\,deg$ and the same frequency and freestream parameters as for the two-dimensional case.

# 3 Results

Figure 6 presents mean velocity vectors of the unsteady flowfield around the unswept wing for the phase angles $t/T = 30, 37, 43, 49, 54$ and 60%. The phase angle is defined by the actual time $t$ beginning with the minimum angle of attack divided by the time for one oscillating periode $T$. The NACA 0012 profile is mapped on the lower focal plane in the pictures. A region with reverse flow components is marching upstream at increasing phase angle starting at the trailing edge at a phase angle of about 21%. This region does not show the behavior of a separation, the boundary layer thickness does not increase very much and the lift rise remains high.

Reaching the leading part of the profile the upstream marching of the reverse flow region stops and therefore a leading edge vortex, the dynamic stall vortex, occurs approximately at a phase angle of 40%. This vortex accelerates the external flow, the flow in the rear part of the profile reattaches, Fig. 6c, (fully reattached at a phase angle of about 45%) and the lift increases further. The velocity vectors show the reattachment of the flow downstream of the vortex, Fig. 6c,d. As the vortex core grows with increasing phase angle it looses energy and moves slowly rear- and outward and then at a phase angle higher then 50% it suddenly increases very much and bursts, Fig. 6f. Now, the lift breakes down, as seen in [3]. With further increasing phase angle, downstroke motion, the flow reattaches uniformely beginning at the leading edge of the profile. A fully reattached flow is achieved at a phase angle of about 90%. Detailed results according the turbulence and vorticity fields can be found in Ref. [16].

First flowfield observations of a swept wing with a yaw-angle of 15° show a significant different behavior than the two-dimensional case. Fig. 5a,b shows the comparision between both cases at the upper dead center at the same reduced frequency. After the bursting of the first dynamic stall vortex which occurs at a lower angle of incidence compared to the two-dimensional case a second vortex occurs, Fig. 5c, but the flow in the rear section remains separated. The reattachment process therefore is delayed.

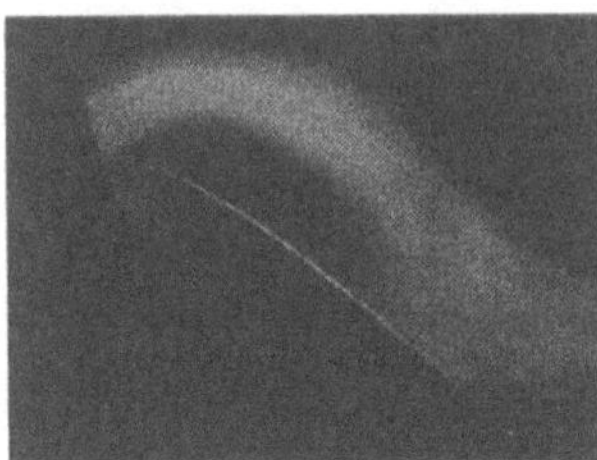

a) Two-dimensional case, upper dead center, $k = 0.3$, $\alpha = 20^o$, H. Ranke [17]

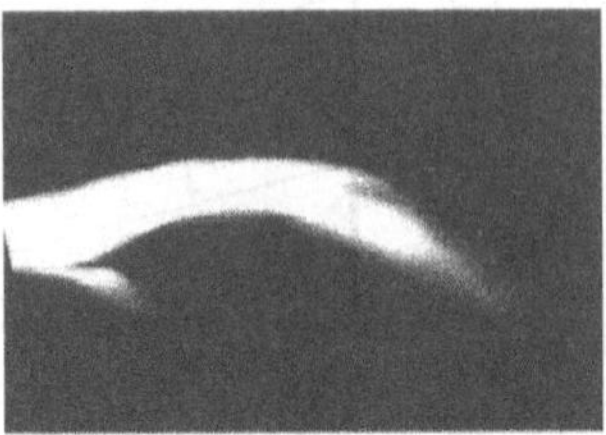

b) Three-dimensional case, upper dead center, $\beta = 15^o$, $k = 0.3$, $\alpha = 20^o$

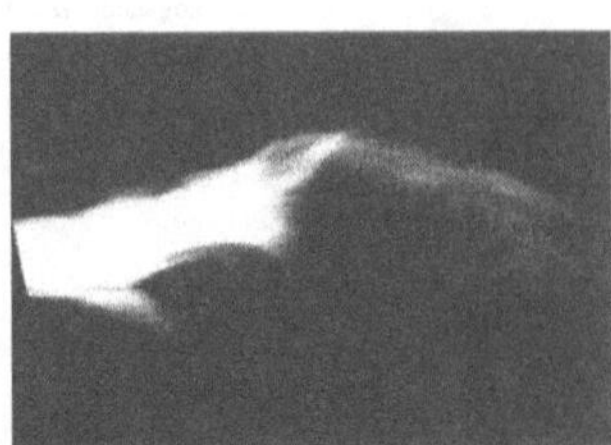

c) Three-dimensional case, downstroke motion, $\beta = 15^o$, $k = 0.3$, $\alpha \simeq 19^o$

Figure 5: Smoke visualisation

# 4 Conclusions

Major results of the dynamic stall investigations are as follows:

1. The dynamic stall results provide strong evidence in the correct operation of the advanced four-wire probe to quantify streamwise and reverse velocity components in two-dimensional flow.
2. A comparison of hot wire, laser light sheet and pressure measurements [3] shows a good agreement.
3. For the test Reynolds number and reduced frequency the onset of the dynamic stall vortex is orginated in a small region of reverse flow marching upstream. When the upstream movement of this reverse flow region becomes slower than the reverse flow velocity components, the dynamic stall vortex occurs.
   Together with calculations of unsteady boundary layer separation a better look into the dynamic stall phenomena can be found.
4. Three-dimensional unsteady boundary layer separation on swept wings show a different behavior than the two-dimensional case. Anemometer measurements results of the swept wing case are expected to give new insights into the unsteady separation process

# References


[1] Markus Raffel. PIV–Messungen instationärer Geschwindigkeitsfelder an einem schwingenden Rotorprofil. Forschungsbericht DLR–FB 93–50, Institut für Strömungsmechanik der DLR Göttingen, 1993.

[2] L. E. Ericsson and J. P. Reding. Fluid mechanics of dynamic stall. Part I. Unsteady flow concepts. *Journal of Fluids and Structures*, 2:pp.1–33, 1988.

[3] R. A. McD. Galbraith, M. W. Gracey, and R. Gilmour. *Collected Data for Tests on a NACA 0012 Aerofoil.* University of Glasgow, Department of Aerospace Engineering, Report 9208, Feb. 1992.

[4] H. Ranke. Unsteady separation in two-dimensional turbulent flow. ICAS 94, 19th Congress of the International Council of the Aeronautical Sciences, Anaheim, California, Paper 94–4.5.1, Sept. 1994.

[5] H. Ranke. *Instationäre Grenzschichtablösung in ebener, kompressibler Strömung.* PhD thesis, Lehrstuhl f. Fluidmechanik, Technische Universität München, 1994.

[6] B. Laschka. On some fundamentals of unsteady viscous flows. In *La Sapienzia.* Universita degli Studi di Roma Dipartimento Aerospaziale, 1994.

[7] L. W. Carr. A review of unsteady turbulent boundary-layer experiments. In *Unsteady Turbulent Shear Flows.* Springer-Verlag, 1981.

[8] M. Raffel, J. Kompenhaus, and P. Wernert. Investigation of the unsteady flow velocity field above an airfoil pitching under deep dynamic stall conditions. *Experiments of Fluids*, Vol.19:pp. 103–111, 1995.

[9] J. Schwab. Dynamic Stall - Messungen an einem schwingenden Tragflügel. FLM 93/42, Lehrstuhl für Fluidmechanik, TU-München, 1993.

[10] B. Laschka, H. Ranke, and C. Breitsamter. Unsteady measurement techniques applied to attached and separated flows. Presented at the International Conference devoted to the 75th Anniversary of TsAGI, Zhukovsky, Russia: Methods and Means for Experimental Investigations in Aeronautics – also avaible as report FLM–6/94, TU-München, Nov.1993.

[11] B. Laschka, H. Ranke, and C. Breitsamter. Application of unsteady measurement techniques to vortical and separated flows. *ZfW Journal*, Vol.19(No.2):pp.90–108, 1995.

[12] B. Laschka, H. Ranke, and C. Breitsamter. Application of unsteady measurement techniques to vortical and separated flows. In *Jahrbuch 1994 der DGLR*, 1994.

[13] F. D. Johnson and H. Eckelmann. A variable angle method of calibration for x-probes applied to wall-bounded turbolent shear flow. *Experiments in Fluids*, Vol.2:pp. 121–130, 1984.

[14] M. Zabat and F. K. Browand. In-situ swinging arm calibration for hot-film anemometers. *Experiments in Fluids*, 12:pp.223–228, 1992.

[15] R. M. Lueptow, K. S. Breuer, and J. H. Haritonidis. Computer-aided calibration of x-probes using a look-up table. *Experiments in Fluids*, pages pp. 115–118, 1988.

[16] A. Raco and H. Ranke. Unsteady Flow-field Hot-wire Measurements on a NACA 0012 Airfoil in Dynamic Stall Motion. In *Jahrbuch 1995 II der DGLR*, 1995.

[17] H. Ranke and J. Schwab. Videoanimation: Messungen am Schwingflügel. FLM, Lehrstuhl für Fluidmechanik, TU-München, 1993.

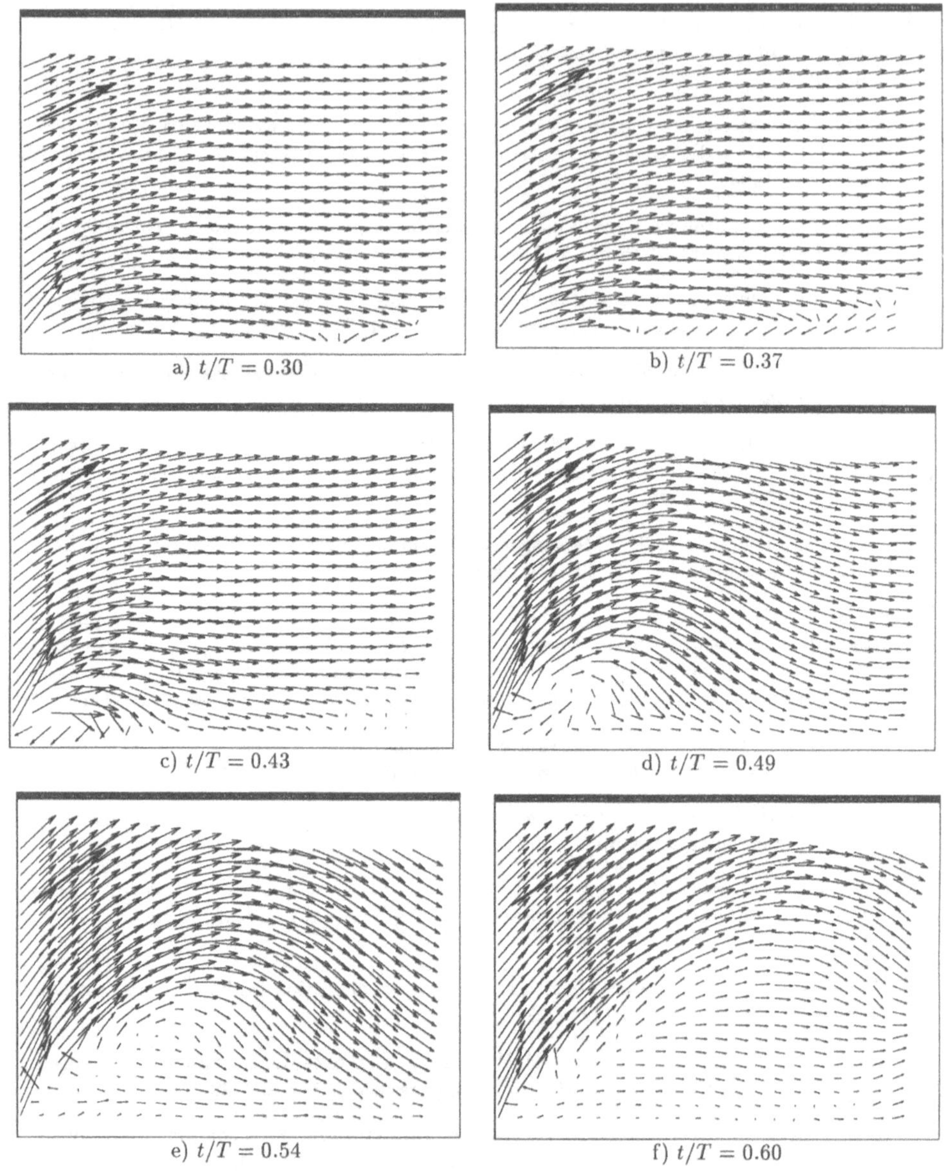


Figure 6: Velocity vectors of the pitching NACA0012 airfoil. The upper airfoil surface is mapped on the lower focal plane

# Numerical Aerodynamic Optimization Study for a Supersonic Aircraft

U. Herrmann, M. Orlowski

DLR, Institut für Entwurfsaerodynamik

Lilienthalplatz 7, D-38108 Braunschweig, F. R. Germany

## Summary

The DLR aerodynamic optimization system MODeM is introduced. Up to now most optimization work of supersonic configurations concentrate on the aircraft's wing. To be able to improve supersonic aircraft even further, DLR aims with this work also to optimize the fuselage of supersonic aircraft. Numerical accuracy of Euler flow solutions in supersonics are investigated by mesh density variations to minimize computational effort with known accuracy. Sensitivities of fuselage design parameters are given. A generic supersonic aircraft is constructed by combining an optimized wing with a generic fuselage. The most sensitive fuselage design parameters are selected to optimize this aircraft aerodynamically. The inviscid optimized geometry is recalculated with DLR's Navier-Stokes method. The aircraft's shape and off-design behaviour is confirmed with the viscous method.

## Introduction

Economical reasons and social developments indicate growing air traffic in the near future. As a response to these forecasts ultra high capacity aircraft programs are activated and intensified. Other concepts try to increase transport performance by higher cruise speeds. They result in proposals for a Concorde successor type aircraft. Realization of such supersonic civil transports (SCT) is highly dependent on substantial improvements compared to Concorde's performance. Improvements have to be gained in aerodynamic performance, engine efficiency, structural weight, pollution and noise. This will ensure economic viability and fulfil current and expected environmental constraints.

The necessary improvement of second generation SCT's in aerodynamic performance is generally estimated as a 30% increase of the lift-to-drag ratio (L/D). Sonic boom and noise aspects demand subsonic cruise over populated areas. Therefore, future SCT's will be dual-point designs, i.e. the aerodynamic performance has to be optimal at supersonic and transonic cruise conditions. Noise reduction requirements at take off and landing exist because of FAR constraints. They steer the low speed high lift system design towards low drag at large L/D conditions. It is believed that only attached flow will allow this but at the expense of vortex lift. The aircraft's planform area increases to compensate for this. A larger planform influences aerodynamic performance also at trans- and supersonic cruise conditions. This highly complex design task is viewed solvable only with optimization techniques.

Classical optimization approaches are based on controlled shape variations. Either the shape of the aircraft in general or aircraft parts are varied. Analysis codes coupled with numerical optimizers judge aerodynamic effects of the aircraft's changing shape. The optimizer determines the new shape variation of a datum aircraft based on results of previous attempts and given constraints. It finally establishes variations that design a better aircraft shape with respect to given requirements and constraints. The optimal aircraft shape is found if the absolute maximum of the (constraint) design space is found.

To contribute to the challenging goal of considerable increasing overall aerodynamic efficiency of second generation SCT's DLR is working on an optimization tool called Multi-point Optimization Design Method (MODeM). It couples CFD and optimization methods based on previous experience with super- and hypersonic configurations [1] and the optimization of transonic airfoils and wings [2]. Attention is paid to enable the tool to treat unconventional configurations such as waveriders and oblique flying wings as well as conventional wing-body configurations. The present work continues previous steps towards the optimization tool MODeM [3]. Here we demonstrate the increase in aerodynamic performance (L/D) by optimization of a generic wing-body supersonic civil transport. Aiming at further improving a partially optimized generic SCT (a previous optimized supersonic wing is combined with a generic fuselage) we concentrate on aerodynamic optimization of the fuselage. Optimizing the whole configuration would require additional input from other disciplines which is not available so far. The optimized geometry is validated by viscous recalculations.

## Optimization Tool MODeM

Any optimization needs a geometric representation which allows variations of the configuration with a minimum number of parameters. MODeM's geometry generator is based on a set of mathematical functions introduced by Sobieczky [4] and has been extended to treat conventional and unconventional configurations in the same manner. The mesh generation (surface mesh, farfield and volume mesh) for this investigation is done with an algebraic mesh generation tool based on [4],[5],[6].

Optimization procedures need fast and robust flow solvers. Effects of incremental geometry variations have to be judged on relative coarse meshes. High accuracy of the flow solution has to be guaranteed. High accuracy gives confidence that aerodynamic effects through geometric variations are resolved at minimal computational expense. Mesh inherent discretization errors can never be prevented and increase with coarser meshes. But it is essential that throughout geometric variations the distribution of the unpreventable inherent local discretization error stays constant. This keeps results comparable. The DLR flow solver [7] fulfils these accuracy requirements. It solves either the Euler- or Navier-Stokes equations in integral form based on finite volumes. The spatial discretization is an improved version of the AUSM flux vector splitting formulation [8]. Integration in time is done by an explicit Runge-Kutta multistage scheme. Convergence to steady state can be accelerated by using multiple grids or local time stepping techniques and implicit residual smoothing.

Most CPU-time within an optimization is spent on the flow solver because each geometry variation requires a new flow solution. Thus, an optimization algorithm with a minimal number of calls to the flow solver is necessary. Finite-difference and adjoint equation methods are considered at DLR. A finite-difference method using parabolic interpolation is implemented here. It is believed that minimal drag at supersonic and transonic onflow conditions will require attached flow. Therefore, in order to reduce numerical effort, inviscid flow solutions are calculated. Viscous drag is taken into account by a surface area dependent flat-plate analogy Turbulent flat plate drag). The validity of this procedure will be proven.

The control structure of DLR's optimization tool MODeM is sketched in **Fig 1**. MODeM [3] is entirely steered by a task broker. This approach enables to implement the tool as a combination of independent modules. Thus, maintenance is easy and updates of the modules can easily be carried out. As indicated in Fig 1, the approach enables also to exploit the inherent parallelity of the problem.

## Fuselage design

Primary demand on any SCT is a sufficiently large volume for payload. Here we require the transportation of 250 passengers at supersonic cruise conditions (M=2; $c_L$=0.1). The necessary volume can be concentrated in a fuselage or distributed in a blended configuration. Unconventional, blended configurations are expected to outperform aerodynamically conventional wing-body configurations [1]. Nevertheless experience in aerodynamic optimization is gained in this investigation by concentrating on the fuselage of conventional wing-body configurations. Furthermore restricting the geometric changes to relative small parts of the aircraft requires accurate drag prediction. Typical planforms of investigated generic SCT wing-body configurations are shown in **Fig 2**.

The parts of the generic SCT are depicted in **Fig 3**. The wing geometry is fixed and has a half-span of 21 meters. The wing profiles and their twist distribution have been optimized [9] for a similar wing planform. The generic fuselage consists of a front cap, a cylindrical part and a rear cap, all being variable in length and diameter. The three parts form a fuselage of constant volume for 250 passengers. In this investigation the fuselage of a generic SCT is defined by five general design variables:

- the fuselage cap length (front and rear cap are of identical length)
- the fuselage incidence angle (relative to the wing and the onflow direction)
- the relative horizontal position between wing and fuselage
- the fuselage diameter
- the relative vertical position between wing and fuselage.

Several constraints have to be met. Due to structural reasons the fuselage cap lengths is not allowed to exceed 30m. The incidence angle between fuselage and wing is limited to ±1.5°. This prevents problems caused by a wing completely crossing the cabin. Existing regulations for cabin layout drove the choice of a 3m resp. 3.5m diameter cylindrical fuselage section, corresponding to 5 resp. 6 seats per row. The length of the cylindrical fuselage part is determined by the volume requirement mentioned above. Utilizable volume inside the front and rear fuselage caps is taken into account. This reduces the overall fuselage length by shortening the cylindrical part. Furthermore the vertical and horizontal wing/fuselage position are geometrically restricted by the outline of the fuselage.

Flow fields are calculated for the generic wing body configurations using the DLR flow solver in it's Euler variant. For simplicity reasons viscous drag is determined from flat plate turbulent viscous drag at M= 2; Re= $145*10^6$ and scaled with the aircraft's wetted surface. The aerodynamic performance is the primary aspect of this investigation. It is characterized by the lift-to-drag ratio with the drag as the sum of pressure and viscous drag.

Results are given first for numerical accuracy aspects. It follows the investigation of the sensitivity of the fuselage design variables. Aerodynamic coefficients of three different configurations are calculated on three differently dense meshes each. The three configurations have identical fuselage diameter (radius of 1.5 meter). At constant volume their fuselage cap lengths are 15, 19, and 23m. In **Fig 4** the mesh dependence of the aircrafts inviscid drag (left hand side) and the lift-to-drag ratios (right hand side) are given. The coarse mesh resolves the flowfield with 24x12x26 cells in longitudinal, body normal and circumferential direction. The medium and fine meshes have 48x24x52 and 96x48x104 cells respectively. Mesh density is given in Fig 4 as the inverse of the cell number power two third. Linear variation of aerodynamic coefficients is expected for second order codes when varying the mesh density by doubling the cell numbers in all three directions.

The three fuselages of different length show comparable inviscid drag change with mesh density. The coarse mesh is not able to capture all flow features for an accurate drag prediction. However the differences between medium and fine mesh are about 3% with respect to the drag and 0.4% with respect to the lift-to-drag ratio. These small differences, especially in the lift-to-drag ratio makes it possible to use the medium mesh for further investigations and optimization purposes. The advantage is obvious: reduced numerical effort compared to the fine mesh.

The convergence behaviour of the aerodynamic coefficients is shown in **Fig 5** for the fine mesh. Both, lift and pressure drag for the different fuselages from Fig 4 are plotted against the residual reduction of the density equation. While the calculated pressure drag does not change significantly after two orders magnitude of residual reduction, the lift prediction needs three orders magnitude residual convergence. In conclusion the numerical effort can additionally be reduced by converging numerical solutions during the optimization only for three orders of magnitude.

The generic aircrafts viscous lift-to-drag ratio sensitivities (pressure drag calculated with the Euler equations and the viscous drag estimated through a turbulent flat plate analogy) with respect to the fuselage design variables are given next.

**Fig 6** shows the influence of the fuselage cap lengths C for two fuselage radii (corresponding to 5 and 6 seats per row) on pressure drag (right hand side) and on lift-to-drag ratio. The overall length of the wider fuselage varies from 74 to 83 m, the slender fuselage from 84 to 94 m. The slender one (R=1.5m) generates less pressure drag and approaches good lift-to-drag ratios with cap lengths C greater than 23m. The wider fuselage shows higher sensitivity to the fuselage cap length. The greater flow turning angle due to the increased fuselage radius raises the pressure drag level compared to the slender body and increases the fuselage cap length sensitivity. Increasing the fuselage diameter by 16% (R=1.5m → R=1.75m) approximately doubles the sensitivity of the cap length variable. The aircrafts lift-to-drag ratio reduces only by ~1% ($C_L/\Sigma C_D$= 9.256 > 9.161) through the diameter increase. Structural advantages of a shorter aircraft drove the decision to stay with the wider fuselage.

In **Fig 7** the sensitivity of the fuselage incidence angle is shown. This design variable shows a higher sensitivity. Because the inclined fuselage generates lift in supersonic flow, the lift-to-drag ratios are improved, although the pressure drag increases with fuselage inclination.

The remaining two sensitivity studies focus on the relative position between fuselage and wing. The sensitivity of the horizontal position of the wing relative to the fuselage middle position is given in **Fig 8**. **Fig 9** depicts the sensitivity of the vertical position. Negative vertical positions correspond to a low wing aircraft. Mesh generation imperfections along the intersection line between wing and fuselage restricted the range of vertical wing position. Both sensitivities are less strong than the fuselage cap length or the fuselage incidence angle. Tendencies of both figures indicate that the aircraft should be a low wing type aircraft with the wing pushed as far to the front as area ruling indicates.

## Optimization and optimum validation

The investigated sensitivities of the fuselage design parameters indicate a relatively smooth design space. MODeM's capability of constraint optimization is demonstrated by increasing the aircraft's aerodynamic performance at supersonic cruise conditions. From the investigated fuselage design variables only three are choosen as optimization variables. They are: the fuselage cap length, the fuselage incidence angle and the relative horizontal wing position. The fuselage radius is fixed at 1.75m. The fuselage provides volume for 250 passengers. The previous mentioned constraints are still valid for the optimization. The lift-to-drag ratio of the generic aircraft will be optimized. The history of the optimization is depicted in **Fig 10**. A substantial increase in the lift-to-drag ratio of nearly 10% is reached within twenty steps. Comparison of these changes is made possible by dividing their values by their admissible variation. During the optimization the fuselage cap length and the fuselage incidence angle are increased immediately. Then the wing is moved backwards and the cap length is temporarily reduced. This increases the fuselage diameter at the front most wing / fuselage intersection allowing further increase of the fuselage incidence angle. Note that two design variables reach their limits (100%).

**Fig 11** compares the polars of the initial and the optimized SCT configuration. Results are still gained by Euler calculations with the viscous drag estimated through a turbulent flat plate analogy. Both aircraft geometries are given in Fig 2 with the shorter one being the initial configuration. The optimized configuration outperforms the initial aircraft not only at the design angle of attack ($\alpha$=3°) but also at off-design conditions. Furthermore the calculated aircraft performance of both configurations is validated by viscous recalculations using the DLR Navier-Stokes Code [10]. The agreement between the Navier-Stokes results and the polar maxima is good because attached flow is found at supersonic cruise conditions. This confirms the initial assumption which enabled the inviscid optimization approach. The inviscid calculated optimization gain of 9.5% in the lift-to-drag ratio is confirmed by turbulent Navier-Stokes recalculations to 6%. This reduction is due to an inviscid underestimation of the lift-to-drag ratio of the initial SCT. The viscous lift-to-drag ratio of the optimized configuration is lower mainly because of the configuration's higher viscous interference drag. Even off-design behaviour of the inviscid optimized configuration at higher incidences is also confirmed well through viscous recalculations, see Fig 11.

**Fig 12** shows mesh convergence of the aerodynamic coefficients of the Navier-Stokes recalculations for the optimized SCT. Lift (multiplied by 100 to fit into the graph), pressure and viscous drag as well as lift-to-drag ratios are plotted against the inverse of the overall mesh cell number N power two third (compare explanation of Fig 4). Lift and pressure drag show nearly perfect second order accurate behaviour. Viscous drag is slightly dependent on mesh density. That is expected because mesh coarsening in wall normal direction degrades the calculation of boundary layer gradients. Increasing mesh density raises viscous drag most resulting in lift-to-drag ratios degrading with mesh density as given in Fig 12.

## Conclusion

Structure and modules of the DLR multi-point optimization design method (MODeM) are described. Sensitivities of lift-to-drag ratios of a generic SCT wing/body configuration with respect to fuselage design variables are studied. Numerical investigations indicate minimal required mesh sizes and necessary convergence levels for this task. The sensitivity of five fuselage design variables is investigated. Two variables, fuselage cap length and relative fuselage incidence show relatively high sensitivity. Each variable changes the lift-to-drag ratio of the configuration in the order of 3%. Weaker sensitivities are found for the relative horizontal and vertical wing position. Three fuselage variables are selected to perform a constrained SCT optimization. The inviscidly calculated gain in the aerodynamic performance of 9.5% is confirmed by viscous recalculation of the optimized geometry, The accuracy of this turbulent Navier-Stokes solutions is investigated by studying the influence of the mesh density. Not only confirmation of the inviscid optimized SCT geometry is obtained. The inviscid predicted off-design behaviour of the configuration is confirmed. Here viscous flow solutions indicate a 30% loss of the aerodynamic performance at supersonic cruise conditions of the optimized configuration compared to inviscid results.

## References


[1] Eggers, Th.; Strohmeyer, D.; Nickel, H.; Radespiel, R.:
*Aerodynamic Off-Design Behavior of Integrated Waveriders From Take-Off up to Hypersonic Flight.*
AIAA Paper 95-6091, 1995.

[2] Bartelheimer, W.:
*An Improved Integral Equation Method for the Design of Transonic Airfoils and Wings.*
AIAA Paper 95-1688, 1995.

[3] Orlowski, M.; Herrmann, U.:
*Aerodynamic Optimization of Supersonic Transport Configurations.*
ICAS-96-4.3.3; 20th Congress of the International Council of the Aeronautical Sciences, Sorrento, Italy, 1996.

[4] Sobieczky, H.; Choudhry, S. I.; Eggers, Th.:
*Parameterized Supersonic Transport Configurations.*
7th European Aerospace Conference, Toulouse, France, 1994.

[5] Abolhassani, J.S.; Stewart, J.E.:
*Surface Grid Generation in a Parameter Space.*
J. of Comp. Physics, 113, 1994, pp. 112-121.

[6] Herrmann, U.:
*IMESH - An Interactive Mesh Generation Package for Graphics Super Workstations.*
3rd Int. Conference on Numerical Grid Generation in Computational Fluid Dynamics and related Fields, Barcelona (Editor: A.S-Arcilla et al.), Spain, 1991, pp. 467-478.

[7] Kroll, N.; Radespiel, R.; Rossow, C.-C.:
*Accurate and Efficient Flow Solvers for 3D Applications on Structured Meshes.*
AGARD FDP/VKI Special Course on Parallel Computing in CFD, Brussels, Belgium, AGARD R-807, 1995, pp. 4-1 - 4-59.

[8] Kroll, N.; Radespiel, R.:
*An Improved Flux Vector Splitting Discretization Scheme for Viscous Flows.*
DLR-FB 93-53, 1993.

[9] Doherty, J. J.; Parker, N. T.:
*Dual-Point Design of a Supersonic Transport Wing using a Constrained Optimization Method.*
7th European Aerospace Conference, Toulouse, France, 1994.

[10] Herrmann, U.;
*Analysis of a Supersonic Civil Transport Configuration (ATSF-1) at Subsonic, Transonic and Supersonic Mach Numbers Using a Navier-Stokes Solver.*
DLR FB 96-34, 1996.

# Figures

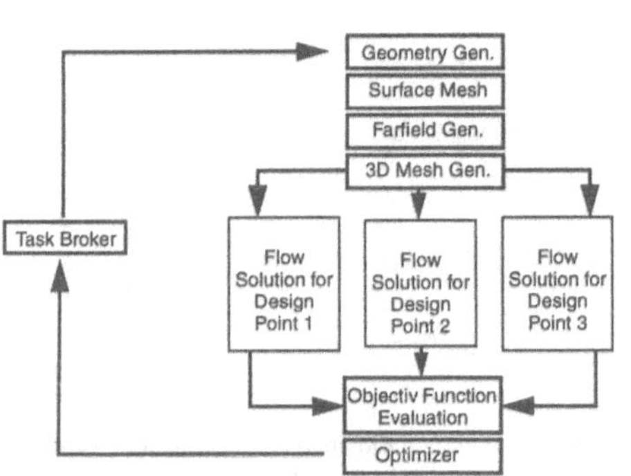


**Fig 1** Control structure of the Multi-point Optimization Design Method (MODeM).

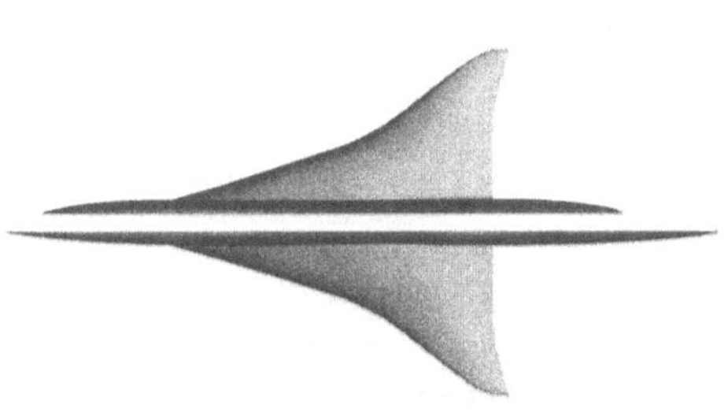

**Fig 2** SCT's resulting from different fuselage designs.

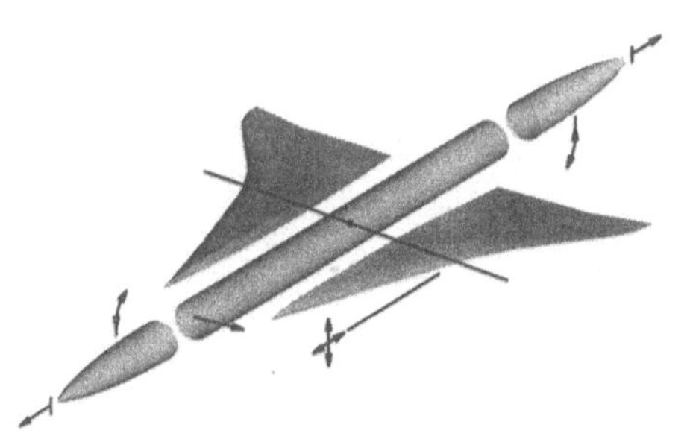

**Fig 3** Generic fuselage design parameters: fuselage cap length, fuselage radius, fuselage incidence angle, horizontal and vertical wing position.

$C_{DP}$ R=1.5; C=15
$C_L/C_{DP}$ R=1.5; C=15
$C_{DP}$ R=1.5; C=19
$C_L/C_{DP}$ R=1.5; C=19
$C_{DP}$ R=1.5; C=23
$C_L/C_{DP}$ R=1.5; C=23
$C_{DP}$
$C_L/C_{DP}$
0.0068 0.0066 0.0064 0.0062 0.0060 0.0058 0.0056
18.0 17.0 16.0 15.0
0.0000 0.0005 0.0010 0.0015 0.0020 0.0025 0.0030
$1/[N^{2/3}]$

**Fig 4** Prediction variation of the aerodynamic coefficients with mesh density.

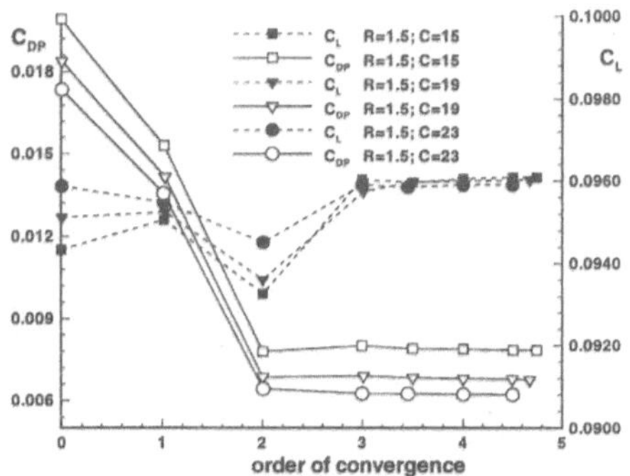


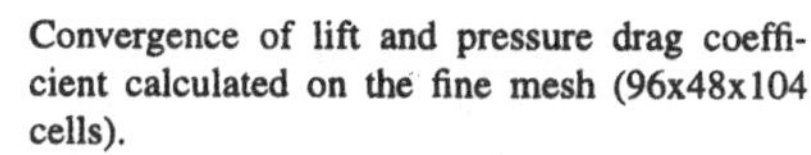

**Fig 5** Convergence of lift and pressure drag coefficient calculated on the fine mesh (96x48x104 cells).

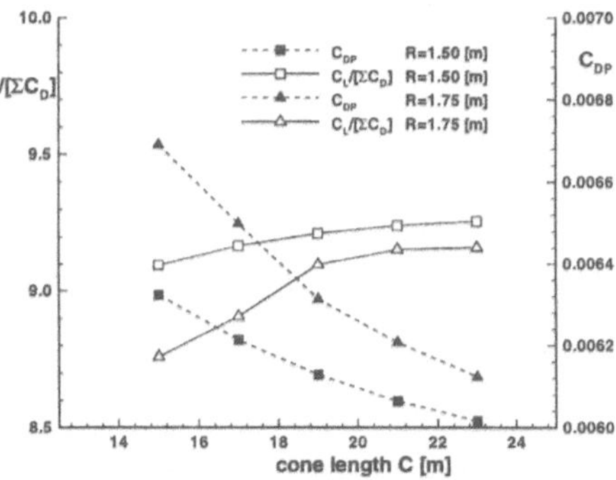


**Fig 6** Lift-to-drag ratio and pressure drag variation with fuselage cap length for two fuselage radii.

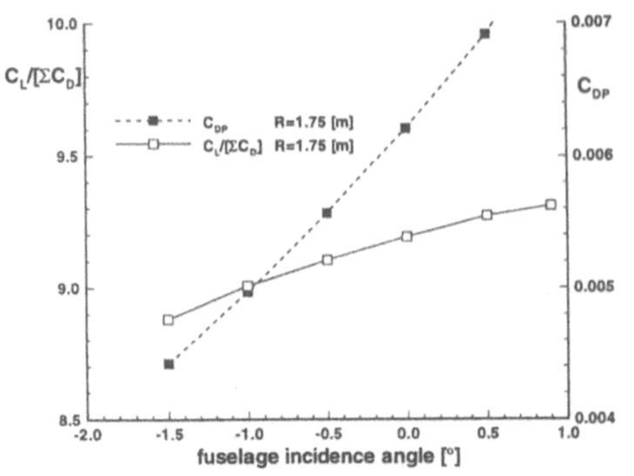


**Fig 7** Lift-to-drag ratio and pressure drag variation with fuselage incidence angle (R=1.75m, fuselage cap length=21m).

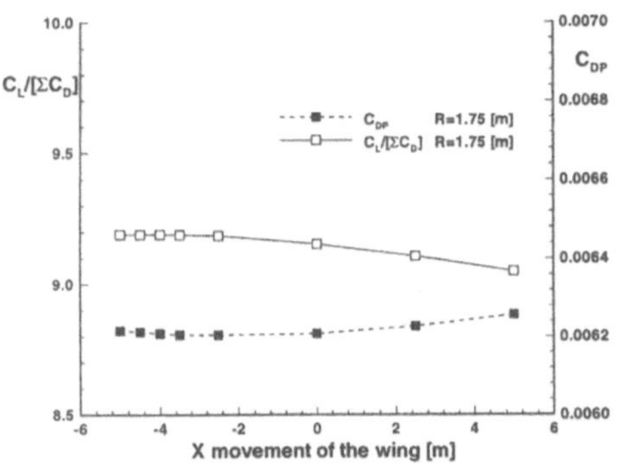


**Fig 8** Lift-to-drag ratio and pressure drag variation with wing horizontal position relative to the fuselage (R=1.75m; fuselage cap length=21m) middle.

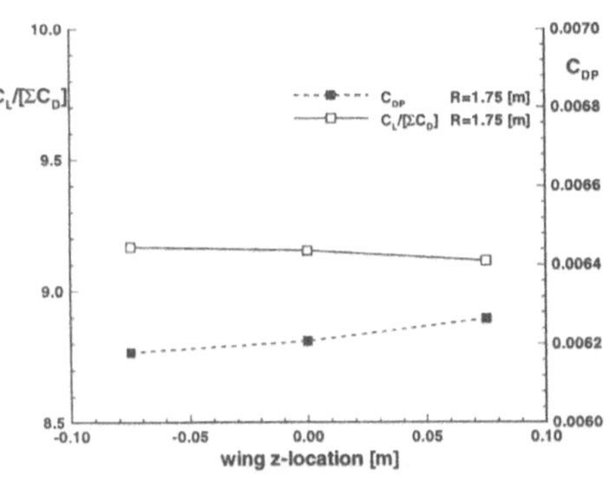


**Fig 9** Lift-to-drag ratio and pressure drag variation with wing vertical position relative to the fuselage (R=1.75m; fuselage cap length=21m) middle.

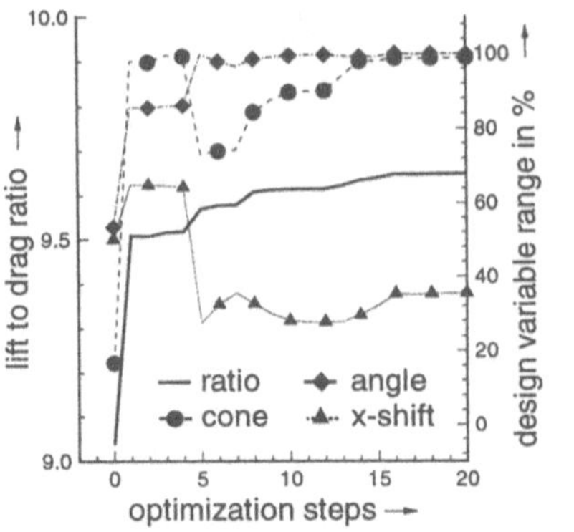

**Fig 10** Lift-to-drag ratio increase and design variable variation during the optimization.

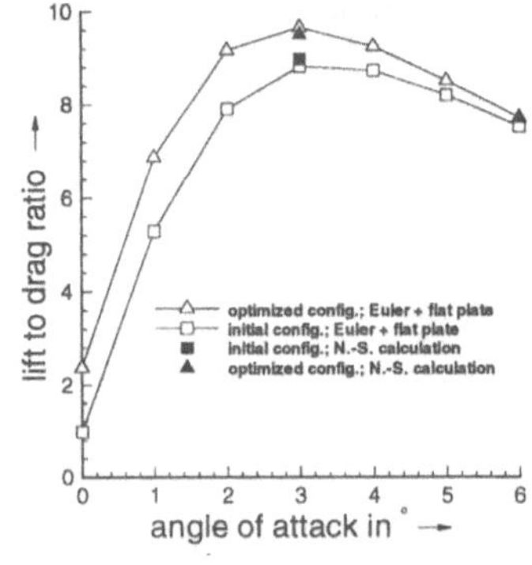


**Fig 11** Validation of initial and optimized SCT configuration at M=2.

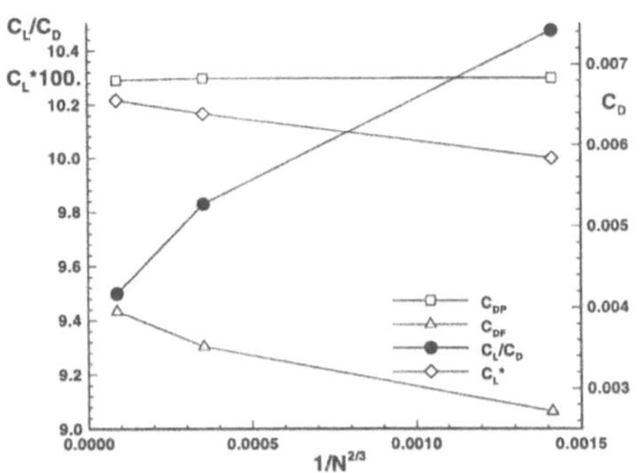


**ig 12** Mesh dependence of viscously calculated aerodynamic coefficients of the optimized SCT configuration at M=2; $\alpha$=3°; Re=$145*10^6$.

# Investigation on a Transonic Transport Aircraft Wing with an Optimized Trailing Edge

H. Köster, G.Wichmann

Deutsche Forschungsanstalt für Luft- und Raumfahrt (DLR) e.V.
Institut für Entwurfsaerodynamik
Lilienthalplatz 7, D-38108 Braunschweig

## Summary

After successful preliminary theoretical and experimental studies on several airfoils with and without optimized trailing edges by DRA, ONERA and DLR an investigation of a clean wing with such a modified trailing edge has been carried out. As a first step, optimized trailing edges for three airfoils at different span stations of this wing have been developed and investigated. The results showed a drag reduction for the optimized airfoils at the prescribed lift coefficients. Therefore in a second step the wing was also equipped with such a trailing edge optimization (TEO) modification. Pressure and friction drag calculations demonstrate that the modified wing provides some advantages compared with the original wing. The wave drag has been reduced so that a total drag reduction of up to 3.2% could be achieved but on the other hand the negative pitching moment has increased producing some additional trim drag. Summarizing the TEO modification of the original wing it could be stated that TEO seems to be a means to improve the performance of a transonic wing especially when the lift coefficient shall be increased.

## Introduction

In 1989 investigations on a new concept for the design of supercritical airfoils were published [1],[2]. The special features of this concept are: First the trailing edge of the airfoil must have a finite thickness which is already often the case at modern supercritical airfoils with rearloading, second the lower surface of the trailing edge diverges away from the upper side and third the concave camber at the lower side has to increase strongly in the direction of the trailing edge. By using this concept first the camber of the airfoil very near the trailing edge is remarkably enlarged leading to a load shifting to the rear part of the airfoil with an increased nose-down pitching moment and second the base pressure is decreased reducing the pressure gradient on the upper side but increasing the base drag. Consequently the trailing edge thickness must not become too large in order to avoid too large additional base drag.

In 1991 Airbus Industrie initiated a research programme to be performed by DRA, ONERA and DLR for investigating airfoils with thick and strongly cambered trailing edges [3]. After successful preliminary experimental and theoretical studies on several Airbus airfoils [4] which showed a reduced wave drag especially at higher lift coefficients it was decided to develop an optimized trailing edge for an Airbus wing without fuselage in two steps. In the first step the trailing edges of three airfoils normal to quarter chord line at different span stations of this wing were modified with an optimized camber, slope and additional trailing edge thickness and then investigated at the corresponding Mach number normal to quarter chord line M = 0.71 by DRA, ONERA and DLR . In the second step the original wing was then reshaped with the optimized trailing edge of the airfoils in order to perform numerical three-dimensional studies at cruise flight Mach number M = 0.78 with different numerical methods by DRA, ONERA and DLR.

The present paper describes the DLR results of the numerical investigations of the flow around the original and modified airfoils and wings .

## Airfoils, wing and their modifications

The considered Airbus wing is a swept back transonic wing the planform and computational surface grid of which are shown in Fig. 1. The aspect ratio is $\Lambda = 9.13$ and the leading edge sweep angle is $\varphi = 27.4^{o}$. The optimized trailing edge shapes for the three wing sections normal to quarter chord line were developed by ONERA for minimum drag under the constraint that the difference of pitching moment coefficients should not exceed smaller values than $\Delta c_m = -0.025$. The equation

$$\Delta Z = -\Delta Z^{te}\left[\frac{X - X_o}{1 - X_o}\right]^m \quad X_o \leq X \leq 1$$

with $$m = -\tan(\alpha^{te})\frac{1 - X_o}{\Delta Z^{te}}$$

was used for varying the trailing edge at the lower surface (see Fig. 2).

The coordinates of the airfoils and their modifications at the relative span stations $\eta = 0.20$, 0.45 and 0.70 (airfoils at $\eta = 0.45$ see Fig. 3) as well as of the original wing A2 and the modified wing TEO2 were provided by ONERA [ 5]. A detailed figure of the wing shape near the trailing edge is shown in Fig. 4.

## Numerical methods

For carrying out the two-dimensional calculations an Euler-method [6] was used which was coupled with an integral boundary layer method which solves the boundary layer equations simultaneously. The boundary layer displacement thickness is then added to the airfoil surface. The transition point between laminar and turbulent boundary layer is determined by a simplified $e^N$-method and the wake is taken into account by an empirical statement.

For the calculation of the flow field of the original and the TEO wing the DLR EULER code CEVCATS [7] was used. In this method a sufficient resolution of the trailing edge flow is ensured. The three-dimensional computational grids were generated using a method based on a C-H type mesh generator with a C-structure in streamwise, see Fig. 5, and an H-structure in spanwise direction [8]. In order to take into account the viscosity, boundary layer calculations with a three-dimensional boundary layer integral method [9] were carried out, in which at a prescribed transition location calculation of the turbulent boundary layer is started. For the laminar part, an analytical model of the displacement surface is used, which is not part of the solution of the boundary layer equations. Within an iterative coupling procedure applying the displacement concept, the resulting boundary layer displacement thickness was then added to the wing surface followed by the corresponding flow analysis using the Euler code. Compared to non-conservative potential flow solvers, the calculated pressure distributions from Euler methods show stronger shocks terminating the supersonic region in transonic flow, stronger adverse pressure gradients near the trailing edge, and local influences of the tip vortex on the pressure distribution. In most cases here the domain of validity of the boundary layer code is exceeded. Due to these reasons the upper surface boundary layer displacement thickness was determined using the DLR FLO22VIS code [10] in which the non-conservative full potential code FLO22 of Jameson [11] is coupled with the above mentioned integral boundary layer code of Stock [9]. The resulting displacement surface of the upper side added to the wing contour then serves as input for the final Euler flow analysis described in [7].

## Results

### Two-dimensional flow

For the three original and three optimized airfoils two-dimensional calculations were carried out at Mach number M = 0.71. For the prescribed lift coefficients, drag reductions of nearly 4% could be obtained whereas the negative pitching moment increased by $\Delta c_m$ = -0.02 to -0.03. Fig. 6 shows as an example the comparison of the drag polars for the A2-45 airfoil and the optimized airfoil TEO2-45 for the given Reynolds number with fixed transition at 2% of local chord on upper and lower surface. In the higher lift range the optimized airfoils generally yield higher lift coefficients between $\Delta c_l$ = 0.06 and 0.08 for a constant drag coefficient. On the other hand there is a small increase in drag of up to $\Delta c_d$ = 0.0001 for lower lift coefficients. Lift coefficient versus angle of attack and versus pitching moment coefficient are presented in Figs. 7and 8 for the same example. Generally the $c_l(\alpha)$ curves of the TEO airfoils are shifted to smaller angles of attack by $\Delta\alpha = 0.6^o$ to $0.7^o$ and show a higher maximum lift coefficient with increments of $\Delta c_{l\ max}$ = 0.06 to 0.07. Because of higher rear loading of the TEO airfoils the negative pitching moment is increased as already indicated by $\Delta c_m$ = -0.02 to -0.03. Comparing the pressure distributions of A2-45 and TEO2-45 airfoils for the prescribed lift coefficient in Fig. 9 one can see that for the latter one the maximum local supersonic Mach numbers as well as the size of the supersonic region have been decreased thus reducing the shock strength and the resulting wave drag. It can also be seen that the load distribution is shifted downstream by the higher camber at the trailing edge thus increasing the negative pitching moment (Fig. 8).

### Three-dimensional flow

Three-dimensional wing investigations were carried out at cruise flight Mach number M = 0.78 and Reynolds number Re = $26 \times 10^6$ based on the aerodynamic mean chord with fixed boundary layer transition at 8% of local chord on upper and 5% on lower surface . The calculation of the wing drag coefficients was done without considering the base drag of the trailing edge.

The comparison of the results of the pressure calculations for the lift coefficient $c_L \approx 0.55$ are presented in Fig. 10. It already shows that the supersonic region of the TEO2 wing which increases from root to tip is smaller than that of the original A2 wing. The reduced strength of the shocks for the TEO2 wing leads to smaller wave drag as shown below. This trend increases with increasing lift coefficients. A more detailed comparison for the wing station $\eta = 0.48$ is given in Fig. 11. The small change of the contour at the trailing edge of the TEO2 wing produces a shift of the load to the rear part of the wing thus reducing the velocities on upper side in the nose region and consequently the shock strength but increasing the nose down pitching moment. For the wing lift coefficient $c_L \approx$ 0.55 the spanwise distributions of the local lift and pitching moment coefficients are presented in Figs. 12 and 13. The comparison shows that there is insignificant difference between the lift distributions whereas the nose down pitching moment is remarkably enlarged due to the above mentioned change of the load distribution.

This behavior causes an additional trim drag and thus partly offsets the drag reduction obtained by the TEO trailing edge .

The breakdown of the total drag $c_D$ in Fig. 14 shows that the friction drag $c_{Df}$ for both wings is nearly equal whereas the pressure drag $c_{Dp}$ is smaller for the TEO2 wing because of reduced wave drag. That leads to a total drag reduction of up to 3.2% assuming an additional base drag of the order of one drag count for the TEO2 wing. The considerable shift of the pitching moment coefficient (Fig. 15) resulting in a corresponding nose down effect is connected with a relatively small shift of the location of the center of pressure (Fig. 16 ).

## Conclusions

Three airfoils of the A2 wing at the span stations $\eta = 0.2$; 0.45 and 0.70 were modified with optimized camber, slope and additional thickness at the trailing edge of the lower surface. Two-dimensional calculations then show that at the Mach number normal to quarter chord line $M = 0.71$ in the higher lift range an additional lift $\Delta c_l = 0.06$ to 0.08 at constant drag coefficient could be obtained. Accordingly for fixed lift coefficients a reduction of the total drag was achieved due to the reduction of wave drag of the three reshaped airfoils.

The A2 wing was then reshaped using the optimized airfoil trailing edges. Calculations carried out for that wing with a three-dimensional EULER method at cruise flight Mach number $M = 0.78$, combined with a three-dimensional boundary layer method, show for a given lift coefficient pressure distributions on the modified wing with reduced supersonic velocities and weaker shocks which result in a reduced wave drag. On the other hand, the modification causes a shift of the air loads in the downstream direction producing a larger negative pitching moment and an additional trim drag. Nevertheless, this optimized trailing edge can be used to improve the performance of the A2 wing for higher lift coefficients.

## References


[1] Henne, P.A.; Gregg, R.D.: *New Airfoil Design Concept.* Journal of Aircraft, Vol. 28, pp. 300-311, 1991.

[2] Gregg, R.D.; Hoch, R.W.; Henne, P.A.: *Application of Divergent Trailing Edge Airfoil Technology to the Design of a Derivative Wing.* Aerospace Technology Conference and Exposition; Anaheim, California, Sept. 25-28, 1989. SAE Technical Paper 892288.

[3] Ashill, P.; Thibert, J.J.; Bézard, H.: Private Communication.

[4] Köster, H.: *Investigations of Airfoils with Cambered Trailing Edges.* Deutscher Luft- und Raumfahrtkongress 1993, 28. Sept. - 1. Okt. 1993, Göttingen. DGLR Paper 93-03-83.

[5] Bézard, H.: Private Communication, 1994.

[6] Drela, M.; Giles, M.B.: *Viscous-Inviscid Analysis of Transonic and Low Reynolds Number Airfoils.* AIAA Journal, Vol. 25 (1987), pp. 1347 - 1355.

[7] Radespiel,R; Rossow, C.-C.; Swanson,R.C.: *Efficient Cell-Vertex Multigrid Scheme for the Three-Dimensional Navier-Stokes Equations.* AIAA-Journal, Vol. 28, No.8, (1990), pp 1464 - 1472.

[8] Leicher, S.; Fritz, W.; Grashof, J.; Longo, J.: *Mesh generation Strategies for CFD on Complex Configurations.* Lecture Notes in Physics, Vol. 170, Eight International Conference on Numerical Methods in Fluid Dynamics. Berlin/Heidelberg: Springer Verlag, 1982, S. 329 - 334.

[9] Stock, H.W.: *Integral Method for the Calculation of Three-Dimensional laminar and Turbulent Boundary Layers.* NASA TM 75320, (1978).

[10] Wichmann, G., Zhang Zhong-Yin: *Calculation of Three-Dimensional Transonic Wing Flow by Interaction of a Potential and an Integral Boundary layer Method.* DLR IB 129 - 89/12, (1989).

[11] Jameson, A.; Caughy, D.A.: *Numerical Calculation of the Transonic Flow past a Swept Wing.* ERDA report C00-3077-140, New York University, (1977).

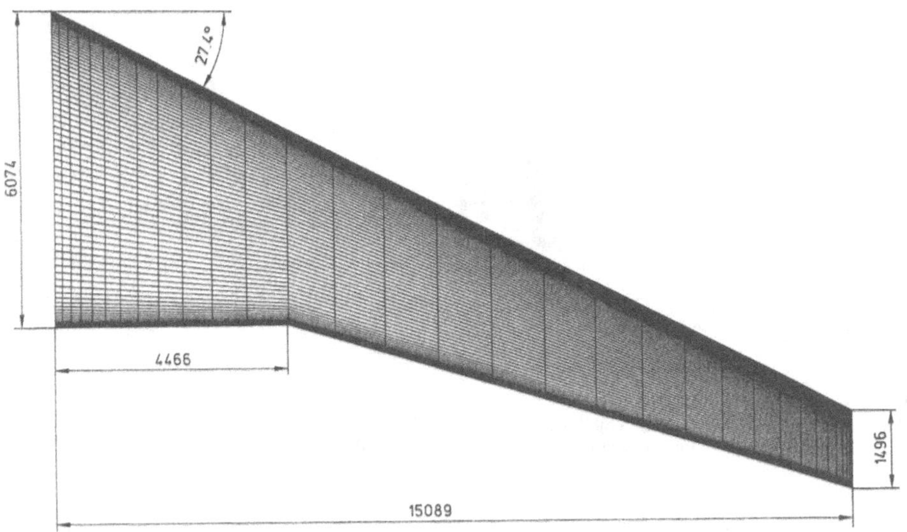


**Fig. 1** Wing planform and grid

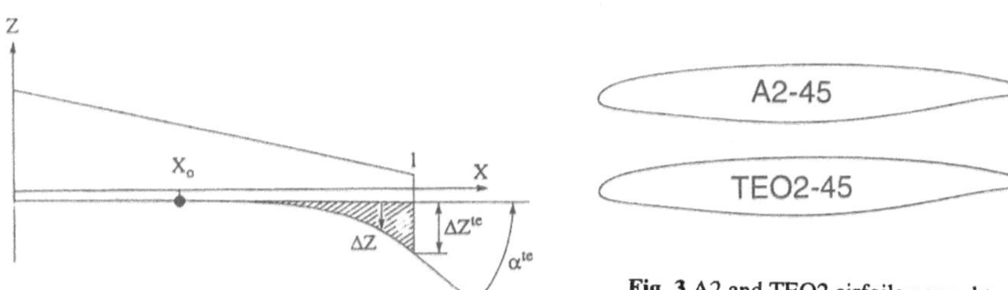


**Fig. 2** Trailing edge modification on lower surface

**Fig. 3** A2 and TEO2 airfoils normal to quarter chord line at $\eta = 0.45$

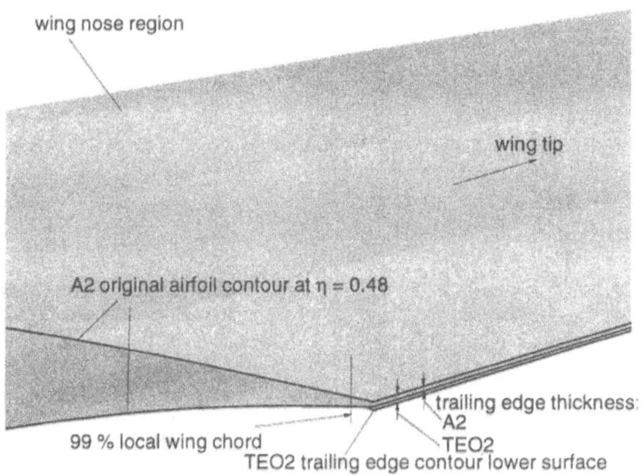


**Fig. 4** TEO2 modification on A2 wing

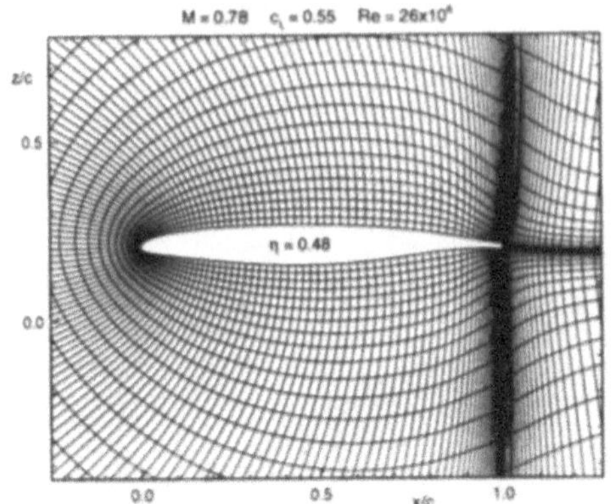


**Fig. 5** Grid of TEO2 wing section with boundary layer displacement thickness at $\eta = 0.48$

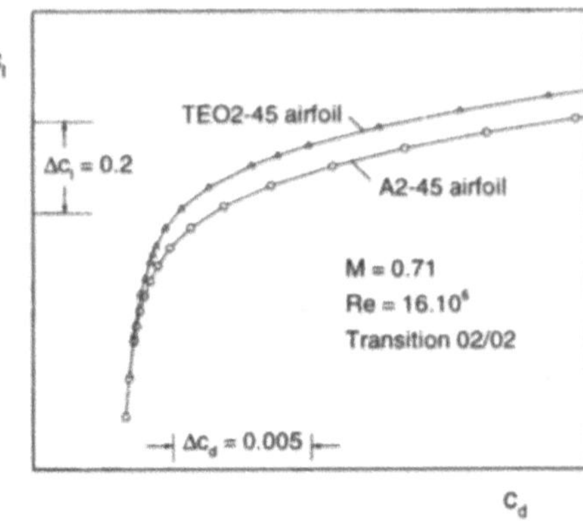


**Fig. 6** Comparison of drag polars of airfoils A2-45 and TEO2-45

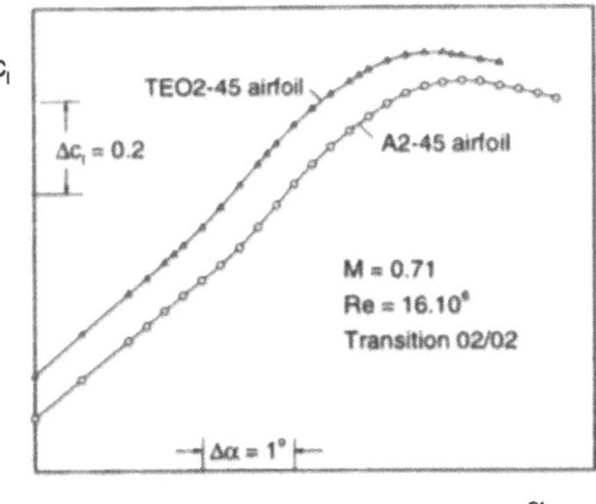


**Fig. 7** Lift coefficient versus angle of attack for airfoils A2-45 and TEO2-45

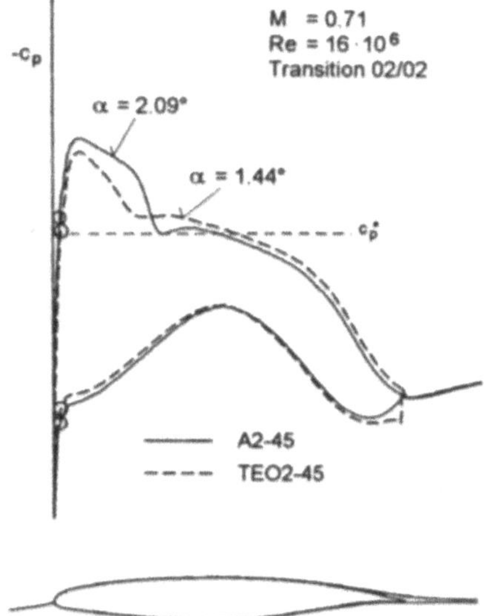


**Fig. 9** Comparison of two pressure distributions of airfoils A2-45 and TEO2-45 at $c_l = 0.66$

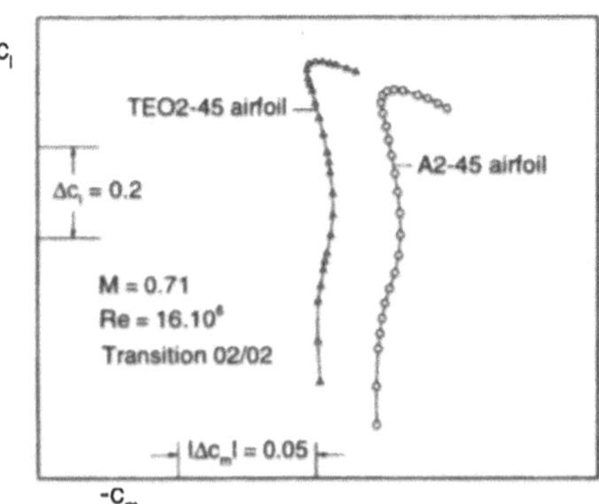


**Fig. 8** Comparison of pitching moment coefficients of airfoils A2-45 and TEO2-45

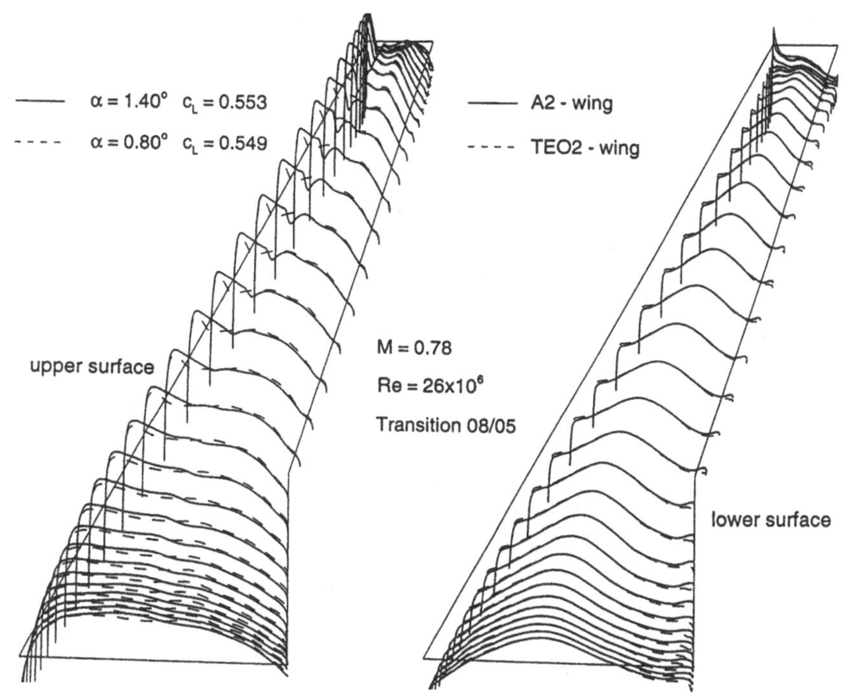


**Fig. 10** Comparison of wing pressure distributions of A2 and TEO2 wings

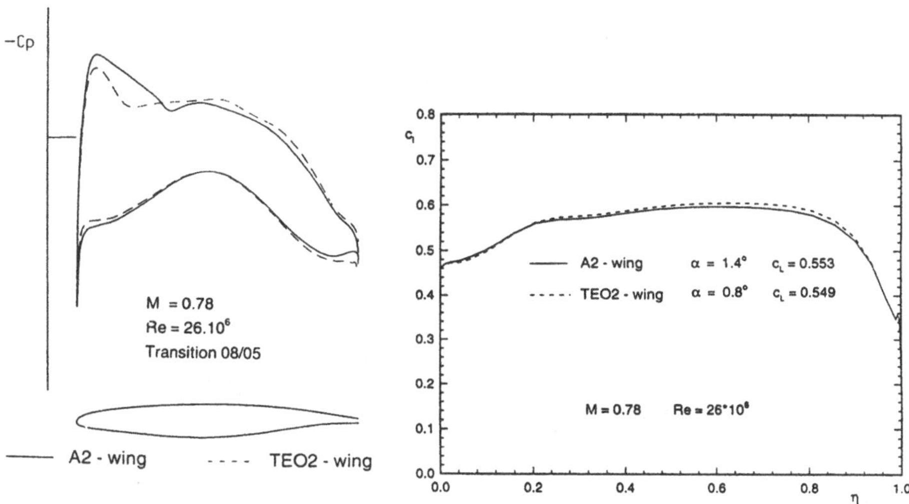


**Fig. 11** Comparison of pressure distributions of A2 and TEO2 wing sections at $\eta = 0.48$

**Fig. 12** Spanwise local lift coefficients of A2 and TEO2 wings

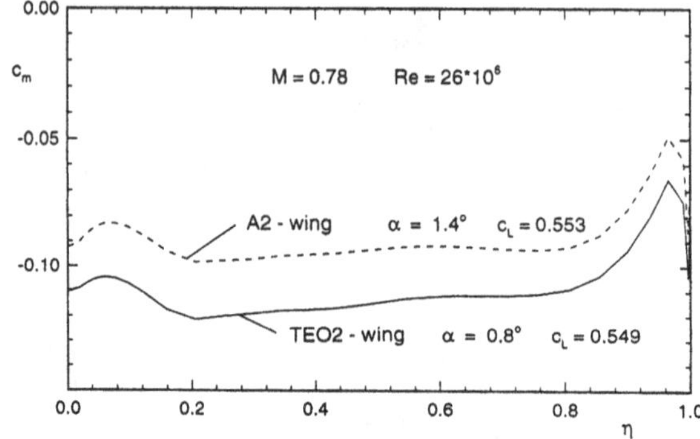


**Fig. 13** Spanwise local pitching moment coefficients of A2 and TEO2 wings

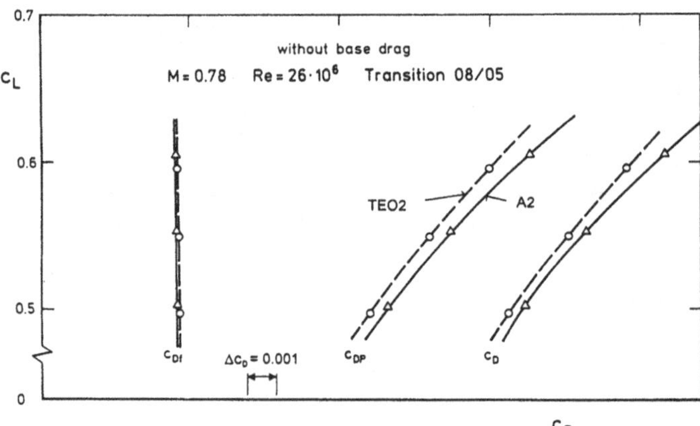


**Fig. 14** Drag polars of A2 and TEO2 wings

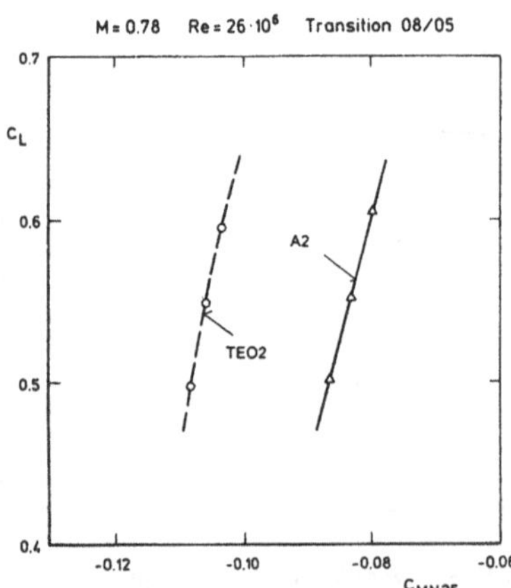


**Fig. 15** Pitching moment coefficients of A2 and TEO2 wings

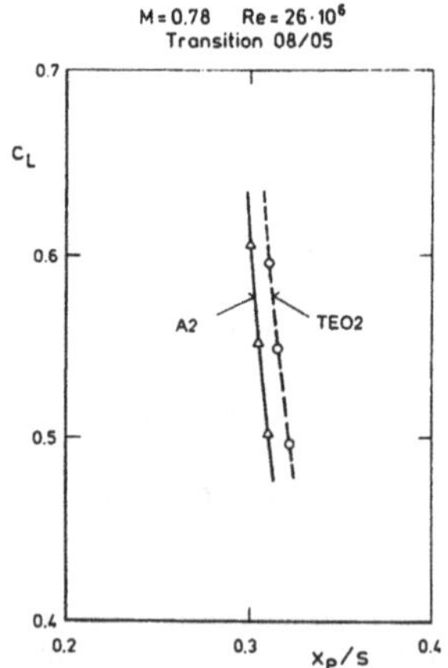


**Fig. 16** Center of pressure of A2 and TEO2 wings

# CALIBRATION OF CODES FOR COMPUTING BI-CONIC FLOWFIELDS

**Alois F. Kreins, Matthew C. Towne, Chad M. Morgan**
U. S. Air Force Academy, CO 80840-6222, USA

**Jim Trolier, Natesan Rajendran**
Science Applications International Corporation, Wayne, PA 19087-1803, USA

## Summary

More and more, the designer of a new reentry vehicle relies on laboratory simulations to evaluate the designs. Numerical simulations of the configuration's trajectory incorporate data bases for the aerodynamic characteristics of the configuration. These aerodynamic data bases are developed using computations made with state-of-the-art design codes, which are verified by comparisons with wind-tunnel test data. Comprehensive surface pressure and force and moment measurements were obtained on hypersonic configurations at Mach 4.28 for validation of computational tools. The experiments were carried out in the United States Air Force Academy's Tri-Sonic Wind Tunnel for sharp and blunted bi-conics (which are representative of some reentry vehicle configurations). Data were obtained over an angle-of-attack range of $\pm 10°$. These data are compared to results of computed flow models of various sophistication levels.

## 1. Introduction

Bi-conic configurations are found in orbit-transfer vehicle applications possessing moderate lift-to-drag ratios such as a Mars return-sample vehicle or other volumetrically efficient re-entry vehicles. According to Miller and Gnoffo [8], the bi-conic shape is widely used because it offers the best compromise between high lift-to-drag ratio (for insertion accuracy) and high volumetric efficiency with a suitable hypersonic ballistic coefficient. The flowfield over a bi-conic configuration at moderate angles of attack often contains viscous/inviscid interactions, which include free-vortex separations with reattachment in the leeward plane of symmetry (Stetson [12], Rainbird [9], Feldhuhn et al. [5]).

Advanced Computational Fluid Dynamics (CFD) codes are used to generate solutions for comparing with experimental data and for generating flowfield solutions at flight conditions. These CFD solutions contain approximations in numerical algorithms, flow models (such as turbulence models), and flow field discretization (i.e. levels of grid refinement) that may result in significant error. Thus, it is critical that systematic studies be conducted to validate the CFD tool for the problem type on which it is employed.

In this paper, experimental data obtained in the Tri-Sonic Wind Tunnel (TWT) at the United States Air Force Academy (USAFA) are compared with CFD results. The computations will include flow models of varying degrees of rigor. Specifically, the measurements are compared with numerical results generated using modified Newtonian flow, an Euler solver, plus Parabolized Navier Stokes (PNS) and Navier Stokes solvers (NS). The principal objective of the study is the determination of the level of rigor needed in the numerical flowfield models in order to obtain a given level of fidelity in the computed solutions.

For the experimental program, pressure distributions as well as force and moment measurements for bi-conic configurations were obtained. One of the configurations has a sharp nose. However, since severe aerothermal conditions exist in the stagnation region for this pointed configuration flying at hypersonic speeds, a bi-conic configuration having a spherically-blunted tip was also investigated (the stagnation-point heating varies inversely with the square root of the nose radius [4]). These configurations are typical of many theater missiles today. Data were obtained at a free-stream Mach number of 4.28, with the free-stream Reynolds number of $11.82 \times 10^6$ / ft ($38.81 \times 10^6$ / m), over an angle-of-attack range from $-10^0$ to $+10^0$.

## 2. Experimental set-up and test procedures

### *2.1 Wind tunnel and test conditions*

The experimental investigations were conducted in the TWT located at the USAFA in Colorado. This "blow down" facility (test gas is air) is capable of producing nominal test section (square) Mach numbers from 0.14 to 4.38 and test-section Reynolds numbers in the range of $5\times10^6/\text{ft}$ $(1.64\times10^7/\text{m})$ to $30\times10^6/\text{ft}$ $(9.84\times10^7/\text{m})$. The mean wind-tunnel test conditions for the experimental investigation are summarized in **Table 1.**

**Table 1**
Mean wind tunnel test conditions

| $M_\infty$ | $p_{t1}$ [psia] ([bar]) | $T_{t1}$ [F] ([K]) | $p_\infty$ [psia] ([bar]) | Re $[10^6/\text{ft}]$ $([10^6/\text{m}])$ |
|---|---|---|---|---|
| 4.28 | 150 (10.34) | 85.9 (303.1) | 0.683 (0.047) | 11.83 (38.81) |

Surface pressure distributions were measured at $0°, 5°$ and $10°$ angles of attack for roll angles in the range of $0^0$ to $180^0$ in $30^0$ increments. Force and Moment data were obtained for angles of attack of $0^0$, $5^0$, and $10^0$. In order to make proper computations, the state of the boundary layer, whether it is laminar or turbulent, should be known. Beckwith [3] gives correlated transition Reynolds number data on sharp cones in wind tunnels and flight. Based on his results, for the conditions tested a laminar boundary layer is expected at least for the forecone.

### *2.2 Models*

For the experimental investigations two different bi-conic configurations were used. The first model was fabricated with a sharp apex, while the second model was designed with a slightly blunted nose. Scale drawings with dimensions of the bi-conic models are shown in **Fig. 1**.

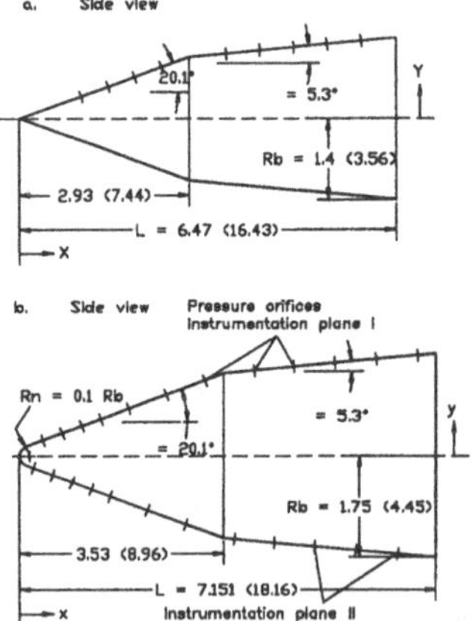


**Figure 1:** Side view of sharp and slightly blunted bi-conic model (dimensions in inches and (cm))

The pressure model was designed to permit rotation about its axis (x-axis) of revolution. Thus, the instrumentation plane (see Fig. 1) could be repositioned between runs, so that circumferential distributions of the pressure distribution could be obtained at each angle-of-attack. **Fig. 2** shows the angle convention when viewing the model head-on (i.e., facing the nose). For the measurement of forces and moments two additional, geometrically identical bi-conic-models one for the sharp

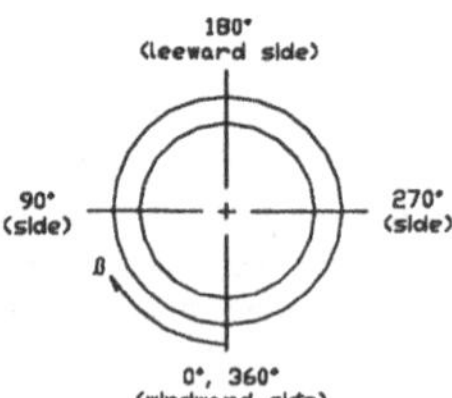


**Figure 2:** Head-on view of model illustrating pressure orifice circumferential distribution

and a second for the blunt configurations were used. A six component balance (internal to the model) was used to measure the forces and moments.

### *2.3 Instrumentation*

Surface pressure data were obtained by using a differential pressure transducer within a Scanivalve® pressure sampling scanner having a ±15 psid measuring range. The measured pressure was converted to an electrical signal by the transducer. This signal was amplified and then transferred to an Hewlett-Packard model 3852A data acquisition control system that was interfaced to a personal computer. In order to further support these measurement results, schlieren photos of the flowfield were also taken.

### 2.4 Uncertainty of Measurement

Quality measurements from wind tunnels and comprehensive error uncertainty considerations are necessary for meaningful computation validations. The total fixed errors are estimated using the root-sum-of-squares method by Kline and McClintock [7]. To demonstrate the accuracy of the experimental data, supersonic flow over a sharp cone is considered in this section. Since the flowfield over the forecone is not affected by the aft conic flow field, the pressure measurements on the forecone should correspond with the values presented in the NACA Report 1135 [1]. **Fig. 3** shows the pressure ratio $p/p_\infty$ as a function of x/L with the fixed error associated to each data point for the sharp bi-conic at 0° angle-of-attack in a Mach 4.28 stream. Observe the measured values are in reasonable agreement with the predictions. This plot also demonstrates the repeatability of the wind tunnel results.

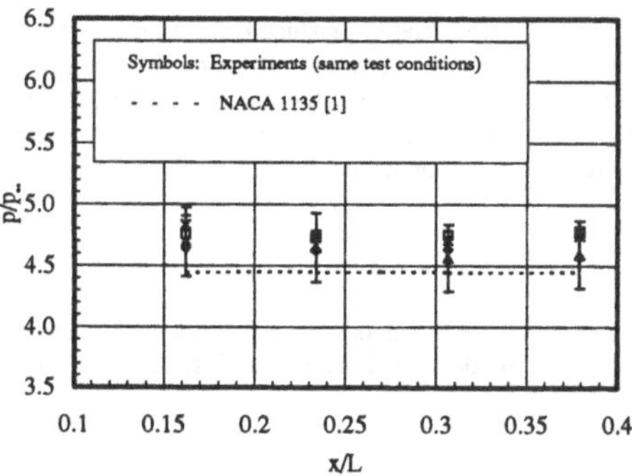


**Figure 3:** Measured and predicted [1] pressure distribution on the forecone of the sharp bi-conic configuration for several test runs and estimated error bands locations ($M_\infty = 4.28$, $Re_\infty = 11.82\times10^6$ / ft, $\alpha=0°$)

## 3. Computational tools

Various levels of computational sophistication are used to compute the flowfield about the cones. The most rigorous computational approach used is the solution of the Navier Stokes equations. After that is the solution of the Parabolized Navier Stokes equations and Euler-equation solutions. Modified Newtonian represents the simplest model used to estimate the pressure distribution. This section presents a brief description of each computational technique used to accomplish the computations.

### *3.1 Navier Stokes solver*

The General Aerodynamic Simulation Program (GASP), v3.0, CFD code [6] was employed for 3D Navier-Stokes solutions. This code numerically solves the governing equations using a finite-volume formulation. A fixed grid must be supplied to GASP, thus, the code captures the shocks produced about bodies in high-speed flow. The algebraic turbulence model of Baldwin and Lomax [2] was used for the entire sharp-nosed bi-conic solutions. For the blunt-nosed bi-conic solution, laminar flow was assumed for the rounded nose-radius region, then the rest of the solution aft of this location employed the Baldwin-Lomax turbulence model. A plane from each of the grids used for the GASP solutions (not only for NS solution but also for GASP PNS and GASP Euler explained below) about the sharp and blunt bi-conics is shown in **Fig. 4**. Each grid has 80 cells in the axial direction, 60 cells radially and 36 cells circumferentially for half the geometry.

### *3.2 Parabolized Navier Stokes solvers*

The Window Cooling (PANSWIC) flow field code ([10]) was employed for Parabolized Navier-Stokes (PNS) solutions. The PANSWIC code solves the Parabolized Navier-Stokes equations for supersonic/hypersonic viscous flow around arbitrary three dimensional geometries. The flow field solutions are generated by marching in space from an initial condition using the Roe upwind algorithm. The initial plane of

a.

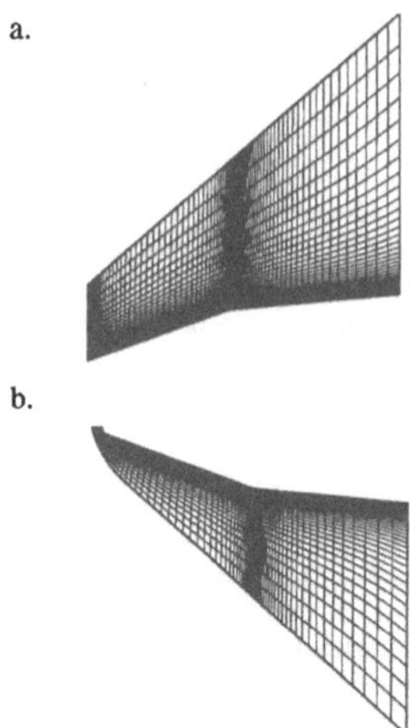

b.

**Figure 4:** Grid used by GASP code: (a) x-z plane for 80×60×36-cell grid (right half of geometry) used in GASP sharp-cone solution, (b) x-(-z) plane for 80×60×36-cell grid (right half of geometry) used in GASP blunt-cone solution

the flow field is obtained using either a Viscous Shock Layer (VSL) code or using a time dependent Navier-Stokes code to define the flow field in the blunt nose region. The outer bow shock is fitted and is computed as part of the solution. The algebraic turbulence model developed by Baldwin and Lomax turbulence model was used for the computations.

PNS solutions were also accomplished using GASP, allowing for a comparison between the shock-fitting (PANSWIC) and shock-capturing (GASP) approaches. A test was conducted to determine if the crossflow plane viscous terms were needed for the grid used in this study. Little benefit resulted from retaining the crossflow plane terms, thus only the viscous terms perpendicular to the cone surface were employed for the GASP PNS solutions.

### *3.3 Euler solvers*

GASP was used to produce Euler solutions for the blunt and sharp bi-conics by simply turning off viscous terms used in the PNS calculations and changing the wall no-slip boundary condition to a tangential flow condition.

### *3.4 Modified Newtonian flow*

The modified Newtonian flow model is an impact flow model, which was used to get a quick estimation of the pressure distribution on the fore- and aftcone of the bi-conic models. A more detailed description can be found in [4].

## 4. Results and Discussion

### *4.1 Pressure Distributions*

A comparison of the experimentally measured pressure distribution with the computational predictions using the various methods is presented in **Fig. 5**, for angles of attack of $0^0, 5^0$, and $10^0$ with orifices at $\beta=0°$ (windward side) and $\beta=180°$ (leeward side). The pressure computed using Modified Newtonian model are significantly lower than the measured values on the forecone (e.g., $\Delta p/p_{\infty,max} = 1.2$, Fig. 5c-- $\alpha = 10^0$, forecone, windward side). This is to be expected since the modified Newtonian model is based on an infinitesimally thin shock layer. For the aftcone (e.g., Fig. 5c-- $\alpha = 10^0$, aftcone, leeward side) Newtonian lies right on top of experimental results. For $\alpha = 10^0$, the angle of attack exceeds the inclination angle of the aftcone at $\beta = 180^0$, resulting in the aftcone region lying in the "shadow of the freestream". Therefore, the modified Newtonian prediction, which assumes that the pressure coefficient is equal 0 in shadow regions, would not be expected to be a suitable approximation.

Also shown in Fig. 5 are the PANSWIC PNS solutions (labeled PNS), GASP PNS solutions, and the GASP Euler solutions. All three computations consistently underpredict experimental results (e.g. for GASP CFD results is $\Delta p/p_{\infty,max} \approx 0.4$, Figs. 5a-c-- $0 \le \alpha \le 10^0$, windward and leeward sides) on the forecone. PANSWIC does a little better job of matching the experimental data than GASP does, most likely because of the inherent advantage of fitting the shock to the solution (PANSWIC) over capturing the shock (GASP). Probably, the better grid for shock-fitting produces a better computation of properties across the shock--especially in the forecone region, where the grid is quite refined since the shock is close to the surface. With shock-capturing, the properties may be off somewhat downstream of the shock and will be transmitted aft.

a. $\alpha = 0^0$

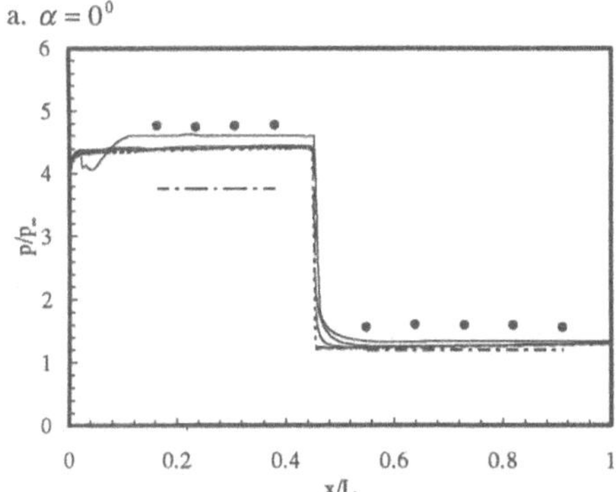

b. $\alpha = 5^0$

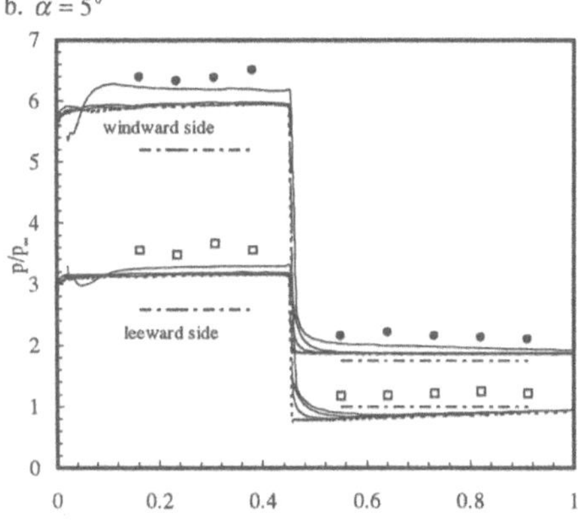


c. $\alpha = 10^0$

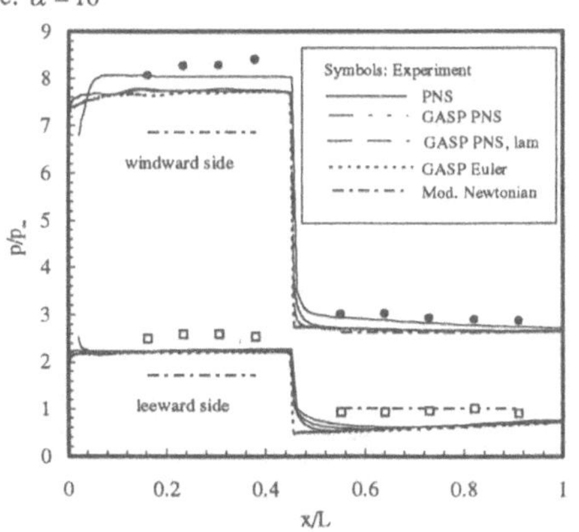


**Figure 5:** Comparison of measured pressure distribution with computational predictions on the sharp bi-conic ($M_\infty = 4.28$, $Re_\infty = 11.82\times10^6$ / ft)

The predicted values calculated by the Euler code also underpredict the measured values. Except from the nose region and the fore- to aftcone transition region, the Euler results slightly underpredict the PNS values, presumably caused by the neglecting viscous terms.

Both, the PNS solutions as well as the Euler solutions provide a reasonably accurate prediction of the pressure distributions on the present sharp bi-conic configuration. Note that the GASP PNS results (laminar and turbulent) and the GASP Euler results are virtually indentical. This fact implies that these bi-conic results are greatly dominated by the inviscid field and there is not yet significant shock-wave/boundary-layer interaction.

**Fig. 6a** shows a schlieren image of the flowfield around the blunted bi-conic. **Fig. 6b** presents GASP's PNS Mach number contours, calculated for the same flow conditions as the schlieren results. A high degree of qualitative agreement exists between experiment and computation.

The pressures computed for the blunted bi-conic are compared with the experimentally measured pressures in **Fig. 7**. Comparisons are made for angles-of-attack of 0°, 5°, and 10°. The streamwise variation of the measured surface-pressures illustrates the overexpansion/recompression of the flow from the spherical nose to the forecone section. The magnitude of the overexpansion peak decreases from the windward to the leeward side--a result which agrees with Miller and Gnoffo [8]. Note that, as in the sharp bi-conic case, the blunt bi-conic GASP results are nearly identical for Euler, PNS, and Navier Stokes solutions implying that the viscous effects are also not important for these flow conditions and geometry. To verify that the crossflow plane physics are being adequately resolved by circumferential grid spacing of $5^0$, the spacing was decreased to $2.5^0$ and the crossflow viscous terms were included--no significant change occurred in the numerical results. The GASP PNS velocity vector distribution in a crossflow plane on the aftcone are presented in **Fig. 8** for an $\alpha$ of 10° and an $\alpha$ of 15°. There is no indication of circumferential reverse flow at an $\alpha$ of 10°, whereas there is reverse flow at an $\alpha$ of 15°. Thus, significant viscous/inviscid interactions would not be expected at angles of attack of 10° or less. Hence, the agreement between inviscid and viscous solutions for $-10° \leq \alpha \leq 10°$ should not be surprising.

a.

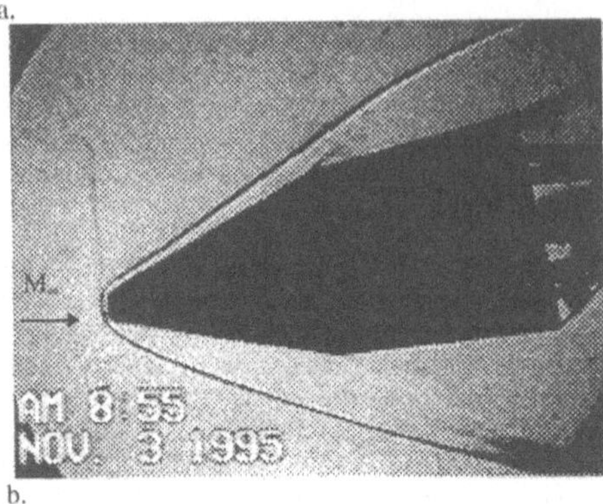


b.

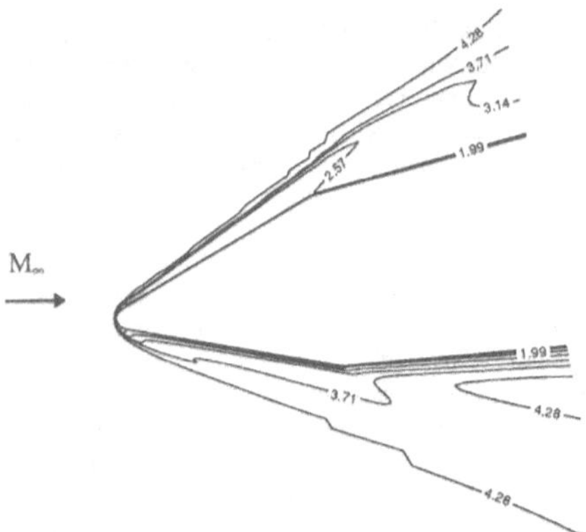


**Figure 6:** (a) Schlieren image of the flowfield around the blunted bi-conic and (b) Mach number contours from PNS calculations by GASP code (both at $M_\infty = 4.28$, $Re_\infty = 11.82\times10^6$ / ft, $\alpha = 10°$ )

#### 4.2 Force and Moment Coefficients

**Figs. 9** through **11** show comparison of PNS predicted values with experimentally measured data of the normal force coefficient ($C_N$), the axial force coefficient ($C_{Ax}$) and the pitching moment ($C_{M_L}$) versus the angle of attack (Note, multiple points are shown for a given $\alpha$ to demonstrate the experimental repeatability). The predicted values of normal force coefficient and pitching moment are in good agreement with the experimentally measured data across the angle of attack range for both the sharp as well as the blunted bi-conic. To allow a comparison of experimental values with PNS predicted results the axial force coefficient only for the forebody is considered.

a. $\alpha = 0°$

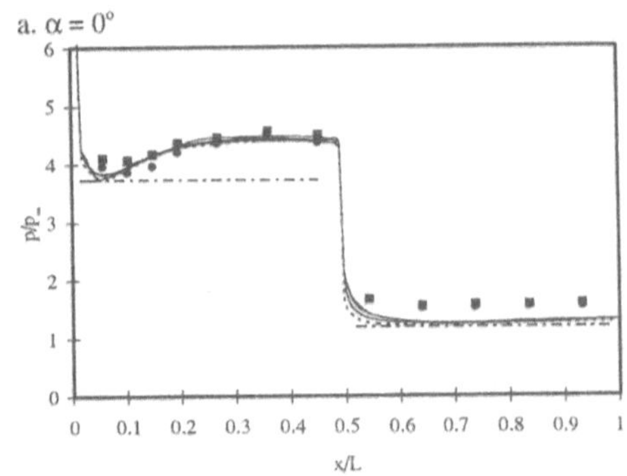


b. $\alpha = 5°$

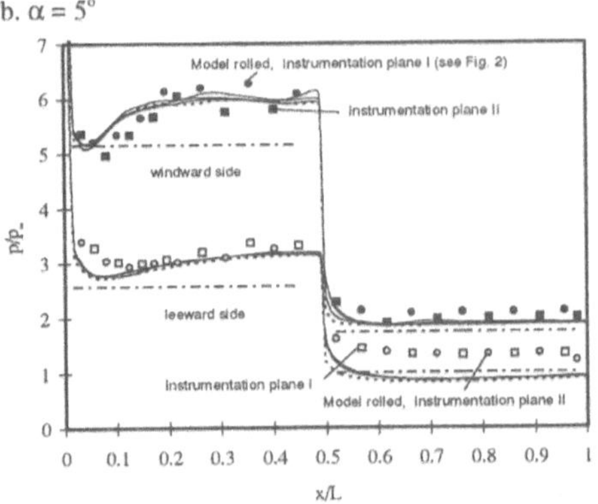


c. $\alpha = 10°$

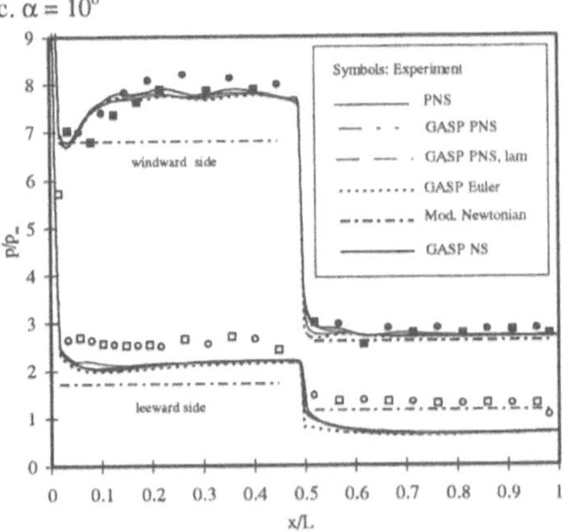


**Figure 7:** Comparison of measured pressure distribution with computational predictions on the blunt bi-conic ($M_\infty = 4.28$, $Re_\infty = 11.82\times10^6$ / ft)

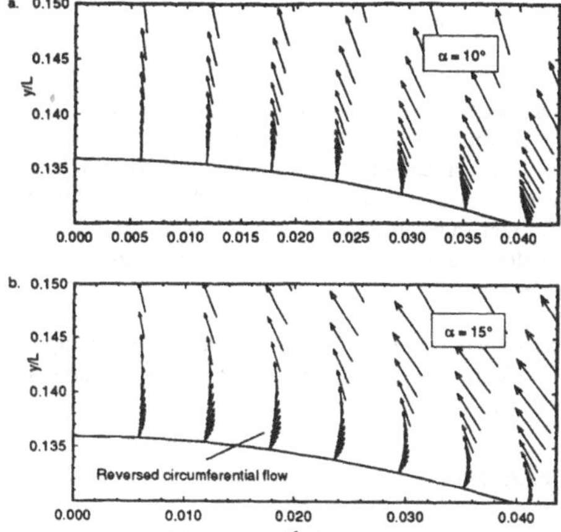


**Figure 8:** PNS-calculated velocity vector distribution in a crossflow plane on the aftcone (x/L = 0.82) for the blunted bi-conic ($M_\infty = 4.28$, $Re_\infty = 11.82\times10^6$ / ft)

The predicted values of axial force coefficient by PNS are in good agreement with the experimental data for the low angle of attack range. With increasing angle of attack, the prediction tends to underpredict the experimental values. This tendency has also been seen in previous studies (Rutledge et al. [11]). Possibly, the difference lies in the inability of the sparse grid to capture streamwise separation.

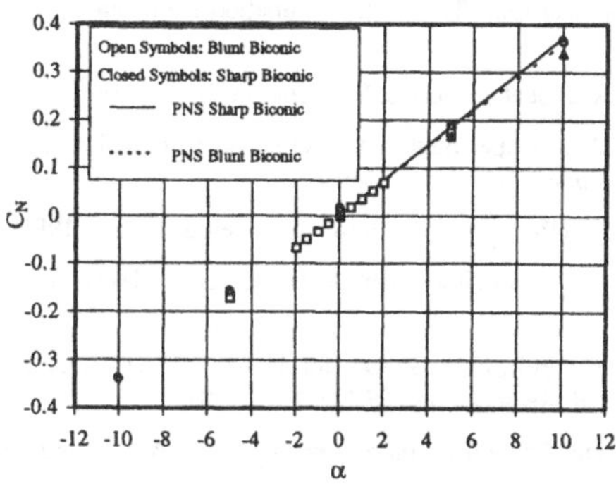


**Figure 9:** Comparison of experimentally measured Normal force coefficient with predictions for the sharp and blunted bi-conic ($M_\infty = 4.28$, $Re_\infty = 11.82\times10^6$ / ft)

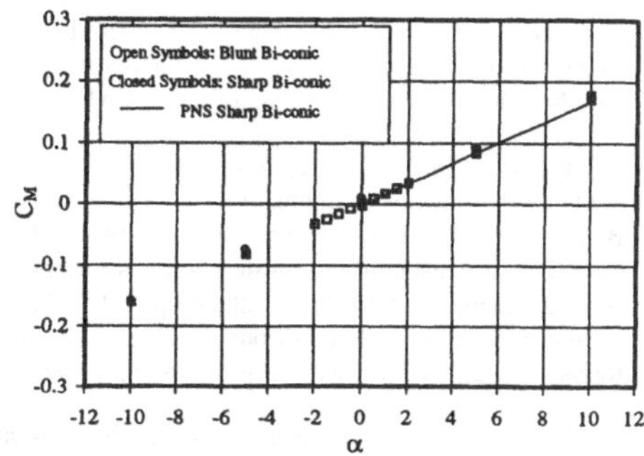


**Figure 10:** Comparison of experimentally measured pitching moment coefficient (for a reference point located at the end of the model (x = L)) with predictions for the sharp and blunted bi-conic ($M_\infty = 4.28$, $Re_\infty = 11.82\times10^6$ / ft)

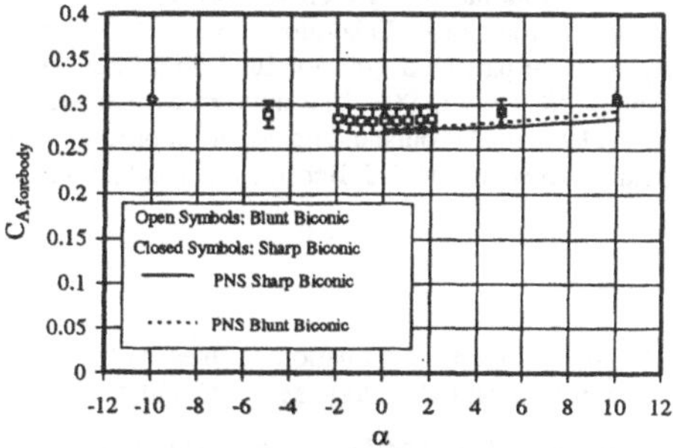


**Figure 11:** Comparison of experimentally measured axial force coefficient with predictions for the sharp and blunted bi-conic ($M_\infty = 4.28$, $Re_\infty = 11.82\times10^6$ / ft)

## 4. Concluding Remarks

Surface pressures and forces and moments have been measured on a sharp and on a blunted bi-conic configuration at a Mach number of 4.28 in the Tri-Sonic Wind Tunnel at the United States Air Force Academy. The experimental uncertainty was found to be acceptable.

The experimental data have been compared with results based on various levels of computational sophistication. As expected, the simple, modified Newtonian flow model provided only a rough approximation of the forebody pressures. However, for the aftcone, it often was in very good agreement with the experimental data and at other times, it was as far off as $\Delta p/p_\infty = 1.2$ for the data presented. The PNS solutions using shock-fitting technique were slightly better (about 2% difference with experiment versus 6% for the shock-capturing PNS and Euler solutions) for the forecone windward pressure predictions on the sharp bi-conic. The shock-captured PNS and Euler solutions gave pressure predictions comparable to the shock-fitted PNS solution for the aftcone region and the forecone leeward region of the sharp bi-conic.

For the blunt bi-conic, windward GASP predictions tended to be in better agreement than leeward predictions: indicating expansions are more difficult for the computational model to predict than compressions. Better leeward side predictions should result with greater grid refinement in that region. For the conditions investigated, reasonably accurate results were obtained from inviscid calculations on the sharp and blunt bi-conic geometries for angles of attack up to 10°. This fact implies that the viscous flow effects are second order for the geometries and conditions considered. At higher angles of attack, higher Mach numbers, and/or lower Reynolds numbers, however, this would not necessarily be the case.

Based on the comparisons between the present measurements and the computed values, the computer codes, which will be used in the design of theater missiles defense systems, should provide suitable results for developing aerodynamic models for use in trajectory simulations.

## Acknowledgment

This work has partly been supported by the Deutschen Forschungsgemeinschaft (DFG) and the Air Force Wright Aeronautical Laboratory "WL/MNGI" through project order ATL-95-629. This work was also supported in part by a grant of HPC time from the DoD HPCC, Naval Oceanographic Office, Stennis Space Center MS, Cray C-90. The authors wish to acknowledge the financial support and guidance of Maj. B. Schneider and the outstanding highly motivated technical assistance of Mr. L. Lamblin. Also, the authors greatly appreciate Dr. J. J. Bertin's technical guidance and editorial suggestions.

## References

[1] Ames Research, "Equation, Tables and Charts for Compressible Flow", NACA Report 1135, Ames Aeronautical Laboratory, Moffett Field, Calif. 1953.

[2] Baldwin, B. S., and Lomax, H., "Thin Layer Approximation and Algebraic Model for Separated Turbulent Flows," AIAA Paper 78-257, Jan. 1978.

[3] Beckwith, I. E., "Development of a High Reynolds Number Quit Tunnel for Transition Research", AIAA Journal, Vol. 13, No. 3, pp. 300-306.

[4] Bertin, J. J., "Hypersonic Aerothermodynamics", AIAA Education Series, Washington, D.C., 1994.

[5] Feldhuhn, R. H., Winkelmann and A. E. Pasiuk, L., "An Experimental Investigation of the Flow Field Around a Yawed Cone", AIAA Journal, Vol. 9, No. 6, June 1971, pp. 1074-1081.

[6] GASP Version 3.0, The General Aerodynamic Simulation Program, User's Manual, AeroSoft, Inc., 1995.

[7] Kline, S. J. and McClintock, F. A., "Describing uncertainties in single-sample experiments", Mech. Engr. 75, 3-8, January 1953.

[8] Miller, C. G. and Gnoffo, P. A., "Pressure Distributions and Shock Shapes for 12.84° / 7° On-Axis and Bent-Nose Bi-conics in Air at Mach 6", NASA Technical Memorandum 83222, Dec. 1981.

[9] Rainbird, W. J., "Turbulent Boundary-Layer Growth and Separation on a Yawed Cone", AIAA Journal, Vol. 6, No. 12, Dec. 1968, pp. 2410-2416.

[10] Rajendran, N., "Users Guide to the PANSWIC Flow Field Code", Science Applications International Corporation.

[11] Rutledge, W. H., Polansky, G. F. and Clark, E. L., "Aerodynamic Design and Performance of a Bent-Axis Geometry Vehicle", AIAA Atmospheric Flight Mechanics Conference, August 17-19, 1987, Monterey, California.

[12] Stetson, K. F., "Boundary-Layer Separation on Slender Cones at Angle of Attack", AIAA Journal, Vol. 10, No. 5, May 1972, pp. 642-648.

# Flow visualization and application of Particle-Image Velocimetry to the hypersonic configuration ELAC 1

Neven Lang, Markus Jacobs
Aerodynamisches Institut
Rheinisch-Westfälische Technische Hochschule (RWTH) Aachen
Wüllnerstraße zw. 5 u. 7
D-52062 Aachen, Germany

## Summary

Experimental investigations of the flow around the hypersonic research configuration ELAC 1 are presented. Results of flow visualization in supersonic flow and measurements of the velocity field in incompressible flow are shown. The visualization was carried out by a combination of the vapour-screen method and oil flow-patterns. The application of these techniques showed the complete vortex structure at the leeward side. A vortex topology for angles of attack $\alpha \leq 10°$ is presented. Furthermore the difference between turbulent and laminar flow was investigated with a large scale model of the first 40 % of ELAC 1. Velocity measurements were carried out with Particle-Image velocimetry. A three-dimensional reconstruction of the flow field is presented. With this the velocity field was established at the leeward side of ELAC 1.

## 1 Introduction

The hypersonic configuration ELAC was developed in the Collaborative Research Center "Fundamentals of Design of Aerospace Planes" as a "Two-Stage-To-Orbit" configuration. The first stage ELAC 1 has a delta wing with round leading edges [2]. The second stage separates at an altitude of approximately 30 $km$ . Due to the integration of the second stage on the upper side of ELAC 1, it is of fundamental importance to obtain a profound knowledge about the flow field at the leeward side of the configuration [10, 6]. The examination of the vortex structure at the front and between the control surfaces is a part of future investigations.

## 2 Flow visualization in supersonic flow

### 2.1 Wind tunnel

The trans- and supersonic wind tunnel of the Aerodynamisches Institut Aachen (AIA) is shown in figure 1. It is a suction type wind tunnel with a test section size of 40 x 40 $cm^2$. During a run the air is sucked from the balloon through the Laval nozzle, then the test section and finally through the diffuser into the vacuum tanks. The adjustable Laval nozzle and diffuser enable $Mach$ numbers ($Ma$) from 0.2 to 4. The testing time is 3 $s$ to 10 $s$ depending on the $Mach$ number. The $Reynolds$ number ($Re$) also depends on the $Mach$ number, it is $Re_L = 8.6 \cdot 10^6$ $(L = 0.72\,m)$ for the first model and $Re_L = 3.6 \cdot 10^6$ $(L = 0.3\,m)$ for the second model at $Ma = 2$.

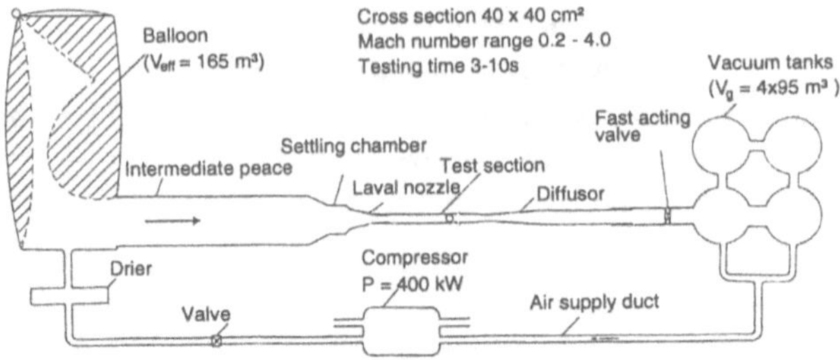


**Figure 1** Trans- and supersonic wind tunnel of the AIA.

## 2.2 Wind tunnel models and test conditions

Two different models were used to investigate the vortex structure at the leeward side of the delta wing configuration. ELAC is a two-stage launch vehicle, the first stage ELAC 1 is a delta-shaped lifting body. The upper and lower sides have semi-elliptic cross sections with different axis ratios and the leading edges are round. Two different models were used to visualize the flow. The first model was used to examine the influence of turbulent in comparison to laminar flow. It is a model of the first 40% of ELAC 1 (scale 1:100). It was applied to achieve a better resolution of the results in spanwise direction. Figure 2 shows a sketch of the second model. Its scale is 1:240, it was used to visualize the flow between the control surfaces.

The $Mach$ number range was $Ma = 1.3$ to $Ma = 2.5$ with angles of attack $\alpha = 6^\circ, 9^\circ$ and $15^\circ$.

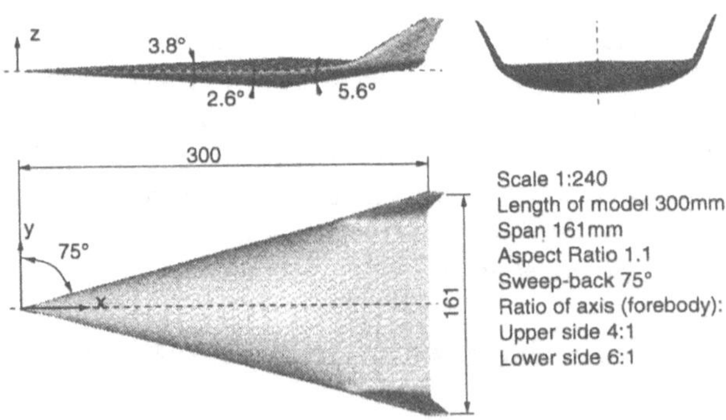


**Figure 2** Model of the first stage of the hypersonic research configuration ELAC 1.

## 2.3 Experimental setup for the vapour-screen method and the oil flow-pattern

The flow visualization was carried out by a combination of vapour-screen and oil flow-pattern. The experimental setup for the vapour-screen technique consisted of a laser-light sheet and a CCD camera, see figure 3. Due to the vapour-screen technique, wet air had to be used for the experiments. Therefore the drier, shown in fig. 1, was turned off at testing time. During a run, the humidity of the air condenses rapidly in the Laval nozzle because of falling temperature. The water droplets are not able to follow the flow in the test section because of their size and higher density. Thus, the droplet concentration in vortex centers is lower than in the outer flow [9]. This slope of droplet concentration was visualized by a continuous argon-ion laser and a light-sheet optic [5]. The light-sheet optic consisted of a focussing optic and a cylindrical lens. In order to get a thin light-sheet, the focussing optics were adjusted according to the position of ELAC 1. The cylindrical lens was used to illuminate a plane perpendicular to the freestream velocity field. The attachment and separation lines of the different vortices were examined by oil flow-patterns. Before a test run, a solution of lamp oil and a fluorescent dye is applied to the model surface. The lamp oil evaporates during the test period. Thus the remaining powder on the model surface shows the pattern of the flow near to the model surface.

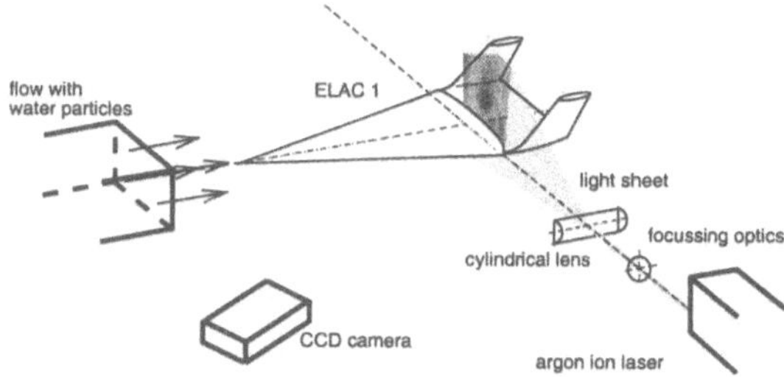


**Figure 3** Experimental setup for vapour-screen with a laser.

# 3 Particle-Image velocimetry in a low-speed tunnel

The measurement of the velocity distribution with Particle-Image velocimetry (PIV) operates by illuminating a particle laden flow in a plane. This light plane must be pulsed to obtain an information about the particle displacement. Several photographs of this light-sheet were taken on a 24x36 $mm^2$ film with different positions of the model. The model was shifted in z-direction between the photographs, see fig. 4a.

## 3.1 Wind tunnel and model

The velocity measurements around ELAC 1 were performed at a low speed wind tunnel of the AIA. The wind tunnel was driven in Eiffel configuration. This enabled a simplified particle seeding for the measurements. A rising particle concentration in the test section was avoided. The size of the cross section is 50x50 $cm^2$, the contraction ratio is 5.7:1. The velocity range of the tunnel is from 0 $m/s$ to 34 $m/s$ with a turbulence level of less than 0.4 %. For the measurements the smaller model of ELAC 1 was used, see fig. 2.

### 3.2 Experimental setup

For application of PIV the flow must be enriched with tracer particles. The particles were added to the flow in front of the rectifier. Oil aerosols with a diameter of $d_p \approx 1\,\mu m$ were used as tracer particles. They were produced by an aerosol generator, which operates with 8 laskin nozzles [4]. Size and density of the used aerosol guaranteed a negligible velocity lag between flow and particles. In order to get a sufficient amount of scattered light, the diameter should not be less than $d_p = 1\,\mu m$ [3]. For avoidance of gaps in particle concentration, a special seeding system was developed, which attained a minimum of 15 particle pairs per interrogation volume.

A ruby laser was used as light source for the double-pulsed laser light-sheet. The wavelength $\lambda$ is 694.3 $nm$, the pulse energy is approximately 400 $mJ$ and the pulse has a duration of 30 $ns$. The time delay between the pulses was 81 $\mu s$. A telescopic optic and two cylindrical lenses were used to form a light-sheet of 50 $mm$ in width with a height of 0.3 $mm$ [12]. The light plane was adjusted perpendicular to the freestream velocity field and the plane of the film in the camera. The photographs were taken from the upper side of the wind tunnel. According to the optical requirements a high-resolution technical pan film and a Leica Makro-Elmarit-R 1:2.8/60 $mm$ were used [13]. Due to the very small depth of focus of $\delta_z < 1\,mm$ [1], the position of camera and light-sheet remain unchanged during testing time. Just the position of the model was changed with the z-traverse in steps of 4 $mm$ during the test period. A sketch of this arrangement is shown in figure 4a.

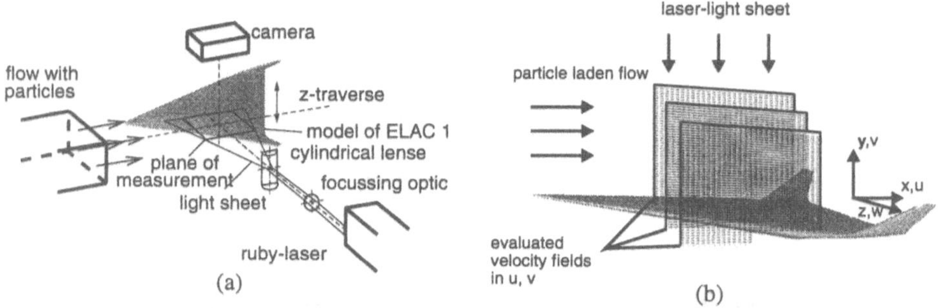


**Figure 4** Experimental setup for the application of PIV to wind tunnels (a), sketch of the used PIV method in incompressible flow with three evaluated velocity planes (b).

### 3.3 Evaluation and three-dimensional reconstruction

The photographs of the different light-sheets in spanwise direction were evaluated with autocorrelation methods. Thus, every photograph was divided in interrogation areas with a diameter of 1 $mm$. The evaluation was carried out with a step size of 0.3 $mm$, the overlap is therefore more than 50 %. A mean velocity was assigned to every of the 27 x 61 interrogation areas by this method. To achieve an optimal evaluation, the interrogation areas had to contain a minimum of 15 particle pictures [1].

After evaluation of all photographs taken at the different light-sheet positions, the in-plane velocity components $u$ and $v$ were validated and interpolated at points of low-grade evaluation as described in [8]. Afterwards the continuity equation was applied for the incompressible flow field (3.1). Figure 4b is representative for the step after validation and interpolation of several light-sheets. The evaluated in-plane velocity components were smoothed by a four-neighbour averaging filter [15]. The derivative of the out-of-plane component was calculated with the help of equation (3.2). The derivatives $\frac{\partial u}{\partial x}$ and $\frac{\partial v}{\partial y}$ were computed by central differences at all evaluated points of each $x$,$y$-plane [14]:

$$div(\vec{u}) = \frac{\partial u}{\partial x} + \frac{\partial v}{\partial y} + \frac{\partial w}{\partial z} = 0, \tag{3.1}$$

$$\frac{\partial w}{\partial z}(x_i,y_j,z) = -\left(\frac{\partial u}{\partial x} + \frac{\partial v}{\partial y}\right)(x_i,y_j,z). \tag{3.2}$$

The out-of-plane velocity was then calculated by integration, starting at the symmetry plane of the delta wing ($z_o = 0$), eqn. (3.3). The stepsize of the integration in z-direction is equivalent to the distance between the measured x,y-planes, it is $\Delta z = 4\,mm$. The sign of the so integrated velocity component is determined by the boundary condition $w(x_i,y_j,z_0 = 0) = 0$:

$$w(x_i,y_j,z) = w(x_i,y_j,z_0) - \int_{z_0}^{z} \left(\frac{\partial u}{\partial x} + \frac{\partial v}{\partial y}\right)(x_i,y_j,z_k)\,dz_k. \tag{3.3}$$

A lowpass Bandworth filter was applied to the resulting three-dimensional velocity field. The vorticity was interpolated by cubic spline interpolation, see fig. 8a.

# 4 Results

## 4.1 Flow visualization

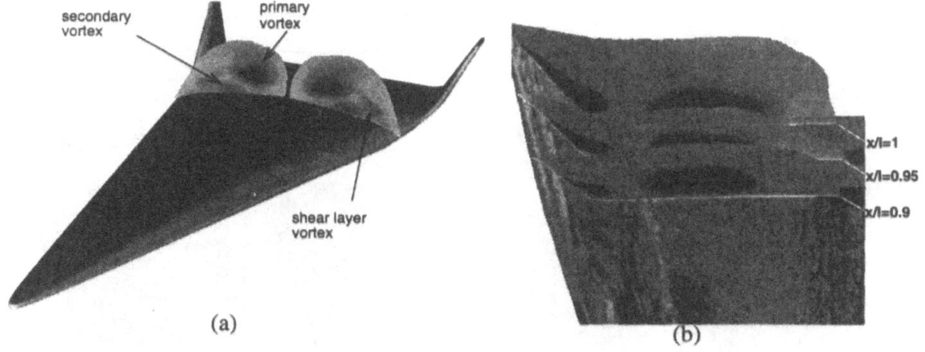


**Figure 5** Part of a vapour-screen at ELAC 1 ($Ma = 2, \alpha = 15°, \frac{x}{l} = 0.66$)(a) and vapour-screen with oil flow-pattern between the control surfaces ($Ma = 2, \alpha = 9°, \frac{x}{l} = 0.9, 0.95, 1$)(b).

Figure 5a shows a part of a vapour-screen image obtained at $Ma = 2$ and $\alpha = 15°$. The positions of the primary and the secondary vortex can clearly be seen. Between the shear-layer and the outer flow a further shear vortex was detected at $\alpha = 15°$. Results from a solution of the parabolized Navier-Stokes equations, see [7], show a good match of position and of size of primary, secondary and shear-layer vortices. As a result from the visualization a topology of the vortex system is suggested in figure 6. At $\frac{x}{l} = 0.66$ primary and secondary separation and attachment lines can be found in $S_1'$ resp. $S_4'$ and $S_3'$ resp. $S_2'$ [11]. Between the control surfaces the topology shows primary and secondary vortices $F_1$ and $F_2$ and a third vortex $F_3$, which was detected in the corner between control surfaces and body. Separation lines are $S_1'$ and $S_4'$, $S$ and $S''$ are saddle points.

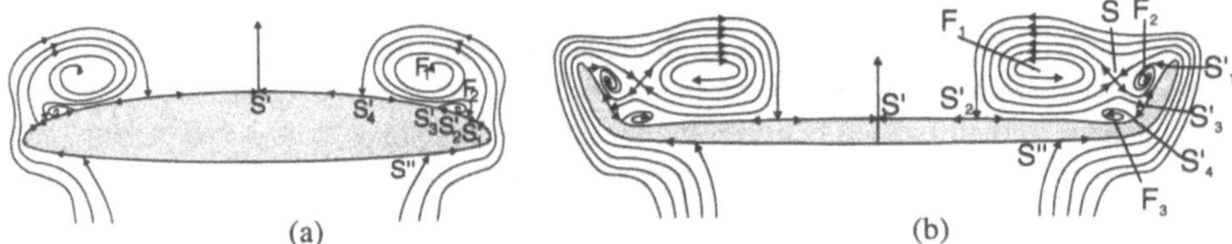


**Figure 6** Estimated vortex topology for $\alpha \leq 10°$ at $\frac{x}{l} = 0.66$(a) and $\frac{x}{l} = 0.95$(b).

Furthermore the difference between turbulent and laminar flow was examined with the help of fixed transition at the leading edge of ELAC 1. In comparison to the laminar flow, figure 7 shows a retarded separation of the turbulent flow. In turbulent flow the attachment line of the primary vortex moves to the centerline, the vapour-screen shows a vortex of increasing width and less height. The separation and attachment lines were measured with the help of oil flow-patterns. The effect of increasing $Mach$ number is shown in figure 7b. Separation and attachment move towards the centerline in both cases.

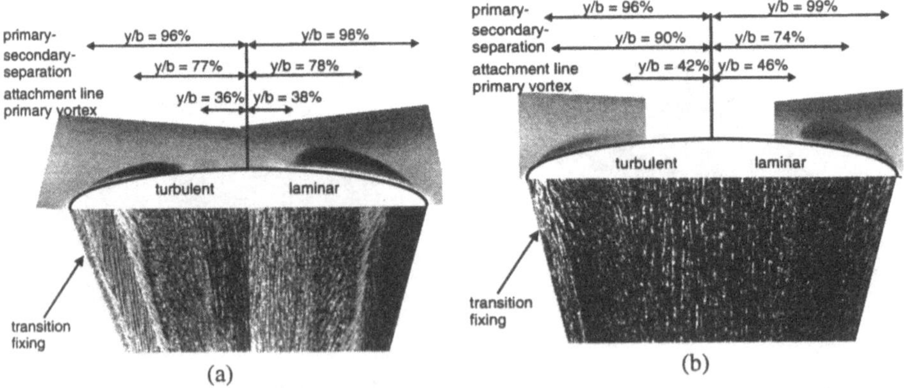


**Figure 7** Vapour-screen (top) and oil flow-pattern (bottom) in laminar and turbulent flow at $\alpha = 9°$, $\frac{x}{l} = 0.3$, $Ma = 2$ (a) and $Ma = 2.5$ (b).

## 4.2 Velocity field

The results of the velocity measurements and the reconstruction of the component $w$ in $z$-direction are shown in figure 8. The velocity vectors present the direction of the flow on the leeward side of the delta wing. Figure 8a shows a cut through the reconstructed flow field perpendicular to the mean flow. In the lower left corner the cross section of the model is represented. The primary vortex can clearly be seen, the printed vorticity indicates the expansion of the vortex. The velocity near to the model is not zero because of the arrangement of the laser light-sheet, the flow and the angle of attack $\alpha = 10°$, see fig. 4b.

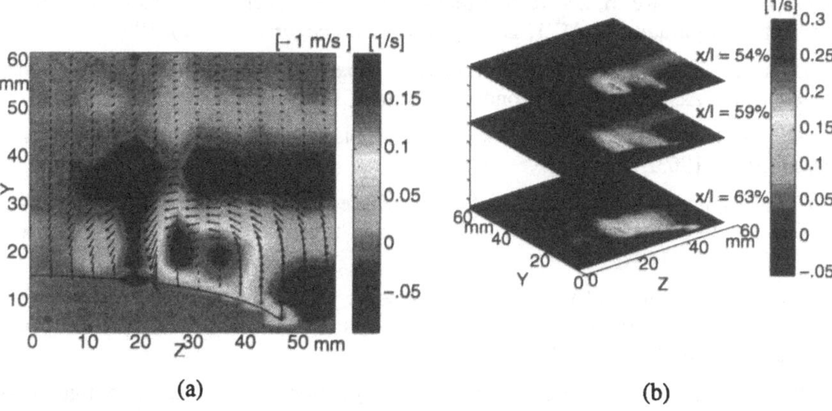


**Figure 8** Velocity field and vorticity at $\alpha = 10°$, $\frac{x}{l} = 0.62$, $u_\infty = 11\ \frac{m}{s}$ ($Re = 2 \cdot 10^5$) (a) and vorticity in three cross sections at $\frac{x}{l} = 0.54, 0.59$ and $0.63$ (b).

Figure 8b shows the main advantage of the used method. After evaluation of all 16 photographs and reconstruction of the third velocity component, the velocity field is available in all three dimensions from $\frac{x}{l} = 50\%$ to $\frac{x}{l} = 66\%$. According to these results the graph shows the vorticity in three cross sections. The line of sight is from the lower side perpendicular to the model in $y$,$z$-direction.

# 5 Conclusion

The flow around the delta-wing configuration ELAC 1 has been examined in supersonic flow with vapour-screen and oil flow-patterns. The vortex system at the leeward side was visualized. Primary, secondary vortices, and a further shear-layer vortex ($\alpha = 15°$) were detected in supersonic flow. An estimated vortex topology is presented. Further the effect of turbulent flow was investigated. Turbulent flow retards separation, the primary vortex develops lower but the width increases in comparison to laminar flow. The velocity distribution was measured with PIV at a low-speed wind tunnel of the AIA. After the reconstruction of the third velocity component in spanwise direction, the velocity distribution at the leeward side of the model was obtained. The data set is well suited for the validation of future numerical investigations. The aim of future investigations is the measurement of the velocity field in the trans- and supersonic wind tunnel of the AIA, see fig. 1.

## Acknowledgement

These investigations were supported by the Graduiertenkolleg "Transportvorgänge in Hyperschallströmungen" and the Sonderforschungsbereich 253 "Grundlagen des Entwurfs von Raumflugzeugen" of the DFG.

## References

[1] R. J. Adrian. Particle-imaging techniques for experimental fluid mechanics. ***Annu. Rev. Fluid Mech.***, 23:261–304, 1991.

[2] F. Decker, G. Neuwerth, and R. Staufenbiel. Low-speed aerodynamics of the hypersonic research configuration ELAC 1. ***Z. Flugwiss. Weltraumforsch.***, 17(2), 1993.

[3] F. Durst, A. Melling, and J. H. Whitelaw. ***Principles and practice of laser–doppler–anemometry***. Academic Press, London, 2. edition, 1981.

[4] W. H. Echols and J. A. Young. Studies of portable air-operated aerosol generators. ***NRL Report 5929***, 1963.

[5] G.E. Erickson and A.S. Inenaga. Fiber-optic-based laser vapor screen flow visualization system for aerodynamic research in larger scale subsonic and transsonic wind tunnels. ***NASA TM-4514***, 1994.

[6] D. Hänel, A. Henze, and E. Krause. Supersonic and hypersonic flow computations for the research configuration ELAC 1 and comparison to experimental data. ***Z. Flugwiss. Weltraumforsch.***, 17(2), 1993.

[7] E. Krause and A. Henze. Recent progress in hypersonics: The ELAC configuration. In ***Proc. 2. Europ. Symp. Aerothermodynamics for Space Vehicles***, pages 567 – 573. ESTEC, ESA SP-347, Nov. 1994.

[8] E. Krause, W. Limberg, A. Abdelfattah, H. Fechner, R. Ortmann, and S. Rotteveel. Experimentelle und numerische Untersuchung der Einlaß- und Kompressionsströmung bei höheren Drehzahlen. ***Kolloquium des SFB 224 Motorische Verbrennung, RWTH-Aachen***, 1996.

[9] J. McGregor. The vapour-screen method of flow visualisation. ***Journal of Fluid Mechanics***, 11(4):481–511, 1961.

[10] C.A. Müller and J. Ballmann. Flow computation for the hypersonic configuration ELAC 1 at low speeds and large incidence. ***Z. Flugwiss. Weltraumforsch.***, 17(2), 1993.

[11] D.J. Peake and M. Tobak. ***Three-dimensional interactions and vortical flows with emphasis on high speeds***, volume AGARD-AG-252. NASA, 1980.

[12] J.P. Prenel, R. Porcar, and A. El Rhassouli. Three-dimensional flow analysis by means of sequential and volumic laser sheet illumination. ***Experiments in Fluids***, 7:133 – 137, 1989.

[13] M. Raffel. ***PIV-Messungen instationärer Geschwindigkeitsfelder an einem schwingenden Rotorprofil***. PhD thesis, Deutsche Forschungsanstalt für Luft- und Raumfahrt, Okt. 1993. DLR-FB 93-50.

[14] O. Robinson and D. Rockwell. Construction of three-dimensional images of flow structure via particle tracking techniques. ***Experiments in Fluids***, 14:257 – 270, 1993.

[15] G.R. Spedding and E.J.M. Rignot. Performance analysis and application of grid interpolation techniques for fluid flows. ***Experiments in Fluids***, 15:417 – 430, 1993.

# Simulation of the Flow Past a Circular Cylinder Using Local Block Refinement

**C. F. Lange, M. Breuer, F. Durst**
Department of Fluid Mechanics
University of Erlangen–Nürnberg
Cauerstr. 4, D-91058 Erlangen, Germany

## Summary

*The crossflow past a heated cylinder was investigated in the Reynolds number range of 0.001–40 and for temperature loadings 1.003 to 1.5. For the simulation a new blockwise local refinement procedure was employed. Comparison of this procedure with computations on conventional block structured grids demonstrates the advantages of the new method. The numerical investigation revealed a strong dependence of the drag coefficient on the size of the computational domain. The analysis of the Nusselt number results led to the suggestion of a different representation of this quantity.*

## Introduction

In the case of flow around an immersed body, the main region of interest for the computation is usually the vicinity of the body and the wake region behind it. In spite of this, numerical predictions show that the computational domain has to be extended to a certain distance from the body so that the boundary conditions do not disturb the solution in the region of interest, i.e. the free flow solution sought. High velocity gradients require very fine grids in the immediate vicinity of the body, whereas at larger distances coarser grids may be employed. Hence, flow around immersed bodies is a typical problem where local refinement of the grid is needed to yield more accurate flow predictions with the available computational resources.

Assuming a structured discretization of the computational domain, it is always possible to divide this domain into blocks, embedding the region to be refined within a few of them. To permit a local refinement of the mesh, the block structure has to be generalized, allowing for discontinuities of the grid lines through the block interfaces. This approach is called *patched grids* [1] or *zonal grids* [2]. Patched grids have been used so far mainly in aerodynamics to solve the Euler equations or a combination of Euler/thin-layer Navier-Stokes equations [3, 4, 5, 6]. For the present work, a new local block refinement procedure based on the patched grid approach was implemented in an existing parallel flow simulation code, called FASTEST. Under extensive use of the local block refinement, it was possible to carry out a detailed investigation of the crossflow past a heated cylinder at low Reynolds numbers.

In the next two sections, brief descriptions of the basic equations, the discretization and the refinement procedures are presented. Subsequently the problem of a cylinder in crossflow is described and interesting results are presented, analyzing the drag coefficient and the Nusselt number as function of the Reynolds number.

## Basic Equations and Discretization Procedure

The governing equations for an incompressible Newtonian fluid in a two-dimensional geometry expressing the conservation of mass, momentum and energy are given by

$$
\begin{aligned}
\frac{\partial}{\partial x_j}(\rho U_j) &= 0 \,, & (1)\\
\frac{\partial(\rho U_i)}{\partial t} + U_j \frac{\partial(\rho U_i)}{\partial x_j} &= -\frac{\partial P}{\partial x_i} - \frac{\partial}{\partial x_j}\left[-\mu\left(\frac{\partial U_i}{\partial x_j} + \frac{\partial U_j}{\partial x_i}\right)\right] \,, & (2)\\
\rho c_p \left[\frac{\partial T}{\partial t} + U_j \frac{\partial T}{\partial x_j}\right] &= \frac{\partial}{\partial x_j}\left[k \frac{\partial T}{\partial x_j}\right] + \Phi + q_T \,, & (3)
\end{aligned}
$$

where $U_i$ and $x_i$ $(i = 1, 2)$ are the Cartesian velocity components and coordinates, respectively, $t$ is the time, $\rho$ is the density, $\mu$ is the dynamic viscosity, $P$ is the pressure, $c_p$ is the specific heat at constant pressure, $T$ is the temperature, $\Phi$ is the viscous dissipation function and $q_T$ is a distributed heat source.

The discretization procedure employed in this work followed the typical finite volume scheme with flux balances for each control volume (CV) forming a nonlinear algebraic system of equations based on eqs. 2 and 3. Instead of eq. 1, a pressure correction equation was solved iteratively utilizing the SIMPLE algorithm with a colocated arrangement of variables [7]. The scheme is second order accurate in space (central differences) as well as in time (Crank-Nicolson). An iterative ILU decomposition method was used for the solution of the sparse linear systems of the resultant equations and a nonlinear multigrid scheme was employed for convergence acceleration [8]. For parallel computation, a block structured grid partitioning and a message passing strategy were used [9]. These procedures were implemented in a multi-platform code and formed the background for the implementation of the refinement strategy, described below.

## Local Block Refinement

Similarly to the idea of grid partitioning for parallel processing, for the computations in the present work the grid was divided into blocks that might have different mesh densities. For instance, if two blocks were at different processors, the information needed by one processor from its neighbor was limited to a layer of auxiliary CVs along the interface. This information was used as a boundary condition for the block and was updated after each iteration of the solver. For each CV in the vicinity of an internal boundary, an auxiliary CV on the other side of the interface existed and contained the data of an equivalent CV in the neighboring block. Since the point distribution along the interface did not necessarily coincide, patched grids required an additional data structure and an interpolation procedure for each block in order to provide the information expected by its neighbor (see Fig. 1).

To supply the information to the neighbor, an array of virtual CVs fitting the point distribution of the neighbor was computed at each interface segment. This provided a

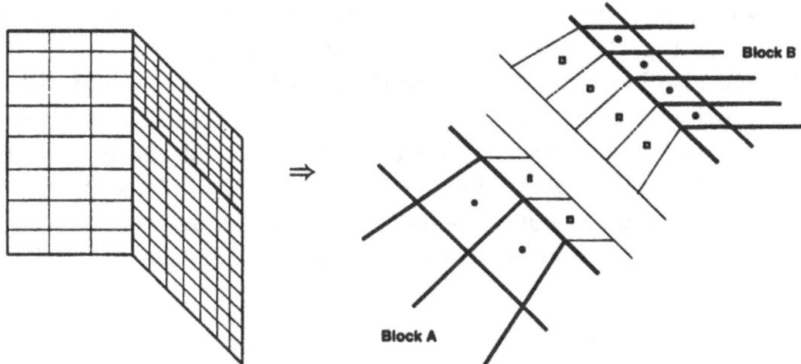


Figure 1: Example of patched grids and blocks with auxiliary CVs.

data structure identical with that expected by the neighboring block. This structure was composed of two kinds of data, namely geometric data and cell centered variables.

The geometric data described the virtual CVs, the position of their corners, centers and their side lengths and volumes. They needed to be computed and exchanged only once since they did not change during the computation. On the other hand, dependent variables, matrix coefficients and other cell centered quantities had to be interpolated and exchanged at each new iteration. A linear interpolation procedure was employed and proved to be very robust and efficient. For more details about the refinement procedure, see ref. [10].

## Cylinder in Crossflow

The numerical method described in the previous sections was employed to investigate the crossflow past a heated cylinder at low Reynolds numbers. The dimensions and boundary conditions of the problem were similar to the case of a hot-wire anemometry probe. The fluid that was considered in the study was air having a temperature of 20 °C at the inflow. All properties of the fluid were treated as temperature dependent. The diameter of the cylinder was 5 $\mu$m and its temperature, assumed to be constant, was computed with values between 21 °C and 166 °C. This corresponds to a range of temperature loadings ($TL$) from 1.003 to 1.5 . $TL$ is defined as

$$TL = \frac{T_w}{T_\infty} \,, \tag{4}$$

where $T_w$ is the wire (cylinder) temperature and $T_\infty$ is the fluid temperature in the region far away from the cylinder, both in absolute units (K). The inflow velocity was adjusted in order to cover a range of Reynolds numbers based on the diameter ($Re$) between $10^{-3}$ and 40. Above these values, velocities were very high and incompressibility could no longer be assumed.

In order to exemplify the gains obtained by the local block refinement procedure used in this work, the authors analyzed the case when the computational domain was 100 times wider than the cylinder diameter. Fig. 2 shows a comparison between grids with and without local block refinement. The locally refined grid was chosen after extensive tests in order to minimize the computing time without accuracy losses. A zoom of both grids can be seen in Fig. 3, where the identity of the point distribution in the vicinity of the cylinder can also be recognized. Note that the figures show only the coarsest grid of a five level multi-grid procedure. The solution was actually computed on the fifth grid. The locally refined grid had a total of 59,400 CVs on the fifth grid level, while the conventional grid had 121,900 CVs, both with 450 CVs along the two-dimensional cylinder surface. The memory requirements followed the same proportion as the number of CVs. The conventional grid needed twice as much memory as the locally refined one.

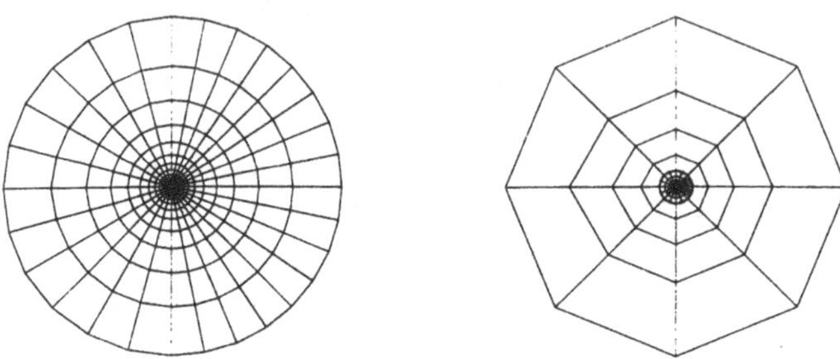

Figure 2: Conventional and local block refined O-grid around a cylinder.

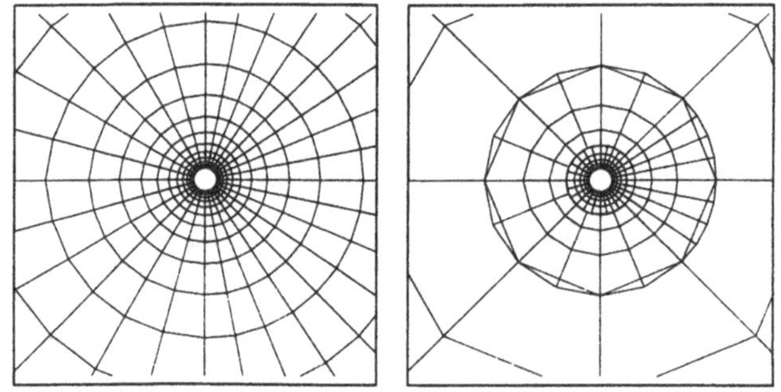

Figure 3: Detail of a conventional and a locally refined grid.

The necessity to subdivide the single block into three blocks for local refinement caused a relative worsening of the convergence compared with the conventional case. In spite of this, the computations with local block refinement (LBR) were 3 to 4 times faster, providing the same level of accuracy as the conventional grid (see Fig. 4).

## Drag Coefficient and Nusselt Number

The most interesting numerical aspect that arose from this work was the strong dependence of global values, such as the drag coefficient, on the relative size of the computational domain. At very small Reynolds numbers ($Re < 1$) the region of influence of the cylinder tended to grow nearly inversely proportional to the value of $Re$. If the boundary of the

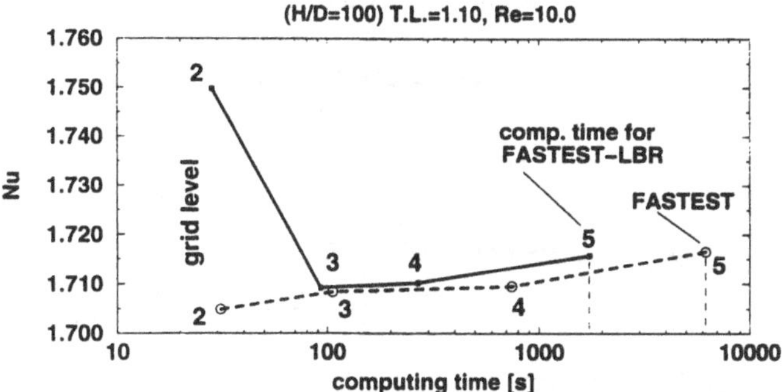


Figure 4: Value of Nusselt number and computing time for different grid levels.

computational domain did not accompany this expansion for decreasing Reynolds numbers, the error caused by the artificial boundary conditions (that are valid actually only at infinity) disturbed the solution, affecting even the vicinity of the cylinder. By means of Fig. 5 it is easy to verify the problem. In both cases the velocity vector $\vec{U}$ is prescribed at the inlet (left half of the outer boundary). The normal gradient of the velocity vector $\frac{\partial \vec{U}}{\partial \vec{n}}$ is set to zero as outflow boundary condition (right half of the outer boundary). In the case of $Re = 1$ the isolines of the velocity component $U$ show that the region strongly perturbed by the cylinder is enclosed in the computational domain, whereas for $Re = 0.1$ this region of influence is larger than the domain and therefore highly disturbed by the outer boundary conditions. When the domain was expanded ten times, similar pictures were obtained by comparing the results of $Re = 0.1$ and $Re = 0.01$.

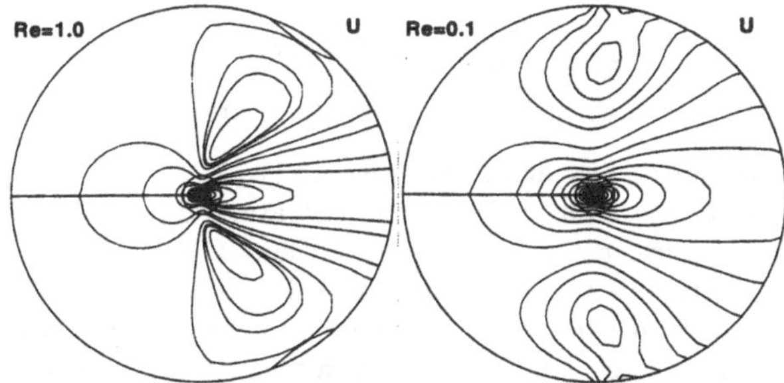


Figure 5: Isolines of velocity component $U$ for a domain 100 times wider than the cylinder diameter at two different Reynolds numbers.

The influence on the drag was similar to the effect of the presence of walls at some distance from the cylinder, namely an increase in the coefficient. This effect was reported previously by Sucker and Brauer [11]. In Fig. 6 the present results are plotted for different sizes of the computational domain to quantify its influence on the drag coefficient. For the sake of comparison, the analytical value obtained with the full Oseen expansion calculated by Tomotika and Aoi [12] is also plotted. Very good agreement with experimental data was also found for larger Reynolds numbers, where the analytical solution fails.

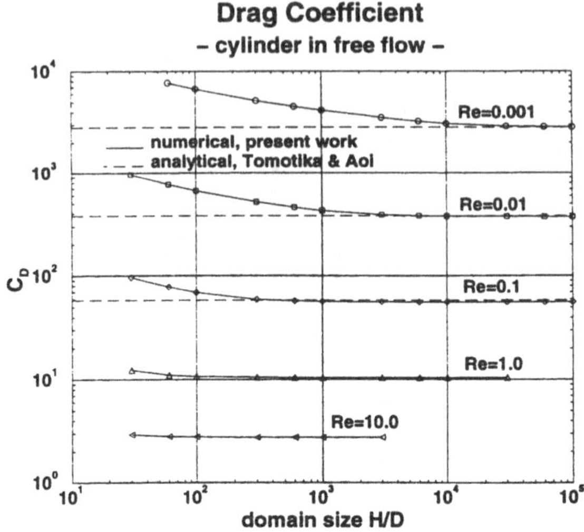


Figure 6: Influence of the size of the computational domain on the drag coefficient.

A further topic analyzed in the present work was the convective heat transfer from the cylinder to the flowing air. A nondimensional quantity that characterizes the convective heat transfer is the Nusselt number. Since the Nusselt number is not a characteristic of the flow, like the Reynolds number, but an auxiliary quantity, there is not a unique definition of the temperature at which the thermal conductivity is determined. In a general form it can be defined as

$$Nu = \frac{q_w}{q_{ref}} = \frac{k_w}{k^*} \left. \frac{\partial \left( \frac{T_w - T}{T_w - T_\infty} \right)}{\partial \left( \frac{r}{D} \right)} \right|_w , \tag{5}$$

where $q_w$ is the actual heat flux at the cylinder wall, $q_{ref}$ a reference heat flux, $k_w$ the thermal conductivity of the fluid at cylinder temperature, $T_w$ the cylinder temperature, $T_\infty$ the inflow temperature, $r$ the radial direction and $D$ the cylinder diameter. $k^*$ is the thermal conductivity at a prescribed temperature. Owing to the large scatter of the experimental data, most investigators try to take into account the influence of the temperature loading on the Nusselt number, calculating the fluid properties at the arithmetic mean temperature, also called *film temperature*: $k^* = k_f$. Although it helps to group the data, Collis and Williams [13] showed that a systematic difference between the curves for different temperature loadings remains. They proposed a correction with an empirical factor and succeeded in obtaining a unique curve for air in the range of their experiments. A comparison of the present results with the data of Collis and Williams based on their correction showed very good agreement, attesting to the quality of the computations (see Fig. 7).

Because of the lack of universality of empirical corrections and the general aim of returning to the basic definition of the Nusselt number as a nondimensional temperature gradient,

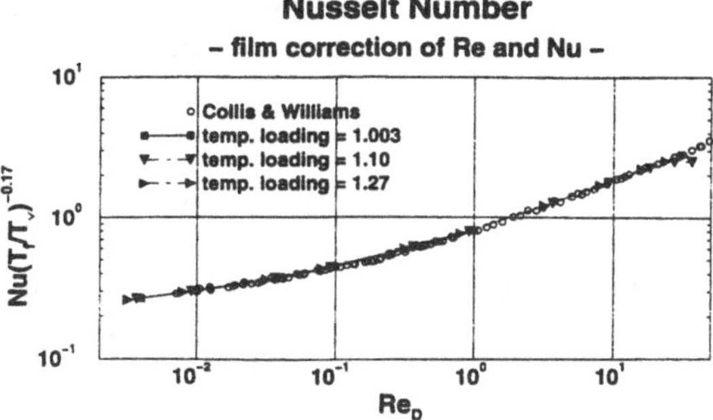


Figure 7: Comparison of computed with experimental data.

the authors propose a different form of correction to account for the temperature loading: both the Nusselt number and the Reynolds number should be calculated with properties at the cylinder temperature $T_w$. In this case $k^*$ is equal to $k_w$ in eq. 5. Applying this concept to the present results, the authors obtained good coincidence of the curves without any empiricism, as can be seen in Fig. 8.

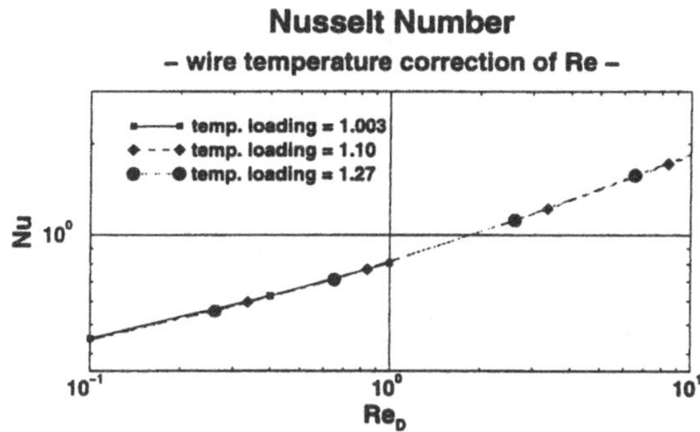


Figure 8: Proposed representation of the curve $Re \times Nu$: all properties at wire temperature.

# Conclusion

The authors applied a new local block refinement procedure to the simulation of the viscous flow around a heated cylinder. The procedure proved to be very robust and well suited for practical applications. Gains of 70–80% in computing time and about 50% in memory requirements compared with the conventional method made it possible to perform a more detailed and accurate analysis.

The simulation of the crossflow past a circular cylinder at low Reynolds numbers showed very good agreement with both experimental and analytical data. An interesting numerical aspect revealed by the investigation is the strong dependence of the drag coefficient on the relative size of the domain leading to overestimation, if the domain is too small.

Finally, a different representation of the Nusselt number is proposed, that takes into account the influence of the temperature loading without any empirical correction.

## Acknowledgement

This work was supported by a fellowship from the German Academic Exchange Program (DAAD) for C. F. Lange and by a research grant of the Bavarian Consortium for High Performance Scientific Computing (FORTWIHR), which are gratefully acknowledged.

## References

[1] B. Børresen Jr. A multi-block methodology for solving flow problems in complex geometries. In S. Wagner, J. Periaux, and E. H. Hirschel, editors, *Second European Computational Fluid Dynamics Conference, Stuttgart*, pages 377–382, Wiley, Chichester, 1994.

[2] M. M. Rai. A conservative treatment of zonal boundaries for Euler equation calculations. *J. Comput. Phys.*, 62:472–503, 1986.

[3] M. M. Rai. An implicit, conservative, zonal-boundary scheme for Euler equation calculations. *Comput. Fluids*, 14(3):295–319, 1986.

[4] M. M. Rai. Navier–Stokes simulations of rotor/stator interaction using patched and overlaid grids. *J. Propulsion*, 3(5):387-396, Sept.-Oct. 1987.

[5] A. Kassies and R. Tognaccini. Boundary conditions for Euler equations at internal block faces of multi–block domains using local grid refinement. paper 90-1590, AIAA, 1990. AIAA 21st Fluid Dynamics, Plasma Dynamics and Lasers Conference, 1990, Seattle, WA.

[6] M. Amato and L. Paparone. On the application of the multizone modelling and the local grid refinement technique for high–lift airfoil analysis. In S. Wagner, J. Periaux, and E. H. Hirschel, editors, *Second European Computational Fluid Dynamics Conference, Stuttgart*, pages 370–376, Wiley, Chichester, 1994.

[7] M. Perić, R. Kessler, and G. Scheuerer. Comparison of finite–volume numerical methods with staggered and colocated grids. *Comput. Fluids*, 16:389–403, 1988.

[8] M. Hortmann, M. Perić, and G. Scheuerer. Finite volume multigrid prediction of laminar natural convection: Bench–mark solutions. *Int. J. Num. Methods Fluids*, 11:189–207, 1990.

[9] F. Durst and M. Schäfer. A parallel blockstructured multigrid method for the prediction of incompressible flows. *Int. J. Num. Methods Fluids*, 22:549–565, 1996.

[10] F. Durst, C. F. Lange, and M. Schäfer. Local block refinement using patched grids for the parallel computation of viscous flows. report LSTM 445/N, LSTM, Universität Erlangen-Nürnberg, 1995.

[11] D. Sucker and H. Brauer. Fluiddynamik bei quer angeströmten Zylindern. *Wärme- und Stoffübertragung*, 8:149–158, 1975.

[12] S. Tomotika and T. Aoi. An expansion formula for the drag on a circular cylinder moving through a viscous fluid at small Reynolds numbers. *Q. J. Mech. Appl. Math.*, 4(4):401–405, 1951.

[13] D. C. Collis and M. J. Williams. Two-dimensional convection from heated wires at low Reynolds numbers. *J. Fluid Mech.*, 6:357–384, 1959.

SECONDARY INSTABILITIES IN A LAMINAR SEPARATION BUBBLE

U. Maucher, U. Rist, S. Wagner
Universität Stuttgart, Institut für Aerodynamik und Gasdynamik
Pfaffenwaldring 21, D-70550 Stuttgart, Germany

## SUMMARY

If a strong adverse pressure gradient is imposed on an initially laminar boundary layer, it is decelerated and eventually separates. In this case, disturbances are strongly amplified and finally transition to turbulence takes place. Thus, a strong exchange of high-momentum fluid far away from the wall and low-momentum fluid near the wall is forced and reattachment may occur. Besides this more general understanding of the physics of laminar separation bubbles (LSB), detailed knowledge of the transition to turbulence in a LSB is still missing. In Direct Numerical Simulations (DNS) artificial 2-D and 3-D disturbances are excited. If the size of the bubble is sufficiently large, DNS with realistic 2-D amplitude but very low initial 3-D amplitude indicate temporal amplification of the 3-D disturbances. Depending on the spanwise wavenumber, the self excited 3-D disturbances are either subharmonic or fundamental with respect to the 2-D disturbance. Subsequently, the spatio-temporal development of the 3-D disturbances is analyzed in order to find the source of their amplification.

## NUMERICAL METHOD

The numerical method was developed in the research-group "Transition and Turbulence" of the Institut für Aerodynamik und Gasdynamik IAG [5],[3]. It is based on the complete, incompressible Navier-Stokes equations in vorticity-velocity formulation. A flat-plate boundary-layer is subjected to an adverse pressure gradient by prescribing the potential velocity $u_p$ at the upper boundary of a rectangular integration domain (figure 1). At the inflow boundary there is stationary flow. In a disturbance strip at the wall, artificial 2-D and 3-D disturbances can be excited by periodic, wall-normal suction and blowing. At the outflow boundary boundary-layer assumptions are made. Upstream of the outflow boundary a buffer domain is applied, which damps the unsteady component of the flow smoothly to zero [4]. Thus, reflections at the outflow boundary can be suppressed.
All dimensional variables (marked by bars) are non-dimensionalized in the usual manner by a length length scale $\overline{L}$, the velocity $\overline{U}_\infty$, and the Reynolds number $Re = \frac{\overline{U}_\infty \overline{L}}{\overline{\nu}}$.
The time-integration of the non-dimensionalized Navier-Stokes equations in vorticity-transport formulation

$$\frac{\partial \omega_x}{\partial t} + \frac{\partial}{\partial y}(v\omega_x - u\omega_y) + \frac{\partial}{\partial z}(w\omega_x - u\omega_z) = \frac{1}{Re}\frac{\partial^2 \omega_x}{\partial x^2} + \frac{\partial^2 \omega_x}{\partial y^2} + \frac{1}{Re}\frac{\partial^2 \omega_x}{\partial z^2} \quad (1)$$

$$\frac{\partial \omega_y}{\partial t} - \frac{\partial}{\partial x}(v\omega_x - u\omega_y) + \frac{\partial}{\partial z}(w\omega_y - v\omega_z) = \frac{1}{Re}\frac{\partial^2 \omega_y}{\partial x^2} + \frac{\partial^2 \omega_y}{\partial y^2} + \frac{1}{Re}\frac{\partial^2 \omega_y}{\partial z^2} \quad (2)$$

$$\frac{\partial \omega_z}{\partial t} + \frac{\partial}{\partial x}(u\omega_z - w\omega_x) + \frac{\partial}{\partial y}(v\omega_z - w\omega_y) = \frac{1}{Re}\frac{\partial^2 \omega_z}{\partial x^2} + \frac{\partial^2 \omega_z}{\partial y^2} + \frac{1}{Re}\frac{\partial^2 \omega_z}{\partial z^2} \quad (3)$$

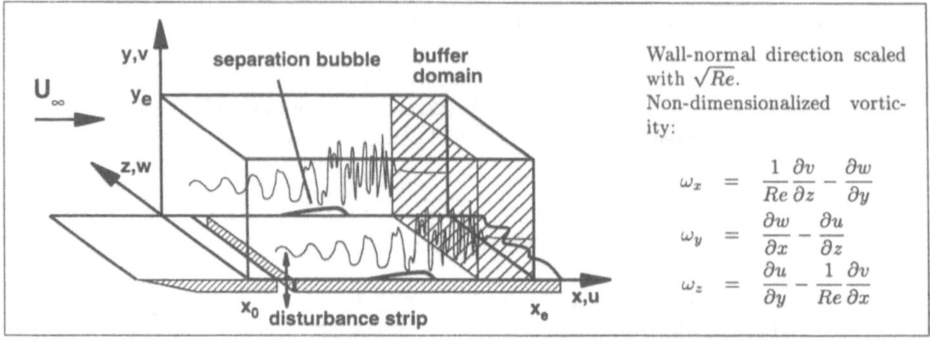


Fig. 1 Integration Domain

is performed with a fourth-order accurate Runge-Kutta scheme. The four stages per time step are coupled with an alternating upwind-downwind (and vice versa) discretisation of the $x$-convection terms. Thus, the scheme is of 4th-order accuracy in time and space and insufficiently resolved oscillations (wiggles) are numerically damped.
In spanwise direction a spectral approach which exploits periodicity is used:

$$f(x,y,z,t) = \sum_{k=-K}^{K} F_k(x,y,t)e^{ik\gamma z}, \tag{4}$$

where $\gamma$ is the basic spanwise wavenumber defined by $\gamma = 2\pi/\lambda_z$. In experiments this assumption has been shown to be in good agreement with physics of boundary layer flows [2]. Assuming spanwise symmetry of the 3-D disturbance development, only the spectral modes from $k = 0$ (2-D) to $k = K$ have to be regarded.
The velocity components can be obtained by solving three Poisson-equations:

$$\frac{\partial^2 u}{\partial x^2} + \frac{\partial^2 u}{\partial z^2} = -\frac{\partial \omega_y}{\partial z} - \frac{\partial^2 v}{\partial x \partial y}, \tag{5}$$

$$\frac{\partial^2 w}{\partial x^2} + \frac{\partial^2 w}{\partial z^2} = \frac{\partial \omega_y}{\partial z} - \frac{\partial^2 v}{\partial y \partial z}, \tag{6}$$

$$\frac{1}{Re}\frac{\partial^2 v}{\partial x^2} + \frac{\partial^2 v}{\partial y^2} + \frac{1}{Re}\frac{\partial^2 v}{\partial z^2} = \frac{\partial \omega_x}{\partial z} - \frac{\partial \omega_z}{\partial x}. \tag{7}$$

Central finite differences with 4th order accuracy are applied for the $x$- and $y$-derivatives. A penta-diagonal system coupled only in $x$ is solved for each line $y = const.$ to obtain $u$ (equation 5) and $w$ (equation 6). A line Gauss-Seidel scheme accelerated by multigrid is applied to effectively solve the $v$-Poisson equation. Nevertheless, the iteration of the $v$-velocity requires 60% of the total CPU-time.
The 2-D component of the downstream velocity $u$ is integrated from the continuity equation:

$$u_{k=0}(x,y) = u_{k=0}(x_0,y) - \int_{\xi=x_0}^{x} \frac{\partial v_{k=0}}{\partial y} d\xi. \tag{8}$$

The numerical code is well adapted to the computer architecture used for the calculations (NEC SX-4: 32 CPUs, vector architecture, shared memory, peak-performance/CPU = 2

GigaFLOPS). In serial mode the performance of the code is approximately 1150 MFLOPS, i.e. 58 % of the peak performance. Optionally, the nonlinear terms are calculated in parallel on lines with $y = const.$ Apart from these terms, the $k$-harmonics of the spectral ansatz are independent from each other and can be treated in parallel. In parallel computations, the performance per CPU decreases only slightly to $\approx$ 1100 MFLOPS/CPU. Calculating with K=10 (11 CPUs) a speedup of about 10 is achieved. This is important for the parametric studies performed further down.

## DIRECT NUMERICAL SIMULATION (DNS), 2-D BASE FLOW

The DNS is performed in two steps. In a first step, a "base flow" with separation bubble is calculated in a 2-D simulation. Spatially amplified Tollmien-Schlichting waves (frequency ($\beta_0 = 10$) are excited by periodic suction and blowing at the wall. In the second step, the development of 3-D disturbances in this two-dimensional, periodic flow is examined. The 2-D "base flow" is addressed in this section, the 3-D disturbance development in the following sections.

In experiments in the Laminar Wind Tunnel of the IAG [7], a LSB on a wing section with a chord-length of $\bar{c} = 0.615\ m$ was investigated. The chord Reynolds number was $Re_c = 1.2 * 10^6$. The 2-D DNS aims at calculating a LSB with comparable physical properties. The free-stream velocity $\overline{U}_\infty$ is $29.3\frac{m}{s}$. In DNS the reference length $\overline{L}$ is chosen to be 6.15 $cm$. Hence, the Reynolds number is 120000, the chord length $c = \bar{c}/\overline{L} = 10$. In 2-D DNS the generation of turbulence and the related fine-scale eddies is suppressed by the absence of three-dimensionality. However, saturated large-scale vortices effect the momentum transport towards the wall, which is necessary to force re-attachment. The size of the bubble is influenced by the initial disturbance amplitude. The lower it is, the later disturbances gain nonlinear amplitude and subsequently the separation bubble grows in length and reverse-flow intensity.

In the experiments, two velocity distributions $u_e(x)$ at the edge of the boundary layer have been measured (figure 2a). The first one (squares) refers to a flow with separation bubble. For the second, turbulent one, separation has been suppressed by fixing a turbulator upstream of the separation bubble (triangles). In DNS, the upper boundary of the integration domain is far away from the boundary layer and is situated in the potential flow. In former simulations, it turned out that due to displacement effects caused by the separation bubble, the velocity distribution at the edge of the boundary layer yields a

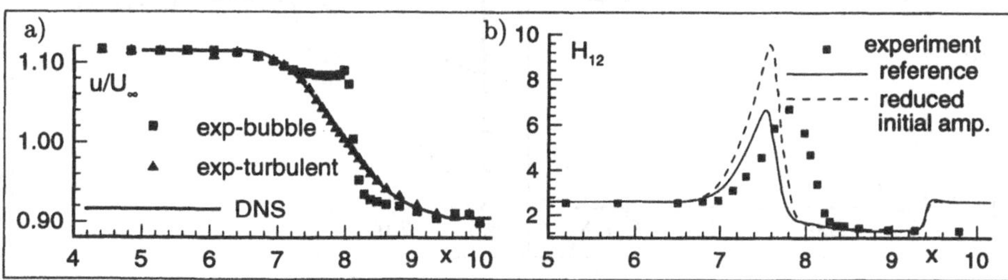


Fig. 2 Comparison of a) the edge-velocity distribution and b) the shape parameter in the experiment (symbols) and the DNS.

typical "separated" pressure plateau at the location of the bubble, and differs from the upper boundary condition $u_p$. For that reason, the turbulent distribution was chosen as upper boundary condition for the DNS (solid line).
In the present calculations however, a very fine grid is necessary to accurately resolve the flow. Therefore the height of the integration domain was restricted by memory and CPU-time requirements. Thus, the upper boundary is rather close to the boundary layer and has strong impact on re-attachment. Consequently, the edge velocity in the simulation is only slightly varied by displacement effects but strongly influenced by the upper boundary. The major interest of the calculations presented here, is to examine stability properties of the spatio-temporal disturbance development. Differences between the experimental and numerical mean-flow have been accepted. In figure 2b, the shape parameter $H_{12}$ for DNS with two different initial disturbance amplitudes is compared to the experimental data (symbols). The computed separation occurs upstream of the location in the experiment and the bubble is shorter, due to the aforementioned suppression of displacement effects. The length of the bubble changes only moderately if the initial disturbance amplitude is decreased (from $u' \approx 10^{-4}$ (solid line) to $\approx 10^{-5}$ (dashed), for example). In contrast, with lower disturbance-amplitude the maximum of $H_{12}$ dramatically increases. The $H_{12}$-maximum of the solid curve is quite close to the experimental maximum. Therefore, this case is regarded to be in better agreement with the experiment and is chosen for the 3-D DNS.

## DIRECT NUMERICAL SIMULATION, 3-D DISTURBANCES

The periodic "base flow" is superposed by 3-D disturbances with very small amplitude ($u' \approx 10^{-10}$). Amplification curves (figure 3a) show the spatial development of the 2-D wave (solid line) and the 3-D wave (dashed). The respective higher harmonics are depicted by the dotted lines. Both waves are excited in the disturbance strip. Subsequently they propagate downstream and are convectively amplified by linear instability. As the 2-D wave gains a nonlinear amplitude ($\approx 1\%$ of $U_\infty$) strong nonlinear amplification (secondary

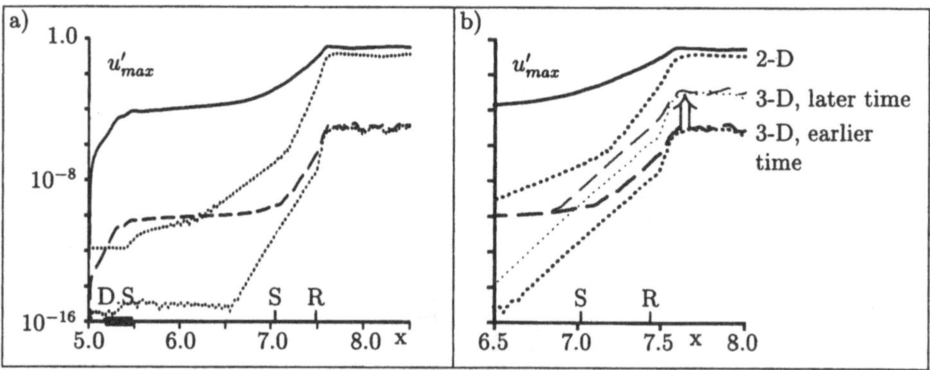


Fig. 3 a) Disturbance amplitude of 2-D (solid) and 3-D (dashed) waves b) vicinity of the bubble. Dotted lines: higher harmonics. Arrow: 3-D temporal growth. S - separation, R - re-attachment. D.S.: disturbance strip.

instability, see Herbert [1]) of the 3-D wave sets in. Finally, the 2-D wave saturates in the re-attachment zone (separation and re-attachment indicated by S and R) and the nonlinear amplification of the 3-D modes vanishes, as well.

A close-up view of the disturbance amplitudes in the vicinity of the separation bubble is plotted in figure 3b. Additional thin lines show the amplitude of the 3-D modes at a later disturbance cycle. For $x < 6.8$, the amplitudes do not differ in both periods shown in the plot. The disturbances propagate downstream in a convective manner. In the separation bubble, however, the amplitude curves exhibit temporal growth of the 3-D modes (arrow). This is in contrast to separation bubbles at lower Reynolds numbers [6]. Indeed, the temporal growth rate is very low compared to the spatial growth rate $\alpha_i$ which is in this domain about 20 – 40 times the temporal rate $\beta_i$.

Interestingly, the nonlinear interaction of the 2-D wave with frequency $\beta_0 = 10$ and a wave-number of $\alpha \approx 20$ and 3-D modes with different spanwise wave-numbers $\gamma$ yields two different temporally amplified 3-D waves, as can be seen in figure 4:

1. "Subharmonic" disturbances (dash-dotted line):
   a 3-D mode with the frequency $\beta = 1/2 * \beta_0$ and the respective higher harmonics ($\frac{2n+1}{2} * \beta_0$) are generated; the spanwise wave number is: $\gamma_{sub} \approx \alpha$.
2. "Fundamental" disturbances (solid line):
   a longitudinal vortex and a 3-D mode with the frequency $\beta = \beta_0$ (and the higher harmonics $n * \beta_0$); the spanwise wave number is: $\gamma_{fun} > \gamma_{sub}$.

The quotes are used to distinguish the underlying mechanisms from the amplification due to fundamental and subharmonic secondary instability in the sense of Herbert [1] which would yield the same kind of 3-D disturbances but without temporal amplification. The calculation for $\gamma = 16$ showed no temporal amplification. Therefore, the dash-dotted line in fact consists of only one point ($\gamma = 20$ and the information that $\beta_{i,\gamma=16} < 0$).

The numerical scheme has been extensively tested. In order to obtain figure 4, 3-D disturbances were excited for only a short period of time in the separation bubble i.e., it is sufficient to force the 3-D disturbances only once to obtain the disturbances mentioned above. In any case, the 2-D wave is continuously excited.

Elimination of possible reflections at the inflow boundary in an additional damping zone between that boundary and the disturbance strip did not affect $\beta_i$. To exclude influences

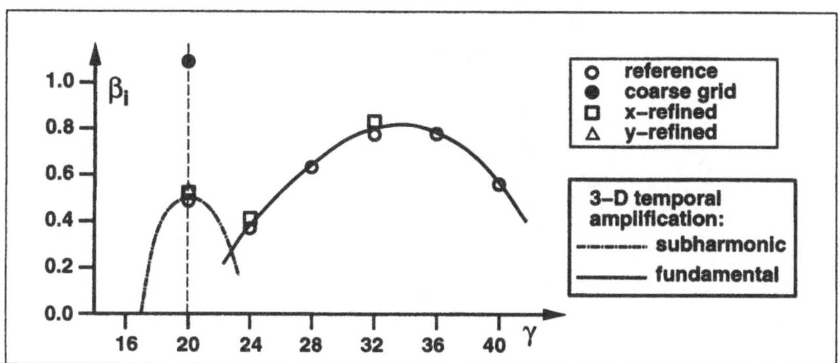


Fig. 4 Dependency of temporal amplification from the spanwise wavenumber $\gamma$. Grid-refinement study at $\gamma = 20$ (dashed line).

of the outflow boundary condition a test calculation with a very long integration domain has been performed. The 3-D disturbances in the bubble exhibit the in the shorter domain observed growth rate $\beta_i$, before the first disturbances reach the outflow boundary. Finally, the influence of the spatial discretisation was investigated for a fixed spanwise wavenumber $\gamma = 20$. The grid was refined from coarse to the reference and an even finer $x$-spacing by a factor of two, respectively. Additionally, the reference discretisation was refined in the wall-normal direction by a factor of two. With the coarse grid (figure 4, filled circle), the amplification rate $\beta_i$ strongly differed from the other cases. In contrast, the refinement of the reference grid showed only negligible influence on the growth rate $\beta_i$ (triangle and square on the dashed line). Comparisons of the amplification rate $\beta_i$ for $\gamma = 24$ and $\gamma = 32$ calculated with the reference grid (circles) and the x-refined grid (squares) confirm this observation. Thus, results with the reference discretisation can be regarded as grid-independent.

To further investigate the origin and the properties of the temporal 3-D growth, the 3-D component of the velocity $u$ is analyzed. Since the amplitude changes spatially and temporally, it is necessary to normalize $u$ to compare the time signals at different streamwise locations. The velocity in the whole $y,t$-domain for a specific $x$-location is normalized with the maximum value in this plane. If the resulting velocity $u_{norm}$ is plotted in a $x,t$-diagram, the spatio-temporal development can be investigated. Figure 5a shows this representation of $u_{norm}$ in the "subharmonic" case ($\gamma = 20$) at a constant distance from the wall ($y = 0.93$) and on the centerline ($z = 0$). Due to the spanwise spectral ansatz, at $z = \pm\lambda_z/2$ the velocity is exactly the same with opposite sign. The spatial $x$-axis is horizontal, the time axis is vertical. Consequently, streaks from lower left to top right (positive gradient) denote convective (downstream) disturbance motion, and streaks with negative gradient indicate upstream influence. The time signals of four characteristic streamwise positions, marked by bold lines, are plotted in figure 5b.

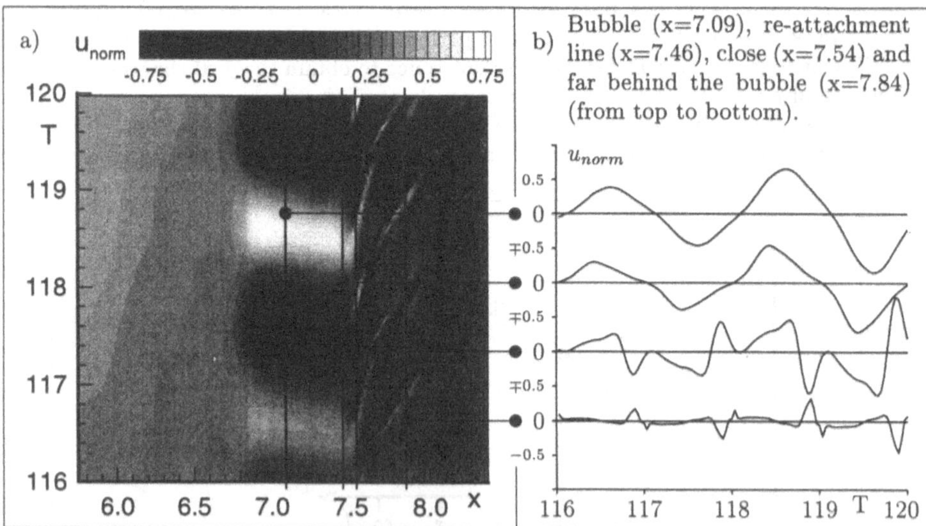


Fig. 5 'Subharmonic' growth ($\gamma = 20$, $y = 0.93$, $T$: period of the 2-D waves): a) temporal-spatial development of the 3-D velocity in constant wall-distance, b) time-signals.

In the bubble ($7.04 < x < 7.46$) the $x,t$-gradient is negative. Disturbances propagate upstream with high velocity. The stronger contrast in the upper part of the plot highlights the temporally increasing amplitude. For $x > 7.5$, streaks with positive gradient are apparent. In this region, the flow is dominated by convection. The time signals significantly change when the re-attachment line ($x = 7.46$) is passed. In the bubble and at the re-attachment line, the subharmonic frequency dominates (two top curves). The nonlinear interaction between 2-D and 3-D modes and correspondingly the generation of higher harmonics is weak. Behind the bubble, however, the amplitude of the higher harmonics is very large due to the saturated 2-D wave (figure 3).

The phase and amplitude of the disturbance profiles of the "subharmonic" case ($\gamma = 20$) and the "fundamental" case ($\gamma = 32$) are plotted in figure 6. Since the time signal is normalized to one, the contribution of the Fourier harmonics to the total time signal can be easily recognized. In the "subharmonic" case (solid lines) the bubble is dominated by the subharmonic oscillation. Vortex structures (weak phase shift), with a rotation centerline almost parallel to the $x$-axis, expand with large velocity from the re-attachment zone into the bubble and change their sign periodically. Downstream of the bubble, the subharmonic oscillation looses influence. Finally, at $x = 7.84$ its amplitude is less than 0.2 related to the local disturbance amplitude. In contrast to the "subharmonic" case, the longitudinal vortices (dotted line) have large impact on the "fundamental" mechanism. In the bubble, their amplitude is approximately three times the amplitude of the fundamental oscillations (dash-dotted). Pairs of counter-rotating longitudinal vortices dominate the bubble.

Typical eigenfunctions (profiles) of classical linear stability theory and of secondary stability theory are characterized by changes of the sign of the amplitude and a corresponding phase shift of $\pi$. While the disturbance profiles in the "subharmonic" do not indicate any such profiles, they are present in the "fundamental" case in the close vicinity of the re-attachment line ($x = 7.46$, $x = 7.54$). Nevertheless, this region is so small, that it seems very unlikely that amplification according to classical linear or secondary instability theory is involved in the observed phenomena. At $x = 7.84$, the profiles of the "fundamental" mechanism are quite similar to the "subharmonic" profiles. In any case, the 3-D behavior is dominated by the 2-D periodic "base flow".

The observed 3-D temporal amplification very likely originates from a new type of sec-

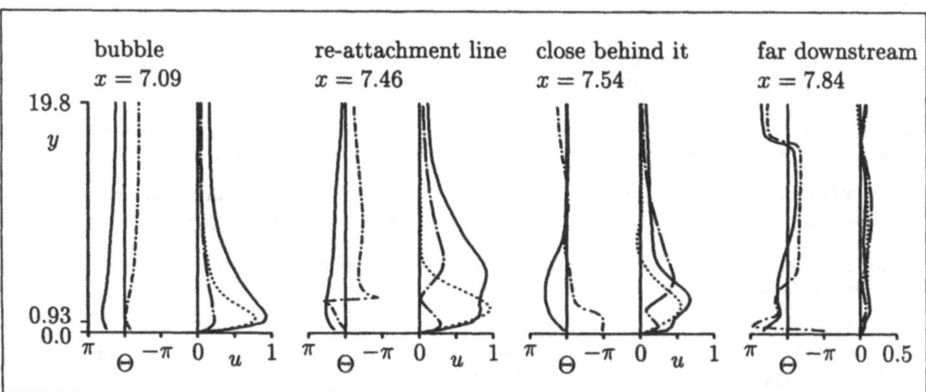


Fig. 6 Comparison of 3-D profiles of the "subharmonic" (solid lines) and the "fundamental" (wave: dash-dotted and longitudinal vortex: dotted) 3-D disturbances.

ondary instability which leads to a spanwise deformation of the large-amplitude TS-wave or "vortices" in the re-attachment region. Thus, the re-attachment line is oscillating in the "subharmonic" case, whereas the increasing deformation due to the longitudinal vortices is only modulated by the fundamental wave in the "fundamental" case.

## CONCLUSION

DNS of a laminar separation bubble exhibit temporal amplification of 3-D modes by interaction with a nonlinear 2-D wave. Compared to the 2-D frequency, this growth can be subharmonic or fundamental. Disturbance waves are only observed in the "fundamental" case, and only in a very localized region near the re-attachment line. There are no hints on amplification due to classical linear stability theory or secondary stability theory. The re-attachment region is the origin of disturbances propagating upstream as well as of waves convecting downstream. Referring to the present knowledge, the re-attachment region therefore has to be considered as the source of the temporal growth of the 3-D modes. To obtain better insight into the wake of laminar separation bubbles, the simulations of both mechanisms will be continued until the temporal amplification leads to transition. Longitudinal vortices behind the bubble for instance, which have been observed in many experiments, might be due to the "fundamental" mechanism observed here.

## ACKNOWLEDGEMENTS

This research is supported by the Deutsche Forschungsgemeinschaft DFG under contract number Ri 680/1-1,2.

## REFERENCES

[1] HERBERT, T.: "Secondary instability of boundary layers", Ann. Rev. of Fluid Mech., 20 (1988), pp. 487–526. 1988.

[2] KACHANOV, Y. S.: "Physical mechanisms of laminar-boundary-layer transition" Ann. Rev. Fluid Mech., 26 (1994), pp. 359–391.

[3] KLOKER, M.: "Direkte Numerische Simulation des laminar-turbulenten Strömungsumschlages in einer stark verzögerten Grenzschicht", Dissertation, Universität Stuttgart, 1993.

[4] KLOKER, M., KONZELMANN, U., FASEL, H.: "Outflow boundary conditions for spatial Navier-Stokes simulations of transition boundary layers", AIAA J., 31(4) (1993), pp. 620–628.

[5] RIST, U., FASEL, H.: "Direct numerical simulation of controlled transition in a flat-plate boundary layer", J. Fluid Mech., 298 (1995), pp. 211–248.

[6] RIST, U., MAUCHER, U., WAGNER, S.: "Direct Numerical Simulation of Some Fundamental Problems Related to Transition in Laminar Separation Bubbles", Computational Fluid Dynamics '96, John Wiley & Sons Ltd., 1996, pp. 319–325.

[7] WÜRZ, W., WAGNER, S.: "Experimental Investigations of Transition-Development in attached Boundary-Layers and Laminar Separation Bubbles", NNFM, this volume.

# Flexibility and Efficiency of a Transport-Equation Turbulence Model for Three-Dimensional Flow

Erik Monsen, Ralf Rudnik, Hans Bleecke
Deutsche Forschungsanstalt für Luft- und Raumfahrt (DLR) e.V.
Institut für Entwurfsaerodynamik
Lilienthalplatz 7, D-38108 Braunschweig, Germany

## Summary

For the prediction of the aerodynamic performance of aircraft or aircraft components, the modeling of the viscous, in particular the turbulent effects is of ever increasing importance. In order to improve the quality of numerical simulations of complex configurations, a more general description of the fluid dynamics as well as a high flexibility in relation to the topology of the computation and high numerical efficiency is required. In this work an algebraic turbulence model, the Baldwin-Lomax model, and a transport equation turbulence model, the two-equation k-ω model of Wilcox, are used for the simulation of the flow around a realistic 3-D Wing-Body configuration. Using single and multiblock versions of the same grid, it is shown that the k-ω model delivers the same results, independent of the number or structure of the blocks, whereas the Baldwin-Lomax model does not. In terms of additional costs, the k-ω model required, for the same number of iterations on the same grid, approximately 40% more memory and 50% more time than the Baldwin-Lomax model. Additionally, issues of grid convergence and a surface roughness boundary condition for the k-ω model are also discussed.

## 1 Introduction

Currently the aerospace industry is increasingly relying on numerical or CFD simulations in the early aircraft design phases. Their requirements for the numerical prediction of aircraft performance include a more general description of the fluid physics, a high flexibility with regard to the computational domain and a high numerical efficiency. To better describe the viscous and turbulent effects, turbulence models playing an ever more important role in achieving an accurate numerical simulation. Algebraic turbulence models, such as the Baldwin-Lomax model [1], are currently widely used by industry for the calculation of realistic problems, such as wing-body configurations [4]. Unfortunately, the range of application of these models, in terms of both the physical accuracy and the numerical handling, is quite limited. Transport-equation models, such as the k-ω model of Wilcox [12], are not as widely used, principally because of the significantly higher cost of using them. These models have distinct advantages, including a wider range of physical applicability, i.e. attached and separated flows, as well as, when properly implemented, neither geometric nor topological limitations.

The k-ω 2-equation model of Wilcox has the potential to meet these requirements. The ability of this model to predict attached and strongly separated flows around airfoils and wings has already been demonstrated in the literature [7,11]. In this work the flexibility and efficiency of the k-ω turbulence model will be investigated. In addition, the algebraic model of Baldwin and Lomax will be used as a baseline for comparison. The flexibility, with regards to multiblock grid topologies, which are necessary for the optimal use of a parallel computer system, will be demonstrated using a 3D wing-body configuration. A single-block grid will be divided into a multiblock grid, in order to prove that the solution is independent of the number as well as structure of the blocks.

To visualize this in 2D, a grid around a RAE 2822 profile has been split at critical positions (Fig. 1). The k-ω model and this split grid were then used to compute a transonic case (Case 10, as described in [2]) with a strongly separated flow (Fig. 2). It was thus proven that a block can be split at the suction peak (Block 2/3), directly behind the shock (Block 3/4) and through the boundary layer (Block 3/4/5), without any visible loss in accuracy. This paper will attempt to prove that this is just as valid in 3D as in 2D.

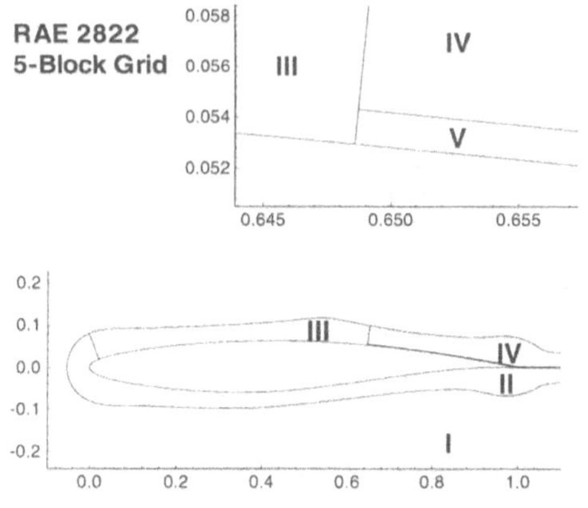


Fig. 1 RAE 2822 5-Block Grid Topology

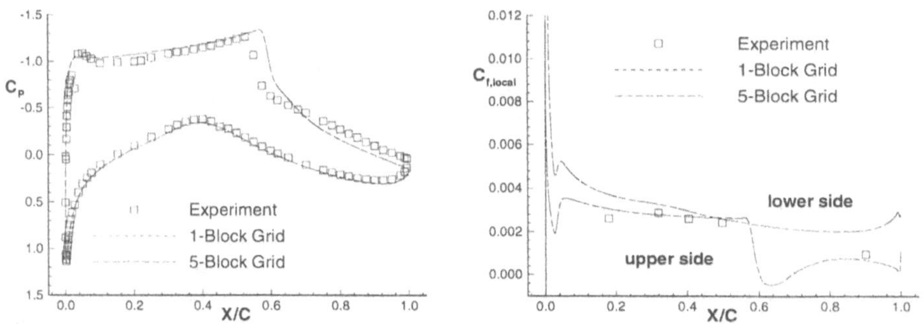


Fig. 2 RAE 2822 Case 10 Pressure Distribution and Skin Friction Distribution

To determine the additional cost of using this model in comparison to algebraic models, the efficiency of the model, measured in terms of computational time and memory requirements, will be simultaneously analyzed.

Additionally, to demonstrate that these results are not just chance, lift and drag curves have been calculated for this case on the standard grid and on a refined grid. An additional degree of flexibility can be added to the k-ω model with the implementation of a surface roughness boundary condition for the ω equation. This boundary condition, which has been used for all of the calculations, will be presented and results of a parameter study will be discussed.

Based on these observations, conclusions, from the engineers point of view, will be drawn as to the range of application and the limitations of both models.

## 2 Numerical Environment

The calculations presented in this paper were computed on the DLR's NEC SX-4/8 vector supercomputer in Göttingen. The system has 8 CPUs, each with a peak performance of 2000 MFlop/s. There are 2 GBytes main memory (MMU) with a additional 4GBytes of extended memory (XMU). In the current queueing configuration, a single user can access up to 1.2 GBytes main memory and 2.9 GBytes extended memory. As this limits grids to about 3 million cells, which represents a medium resolution grid for a complex 3D viscous calculation, a system with more memory is desirable. One such system is the DLR's IBM SP2 cluster in Köln. The core of this parallel system is a cluster of 58 nodes, each with 267 MFlop/s peak performance and 256 MBytes local memory, for a total of 14.5 GBytes and a comparable level of performance. This study aims to prove that computations on this and other parallel computers will be worthwhile.

The flow solver used in this study was FLOWer [5]. The FLOWer code solves the compressible, three-dimensional Reynolds-averaged Navier-Stokes equations. It uses a finite volume formulation on block-structured meshes and it is designed to simulate flows around complex aerodynamic configurations. It is a highly portable code that currently runs on a variety of scalar, vector and parallel computers [3]. FLOWer has directly evolved from the DLR-CEVCATS code and is being further developed in close cooperation with the German aeronautical industry and universities in the framework of the MEGAFLOW project [8].

## 3 Turbulence Models

The Baldwin-Lomax turbulence model [1] is an algebraic relation for the turbulent eddy viscosity $\mu_t$. This is a zonal model, which is subdivided into an inner and outer layer. A special treatment has been added for the wake region of the flow. The main drawback of this model is that the calculation has to be carried out along grid lines normal to the no-slip wall, which prohibits one from subdividing blocks parallel to walls, especially in the boundary layer. This is an extreme limitation, as it can prohibit the optimal load balancing of a parallel calculation.

The k-ω model of Wilcox [12,13] involves the solution of 2 additional transport equations, for the turbulent kinetic energy k and the specific dissipation rate ω, whose ratio defines the eddy viscosity to $\mu_t$. Additionally, there are source terms for the production and destruction of turbulence. The main advantage of this model is that the calculation of the turbulent eddy viscosity is carried out pointwise, e.g. locally at a point in the grid, independent of the mesh topology or the surface geometry. This is a significant advantage, as the automatic parallel load balancing tool is now free to optimally subdivide the grid.

This implementation of the k-ω model offers two upwind discretizations, following an AUSM or a Roe approach, for the convective fluxes of the turbulence terms. To reduce the stiffness of the source terms, a point implicit formulation for the turbulent terms has been added to the Runge-Kutta time marching scheme [6]. Additionally, the multigrid approach has been extended for the turbulence equations, either V- or W-cycles, which results in a factor 5 speed-up in wall-clock turn-around time. As for the main equations, additional boundary conditions are added for k and ω. In the case of the no-slip wall boundary condition, the turbulent kinetic energy k is set to 0 and the specific dissipation rate ω is set proportional to the surface roughness and shear stress [13]:

$$\omega = u_\tau^2/\nu \cdot S_R \qquad S_R = \begin{cases} \left(50/k_R^+\right)^2, & k_R^+ < 25 \\ \left(100/k_R^+\right), & k_R^+ \geq 25\,. \end{cases}$$

The parameter $k_R^+$ needs only be small enough so that the surface is hydraulically smooth:

$$k_R^+ = u_\tau k_R / \nu < 5 .$$

## 4 Computations for the F4 Wing-Body Configuration

The 3D configuration used in this study is the DLR F4 wing-body configuration. Experimental measurements for this model were made in 3 wind tunnels, the NLR-HST tunnel, the ONERA-S2MA tunnel and the DRA 8ftx8ft tunnel, and results were reported by Redeker and Müller in [10]. The selected test case has a Mach number of 0.75, a Reynolds number of $3 \times 10^6$ and an angle of attack of $0.93°$. The computations were carried out on a grid with 192x64x64 or 786,482 cells, which was generated by Daimler-Benz Aerospace, EFV in Bremen. It should be noted that this grid has only 120 cells around the wing and 16 cells normal to the wing in the boundary layer. The Baldwin-Lomax and k-ω models predict wing pressure distributions which are close to the experiment (Fig. 3), but predict a shock to far downstream and fail to meet the suction peak. The coarseness of the grid is most likely the cause of these differences, and will be investigated in section 4.3. However, this should not interfere the analysis of the flexibility and efficiency of the models and will, indeed, be an additional challenge for them.

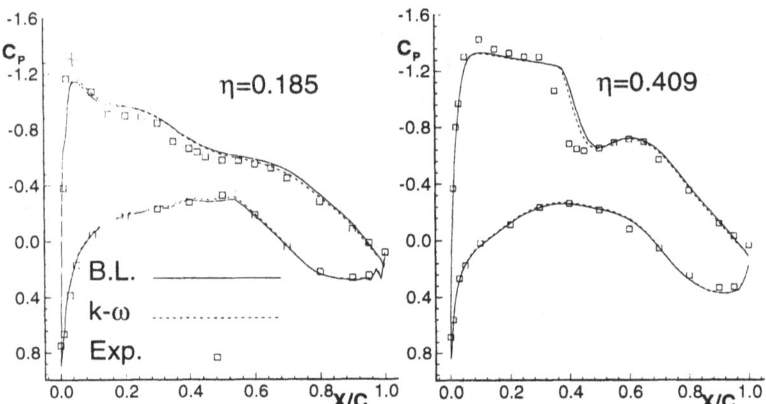


Fig. 3 DLR F4 Wing-Body, $Ma_\infty=0.75$ $Re=3 \times 10^6$ $\alpha=0.93°$, Experiment, Baldwin-Lomax, k-ω

### 4.1 Variation of the Block Topology

Next the original grid was subdivided into 16 equally sized blocks of 48x32x32 cells. A comparison of the 1-Block and 16-Block results for the Baldwin-Lomax model (Fig. 4) shows not only that the convergence history is less stable, but also that the lift and drag coefficients are clearly different. It could be expected, that if the grid is further subdivided, the solution could differ even more. In any case, this proves that the Baldwin-Lomax model is not reliable for use with multi-block grids, which must be generated to optimally load-balance a parallel computation.

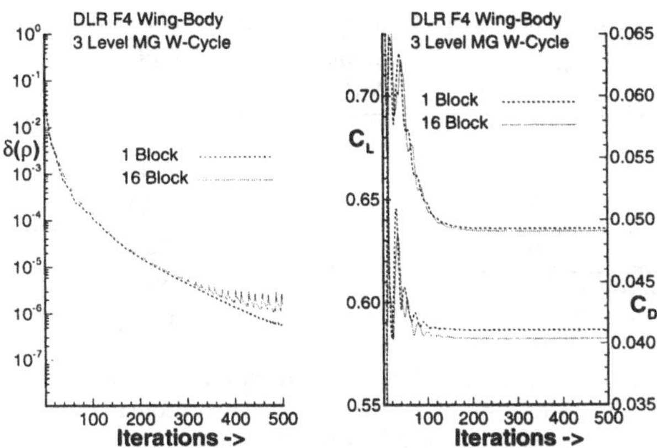


Fig. 4 Baldwin-Lomax, 1 & 16 Blocks, Convergence History of Density Residual, Lift and Drag

A comparison of the results for the k-ω Model (Fig. 5) shows that the convergence of the solution, as well as the lift and drag coefficients, are essentially unchanged by the different grid topology. This agrees with the results already observed in 2D and supports the hypothesis that the k-ω model will deliver the same results independent of the number of structure of the blocks in a multiblock grid, and is therefore flexible enough to be used for optimally load-balanced simulations in a parallel computing environment.

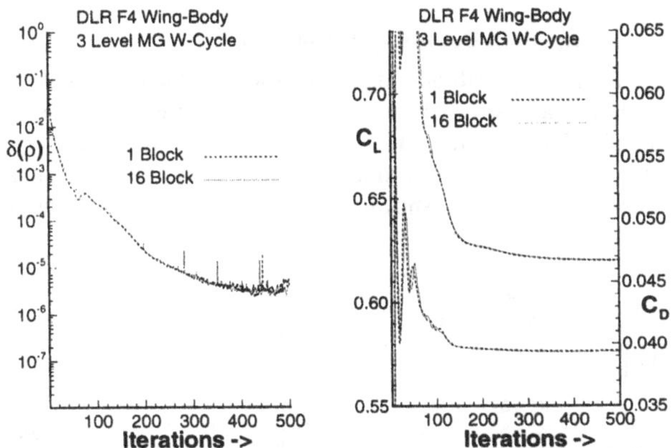


Fig. 5 k-ω of Wilcox, 1 & 16 Blocks, Convergence History of Density Residual, Lift and Drag

However, it is not enough that the k-ω model is topologically flexible. The model must also be numerically efficient. As seven, instead of five, equations must now be solved, it can be estimated that the computational costs will be 40% higher. A comparison of the memory and time requirements for these four runs (Table 1) shows that the k-ω model (with the AUSM discretization) requires almost 40% more memory, but unfortunately over 80% more time.

Table 1: Comparative resource usage for Baldwin-Lomax and k-ω/AUSM, 1 & 16 Blocks

| Model | Blocks | Memory MBytes | CPU Seconds | MFLop/s |
|---|---|---|---|---|
| B.L. | 1 | 522 | 3475 | 761 |
| B.L. | 16 | 546 | 7147 | 382 |
| k-ω / AUSM | 1 | 711 | 6391 | 888 |
| k-ω / AUSM | 16 | 758 | 13202 | 451 |
| How much more does k-ω cost compared to B.L.? | | +38% Memory | +84% Time | +17% |

However, when the convective fluxes of k and ω are discretized with a Roe-like scheme, the computations require just over 50% more time than Baldwin-Lomax (Table 2) and there is no significant change in the results. This result is much more in line with the 40% estimate and promise, with further optimization of the programming, to be even less.

Table 2: Comparative resource usage for Baldwin-Lomax, k-ω/AUSM and k-ω/Roe, 1 Block

| Model | CPU Seconds | Time +% | MFLop/s |
|---|---|---|---|
| B.L. | 3337 | | 817 |
| k-ω / AUSM | 6198 | +86% | 939 |
| k-ω / Roe | 5024 | +51% | 906 |

### 4.2 Variation of the No-Slip Wall Boundary Condition

As mentioned earlier, the no-slip wall boundary condition for the ω equation is a function of the surface roughness and the shear stress. Fig. 6 demonstrates the effect of a smoother wall ( < 5 ) and a rougher wall ( > 5 ). The most dramatic changes can be seen in the lift coefficient, which is higher for the smoother, less dissipative wall and is lower for the rougher, more dissipative wall. The change is the result of the shock shifting slightly downstream for the smoother wall and slightly upstream for the rougher wall.

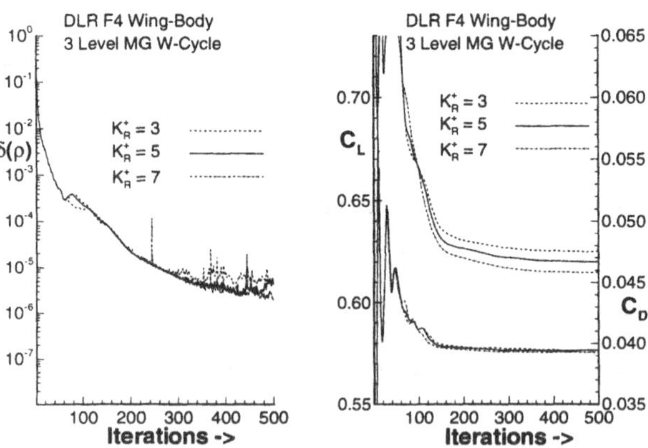


Fig. 6 Variation of Surface Roughness $k_R^+$, Convergence of Density Residual, Lift and Drag

## 4.3 Grid Refinement

A necessary part of any numerical investigation is a grid refinement study. A second grid with 256x88x48 or 1,081,344 cells was generated by Daimler-Benz Aerospace, EFV in Bremen. This grid includes refinements in the resolution of the leading and trailing edges, as well as more cells in the boundary layer. This grid has now 160 cells around the wing and 30 cells normal to the wing in the boundary layer. Comparing wing pressure distributions (Fig. 7), it can be seen that with grid refinement, the suction peaks are now better resolved.

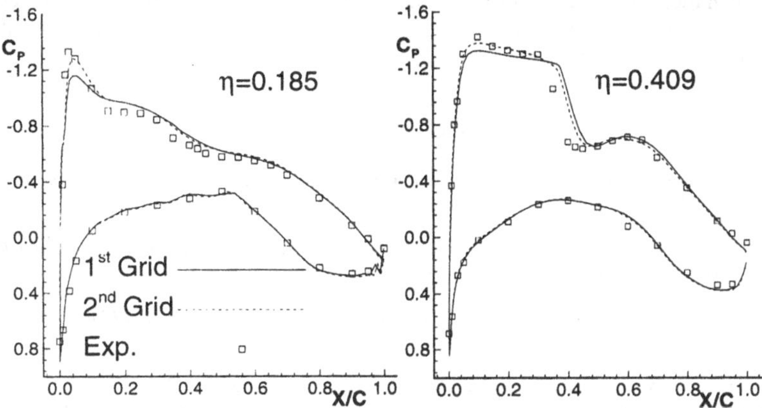


Fig. 7 DLR F4 Wing-Body, $Ma_\infty$=0.75 Re=$3x10^6$ $\alpha$=0.93°, k-ω Results with Grid Refinement

It is also important to compute the flow for a variety of flow conditions, in order to demonstrate that the results were not just by chance. In addition to $\alpha$=0.93°, the flow was computed on the first grid for $\alpha$=0.0°, 2.0° and 4.0° (Fig. 8). Although the initial cases matched the measured lift fairly well, the drag predictions for all angles is much to high. In addition, the lift for the stalled case ($\alpha$=4.0°) is much too low, indicating that the separated flow is not properly resolved. Using the refined grid, the $\alpha$=0.93° and 4.0° cases were recalculated and resulting lift and drag coefficients agree much better with the measurements. This suggests that further grid refinement should lead to even better results.

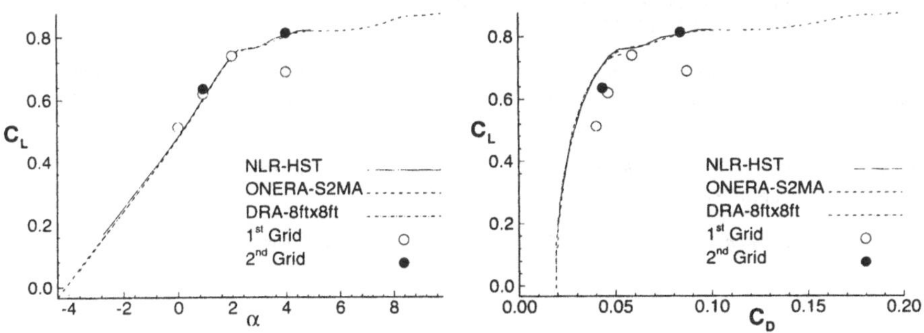


Fig. 8 DLR F4 Wing-Body, $C_L$ vs. $\alpha$ and $C_L$ vs. $C_D$, k-ω Results with Grid Refinement

## 5 Conclusions

It has previously demonstrated in the literature that the k-ω model of Wilcox, a transport equation turbulence model, provided a much more general description of the flow physics than algebraic models, such as the Baldwin-Lomax model. This investigation has demonstrated that the k-ω model is also topologically flexible. The computed results presented were the same, independent of the multiblock structure. The current implementation is also numerically efficient. The solution of 40% more equations, seven instead of five, requires 40% more memory and 50% more time. It can therefore be concluded that the k-ω model should be able to better meet the industrial requirements for the prediction of aerodynamic aircraft performance on current vector and parallel high-performance computers.

## References


[1] Baldwin, B., Lomax, H., *Thin Layer Approximation and Algebraic Model For Separated Turbulent Flows*, AIAA-Paper 78-0257 (1978).

[2] Cook, P.H., Mc Donald, M.A., Firmin, M.C.P., *Aerofoil RAE 2822 - Pressure Distributions and Boundary Layer and Wake Measurements*, AGARD-AR-138 (1979).

[3] Eisfeld, B., Bleecke, H.-M., Kroll, N., Ritzdorf, H., *Parallelization of Block Structured Flow Solvers*, AGARD R-807, pp 5.1-5.20, 1995.

[4] Elsholz, E., Longo, J.M.A., *Navier-Stokes Simulation of a Transonic Wing-body Configuration.*, 1993 European Forum 'Recent Developments and Applications in Aeronautical CFD', Sept. 1-3, 1993, Bristol, UK.

[5] *Installation and User Handbook for the Project FLOWer*, Internal Doc. No. QS-FLOWer-3008.

[6] Mavriplis, D.J., Martinelli, L., *Multigrid Solution of Compressible Turbulent Flow on Unstructured Meshes Using a Two-Equation Model*, NASA CR 187513 (1991).

[7] Monsen, E., Rudnik, R., Investigation of the Blunt Trailing Edge Problem for Supercritical Airfoils, AIAA-Paper 95-0089 (1995).

[8] *Projecktbeschreibung MEGAFLOW*, DLR-IB 129-96/8 1996.

[9] Radespiel, R., Rossow, C.-C., Swanson, C., *Efficient Cell-Vertex Multigrid Scheme for the Three-Dimensional Navier-Stokes Equations*, AIAA-Journal, Vol. 28, No. 8, pp. 1464-1472 (1990).

[10] Redeker, G., Müller, R., *A Comparison of Experimental Results for the Transonic Flow around the DFVLR-F4 Wing Body Configuration*, DFVLR-IB 129-83/71 (1983).

[11] Rudnik, R., Ronzheimer, A., Schenk, M., C.-C. Rossow, *Berechnung von 2- und 3-dimensionalen Hochauftriebskonfigurationen durch Lösung der Navier-Stokes Gleichungen*, DGLR-Jahrestagung, 24.09.-27.09.1996, Dresden.

[12] Wilcox, D.C., *Reassessment of the Scale-Determining Equation for Advanced Turbulence Models*, AIAA Journal, Vol. 26, pp. 1299-1310 (Nov. 1988).

[13] Wilcox, D.C., *Turbulence Modeling for CFD*, DCW Industries, Inc., La Cañada, CA, 1994.

# VISUALIZATION OF HIGH-SPEED BOUNDARY-LAYER TRANSITION WITH FPA INFRARED TECHNIQUE

L. Müller and A. Henckels
DLR, Hauptabteilung Windkanäle, Abteilung Köln-Porz
Linder Höhe, D-51147 Köln

## SUMMARY

*A modern "Focal Plane Array" (FPA) infrared system was used to visualize natural transition phenomena on a 2.9 degree PTFE cone model, exposed to the hypersonic flow field of the H2K facility at flow conditions of Mach number 6 and Reynolds numbers up to $2{\cdot}10^7$ /m. To study bluntness influence, the model was alternately equipped with four different nose segments of 1 to 15 mm nose radius. Visualized transition locations, measured temperatures, and evaluated heat fluxes were found to be in good agreement with approximate engineering relations taken from literature. By that the infrared system has been qualified for high speed boundary-layer transition detection and the technical performance of such advanced infrared systems was demonstrated, in particular to provide experimental data to validate numerical codes for hypersonic transition prediction.*

## INTRODUCTION

The knowledge of boundary-layer transition phenomena plays an eminent role in aircraft design, because the transition influences drag and by that the required thrust. Moreover, in case of hypersonic flight speeds, the designer of the heat protection system needs precise transition data since the heat loads are essentially controlled by this phenomenon. Today, design data are generated by CFD codes, which need information on transition to implement turbulence models. Also the experimentalist is interested in reliable information about the transition location on his wind-tunnel model for the interpretation of results.

Models being equipped with thermocouples are known as costly and provide only local information. To install gauges at optimum position, the particular flow field has to be known in advance. Therefore, infrared cameras providing field information have been established in aerothermal testing and transition detection as non-intrusive measuring devices (e.g. [2], [3], [6]). For most of the applications single-detector serial-line scanners were used, even though multi detector FPA (focal plane array) infrared imaging systems have a much higher spatial resolution, because all the array elements stare simultaneously when collecting infrared energy [4].

In the past FPA systems were relatively expensive and a consistent temperature calibration of its high number of detector elements was difficult to achieve. Recently this situation changed by technical progress as a number of distributors offer FPA measurement systems at moderate prices. The intention of the present study is to demonstrate the measurement capabilities of such an advanced system for hypersonic transition research, i.e. in an aerodynamic field still being short of experimental data.

## EXPERIMENTAL ARRANGEMENT AND DATA EVALUATION

Cone transition experiments have been conducted at the H2K facility, a conventional blow down hypersonic wind-tunnel. Electrically heated air expands through an axisymmetric nozzle with 60 cm exit diameter, providing free-stream flow conditions of Mach number 6 and Reynolds numbers up to $2 \cdot 10^7$ /m. Inside the test chamber the 2.9 degree sharp cone model was installed. Its geometric length of 1.2 m correspondingly decreases by mounting different tip segments having 1, 5, 10 and 15 mm blunt radius. In order to reduce internal heat fluxes the cone was machined of PTFE (Teflon) material and its surface was sprayed with black paint to achieve a defined emissivity value of 0.9 in the infrared regime.

During the tests an AVIO TVS-8000 infrared thermal video system was looking from the top of the test chamber through a $CaF_2$ window with 97.5% transmittance in the spectral range of the IR-system (Fig. 1). This IR-system, which was developed by Cincinnati Electronics Corporation and Nippon Avionics Company, Ltd. of Japan, is equipped with a Sterling cooled InSb array of 160(H) x 120(V) elements, which are sensitive to radiation in the 3-5 μm region and allow to resolve temperature differences up to 0.025°C. At a distance of 1.5 m between camera and model the spatial resolution of the system is about 1.5 mm.

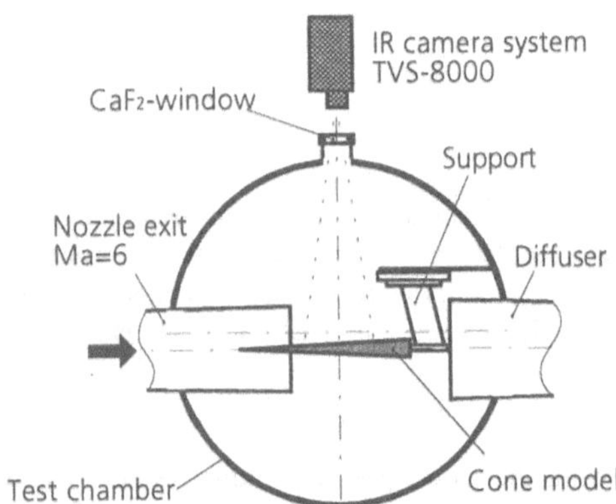


Fig. 1: Position of FPA camera, window and model at the H2K test chamber.

Even though the camera manufacturer already calibrated the system, a calibration check was carried out by placing a reference source inside the test chamber, being at the same pressure level as under flow condition. By turning the surface of the reference source, it was proven that the temperature measurement error due to surface inclination does not affect the temperature measurements near to the cone´s symmetry line. During one test 512 frames in a 0.1 seconds interval were saved on optical disk for post processing. After the test heat fluxes were evaluated by a special procedure taking into account the relatively long flow exposure time of the model, as being typically for blow down facilities [5].

After carefully aligning the cone in flow direction, infrared images were taken to visualize the transition zone, its extension and its dependency on the nose radius as well as on the free stream Reynolds number. Figure 2 demonstrates, how the transition zone clearly became evident from the infrared images. Only slight misalignments of the cone caused asymmetrical heating patterns (Fig. 3). As is known from boundary-layer stability theory [1], the transition location moves downstream on the windward and upstream on the leeward side of the cone.

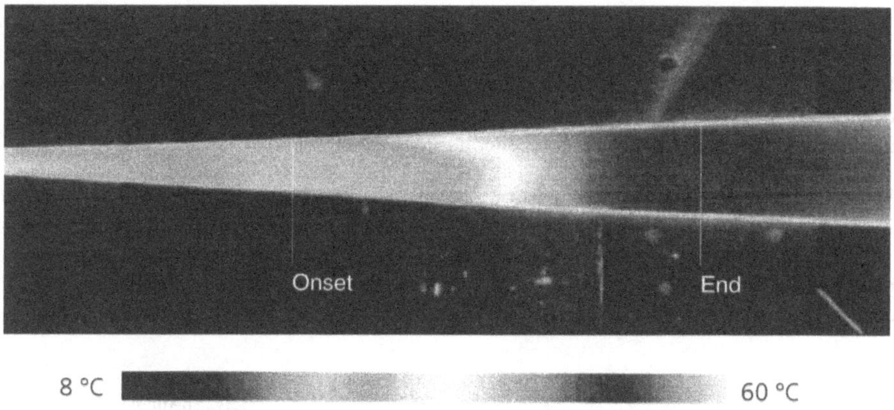


Fig. 2: Temperature distribution after 10 seconds test time on a blunted cone with 5 mm nose radius at Ma = 6 and Re = $16 \cdot 10^6$ /m.

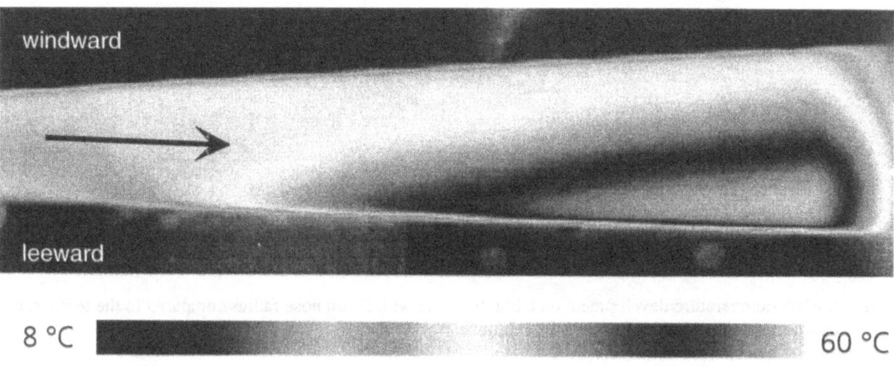


Fig. 3: Temperature distribution on a blunted cone with 5 mm nose radius at 1° angle of yaw (flow direction indicated by arrow, same flow condition as Fig. 2).

For the adjustment of the camera´s temperature measurement range some knowledge about the expected surface temperature-rise along the transition zone is required in advance. Considering an instationary heating process including radiation effects, this temperature-rise considerably deviates from the difference between laminar and turbulent recovery temperatures. In order to predict the surface-temperature history on the cone, heat flux calculations were performed using the reference temperature method [1]. These results served as boundary conditions to the one-dimensional heat equation, which was solved numerically in order to get some approximate values for the surface-temperature development in the experiment.

The definition of the transition zone slightly varies in literature. In this paper, transition onset is indicated as the point of increasing temperature or heating rate with respect to the laminar flow. The transition ends at the maximal temperature or heating rate. Even though the onset of transition could have been predicted by engineering calculations, they do not provide information about the size of the transition zone.

Figure 4 exemplarily shows a comparison between predicted and measured temperature profiles at 1, 5 and 20 seconds test duration. Due to the high velocities of the hypersonic boundary-layer flow the transition zone is found to be significantly larger compared to low speed transition. Even though the temperature-rise over the transition region increases with time, the location of transition onset and end is found to be not affected by the surface-temperature.

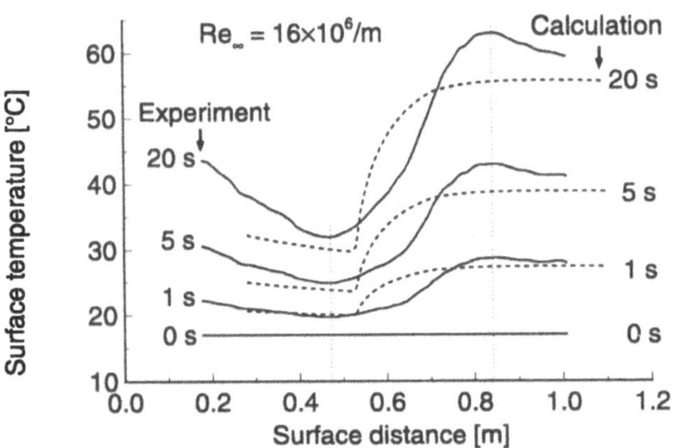


Fig. 4: Surface-temperature development on a blunted cone with 5 mm nose radius compared to the temperature distribution calculated by engineering relations.

From the measured temperature-rises heat-flux values are evaluated and plotted at certain time intervals as Stanton number distributions along the cone axis (Fig. 5). Due to flow establishment the Stanton number values take more than a second to stabilize. The final distribution then agrees fairly well with the calculated one for sharp cones. Transition onset, denoted by $x_{T,B}$ and transition end, denoted by $x_{T,E}$ become evident also from the Stanton number distribution.

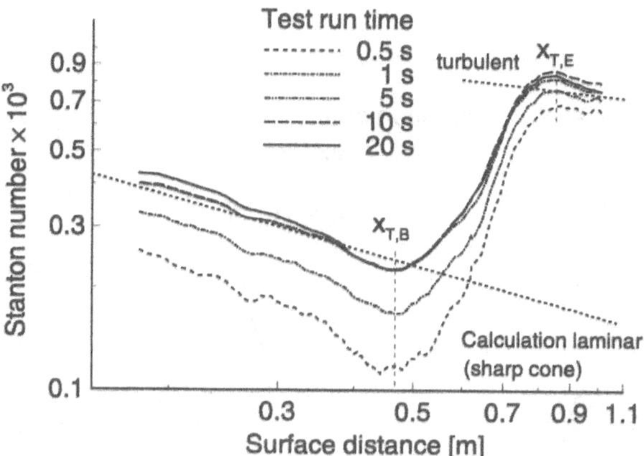


Fig. 5: Stanton number distribution on the axis after certain time intervals for the blunted cone with 5 mm radius at Ma = 6 and Re = $16 \cdot 10^6$ /m.

In order to estimate the measurement error, the heat-flux evaluation procedure was performed for slight variations of its input parameters. The measured temperature-rise, the flow condition parameters, the influencing material data as thermal diffusivity and surface emissivity, as well as the test duration were changed within their uncertainty margins. This results in a heat-flux error of less than 12%, a range which is comparable to conventional gauge measurements in short-time facilities.

## NOSE-TIP BLUNTNESS

Among others, Stetson [7] reported that nose-tip bluntness has a large influence on cone transition phenomena. The curved bow shock in front of a blunted nose generates an entropy layer, which interacts with the viscous boundary-layer and influences the laminar/turbulent transition process to a large extent.

Results of a H2K test series with the cone equipped with four noses of different bluntness are shown in Fig. 6. The Stanton number distribution indicates that the bluntness becomes apparent in a shift of the onset location. For nose radii of 1, 5, and 10 mm the entropy layer on the cone is partly swallowed and therefore transition onset moves downstream with increasing bluntness. For 15 mm nose radius transition is dominated by the entropy effects of the nose region and therefore significantly moves upstream again.

This phenomenon, denoted as "blunt-nose paradox", also becomes evident from Fig. 7 by plotting measured transition locations. Additionally, for larger radii an increase of the size of the transition zone is observed. Nevertheless, a comprehensive physical explanation on the influence of the entropy layer effect on the instability mechanism of hypersonic boundary-layers is still missing.

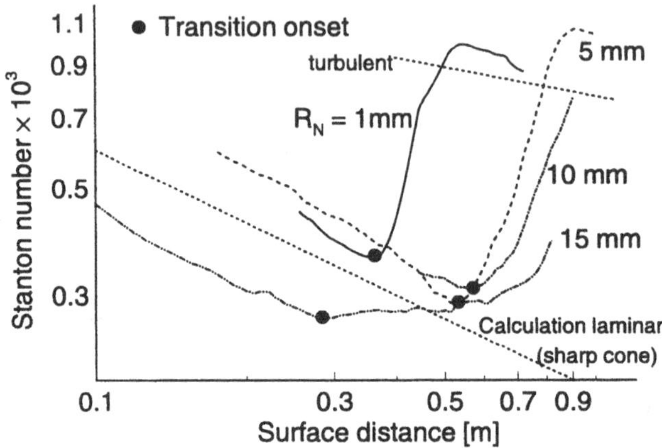


Fig. 6: Stanton number distribution for different nose radii for Ma = 6 and Re = $12 \cdot 10^6$ /m (circles denote transition onset).

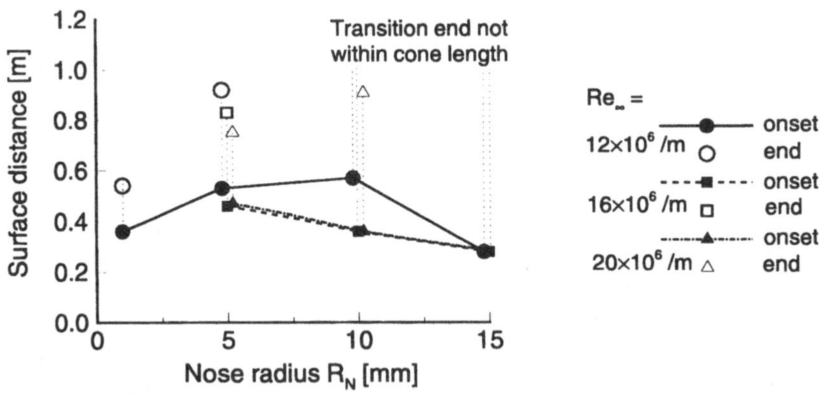


Fig. 7: Bluntness effect on transition onset and completion at Ma = 6.

## FORCED TRANSITION

It is well known that surface roughness influences the transition process as well. In wind-tunnel experiments, trigger elements are sometimes used to fix transition to a certain location. In addition to the test series with natural cone transition, transition triggering is achieved by the help of a cylindrical element being located some distance downstream from the blunt nose-tip of 15 mm radius. Due to vortices which are generated by the trigger element, a turbulent wedge spreads with the flow along the cone surface. Figure 8 shows the resulting infrared images for trigger elements of different sizes.

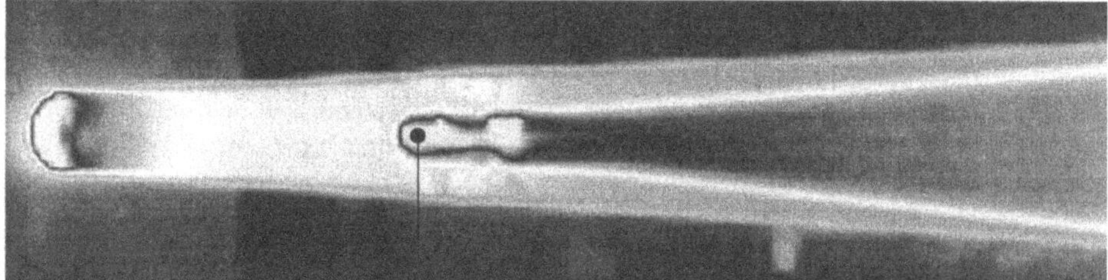

Cylinder, d = 3 mm, h = 2 mm

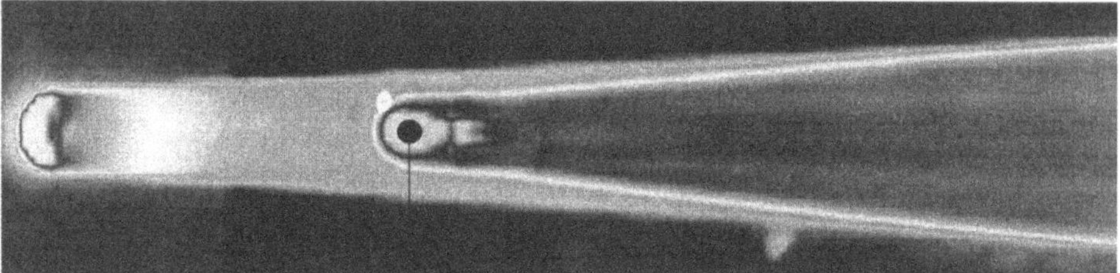

Cylinder, d = 6 mm, h = 5 mm

Fig. 8: Infrared images of the transition wedge downstream a cylindrical trigger element of 3 mm and 6 mm diameter after 10 seconds test duration.

## CONCLUSIONS

At the H2K facility, transition experiments were performed on cones with different bluntness. Transition onset and end were visualized by an advanced focal plane array infrared system, providing exceptionally high spatial resolution and temperature sensitivity. In particular the following results were obtained:

- The locations of transition onset and end were observed to be fixed during the test run, i.e. it was not affected by the rise of surface-temperatures. During the 20 seconds of the test run the equilibrium temperature on the PTFE cone was not reached. Furthermore, a weak unit Reynolds number effect was recognized for this facility.
- Heat-flux values were evaluated from the measured temperature histories and the Stanton number distribution along the cone´s symmetry line was successfully presented in accordance to calculations using the reference temperature method.
- The influence of different nose radii on the transition location and the size of the transition regime was investigated. The gained experimental data may support the validation of numerical codes for the prediction of transition.
- Forced transition was demonstrated by fixing cylindrical trigger elements behind the blunted nose-tip.

All results confirm the ability of focal plane array infrared systems and their qualification for aerothermodynamic measurement applications in cold hypersonic blow-down wind-tunnels, in particular for hypersonic transition studies.

## REFERENCES

[1] Bertin, J. J.: *Hypersonic Aerodynamics.* AIAA Education Series ed. by J. S. Przemieniecki, AIAA Washington, 1994.

[2] De Luca, L., Carlomagno, G. M., and Buresti, G.: *Boundary layer diagnostics by means of infrared scanning radiometer.* Experiments in Fluids, Vol. 9 (1990), pp. 121-128.

[3] Gartenberg, E., Roberts Jr., A. S.: *Twenty-five years of aerodynamic research with infrared imaging.* J. of Aircraft, Vol. 29 (1992), No. 2, pp. 161-171.

[4] Gaussorgues, G.: *Infrared Thermography.* Microwave Technology Series 5, pp. 234, Chapman & Hall, London, 1994.

[5] Henckels, A., Kreins, A. F., Maurer, F.: *Development and application of infrared- and other flow visualization technique in a hypersonic wind tunnel of DLR Cologne.* European Forum on Wind Tunnels and Wind Tunnel Test Techniques, pp. 11.1-11.12, Southampton, 1992.

[6] Simeonides, G. et al.: *Infrared thermography in blowdown and intermittent hypersonic facilities.* J. Thermophysics, Vol. 4 (1990), No. 2, pp. 143-148.

[7] Stetson, K. F.: *Hypersonic Boundary-Layer Transition.* Advances in Hypersonics, Vol. 1, ed. by J. J. Bertin et al., p. 343, Birkhäuser Boston, 1992.

# Application of PIV in the Large Low Speed Facility of DNW

K. Pengel[#], J.W. Kooi[#], M. Raffel[*], C. Willert[*], J. Kompenhans[*]

[#]German- Dutch Wind Tunnel DNW,Postbus 175, 8300 AD Emmeloord, The Netherlands
[*]Institut für Strömungsmechanik, Deutsche Forschungsanstalt für Luft- und Raumfahrt (DLR)
Bunsenstraße 10, D-37073 Göttingen, Germany

## Summary

A Particle Image Velocimetry (PIV) system has been successfully applied in the large low-speed wind tunnel (LLF) of the German Dutch Wind Tunnels (DNW). The described system was developed for the DNW by the German Aerospace Research Establishment (DLR). The system equipment and the measurement results of a wake vortex investigation behind an Airbus half-model in the DNW are presented in this paper. The PIV-data were compared with five-hole probe data, which were measured at the same downstream location. The comparison of data and the demonstration show that industrial PIV measurements will be possible in large low-speed wind tunnels in the near future.

## Introduction

Laser Doppler Anemometry (LDA) and Particle Image Velocimetry (PIV) are two non-intrusive velocity measurement techniques which have proven in the last fifteen years their capabilities in a number of applications. As far as wind tunnel testing is concerned the applications were mostly limited to small or moderate sized facilities. This is not surprising as use of LDA or PIV in large facilities requires powerful laser systems and high quality long range optics. Meanwhile this equipment has become available which means that the application of LDA and PIV in large facilities as the German Dutch Wind Tunnel DNW is possible. From these two systems DNW has selected PIV to become available on a routine base to the DNW users and DNW has procured the necessary hard- and software. Whenever required DNW can also provide a LDA system through its parent organization DLR. As a matter of fact, this LDA system has been successfully used to measure the flow field of a rotor.

The test case selected for the first application of PIV in the DNW was the measurement of the development of the vortex system downstream of an aircraft model [1]. During landing and take off the trailing vortices from the preceding aircraft can be dangerous to the following aircraft. The air safety regulations require a minimum separation based on the expected intensity of these vortices. Because the flow field of interest extends in streamwise direction over the distance of many spans, in an experimental set-up in a wind tunnel recourse has to be made to relative small scale models installed in the test section as far upstream as possible. For the PIV test a half-span model representing a modern twin-engined transport aircraft was mounted on the floor of the 8 x 6 $m^2$ test section of the DNW. The wing of a half span of 1.25 meter was in a high-lift configuration, which means that the slats and the flaps were deflected. An important reason for the selection of this particular model and test set-up was that the wake vortex at the same configuration had been measured with a five-hole probe rake. With this rake the wing wake flow was measured at 12 different planes at streamwise distances ranging from 0.01 times the span widths downstream of the wing up to 7 times the span widths.

## Test Set-Up

The model, that has been chosen for the investigation of the development of wake vortices was a half-span, twin-engined 1:13.6 scaled Airbus model of the German DASA Airbus GmbH. The half-span width is 1.25 m and the flaps were installed in a high-lift landing configuration. This model was installed in the most up-stream position of the 8 x 6 $m^2$ closed test section of the DNW so that the development and in particular the rolling-up of the wake vortices could be investigated up to 7 span widths behind the model in the 20-m-long test section.

**Figure 1:** Model and camera support system in the DNW test section.

The experimental set-up of the PIV system in the DNW test section consists of several sub-systems developed or composed by the DLR Institute for Fluid Mechanics. First of all tracer particles, generated by an aerosolgenerator, have to be added to the flow. Therefore, three of these generators in parallel and a specially developed low-disturbance, multi nozzle injection device were installed on a traverse inside the settling chamber of the closed loop wind tunnel. This seeding device generates oil droplets of about 1 µm diameter in a very high concentration so that enough tracer particles are present in the cores of the vortices even in region with high rotational velocities.

The particles were illuminated in two selected observation planes perpendicular to the flow and downstream of the wing tip. One plane 0.93 m or 0.37 span widths behind the tip, the other one 5.0 m or 2.0 span widths downstream of the origin. The light sheets were generated by a double-oscillator Nd:YAG pulse laser with an output energy of 320 mJ for each pulse and a typical time delay between the two pulses of about 20 µs. The repetition rate of the two pulses is 10 Hz. The laser system generates two green light pulses ($\lambda$ = 532 nm) with a pulse duration of 16 ns and with the same output energy. In order to generate laser sheets within the observation planes the two laser beams were spread by cylindrical lenses. This optical device generates

a light sheet, which is 1.5 mm thick at the area of interest. The laser and the optical devices were mounted outside of the test section.
Recording of the light scattered by the particles was done in two ways. In the first place a 35 mm photographic camera equipped with a 100-mm-focal-length lens was used [2]. The camera recorded an area of 30 x 30 $cm^2$ at 5m downstream of the wing and 20 x 20 $cm^2$ at 0.93 m behind the model. A rotating mirror in front of the camera lens allows to remove directional ambiguity. The drawback of the photographic recording is the turn around time for development of the film. At the cost of somewhat lower resolution this drawback can be circumvented by the use of a video camera. The video camera incorporates a full-frame interline CCD technology [3]. These sensors are capable of shuttering (exposing) and storing the entire array of pixel (1000 x 1000). The digital information from the camera was transferred to a work station where it was stored in memory and could be viewed almost in real time. Both the photographic and the video images were processed automatically. To this end the developed films were scanned and the images were stored in the same way as the video images. Special algorithms and visual inspection of the results finally validated the results. These two cameras were installed on an in-flow traversing system of DNW, which was mounted about 8m downstream of the model and about 3m respectively 7m downstream of the two observation areas. The optical axis of the cameras were arranged perpendicular to the light sheet. In order to attain data of the entire vortex structure (within the observation plane) the camera was scanned step by step vertically and horizontally by means of the traversing system on which the camera is mounted.

## Measurements and Results

The wake vortices to be investigated were generated at a wind velocity of $V_\infty = 60$ m/s. The two observation planes were generated at the above mentioned locations, perpendicular to the ambient flow. All PIV parameters (laser pulse duration, time interval between the two pulses etc.) were adjusted to the expected flow conditions in the observation plane. Additionally the second-pulse-light sheet was shifted in flow direction by a small amount because otherwise the ambient flow and the axial flow of the vortices would force the tracer particles moving out of the light sheet i.e. the observation plane during the second pulse. They only would be illuminated ones, and the requirements for a PIV- analysis would not be fulfilled.
A typical result obtained 5 m downstream of the model is represented in figure 2. The figure shows a vector map as a result of one double-frame analysis. Structures of two vortices, the tip vortex and the flap vortex can be easily recognized. The averaged velocity vector maps were compared with the results from the five-hole probe measurements, published in [1] and the overall agreement was very good.
The size of the observation area is defined by the flow and the PIV-system parameters: the tracer particle size (1μm), the expected maximum tangential velocity of the vortices ($V_{tang.} \cong$ 50 m/s), the time interval between the two laser pulses ($\tau = 20$ μs), the distance between the camera and the observation plane (d = 3.0m) and the resolution of the video camera (1000 x 1000 pixels). A larger number of pixels of the CCD-chip or a smaller observation distance d would increase the size of this observation area in one video frame but neither the number of pixel nor the observation distance could be changed. Decreasing the observation distance would lead to a disturbance of the flow in the measurement plane, cameras with a higher resolution are not yet available. The location of the aerodynamic center of the model within all represented vector maps is 0 in vertical and horizontal direction.

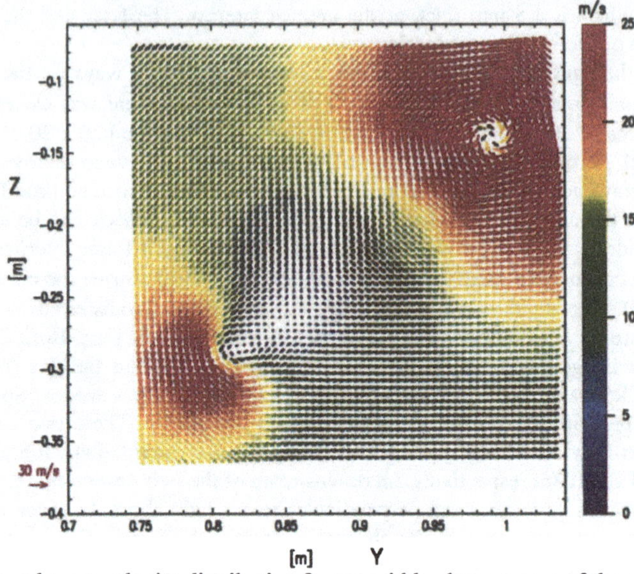


**Figure 2:** Averaged cross velocity distribution 2 span widths downstream of the wing tip of the half model. Wind velocity $V_\infty = 60$ m/s.

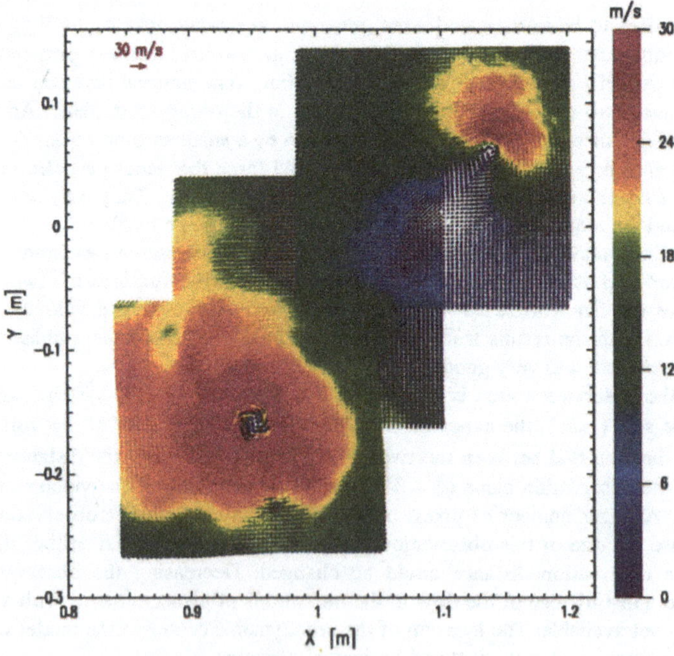


**Figure 3:** Composite of three vector maps measured 0.93m downstream of the wing tip.

Figure 3 shows composed vector maps of the main part of the observation plane at 0.93m downstream of the wing tip. One video frame area of 20 x 20 cm$^2$ is too small to represent the main 2 vortices. The partly superimposed single frames cover the area of the stronger flap vortex and the tip vortex. Also the area between the two vortices where the cross velocities compensate each other is perceptible.

The individual PIV results (before being averaged, i.e. instantaneous flow fields) showed an unsteady vortex (figure 4) which was almost absent in the five-hole probe results and in the averaged PIV results as well (figure 5). The mean induced drag, for example, calculated from this data would be too small.

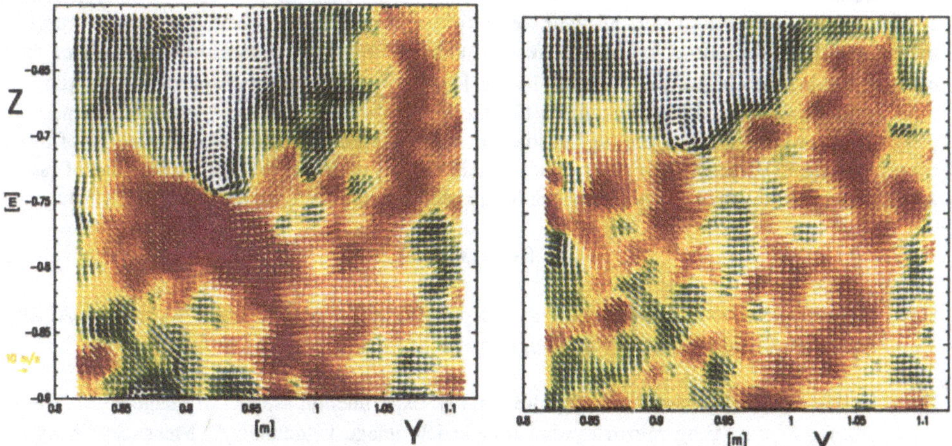


**Figure 4:** Two instantaneous cross velocity distribution at different times 2 span widths down stream of the wing tip in an area underneath of the two main vortices. The locations of the shown vortex center are different .

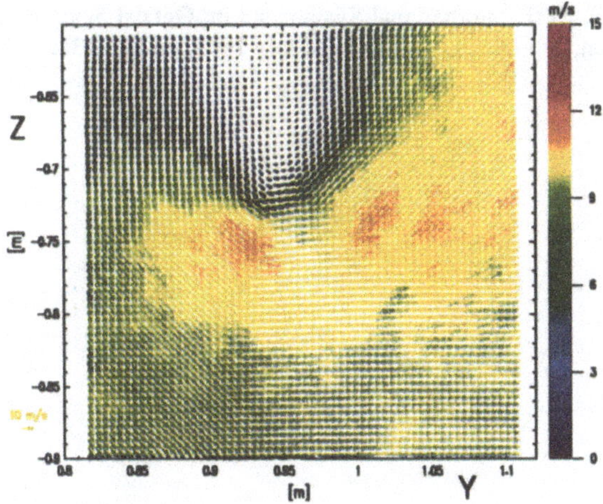


**Figure 5:** Averaged cross velocity distribution of the same vortex as shown in figure 4.

The location of the center of the flap vortex in figure 2 and figure 3 are only marginal different. The vortex is almost stable in his position. However the less strong tip vortex changes its position by rotating around the stronger flap vortex. Results from the five-hole probe measurements, which were also made at several positions in-between 0.93m and 5m emphasize that it is about half a revolution of the tip vortex around the flap vortex. Thus for the investigated wing configuration the roll-up center of this particular wake is the region of the flap vortex.

## Conclusion

The results presented in this paper showed that a two-dimensional PIV can be utilized successfully even when strong out-of-plane velocity components are present in a given experimental set-up. Comparisons of the PIV results with well established measuring techniques as e.g. the five-hole probe technique, published in [1], showed a very good agreement. Requirements of large sized wind tunnel facilities like the DNW can be fulfilled if sophisticated optics, powerful pulse laser systems and high-resolution video cameras are used as components of the corresponding PIV system. However developments of a 3-dimensional PIV system and of fast and long-ranged traversing systems for the light sheet displacement are required for the future.

## References

[1] Huenecke, Klaus :"Structure of a transport aircraft-type near field wake", AGARD FDP Symposium (May,1996), Congress Proceedings-548 pp 5-1 to 5-9.

[2] Raffel, M; Kompenhans, J.: "Theoretical and experimental aspects of image-shifting by means of a rotating mirror system for particle image velocimetry", Meas. Sci. Tech., Vol. 6, pp 795-808, 1995.

[3] Willert, Chr.; Stasicki, B; Raffel, M; Kompenhans, J :"A digital video camera for application of particle image velocimetry in high-speed flows", SPIE Proceedings Article No. 2546-19, SPIE International Symposium on Optical Science, Engineering and Instrumentation, San Diego, U.S.A., 9-14 July 1995, pp 124-134.

# Refined Streamline Patterns for Unsteady Boundary Layer Separation

H. Ranke*, J. Henkner

Lehrstuhl für Fluidmechanik, Technische Universität München,
Boltzmannstr. 15, 85747 Garching, Germany

## Summary

Unsteady boundary layer separation is one of the essential phenomena in fluid mechanics which is still not fully understood. The well known Moore-Rott-Sears criterion (MRS) states that the onset of separation is reached if in a coordinate system moving with the separation point both the velocity parallel to the wall and its normal derivative vanish at one point simultaneously. This is a good approach to identify an upstream moving separation. It can be refined by considering three distinguished points in order to specify the separation. Those are (1) the MRS point, (2) the Onset of Reverse Flow of the separation region (ORF) and (3) the Stagnation point (STG) within the boundary layer. For a downstream moving separation point, however, the up-to-now published streamline pictures seem to be incorrect. A new separation flow pattern for that case as well as a new criterion to describe it will be suggested.

## Nomenclature

**Symbols:**

| | |
|---|---|
| $a, b$ | constants |
| $L$ | characteristic length |
| $p$ | static pressure |
| $p_e$ | external flow static pressure |
| $Re$ | Reynolds number ($\frac{U_0 L}{\nu}$) |
| $t$ | time |
| $u, v$ | flow components in $x$- and $y$-direction |
| $U_e$ | external flow velocity |
| $U_0$ | reference velocity |
| $U_1$ | velocity at the downstream boundary of the computational area |
| $U_w, V_w$ | wall velocities in $x$- and $y$-direction |
| $x, y$ | wall fixed coordinate system |
| $X_1$ | $x$-coordinate at the downstream boundary of the computational area |
| $X_I$ | $x$-coordinate at a specified point of the computational area |
| $\delta$ | boundary layer thickness |
| $\nu$ | kinematic viscosity |

**Abbreviations:**

| | |
|---|---|
| MRS | Moore-Rott-Sears |
| ORF | Onset of Reverse Flow of the separated region |
| STG | Stagnation |
| WJ | Williams-Johnson |

**Superscripts:**

transformed coordinates and flow components

## 1 Introduction

Steady two-dimensional separation is characterized by a semi-bubble. Its region is limited by the wall and by the separation line. At unsteady flow separation instead full bubbles, limited by the streamlines which belong to a free stagnation point, can be observed. Reverse flow components without any indication of separation are possible [1, 2]. Therefore it is not correct to describe the onset of unsteady separation according to the classical definition after L. Prandtl

*now: Process Engineering and Contracting Division, Linde AG, Dr.-Carl-von-Linde-Str. 6-14, 82089 Höllriegelskreuth, Germany

via vanishing wall skin friction. F. K. Moore [3], N. Rott [4] and W. R. Sears [5] proposed independently of each other at the end of the fifties the so called MRS point as the onset of unsteady separation. It states that in a coordinate system moving with the separation point the velocity component along the wall direction and its normal derivative ($\frac{\partial u}{\partial y}$) vanish simultaneously in one point of the flow field. Streamline pictures, achieved e.g. from experiments of G. R. Ludwig [6] and of C. A. Koromilas and D. P. Telionis [7] seem to confirm that hypothesis, but there are some details which are not completely accounted for by the MRS criterion. It is the aim of this paper to focus on those details. A qualitative analysis of the velocity profiles and the instantaneous streamline pictures in a moving coordinate system consequently lead to a more refined description for unsteady separation [8, 9]. This is important for a better understanding of that process and for the possibility to interpret unsteady flow field measurements, e.g. the onset of dynamic stall of a pitching airfoil (see L. Carr and M. S. Chandrasekhara [10]. L. E. Ericsson and J. P. Reding [11] or H. Ranke [8]).

# 2 Fundamentals

## 2.1 Basic Equations And Numerical Code

The boundary layer equations of L. Prandtl for two-dimensional, incompressible, laminar, unsteady flows are given by the following set of well known non-dimensional equations, see e.g.: H. Schlichting [12]:

$$\frac{\partial u}{\partial x}+\frac{\partial v}{\partial y} = 0 \tag{1}$$

$$\frac{\partial u}{\partial t}+u\frac{\partial u}{\partial x}+v\frac{\partial u}{\partial y} = -\frac{\partial p_e}{\partial x}+\frac{1}{Re}\frac{\partial^2 u}{\partial y^2} \tag{2}$$

$$\frac{\partial p}{\partial y} = 0\,. \tag{3}$$

Boundary conditions:

$$At\ \ y=0\ :\ u=U_w\,;\ v=V_w$$

$$At\ \ y=\delta\ :\ u=U_e\,;\ \frac{\partial u}{\partial y}=0$$

$$\Rightarrow\quad \frac{\partial U_e}{\partial t}+U_e\frac{\partial U_e}{\partial x}=-\frac{\partial p_e}{\partial x}\,. \tag{4}$$

The method [13] used to solve these equations (Eqs. (1 – 4)) is a time-marching implicit code. For that purpose the boundary layer-equations have been modified by adding the diffusive term $\frac{\partial^2 u}{\partial x^2}$ which occurs in the full Navier-Stokes equations. It balances the pressure gradient term in the momentum equation in regions of the flow field, where u and $\frac{\partial u}{\partial y}$ become very small, see D. T. Tsahalis [14]. It enables the calculation of flows with small regions of reverse flow or with small separation bubbles.

## 2.2 Williams-Johnson Flow

Certain classes of two-dimensional flows can be transformed such that the unsteady flow field quantities f(x,y,t) become steady f(x*,y*). These so-called semi-similar solutions are e.g. (Refs. [15, 16]) the Tani flow, the Curle Cubic flow, the Howarth linearly retarded flow and the Williams-Johnson (WJ) flow. The velocity of the external flow and the boundary conditions of the WJ flow, Ref. [16], where an unsteady locally and temporarily linear retarded flow is transformed into a steady flow over a moving wall, are as follows:

$$U_e=U_0(1-ax-bt)\,;\ \ U_w=0. \tag{5}$$

The transformation

$$x = x^* + \int_{t_0}^{t} U dt\,;\;\; y = y^*\,;\;\; t = t^*$$

$$u = u^* + U\,;\;\; U = -\frac{b}{a}U_0\,;\;\; v = v^* \qquad (6)$$

applied to Eq. (5) yields to the time-independent form:

$$U_e^* = U_0^*(1 - a\frac{1}{1+b/a}x^*)\,;\;\; U_0^* = U_0(1+\frac{b}{a})\,;\;\; U_w^* = -\frac{U}{U_0^*}U_0\,. \qquad (7)$$

A more detailed description is given in Ref. [16].

Fig. 1 shows the profiles of the u-component of a WJ flow. This figure as well as those published in literature reveal that there are no intersections of the velocity profiles upstream of the MRS point. As a consequence in front of the MRS point the flow component u is decelerated ($\frac{\partial u}{\partial x} < 0$) at any distance from the wall. Because of the continuity equation the flow component v therefore must be everywhere directed away from the wall. When the flow approaches the MRS point v becomes very large.

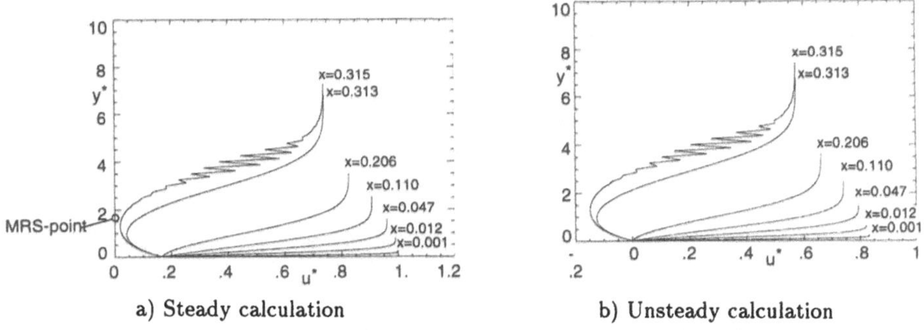


Figure 1: Williams-Johnson Flow, $a = 1$, $b = 0.2$

The original WJ flow and in addition a modified form which allows the calculation of a small separation bubble are studied. Both cases, namely the flows over an upstream and over a downstream moving wall will be considered. Those flows correspond to downstream and upstream moving separation points, respectively.

# 3 Downstream Moving Wall – Upstream Moving Separation Point

## 3.1 Previously Published Streamlines

Fig. 2a is a typical reproduction of streamlines, e.g. taken from D. P. Telionis and M. J. Werle [17]. In this picture or in similar pictures e.g. those by C. A. Koromilas and D. P. Telionis [7] or G. R. Ludwig [18] some inaccuracies occur in the details:

1. At the MRS-point the MRS-streamline must be perpendicular to the wall ($u = 0$; $\frac{\partial u}{\partial y} = 0$): As said before, the $u$-component approaching the MRS point becomes zero whereas the $v$-component must become very large in order to be consistent with the continuity equation. A finite slope of the MRS-streamline would otherwise require a negative $u$-component upstream of the MRS point.

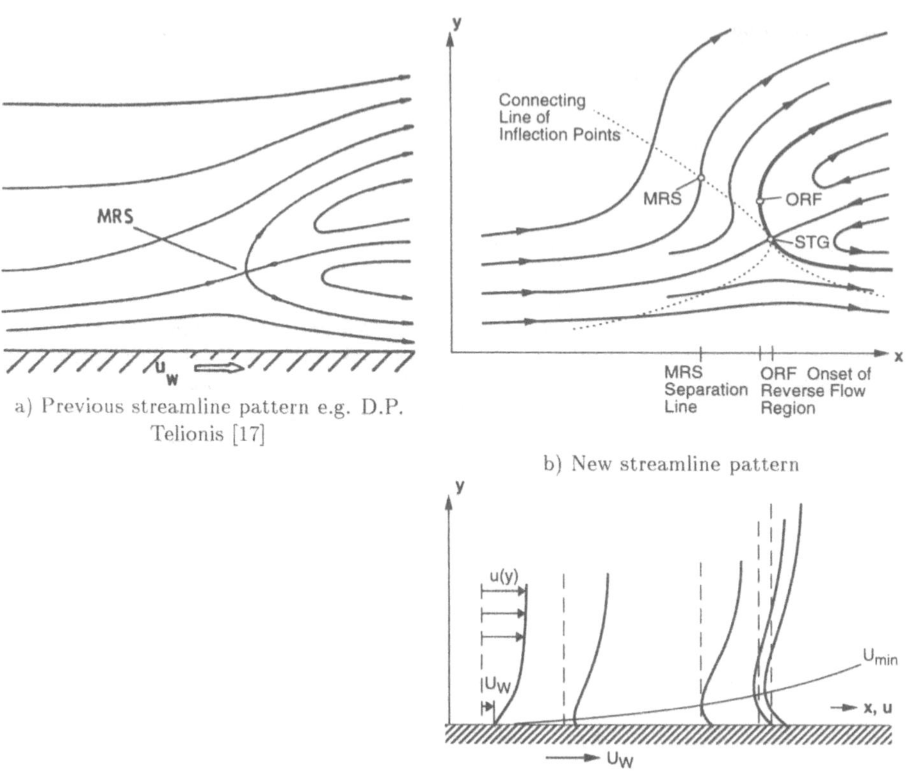

a) Previous streamline pattern e.g. D.P. Telionis [17]

b) New streamline pattern

Figure 2: Downstream moving wall - upstream moving separation

2. The MRS point and the STG point can-not coincide: At the MRS point only the u-component is zero, the v-component is positive. A bifurcation at the MRS point as shown in Fig. 2a is not possible with a nonvanishing $v$-component.

3. The MRS and the STG point streamlines must be different: The STG point streamline must approach the separated region perpendicular to it, as branching streamlines at a free STG point are always normal to each other, Fig. 2a,b.

4. The maxima (in y-direction, compare Fig. 2a and 2b) of the streamlines have to be behind the MRS point because in front of the MRS point always upward (positive) v-velocities exist.

5. The line connecting the minima of the u-velocity component approaches the separated flow region continuously, see Fig. 2c.

As a consequence the MRS point is not a complete description of the separation over a downstream moving wall or an upstream moving separation point, respectively. Moreover, three different points in streamwise direction have to be distinguished (Fig. 2b):

- Moore-Rott-Sears point (MRS): First point where u and $\frac{\partial u}{\partial y}$ vanish.
- Onset of Reverse Flow (ORF): Most upstream point of the separated region.

- Stagnation point (STG): Point where the velocity components u and v vanish.

### 3.2 Calculated Results – Coordinate System Fixed To Separation Point

The transformed locally and temporarily linear retarded WJ flow is modified by setting the longitudinal velocity gradient to zero at the end of the computational area. The result is a boundary layer flow with a separation bubble, see Fig. 3.

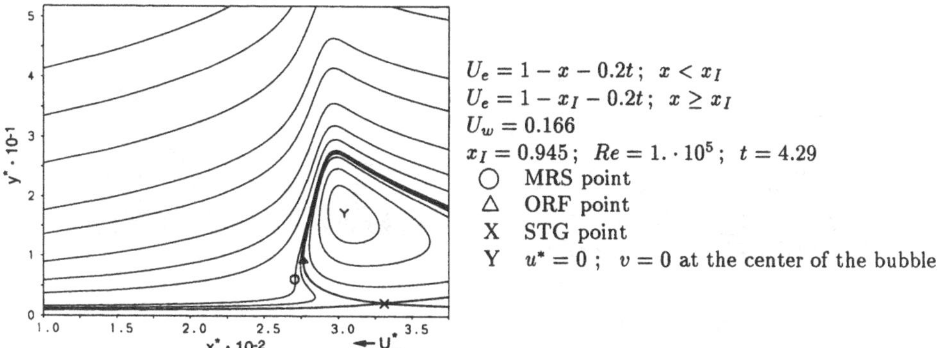


Figure 3: Unsteady result presented in a upstream moving coordinate system

This streamline picture agrees well with the results of O. Inoue [19] who also used reduced Navier-Stokes equations and similar boundary conditions. As expected, the MRS point is clearly located in front of the separated region starting from the ORF point. There is only one STG point underneath the separation bubble. Another point with diminishing velocity components is in the center of the separated region.

## 4 Upstream Moving Wall – Downstream Moving Separation Point

### 4.1 General

When the downstream moving wall decelerates the separated region moves ahead. The STG, the ORF and the MRS point move towards the wall and coincide there, when the wall velocity vanishes. The definitions of the separation according to the present report, Prandtl's criterion and the MRS criterion become identical. However, if the wall velocity is increased upstream the description of Prandtl will fail again. The separated region moves further ahead. In unsteady flow this corresponds to a separation moving downstream. For the streamline picture it has been proposed previously that there still exist a STG point and a MRS point in the flow. In contradiction to this picture it is suggested in this paper that no STG point and no MRS point is available in order to identify separation.

### 4.2 Previously Published Streamline Pictures

Streamline pictures with STG and MRS point, Fig. 4a, may be found e.g. at W. R. Sears, D. P. Telionis [20] and D. T. Tsahalis [14]. They all show a common STG, ORF and MRS point. When the u-velocity profiles pass the zero value they should have a plateau at the MRS point (U=0, $\frac{\partial u}{\partial y}$), which coincides, as mentioned above, with the STG point. It allows the change from positive to negative velocity at vanishing velocity gradient, see Fig. 4b. The development of a plateau at the u-profiles means an acceleration of the flow near the wall. This is not justified in a region where the pressure through the boundary layer is nearly constant. Suitable v-profiles of the velocity, which have not been found in the literature, must show direction changes to comply with the assumed streamlines. Neither published calculation nor experimental results of moving flat plate boundary layers have been found to confirm these streamline pictures. There are only results available for rotating circular cylinders (Refs. [7, 18]). Those results, however, can not be taken for our problem because of the large slope and the curvature of the body contour at the location of separation.

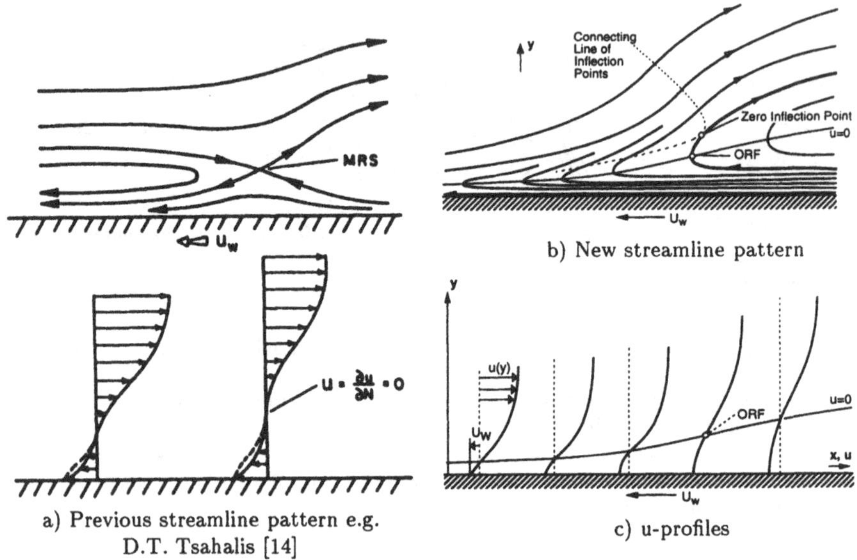


a) Previous streamline pattern e.g. D.T. Tsahalis [14]

b) New streamline pattern

c) u-profiles

Figure 4: Upstream moving wall - downstream moving separation

### 4.3 New Separation Flow Pattern

For the upstream moving wall the three characteristic points describing the separation for a downstream moving wall are not valid anymore, because the MRS and the STG point move far upstream compared to the location of separation. The suddenly very large growing v-component of the velocity accompanied by a very strong increase of the boundary layer thickness, indicate this separation, see Fig. 4b. Separation still begins at a relatively sharp defined position downstream of the forward part of the plate but, of course, upstream of the separation position of a fixed or downstream moving wall. Accordingly the velocity profiles in x-direction, Fig. 4c, and in y-direction contain the same characteristic features as those of a downstream moving wall, Fig. 2c. There is no need for the appearance of plateaus in the u-profiles.

The flow close to the wall moves in opposite direction to the external flow, Fig. 4b. It changes its direction upstream of the separation and aligns with the external flow. When approaching the separation all streamlines coming upstream first are curved outward. Further downstream they align with the main flow, see Fig. 4b. The streamlines close to the wall have one point where the u-component changes its sign followed by two inflection points. The streamlines of the external boundary layer, going downstream only, show no sign change of the u-component and only one inflection point. A line connecting all inflection points, Fig.4b, has one characteristic point in which both inflection points coincide and consequently disappear. This point belongs to a streamline which we propose to be defined as the separation streamline. The most forward point of it gives the Onset of Reverse Flow of the defined separated region (ORF). (For a downstream moving wall the same holds). This definition for the separation line is applicable to all flows above moderate curved surfaces, for accelerated and for decelerated flows, for down- and for upstream moving walls. Therefore an unambiguous and reproducible definition has been found to describe in a streamline picture the location of separation: The separation streamline is the first streamline within the boundary layer without inflection point (Zero Inflection Point).

### 4.4 Calculated Results – Coordinate System Fixed To Separation Point

The calculation procedure for upstream and downstream moving walls is the same apart from the different wall velocity boundary conditions. The existence of a plateau in the u-velocity component, Fig. 4a, cannot be observed. Streamlines and velocity profiles correspond to the description given above. In Fig. 5 such a calculation is presented in a downstream moving coordinate system. Fig. 5 enables the comparison with Fig. 4b, neither MRS nor STG point can be observed near the separation.

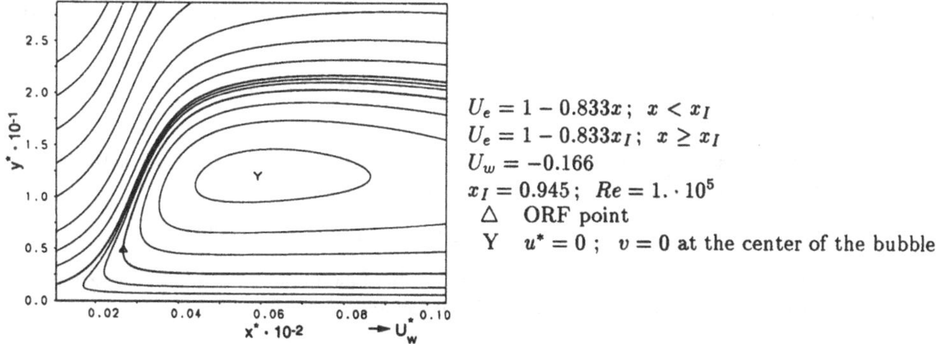


Figure 5: Unsteady result presented in a downstream moving coordinate system

## 5 Conclusion

Instead of using the MRS point a refined description of the unsteady separation process based on velocity profiles and streamline pictures is proposed: The separation streamline is the first streamline without inflection point (Zero Inflection Point). The most upstream point of that line, namely the ORF point, marks the beginning of the separation. For an upstream moving separation point the previously published flow patterns should be refined by including in addition to the MRS-point two other points, namely the Onset of Reverse Flow point (ORF) and the Stagnation point (STG). In the case of a downstream moving separation point neither MRS nor STG point exist near separation.

## References


[1] B. Laschka. Unsteady flows - Fundamentals and applications. In *AGARD CP-386 Conference Proceedings*, pages 1.1–1.21, 1985.

[2] B. Laschka. On some fundamentals of unsteady viscous flows. In *La Sapienza*. Universita degli Studi di Roma Dipartimento Aerospaziale, 1994.

[3] F. K. Moore. On the separation of the unsteady laminar boundary layer. Boundary Layer Research, edited by H. G. Görtler, Springer Verlag, Berlin, 1958.

[4] N. Rott. Unsteady viscous flow in the vicinity of a stagnation point. *Quarterly of Applied Mathematics*, 13(1):pp.444–451, 1955.

[5] W. R. Sears. Some recent developements in airfoil theory. *Journal of the Aeronautical Sciences*, 23(5):pp.490–499, May 1956.

[6] J. C. WilliamsIII. Incompressible boundary-layer separation. *Annual Reviews of Fluid Mechanics*, 9:pp.113–144, 1977.

[7] C. A. Koromilas and D. P. Telionis. Unsteady laminar separation : An experimental study. *J.Fluid.Mech.*, 97(2):pp.347–384, 1980.

[8] H. Ranke. *Instationäre Grenzschichtablösung in ebener, kompressibler Strömung*. PhD thesis, Lehrstuhl f. Fluidmechanik, Technische Universität München, 1994.

[9] H. Ranke. Unsteady separation in two-dimensional turbulent flow. ICAS 94, 19th Congress of the International Council of the Aeronautical Sciences, Anaheim, California, Paper 94–4.5.1, Sept. 1994.

[10] L. W. Carr and M. S. Chandrasekhara. Design and development of a compressible dynamic stall facility. *Journal of Aircraft*, 29(3), May-Jun. 1992.

[11] L. E. Ericsson and J. P. Reding. Fluid mechanics of dynamic stall. Part I. Unsteady flow concepts. *Journal of Fluids and Structures*, 2:pp.1–33, 1988.

[12] H. Schlichting. *Grenzschicht-Theorie*. Verlag G. Braun; Karlsruhe, 1965.

[13] H. Ranke. Phänomene der Grenzschichtablösung bei bewegten Wänden oder instationären Strömungen. FLM 93/16, Lehrstuhl für Fluidmechanik, TU-München, 1992.

[14] D. T. Tsahalis. Laminar boundary-layer separation from an upstream-moving wall. *AIAA Journal*, 15(4), 1977.

[15] J. C. WilliamsIII and T. J. Wang. Semisimilar solutions of the unsteady compressible laminar boundary-layer equations. *AIAA Journal*, 23(2):pp.228–233, Feb. 1985.

[16] J. C. WilliamsIII and W. D. Johnson. Note on unsteady boundary-layer separation. *AIAA Journal*, Vol.12(No.10):pp.1427–1429, Oct. 1974.

[17] D. P. Telionis and M. J. Werle. Boundary-layer separation from downstream moving boundaries. *Transactions ASME Journal of Applied Mechanics*, E40:pp.369–374, Juni 1973.

[18] G. R. Ludwig. An experimental investigation of laminar separation from a moving wall. In *AIAA Paper No. 64-6*, page 12, Jan. 1964.

[19] O. Inoue. MRS criterion for flow separation over moving walls. *AIAA Journal*, 19(9):pp.1108–1111, 1981.

[20] W. R. Sears and D. P. Telionis. Boundary-layer separation in unsteady flow. *SIAM Journal of Applied Mathematics*, 28(1):pp.215–235, 1975.

# Surface Inclination and Heat Transfer Methods for Reacting Hypersonic Flow in Thermochemical Equilibrium

U. Reisch, Th. Streit
Deutsche Forschungsanstalt für Luft- und Raumfahrt (DLR) e.V.
Institut für Entwurfsaerodynamik
Lilienthalplatz 7
D-38108 Braunschweig

## Summary

For the design of hypersonic configurations heat transfer plays an essential role. At high velocities high temperature effects have to be taken into account, since a perfect gas assumption for air leads to major errors concerning the calculation of temperature and heat transfer. In the present paper the application of surface inclination methods and approximate heat transfer methods is studied for chemically reacting air in thermochemical equilibrium. Computations for various geometries including high temperature effects are carried out to validate these procedures. The results are compared to Euler and Navier-Stokes solutions as well as experimental data.

For inviscid flow, the pressure is calculated from surface inclination methods. The respective flow variables are obtained from isentropic relations for perfect gas using piecewise constant gas properties, or from the oblique shock relations, which are solved iteratively. The flow variables are substituted as boundary layer edge values in approximate methods to evaluate the heat transfer, which otherwise are only a function of the wall temperature. Therefore, total coefficients, flow variables and heat loads for complete configurations can be calculated very efficiently and with reasonable accuracy.

## 1 Introduction

The numerical effort solving the Euler and Navier-Stokes equations within the predesign stage of hypersonic vehicles still is too large. As in hypersonic flows aerodynamic heating plays an essential role in the design process [1], the DLR code HOTSOSE [2] using surface inclination methods in combination with approximate methods to determine heat transfer and skin friction has been developed, providing an efficient design tool in case of perfect gas.

Decisive for designing heat protection systems of high speed vehicles is the accurate prediction of skin temperatures and heat fluxes. In a perfect gas, the large freestream kinetic energy is transformed mainly into rotational and translational molecular energy behind a shock and in the boundary layer, resulting in very high temperatures. If the temperature is high enough, vibrational excitation and chemical reactions can occur. In consequence, a part of the freestream kinetic energy is changed into vibrational molecular energy and the zero point energy of the products of the chemical reactions, reducing the energy levels of molecular translation and rotation. Hence, temperatures in high speed flows are significantly lower than in a perfect gas, affecting most aerodynamic and thermodynamic properties. Therefore, the existing code HOTSOSE was extended to applications for chemically reacting air in thermochemical equilibrium [3] using a gas model given in [4] and matching the approximate methods to evaluate boundary layer edge properties, skin friction and aerodynamic heating. The modified code was validated for various geometries, comparing HOTSOSE high-temperature results with Euler and Navier-Stokes solutions as well as experimental data.

## 2 High-temperature effects on gas properties

In high temperature flows vibrational excitation and chemical reactions occur in the gas, affecting its thermodynamic and transport properties. In particular, the specific heats $c_p$, $c_v$, and hence their ratio $\gamma$, as well as the specific gas constant $R$ are variables. For the determination of transport properties, perfect gas relationships are no longer valid, e.g. the Sutherland law of viscosity only applies for temperatures less than 2000K. Therefore, the vectorizable fits for thermodynamic and transport properties according to [4] have been implemented for chemically reacting air in thermochemical equilibrium, using the density $\rho$ and internal energy $e$ as input parameters. The curve fits allow the determination of an effective $\gamma$ to calculate the pressure, the derivatives $\partial p/\partial e$, $\partial p/\partial \rho$ and an equivalent $\gamma_a$ for the evaluation of the speed of sound and the equivalent gas constant $R$ as well as the dynamic viscosity $\mu$ and the heat conductivity $k$.

Along with these output values the same equations as for perfect gas can be applied for equilibrium air.

## 3 Approximate methods for flow analysis

Aerodynamic flow properties as well as the boundary layer thickness, and hence heat transfer, are strongly geometry dependent. Therefore a considered hypersonic vehicle is subdivided into different aerodynamic surfaces, which are blunt nose, blunt leading edge, cone and flat plate. These surfaces are calculated using respective procedures for inviscid and viscous flows.

### 3.1 Surface inclination methods

The inviscid hypersonic flow over a given body can be obtained with little computational effort and with engineering accuracy by using surface inclination methods. These methods predict the surface pressure from the local surface inclination angle $\Theta$ relative to the freestream direction.

In case of a blunt body, the modified Newtonian theory [5] gives the pressure coefficient in terms of $C_{p,max}$ and the surface inclination angle $\Theta$ defined in *figure1*

$$C_p = C_{p,max} \sin^2 \Theta. \tag{3.1}$$

The stagnation pressure coefficient $C_{p,max}$ is iteratively determined from normal shock relations and the gas model $p = p(\rho,e)$, assuming isentropic deceleration at constant gas parameters behind the shock.

Provided that the flow over the surface is isentropic and using piecewise constant gas parameters, all relevant boundary layer edge flow variables may be computed from a start condition and the pressure from a surface inclination method. In other words, to evaluate the inviscid flow properties in a considered point $i$, the gas parameters of the neighbouring upstream point $i-1$ and pressure $p_i$ are used in the isentropic relations.

Complete configurations are calculated along the body streamlines, approximated by the panel rows of the surface grid. Starting at the nose and moving downstream, the pressure on every panel is calculated from the shock expansion method, consisting of Newtonian theory applied to blunt bodies, the Prandtl-Meyer function (see [1]) for expansion flow and the oblique shock theory for compression flow. Newtonian theory may be matched to the Prandtl-Meyer function after the theory given in [6]. For the determination of start values in case of sharp nosed bodies, a tangent cone method or the oblique shock relations are applied.

Further simplifications in the HOTSOSE surface inclination methods are that three-dimensional effects are neglected and the influence of expansion waves, which are in part reflected from the shock wave, is not accounted for.

## 3.2 Aerodynamic heat transfer

Engineering methods to compute local heat transfer and skin friction coefficients are often based on the incompressible boundary layer theory. The incompressible solutions can be adjusted to high speed flows by applying Eckert's reference enthalpy concept [7]

$$h^* = 0.28h_e + 0.5h_w + 0.22h_{aw}, \tag{3.2}$$

with $e$ and $w$ denoting the boundary layer edge and wall conditions, respectively. The adiabatic wall enthalpy $h_{aw}$ is given by

$$h_{aw} = h_e - r_f \frac{u_e^2}{2}. \tag{3.3}$$

The recovery factor $r_f$ takes into account, that the freestream kinetic energy is partly dissipated by viscous effects in the boundary layer. Here, $r_f$ is related to the reference Prandtl number Pr*

$$r_f = (\mathrm{Pr}^*)^n \tag{3.4}$$

with $n$=1/2 for laminar and $n$=1/3 for turbulent flow. Eq.(3.2) defines a reference condition with corresponding density $\rho^*$, temperature $T^*$, viscosity $\mu^*$ and heat conductivity $k^*$, determined by using piecewise constant gas properties as shown above. The start values for this procedure are obtained by iteration.

Local heat transfer is described commonly in dimensionless form in terms of a heat transfer coefficient. Here, the Stanton number has been chosen, defined as

$$St_e = \frac{q_w}{\rho_e u_e (h_{aw} - h_w)}, \tag{3.5}$$

or, related to the freestream

$$St_{0,\infty} = \frac{q_w}{\rho_\infty u_\infty (h_0 - h_w)} \tag{3.6}$$

with $h_0$ being the total enthalpy. Note that the results presented in this paper are given in terms of the Stanton number $St_{0,\infty}$, which is proportional to heat transfer.

All approximate methods to determine aerodynamic heating and skin friction coefficients being used here are only dependent on wall temperature and local boundary layer edge flow properties, and are valid for an isothermal wall. In particular, procedures have been implemented for different geometry options. At the stagnation point, the Fay-Riddell formula [8] is applied, assuming a Lewis number Le=1. Downstream the heat transfer to blunt bodies of revolution is determined after Zoby, Moss & Sutton [9]. Blunt leading edges are treated as infinite swept cylinders, as supposed by Fleming & Krauss [10], while flat plates and cones are calculated with the reference enthalpy method [11] or the formulation of White & Christoph [12]. As some of these methods compute heat transfer and others skin friction, the Reynolds analogy is introduced, relating the local Stanton number $St_e$ to the local skin friction coefficient $c_f$. Note that heat transfer in high temperature flows is mainly driven by enthalpy rather than temperature differences. Therefore, temperature ratios in the above mentioned procedures were replaced by the respective enthalpy ratios.

## 4 Numerical results

The surface inclination methods and the boundary layer approximations including high-temperature effects were checked for several geometries at zero incidence.

One important test case with respect to reentry vehicles are blunt bodies of revolution. *Figure 2* shows for a blunt cone with cone half angle $\Theta_c$=15° the pressure coefficient and temperature distributions at the boundary layer edge, plotted versus the axis of revolution. The curves are obtained from HOTSOSE and Euler calculations for perfect gas and equilibrium air, showing good agreement between the surface inclination methods and the Euler results for both gas models. Note that pressure is affected only slightly by high temperature effects, while temperature drops significantly.

As an example of heat transfer for this type of configuration, the heat loads for the ELECTRE capsule [13] in equilibrium air at a flight velocity of approximately 6000m/s are given in *figure 3* versus the dimensionless surface coordinate. The experimental data from the HEG wind tunnel Göttingen are compared with a Navier-Stokes solution and predictions obtained with the approximate methods implemented in HOTSOSE for the stagnation point (Fay-Riddell) and the body (Zoby et al.), showing good correlation. Considering the nose region, the HOTSOSE results lie even closer to the experiments than the Navier-Stokes computation.

Wings and stabilisers may be computed using the approximate methods for blunt leading edges, geometrically defined in *figure 4*. In the following, results are presented for a fin with a sweeping angle $\Lambda$=75° and a nose radius $r_n$=0.0005m in a plane parallel to the freestream velocity direction. *Figure 5* displays the distributions of $C_p$, Mach number and temperature along the $x$-axis (see *figure 4*) obtained from HOTSOSE for perfect gas and equilibrium air. Due to the large sweeping angle and the, if related to high-speed flows, moderate Mach number $M_\infty$=5.93, the influence of high-temperature effects on the boundary layer edge properties is negligible. Despite these results, the heat transfer to the fin nose clearly increases in equilibrium air, as *figure 6* shows, comparing Stanton number distributions after Fleming and Krauss for a wall temperature $T_w$=999K. *Figure 6* also contains the ratio reference density to boundary layer edge density for the nose region. Since density within the equilibrium bounday layer is increased compared to perfect gas, the boundary layer thickness is decreased, resulting in a larger heat transfer to the surface. Away from the stagnation line, the temperature within the boundary layer drops and the difference in Stanton number between both gas models vanishes.

## 5 Conclusions

The application of surface inclination methods and approximate methods for hypersonic flows to evaluate skin friction and heat transfer for air in thermochemical equilibrium has been studied.

The pressure is calculated from surface inclination methods, while the flow variables at the boundary layer edge are calculated from isentropic gas relations using piecewise constant gas properties or from the exact shock equations, solved iteratively. The gas properties are interpolated from curve fits for equilibrium air.

The boundary layer strongly depends on the surface geometry, and therefore several approximate methods to calculate heat transfer and skin friction coefficients were implemented. As these only depend on the boundary layer edge properties and the wall temperature, the required computational effort is reduced to the solution of closed-form expressions, iteratively solved by implicit formulations and one-dimensional integrals along the streamlines. Different geometry sections in one configuration are accounted for by multiblock treatment. Furthermore, HOTSOSE may be used to calculate the radiation-adiabatic surface temperature of a hypersonic vehicle.

Numerical results obtained from HOTSOSE using the equilibrium air gas model were compared with perfect gas computations, Euler and Navier-Stokes solutions as well as experimental data. Summarizing the considered test cases, the here presented engineering methods predict flow variables at the boundary layer edge and heat loads with reasonable accuracy for air in thermochemical equilibrium.

## References


[1] J. D. Anderson: *Hypersonic and High Temperature Gas Dynamics*. McGraw Hill Book Company, New York, 1989.

[2] Th. Streit, S. Martin, Th. Eggers: *Approximate Heat Transfer Methods for Hypersonic Flow in Comparison with Results Provided by Numerical Navier-Stokes Solutions*. DLR-FB 94-36, 1994.

[3] U. Reisch, Th. Streit: *Surface Inclination Methods and Heat Transfer Methods for Reacting Hypersonic Flow in Thermochemical Equilibrium*. DLR-IB 129-96/10, 1996.

[4] Ch. Mundt, R. Keraus, J. Fischer: *New Vectorized Approximations of State Surfaces for the Thermodynamic and Transport Properties of Equilibrium Air*. ZFW, No.15, 1991, pp. 179–184.

[5] L. Lees: *Laminar Heat Transfer over Blunt-Nosed Bodies at Hypersonic Flight Speeds*. Jet Propulsion, Vol.26, 1956, pp.259–269, 274.

[6] L. G. Kaufman; *Pressure Estimation Techniques for Hypersonic Flows over Blunt Bodies*. Journal of the Aeronautical Sciences, Vol.X, No.2,1963, pp.35–41.

[7] E. R. G. Eckert: *Engineering Relations of Friction and Heat Transfer to Surfaces in High Velocity Flow*. J.Aeronautical Sciences, Vol.22, No.8, August 1955, pp.585–587.

[8] J. A. Fay, F. R. Riddell: *Theory of Stagnation Point Heat Transfer in Dissociated Air*. Journal of the Aeronautical Sciences, Vol.25, No.2, 1958, pp.73–85.

[9] E. V. Zoby, J. N. Moss, K. Sutton: *Approximate Convective-Heating Equations for Hypersonic Flows*. J. Spacecraft, Vol.18, No.1, 1979.

[10] W. J. Fleming, W. E. Krauss: *Aerodynamic Heating from Turbulent Boundary Layers to Swept Surfaces*. Third International Heat Conference, 1966.

[11] G. Simeonides: *Simple Theoretical and Semi-Empirical Convective Heat Transfer Predictions for Generic Aerodynamic Surfaces*. ESTEC Document YPA/1576/GS, Noordwijk, 1995.

[12] F. M. White, G. H. Christoph: *A Simple Theory for the Two-Dimensional Compressible Turbulent Boundary Layer*. J.Basic Eng., Vol.91, 1972, pp.371–378.

[13] K. Hannemann: *Computation of high enthalpy flow past the electre standard model in HEG conditions*. In *Proceedings of the MSTP Workshop 1996 Reentry Aerothermodynamics and Ground-to-Flight Extrapolation, March 25-27, 1996*, Noordwijk, 1996.

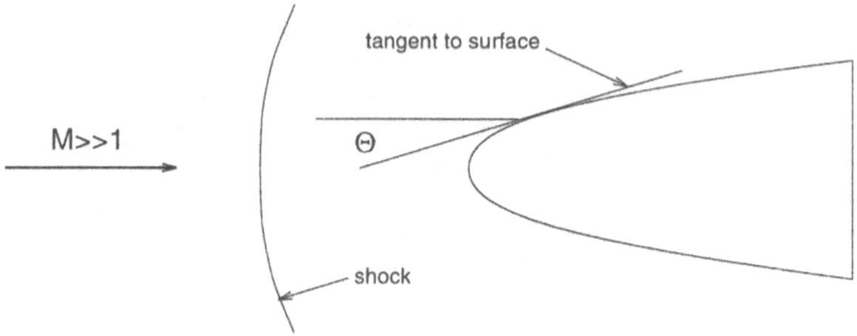


**Figure 1** Definition of the surface inclination angle for Newtonian theory (2-dimensional).

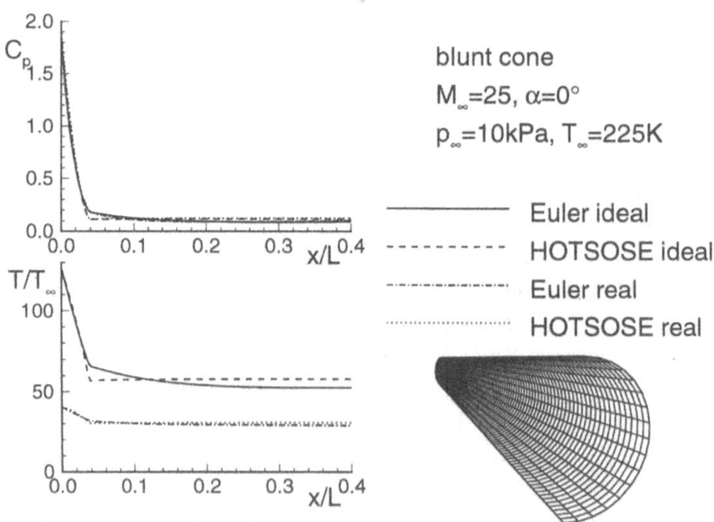


**Figure 2** Pressure coefficient and temperature distributions for a blunt cone along the axis of revolution for perfect gas and equilibrium air.

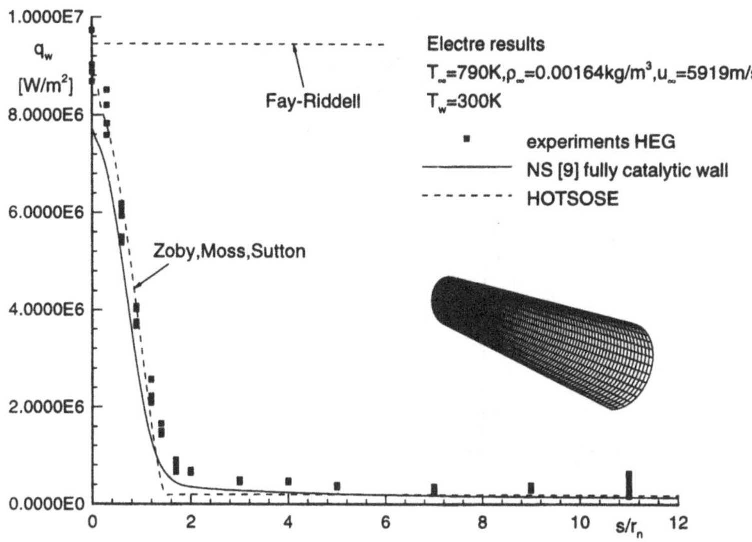


**Figure 3** Heat loads along the surface coordinate of the ELECTRE capsule; comparison of HOTSOSE results with experiments and a Navier-Stokes solution.

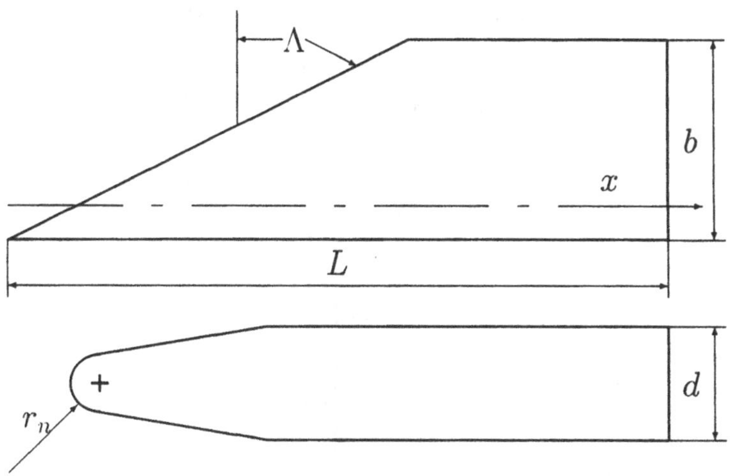


**Figure 4** Geometry definition of a blunt leading edge ($b$, $d$, $L$ not used in the calculation).

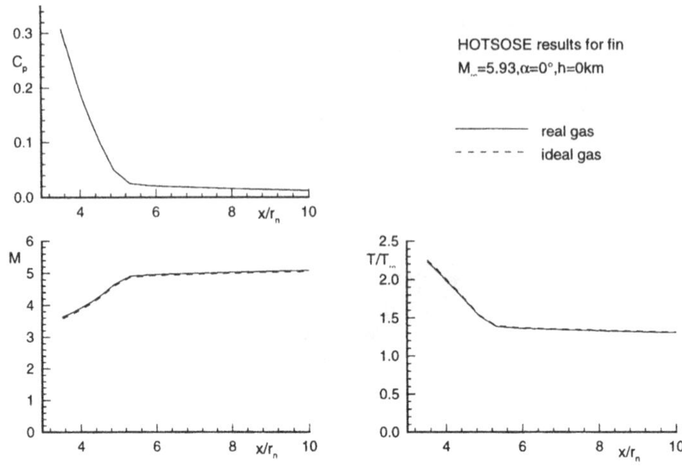


**Figure 5** Pressure coefficient, Mach number and temperature distributions at the boundary layer edge versus the x-axis of a swept fin ($\Lambda$=75°).

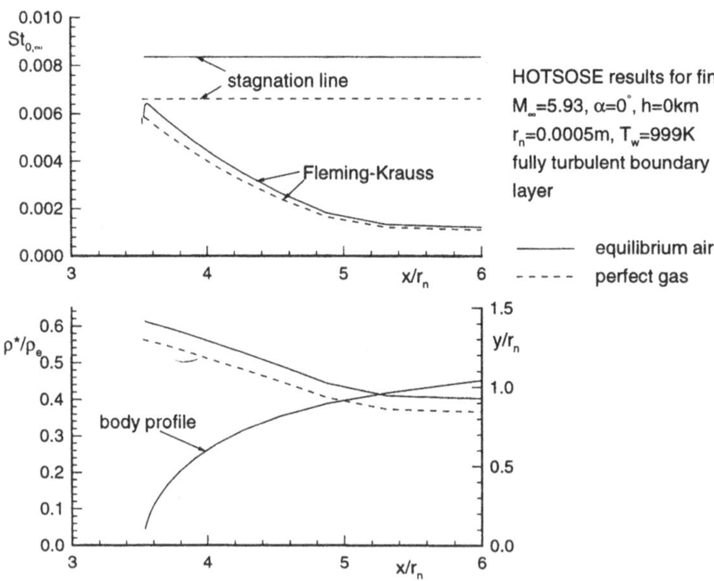


**Figure 6** Stanton number distribution in the nose region versus the x-axis of a swept fin ($\Lambda$=75°)

# Techniques to enhance reliability and efficiency of flow calculations on unstructured grids

C.-H. Rexroth, S. Wittig

Universität Karlsruhe, Lehrstuhl und Institut für Thermische Strömungsmaschinen
Kaiserstraße 12, D-76128 Karlsruhe, Germany

## Summary

If unstructured grids are used for Finite Volume flow calculations, drawbacks in accuracy and computing time tend to outweigh the advantages of geometrical flexibility. To overcome this problem, various techniques enabling an accurate and stable discretization of convection, an error controlled mesh refinement, a rapid solution of the systems of linear equations and a general stabilization of the iterative process are presented. Results for different test cases and a practical application prove the performance of these methods.

## Introduction

When designing components for modern gas turbines, time and costs can be reduced significantly by applying numerical methods. To be successful, these tools have to be reliable and inexpensive in use, and therefore, high accuracy and efficiency are vital.

Employing Finite Volume methods, the most important reasons for poor results are zones of insufficient spatial resolution and inadequate modelling of the convective transport. To describe the convective terms correctly, a stable interpolation scheme of second order accuracy is necessary. On unstructured meshes another possibility to improve accuracy is grid adaption exploiting error sensitive criteria to optimize the spatial discretization.

The amount of computational work required is mainly determined by the convergence and stability properties of the numerical procedure. In this context an efficient algorithm for the solution of the linear equations is essential. Furthermore all kinds of techniques improving numerical stability can contribute to a reduction of processing time.

## Numerical procedure

At the Institut für Thermische Strömungsmaschinen (ITS), Finite Volume methods for the numerical prediction of flows in turbomachinery have been developed and successfully employed for years [1, 2]. In continuing this work, a procedure using unstructured grids to analyse two-dimensional flow fields has been established [3].

### Transport equations and solution algorithm

The transport equations of a turbulent flow are evaluated in a block iterative way. The velocities are calculated from the Reynolds-averaged Navier-Stokes equations. A SIMPLE(C) type correction algorithm yields the corresponding pressure distribution. On the collocated grid a decoupling of the former quantities is prevented by a special treatment of the velocities at the cell faces [4]. An extension covering the transonic flow regime has also been implemented [5]. Turbulence is accounted for by the standard $k, \epsilon$ model. Additionally a transport equation for the enthalpy is solved.

## Adapted spatial discretization

In contrast to structured meshes, unstructured grids offer better geometrical flexibility and adaptivity, because a strict order of mesh cells is not required. In the two-dimensional case, triangles are used to cover domains of arbitrary shape. The grid nodes are defined by the vertices of the triangles. To ensure a strong coupling between the variables, a dual mesh of convex polygons is used for the integration of the transport equations (cf. Fig. 1).

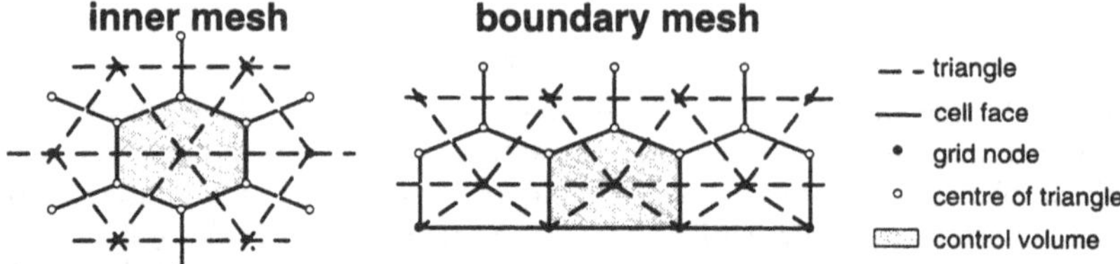


**Fig. 1:** Topology of control volumes

Unlike most structured approaches unstructured grids offer the possibility to modify an existing mesh by inserting or removing grid nodes. This advantage can be exploited to improve accuracy by adapting the spatial discretization to the flow field. To identify areas, where grid adaption is required, error sensitive indicators have to be applied. Normalized gradients $P$ of the transport variables $\phi$ were found to meet this requirement.

$$P = \left[\left|\frac{\partial\phi}{\partial x}\right| + \left|\frac{\partial\phi}{\partial y}\right|\right] \cdot \frac{D}{|\phi_{ref}|} . \tag{1}$$

$P$ is evaluated at the grid nodes from the spatial derivatives of $\phi$, the mean diameter $D$ of the control volume and a reference value $\phi_{ref}$. The expressions $\frac{\partial\phi}{\partial x_i} \cdot D$ may be interpreted as the first order terms of a Taylor expansion for $\phi$. Therefore steep changes of $\phi$ are associated with high values of $P$ whereever the spatial resolution of the mesh is not sufficient. In these regions numerical failures are likely to occur. From this point of view $P$ can be regarded as an indicator for the local discretization error of $\phi$. In zones with high values of $P$ grid refinement is necessary. whereas low values permit coarsening.

The success of the whole adaption process relies on the quality of the coarse grid results. If the flow pattern in the basic solution is not physically correct, adaptive refinement will not be successfull. Especially in turbomachinery applications, the desired accuracy is strongly affected by the modelling of the convective transport.

## Accurate modelling of convective transport

The transport of information in a discretized domain is formulated in terms of transport coefficients $a_i$. $a_i$ consists of a diffusive part $a_{i,diff}$ and a convective part $a_{i,conv}$.

$$a_i = a_{i,diff} + a_{i,conv} . \tag{2}$$

To express the convective exchange $a_{i,conv}$ between adjacent grid nodes $P$ and $I$, the value of the transport variable $\phi$ at the cell face $i$ with area $F$ and fluid density $\rho$ has to be estimated. A common method is the Upwind scheme (UPW), setting $\phi_i$ to the upstream value $\phi_P$ (cf. Fig. 2a).

$$\phi_i^{UPW} = \begin{cases} \phi_P, & (\vec{w},\vec{n}) > 0 \\ \phi_{I,down}, & (\vec{w},\vec{n}) < 0 \end{cases} \quad , \quad a_{i,conv}^{UPW} = max\left[0, -(\rho \cdot F \cdot (\vec{w},\vec{n}))_i\right] . \tag{3}$$

Because the Upwind interpolation is only first order accurate, the polygonal mesh cells are prone to severe discretization errors, better known as numerical diffusion.

The central difference approach (CD) is second order accurate. To determine $\phi_i$, a linear interpolation of $\phi_P$ and $\phi_{I,down}$ is applied (cf. Fig. 2a).

$$\phi_i^{CD} = \frac{\Delta_{down} \cdot \phi_P + \Delta_P \cdot \phi_{I,down}}{\Delta_P + \Delta_{down}} \,, \quad a_{i,conv}^{CD} = -\left(\rho \cdot F \cdot (\vec{w}, \vec{n})\right)_i \cdot \frac{\Delta_P}{\Delta_P + \Delta_{down}} \,. \tag{4}$$

The drawback of the CD scheme are stability problems which arise from violations of the conservation of $\phi$ inside the control volumes.

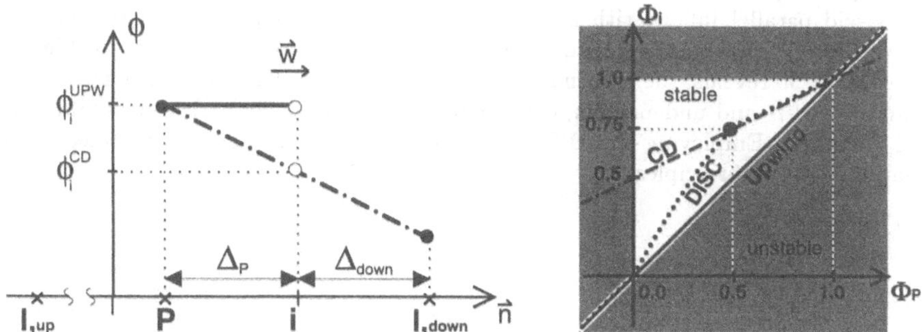


**Fig. 2: (a) UPWIND and CD interpolation; (b) CBC stability diagram**

Second order accuracy and numerical stability are combined with the Derivative based Interpolation Scheme for Convection (DISC) [6]. Its principles are summarized below.

DISC exploits the Convection Boundedness Criterion (CBC) formulated by Gaskell and Lau [7]. The nondimensional CBC parameter $\Phi_P$ depending on the values of $\phi$ at node $P$, the upstream node $I, up$ and the downstream node $I, down$, indicates the actual state of stability at the cell face $i$ (cf. Fig. 2a).

$$\Phi_P = \frac{\phi_P - \phi_{I,up}}{\phi_{I,down} - \phi_{I,up}} \,. \tag{5}$$

The CBC stability diagram then shows suitable values for $\Phi_i$ (cf. Fig. 2b). If for example $\Phi_P$ is less than 0.0 or greater than 1.0 only the Upwind scheme is stable.

$$\Phi_i^{UPW} = \Phi_P \,. \tag{6}$$

Inside a triangle with $\Phi_P$ between 0.0 and 1.0, the CD method is stable, too.

$$\Phi_i^{CD} = \frac{1}{2}\Phi_P + \frac{1}{2} \,. \tag{7}$$

In the CBC diagram, second and third order interpolations meet at the point (0.5, 0.75). These high order schemes (HOS) can be interpreted as linear superpositions of Upwind and CD approach where a weighing factor $\alpha^{HOS}$ controls the CD contribution. The step back to physical quantities $\phi_i$ is straightforward now, because it does not affect $\alpha^{HOS}$.

$$\phi_i^{HOS} = \alpha^{HOS} \cdot \phi_i^{CD} + (1 - \alpha^{HOS}) \cdot \phi_i^{UPW} \,, \quad a_{i,conv}^{HOS} = \alpha^{HOS} \cdot a_{i,conv}^{CD} + (1 - \alpha^{HOS}) \cdot a_{i,conv}^{UPW} \tag{8}$$

The DISC method applies the bare Upwind interpolation outside the stable region. Inside, after following a parabolic curve to the point (0.5, 0.75), it changes to the CD method.

$$\alpha^{DISC}(\Phi_P) = \begin{cases} 0.0\,, & \Phi_P < 0.0 \\ 2\,\Phi_P\,, & 0 \le \Phi_P < 0.5 \\ 1.0\,, & 0.5 \le \Phi_P \le 1.0 \\ 0.0\,, & \Phi_P > 1.0\,. \end{cases} \tag{9}$$

To reduce the computational expense, $\phi_{I,up}$ at the node upstream of P, which is necessary to calculate $\Phi_P$, is determined by expansion into a Taylor series where second order terms are neglected. From this simplification the new scheme was given its name.

$$\phi_{I,up} = \phi_P - \left(\frac{\partial \phi}{\partial n}\right)_P (\Delta_P + \Delta_{down}) + O^2 \ , \ \ \frac{\partial \phi}{\partial n} = (\mathrm{grad}\phi, \vec{n}). \tag{10}$$

To compare Upwind, CD and DISC approach, a square domain has been exposed to an inviscid parallel inflow with a steplike distribution of a passive scalar quantity $\phi^*$ (cf. Fig. 3a). Without numerical diffusion, the same step profile would be visible at the exit. As Fig. 3b reveals, the Upwind result is quite inexact. The CD solution suffers from unphysical over- and undershoots, corresponding to areas of numerical instability in the CBC diagram. Employing the DISC method fairly good accuracy is attained without violating stability principles.

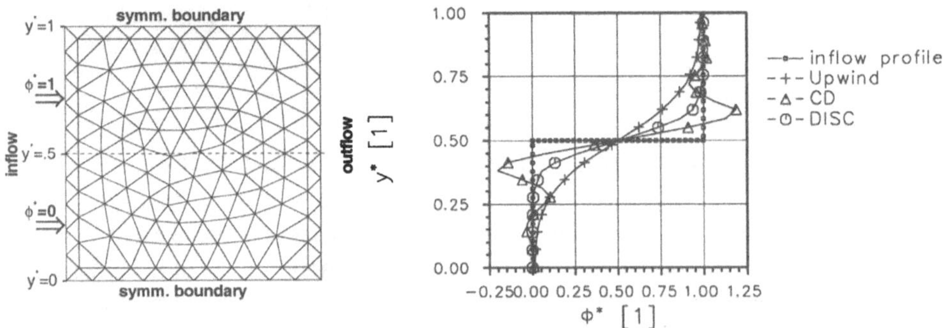


**Fig. 3:** **(a)** Test grid and boundary conditions; **(b)** Outflow profiles of $\phi^*$

## Rapid solution of linear systems

When solving a flow problem numerically, the solution of the systems of algebraic equations arising from discretization requires about 50 % of the total computing time. Furthermore, the stability of the numerical procedure and the number of iterations needed are affected by the properties of the solver. Out of several Conjugate Gradient and Conjugate Residual algorithms, the BiCGSTAB method was found to be most robust and efficient [8]. For increasing the performance of the solver, preconditioning based on an incomplete lower-upper matrix decomposition (ILU) can be applied. The goal of the decomposition is to create an approximation $M$ of $A$, which is easy to invert [9].

$$A \cdot \vec{\phi} = \vec{b} \ , \ \ A \approx L \cdot U = M \tag{11}$$

$$a_{i,j} \neq 0 : \ l_{i,j} = s_{i,j} \ \ (j \leq i) \ , \ \ u_{i,j} = \frac{s_{i,j}}{l_{i,i}} \ \ (j \geq i) \ ; \ \ s_{i,j} = a_{i,j} - \sum_{k=1}^{min(i,j)-1} l_{i,k} \cdot u_{k,j}. \tag{12}$$

Like $A$ itself, the triangular matrices $L$ and $U$ are sparse. Therefore, memory requirements are quite small. However, the effort of computing the inner products $l_{i,k} \cdot u_{k,j}$ contributing to $s_{i,j}$ is prohibitive on unstructured grids. A reduced incomplete decomposition (rILU) overcomes this difficulty.

$$a_{i,j} \neq 0 : \ l_{i,j} = s_{i,j} \ \ (j \leq i) \ , \ \ u_{i,j} = \frac{s_{i,j}}{l_{i,i}} \ \ (j \geq i) \ ; \ \ s_{i,j} = a_{i,j} \, . \tag{13}$$

As an additional advantage of rILU, it is not necessary to store $L$ and $U$ explicitly. Combined with rILU preconditioning, BiCGSTAB is capable of reducing the residual norm of a linear system by more than two orders of magnitude in only one step.

**Stabilizing the outer iteration**

In spite of a stable discretization scheme and a robust solution algorithm, numerical problems may still occur in some critical cases. To cope with these difficulties, additional stabilization techniques have to be applied. Usually, if a calculation does not converge smoothly, one of the following irregularities can be observed:

- The reduction of the residuals of the various transport equations stagnates after a certain number of outer iterations.
- The pressure field starts to oscillate with the amplitude growing constantly.
- Negative values of the turbulent dissipation $\epsilon$ are computed, leading to an instantaneous breakdown of the numerical process.

Trouble with pressure and turbulent dissipation is often initiated by pressure corrections $p'$ lacking a sound physical background at the beginning of a calculation. When the information from the inflow boundaries is still not present in all areas of the domain, the corrections $p'$ tend to be too strong in these zones. The resulting 'low pressure' provokes 'high velocity' and vice versa. Depending on the situation, these pressure-velocity oscillations are not damped and the calculation does not converge. In other cases, the steep velocity gradients destabilize the turbulence model via the production term.

A massflow controlled damping of the pressure field has been developed to prevent overcorrections. It is applied while calculating the actual pressure solution $p^{new}$ from previous values $p^{old}$ and the SIMPLE(C) updates $p'$. The value of the relaxation factor $\alpha(p)^{act}$ rises constantly with the ratio of the massflows $\dot{m}_{out}$ to $\dot{m}_{in}$ entering and leaving the domain.

$$p^{new} = p^{old} + \alpha(p)^{act} \cdot p' \tag{14}$$

$$\alpha(p)^{act} = \min\left\{\alpha(p)^{stand}, \max\left[\alpha(p)^{old}, \frac{\dot{m}_{out}}{\dot{m}_{in}} + 0.01\right]\right\} . \tag{15}$$

From a minimum value of 0.01, assuring a driving pressure force, $\alpha(p)$ increases until the standard value $\alpha(p)^{stand}$ is reached.

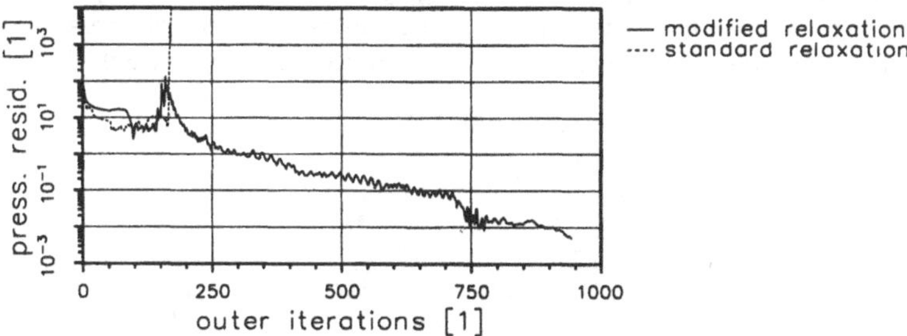


**Fig. 4: Relaxation technique and convergence of step flow calculation**

Fig. 4 displays the convergence history for a flow over a backward facing step [10] in terms of the normalized pressure residual and the number of outer iterations. With the standard approach the residual norm explodes after 172 steps. Applying the modified technique, after a local peak near 160 iterations monotonic convergence is attained. To achieve the stopping criterion of $Res_{max} \leq 5 \cdot 10^{-3}$, 943 block iterations had to be carried out.

Problems with negative values of $\epsilon$ or stagnating residuals can often be traced back to the structure of the transport equation for $\epsilon$ in the standard $k, \epsilon$ model. A part of the source term Sc($\epsilon$) on the right hand side of this equation depends linearly on $\epsilon$ itself. If the values of $\epsilon$ change very fast, a decoupling of the unknown $\epsilon$ and the source term $Sc(\epsilon)$ takes place, leading to the problems mentioned earlier.

In this case, a measure taking into account the particular structure of the equation was found in solving the $\epsilon$-equation with actual coefficients, source terms and boundary conditions a second time immediately after the first pass. The most efficient way to employ this technique is to restart on a solution that stagnated with the common treatment of $\epsilon$. The best results have been obtained using a modified relaxation factor $\alpha(\epsilon)^{double}$.

$$\alpha(\epsilon)^{double} = \sqrt{\alpha(\epsilon)^{single}}\,. \tag{16}$$

As an example, the progress of a calculation of the flow through a cascade of highly loaded gas turbine stator blades [11] is shown in Fig. 5. The conventional approach of solving for $\epsilon$ only once during an outer iteration causes the pressure residual to stay at values of around $10^1$ after 800 steps. Switching to the double solution strategy after 500 iterations, it takes 1689 loops to achieve convergence.

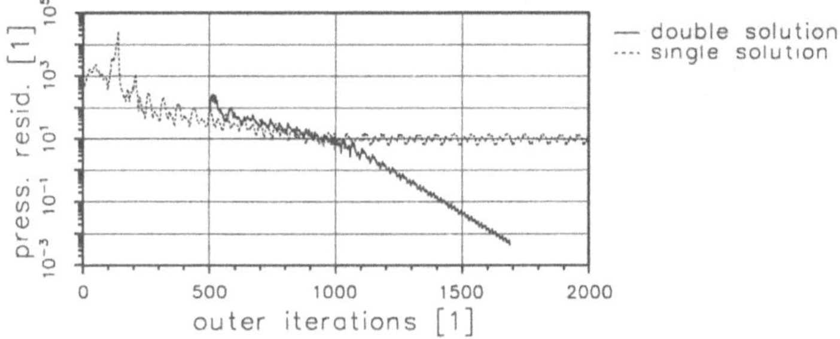


**Fig. 5:** Solution technique and convergence of cascade flow calculation

## Practical application

As a demanding practical application, the highly loaded turbine blade, also discussed in the preceding example, has been selected to demonstrate the effectiveness of all methods presented. The corresponding experiments have been performed by Schiele et al. [11].

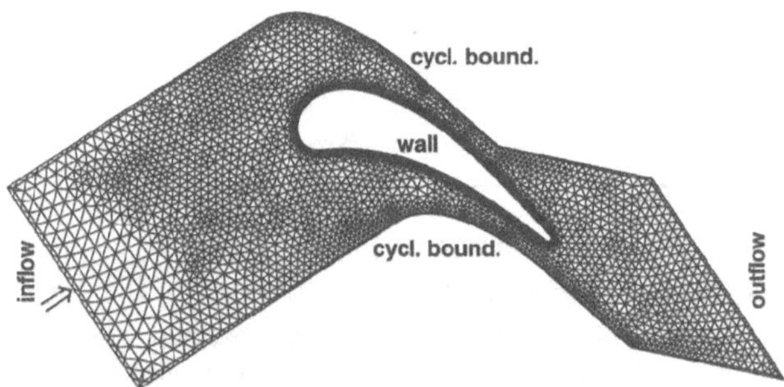


**Fig. 6:** Basic grid for highly loaded turbine blade

For the first step of the numerical analysis a basic grid of 3839 nodes has been used. It can be seen in Fig. 6. To improve the spatial resolution in areas, where inaccuracies may occur, the self-adaptive refinement technique has been applied. The modified grid consists of 4169 nodes. Especially the boundary layer at the leading edge of the blade was resolved in more detail afterwards. On both grids, calculations were carried out at a Reynolds number *Re* of 224 000. At the inlet, total pressure and flow direction were specified as boundary conditions.

Fig. 7 illustrates the reduction of the pressure residual in the course of the numerical process on the basic and the modified mesh. As shown before, on the coarse grid smooth convergence is attained by switching to the double $\epsilon$ solution strategy after 500 outer iterations. Having finished grid refinement, the computation was restarted without encountering problems. This time 838 steps were sufficient to reach the convergence limit.

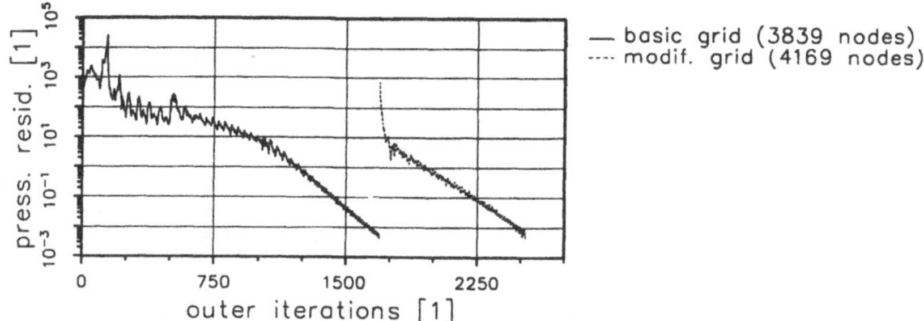


**Fig. 7:** Convergence of cascade flow calculations on different grids

To judge the accuracy of the solution, the pressure distribution calculated along the surface of the blade is compared to the experimental results (cf. Fig. 8). The surface coordinate $s$ has been normalized with the chord length $c$.

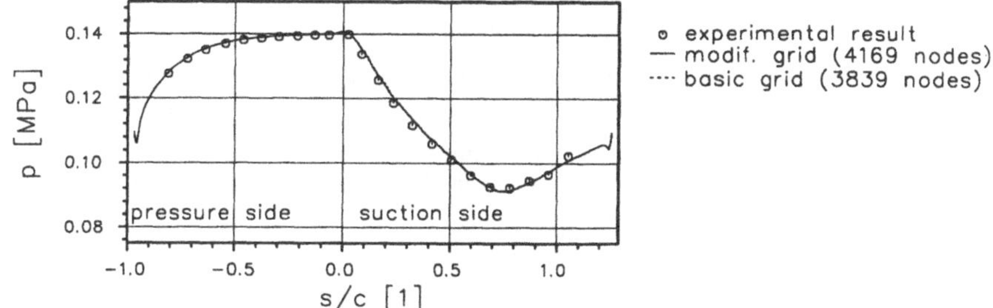


**Fig. 8:** Measured and simulated pressure profiles

Already on the basic mesh, computed and measured values of $p$ agree very well. Due to the grid modification, some minor improvements are visible near the stagnation point on the suction side of the profile. These findings emphasize the good performance of the DISC approach.

## Conclusion

Several new methods for improving the reliability and effectiveness of flow calculations on unstructured grids have been described and demonstrated.

Above all, the DISC approach for modelling the convective transport is capable to assure high accuracy without sacrificing numerical stability. Further improvements are attained through adaptive mesh refinement guided by an error sensing criteria.

To provide an efficient solver for the linear equations, the robust BiCGSTAB algorithm has been combined with matrix preconditioning based on a reduced lower-upper decomposition (rILU). This choice proved to be well suited for unstructured grids.

In order to avoid convergence problems under critical conditions, two techniques are proposed for an additional stabilization of the iterative process. Oscillations of pressure and velocity field are suppressed by a massflow controlled relaxation. A special solution strategy for the transport equation of turbulent dissipation prevents difficulties related to the turbulence model. A practical application for a highly loaded turbine cascade proves the performance of all methods presented.

**Acknowledgements**

This work was supported by a grant of the Deutsche Forschungsgemeinschaft in the context of the Graduiertenkolleg 'Energie- und Umwelttechnik' and the Sonderforschungsbereich 167 'Hochbelastete Brennräume'.

## References

[1] Giebert, D., Hürst, C., Kurreck, M., Rexroth, C.-H., and Wittig, S. Einsatz effizienter Methoden und Höchstleistungsrechner zur Strömungssimulation in Turbomaschinen. In *Deutscher Luft- und Raumfahrt-Kongreß 1995, DGLR-Jahrestagung, Jahrbuch 1995* :941–950, Bonn, 1995.

[2] Noll, B. E. and Wittig, S. A generalized conjugate gradient method for the efficient solution of three-dimensional fluid flow problems. *Numerical Heat Transfer*, 20:207–221, 1991.

[3] Rexroth, C.-H., Kurreck, M. and Wittig, S. Numerische Strömungsberechnung mit unstrukturierten Gittern nach der Methode der Finiten Volumen. In *Deutscher Luft- und Raumfahrt-Kongreß 1994, DGLR-Jahrestagung, Jahrbuch 1994* :437–446, Erlangen, 1994.

[4] Rhie, C. M. *A numerical study of the flow past an isolated airfoil with separation.* Ph. d. thesis, University of Illinois at Urbana-Champaign, 1981.

[5] Karki, K. C. and Patankar, S. V. Pressure based calculation procedure for viscous flows at all speeds in arbitrary configurations. *AIAA Journal*, 27(9):1167–1174, 1989.

[6] Rexroth, C.-H. and Wittig, S. Improved accuracy and effectiveness for Navier-Stokes solvers on unstructured grids. In *Numerical Methods for Fluid Dynamics V*, 1995.

[7] Gaskell, P. H. and Lau, A. K. C. Curvature-compensated convective transport: Smart, a new boundedness-preserving transport algorithm. *International Journal for Numerical Methods in Fluids*, 8:617–641, 1988.

[8] Van der Vorst, H. A. Bi-CGSTAB: A fast and smoothly converging variant of Bi-CG for the solution of nonsymmetric linear systems. *SIAM Journal on Scientific and Statistical Computing*, 13(2):631–644, 1992.

[9] Elman, H. C. *Iterative methods for large, sparse, nonsymmetric systems of linear equations.* Ph. d. thesis, Yale University, 1982.

[10] Eaton, J. K. and Johnston, J. P. Turbulent flow reattachment: An experimental study of the flow and structure behind a backward facing step. Department of Mechanical Engineering, Stanford University, Report MD-39, 1980.

[11] Schiele, R., Sieger, K., Schulz, A., and Wittig, S. Heat transfer investigations on a highly loaded, aerothermally designed turbine cascade. In F. S. Billig, editor, *12th. International Symposium on Air Breathing engines* :1091–1101, Melbourne, Australia, 1995.

# Design of a Laminar-Glove for the A340 and First Results of a High-Speed Wind-Tunnel Test

S. Schmid-Göller, H. Hansen
Daimler-Benz Aerospace Airbus GmbH
Hünefeldstr. 1-5, D-28199 Bremen, Germany

## SUMMARY

This paper describes a first A340 glove design and a high speed wind tunnel test with a 1:35 complete model at the ARA Bedford. The design aim was to integrate the target laminar pressure distribution into the A340 pressure system compensating the 3D wing effects and the influence of pod/pylon. During this work it became obvious, that the geometric design restrictions related to the A340 demonstrator aircraft substantially affect the laminar integration ability. The following high speed wind tunnel test could verify the design predictions concerning the target pressure distribution along wing span and the related transition development. Data for a next, improved design step could be obtained.

## INTRODUCTION

Laminar technology is a key in improving economy and environmental integration of airliners due to the potential of significantly reducing fuel consumption by aerodynamic drag reduction.
At DASA-Airbus, Bremen work is done on design and manufacture preparing a flight test of a HLFC-Glove on a A340 flight demonstrator.
The target is the demonstration of the feasibility of laminar technology for airliners in service. The first essential aspects are high- and low-speed aerodynamics (considering pod/pylon and flap track fairing interference), high lift and suction systems.

## A340 GLOVE DESIGN

### Design Requirements and Restrictions

The design point is A340 cruise, defined by a mach number of .82 and a $C_L$ of .45. At altitudes between 31000 and 39000 ft there are Reynolds numbers up to 45 million at the mid glove station. This area is typical for big airliners and (considering the associated wing sweep) can only be handled by the use of ***hybrid*** laminar technology.
The Glove will be built up on the A340 as a composite structure and consequently has to include the upper and lower side of the A340 original wing.
The trailing edge flap system and the spoilers of the A340 should remain unchanged. Therefore the laminar airfoil contour should join smoothly the original airfoil at the upper and lower side shroud lines.
The nose box will include a suction system and a new high lift system (Krüger flaps).

Following the target to demonstrate the feasibility of laminar technology for a real aircraft the glove will be extended spanwise from inboard to outboard pylon (only on the port or starboard side of the wing). It shall be shown, that the compensation of aerodynamic interference effects of nacelles and pylons by glove design will be possible.
For reasons of manufacture the main part of the glove should be defined by linear lofting.

## Design Process

The laminar 2D pressure target distribution developed for the glove is very different to the A340 pressure distribution. The corresponding airfoils are presented in fig. 1. It demonstrates the slenderness of the laminar airfoil against the original section. This leads to the need of increasing the section chord by 7% to fulfil the requirement for including the whole A340 section (with a certain amount of normal distance). Therefore the leading edge will move forward in the glove region.

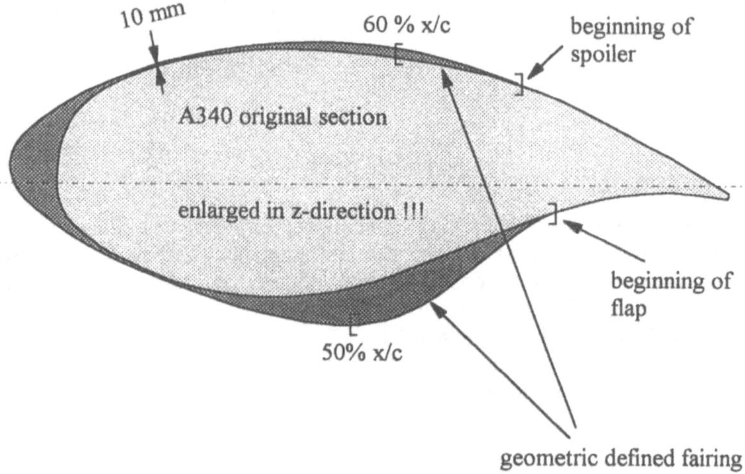


Fig. 1: A340 Glove Section, installed on A340 original geometry

Moreover the laminar airfoil will smoothly convert into the original contour beginning at 60%/50% up to the shroud line. As illustrated in fig. 1, a steep geometric transition follows at the lower side (This situation could only be improved by definitely decreasing laminar/turbulent transition at the lower side).
The glove planform is presented in fig. 2. It consists of three parts: In the center linear region the target distribution shall be introduced. The two outer parts, directly joining to the pylons, are designed to compensate the 3D A340 wing influence and the interference effects of the inboard/outboard pylons and nacelles.
Fig. 3 clearly demonstrates how the target pressure distribution is locally embedded into the A340 system (full potential code, clean wing, glove at an earlier design stage).

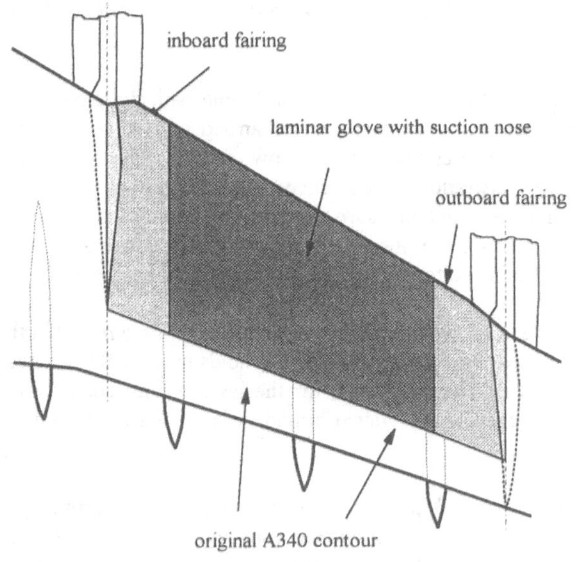


Fig. 2: **A340 Glove Planform**

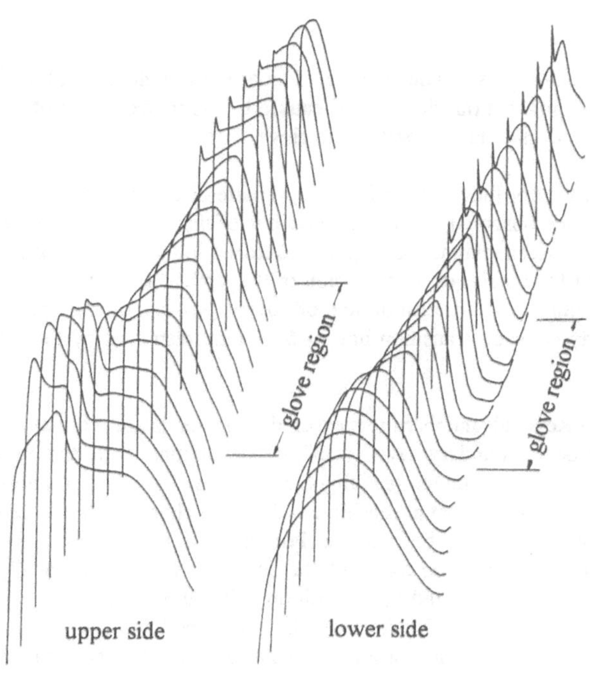


Fig. 3: **A340 Glove Pressure Distribution Overview (FLO22-calc., earlier design stage)**

## RESULTS OF THE HIGH-SPEED WIND-TUNNEL TEST

After the design of the A340 glove a new starboard wing with the glove contour between the inboard and outboard pod was manufactured for an existing A340 windtunnel model. This wing was equipped with four pressure rows as shown in fig.4. The main objective of the high-speed test was the demonstration of the successful integration of a hybrid laminar flow pressure distribution in the existing pressure distribution system of the A340 in the presence of pod and pylon. Furtheron, for the preparation of a flight test with such a laminar glove, the performance and handling characteristics of the aircraft should be checked.

The measurements in the ARA windtunnel were carried out with a model in the scale of 1:35 at a Reynolds number of $Re \approx 2.8 * 10^6$. For this low Reynolds number the boundary layer situation on the glove is not critical. Therefore even for the high leading edge sweep of $\varphi_{L.E.} = 32^0$ no suction was required to stabilise the laminar boundary layer and to maintain a laminar flow on the glove.

A comparison between the target pressure distribution and the measured pressure at the mid glove pressure rows shows a good agreement especially concerning the pressure gradients (glove design case Ma=.82, $C_L$=.45). The upper side of the inboard pressure section deviates from the target pressure distribution after the 35% chord station due to a first small shock. The comparison of the 3D-Euler computations for the fully equipped model (pod/pylon and flap-track-fairings) and the pressure measurements shows a similar effect for the inboard pressure station but the measured small shock is only indicated by the computation as flat part of the pressure distribution.

The analysis of the A340 pressure development inboard of the glove as a function of the Mach number gives some hints that this deviation is caused by the shock system of A340 which is not fully suppressed by the inner fairing part of the glove.

The problems with the not fully achieved target pressure distribution on the inner glove part is the explanation for the observed transition behaviour of the glove. The transition, which was made visible with the Acenaphtene technique, is sketched in fig.4 for the design case of the full equipped model. At the inner part an extension of the laminar flow to only 40% chord can be reached but at the outer part a transition line of 70% is possible on the upper side. The lower side shows a nearly constant transition line of 60% only disturbed by the effect of the flap-track-fairings.

Because of the low Reynolds number the measured transition is not representative for the flight test. For this purpose boundary layer stability calculations with the measured pressure distributions has to be done under real flight conditions. First computations for the target pressure distribution indicated that a transition of 35÷45% chord on the upper side and of 50% chord on the lower side should be possible. For the performance test this computed flight prediction was used for a realistic fixing of the boundary layer. Additionally the effect of a suction system failure was simulated by a transition fixing at 5% chord on both sides. The drag results in comparison to the A340 reference with the unmodified right wing show for the to optimistic free transition case a strong improvement, but even for the realistic flight condition fixing an acceptable improvement is possible. Moreover it has to be taken into account that the

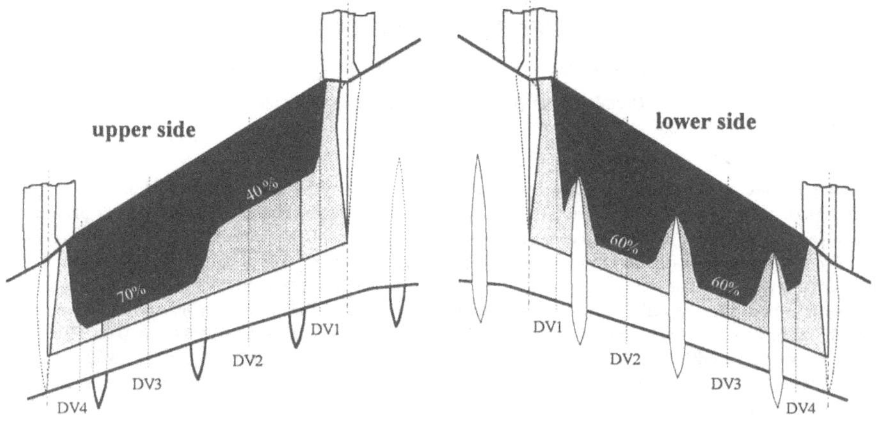


Fig. 4: A340 Glove: Measured transition line, upper and lower side

glove area is only 15% of the total wing area. The turbulent glove is worse than the reference but the pressure distribution shows no indication of a critical trailing edge separation on the upper side for this conditions.

Only a few measured effects could be shown here but in general the results of this first high-speed test campaign can be summarised in the following

## CONCLUSIONS

- The target pressure distribution could be reached especially concerning the pressure gradients. In the inner part of the glove the shock-system of the original A340 could not be damped out sufficiently and has to be improved.
- Under wind tunnel conditions a chordwise extension of the laminar flow of 40-70%c could be reached for the upper side and 60%c for the lower side. For the flight test it can be assumed that the desired transition of 40% on the upper side and 50% on the lower side can be reached.
- Despite the 15% glove area a net-drag reduction (depending on the transition fixing) could be reached in the wind tunnel in comparison to the A340 for the design case. Under turbulent conditions the glove produced more drag than the A340.
- The rolling moment behaviour and clean-$C_{L\,max}$ is not critical.
- The pod/pylon interference drag is in same magnitude as for the A340. Therefore, the aerodynamic integration can be expressed as successful.
- The geometric lofting from the glove region into the original A340 lower side has to be improved (only by reducing the laminar region).

This first HLFC-glove design for the A340 is an excellent base for the preparation of a flight test with an A340 demonstrator aircraft as a major step of introducing HLFC-wings in the next generation of future transport aircraft.

# INVESTIGATIONS OF HYPERSONIC INTAKES

Dieter M. Schmitz    Norbert C. Bissinger    Bert Sander

Daimler-Benz Aerospace AG, D-81663 München

## SUMMARY

The work performed in the BMBF Hypersonic Technology Program included the design of a drone with ramjet propulsion for flight Mach numbers $M_\infty = 3.5 - 6.8$. Due to the fuselage precompression, the Mach number in the intake plane is reduced to $M_0 \approx 6$. For this drone, a 2-D mixed compression intake with a design Mach number $M_{DES} = 6.2$ has been designed. Originally, it was planned to test the full scale intake (capture area $A_C = 0.48 * 0.48\ m^2$), designated ETM3, in the wind tunnel APTU at AEDC. In this wind tunnel, the complete RAM proulsion was to be tested in a second test phase. After the full scale tests have been cancelled, a 1/4.97 scale model of this intake (ETM3M) was defined. For comparison with ETM3M, 2 alternative configurations for the same operating range were designed (model ETM3AL by Dasa, ETM3AT by TsAGI). These intake models were manufactured at TsAGI and tested in the wind tunnel SVS-2 at $M_0 = 3.5 - 6.0$. In addition to the tests at TsAGI, ETM3M was also tested at DLR Cologne in the wind tunnel TMK. Test results are presented and the role of CFD is highlighted.

## MODEL CONFIGURATIONS AND WIND TUNNELS

### Model ETM3M

The intake configuration ETM3 is a 2-D mixed compression intake with two external ramps, designed for an operating range $M_0 = 3.5 - 6.0$. The requirements for this intake have been derived from the operating range of a drone with ramjet propulsion for flight Mach numbers $M_\infty = 3.5 - 6.8$ and Mach numbers in the intake plane $M_0 \leq 6.0$. An essential requirement was the avoidance of a bleed system on the ramps and within the intake. Therefore, the risk of shock induced boundary layer separation had to be reduced as far as possible, limiting the flow deflection at the 2nd ramp $\delta_2$ and at the cowl leading edge $\delta_{lip}$. Especially, the ramp boundary layer is affected by the lip shock and its reflection. The flow deflection at the cowl lip and, thus, the risk of shock induced boundary layer separation is reduced by an increased lip angle. This, however, results in higher lip drag. In accordance with the criteria in [3,4], the maximum allowable flow deflection at the cowl $\delta_{lip}$ is a function of the Mach number ahead of the lip:

$$\delta_{lip} = f(M_2) \leq 7°.$$

To obtain a short intake, the 1st ramp angle is $\delta_1 = 10°$, resulting in lower internal compression and shock/interaction losses, compared with a smaller angle $\delta_1$. The 2nd ramp angle $\delta_2$ is variable. It is limited by the maximum pressure rise to avoid shock induced boundary layer separations (upper limit) and by possible flow expansion at the cowl lip (lower limit). The criterion [2] allows ramp angles $\delta_2 \leq 12°$ in the relevant Mach number range. With the selected ramp angles, the total pressure recovery of the external compression is close to the Oswatitsch criterion. With a cowl lip angle $\alpha_{lip} = 15°$, both criteria for the avoidance of shock induced boundary layer separations can be fulfilled. The minimum ramp angle is $\delta_2 = 5°$ ($\delta_{lip} \geq 0$).

The throat heights $h_{thr}$ for starting the internal compression and for "unstart" have been derived from the intake geometry and the Mach number on the 2nd ramp:

$$(h_{thr}/h_{lip})_{isentropic} = f(M_2) \quad \text{with: } h_{lip} = f(h_C, \delta_1, \delta_2) \text{ and } M_2 = f(M_0, \delta_1, \delta_2).$$

For the definition of the throat height (non isentropic flow between external flow and throat), the isentropic area ratio has been corrected by the total pressure ratio $p_{t,thr}/p_{t2}$ which was derived from previous intake tests [1]. This design work was performed using CATIA. For the calculation of the internal shock system, the 2-D Euler code EU2DIN and the Navier-Stokes code NSFLEX were already applied in the design process. NSFLEX results also provided additional information on shock/boundary layer interactions within the intake with which the final design was validated numerically. For geometrical reasons, it is not possible to obtain the required throat height $h_{thr}$ at a given Mach number $M_0$ and arbitrary ramp angles $\delta_2$ without perturbations (convex or concave corners) of the ramp contour. Modell ETM3M is shown in Fig. 1.

In addition to the baseline tests, by simulating the center body of the ram burner it is possible to investigate its influence on the start and unstart limits (i. e. the contraction ratio $A_{thr}/A_{lip}$) and on total pressure recovery; Fig. 2. With the center body, the measurement plane (AIP) is changed from a circular cross section to an annular cross section. The area of the cross section remains constant.

### Model ETM3AT

Model ETM3AT (Fig. 3) has been designed by TsAGI [9]. In contrast to ETM3M, it features 3 external ramps; the 1st one with a constant ramp angle $\delta_1 = 9.7°$, the 2nd and 3rd ramps variable. With a total external flow deflection $\delta = 26.5° - 29.0°$, the degree of external compression is higher. The lower Mach number ahead of the cowl lip results in a higher area ratio $A_{thr}/A_{lip}$. With a lip angle $\alpha_{lip} = 14°$, the deflection at the cowl lip is significantly above the limits to avoid shock induced boundary layer separation by the lip shock [3,4]. The design Mach number for inviscid flow, i.e. for the *geometric* ramp angle $\delta_1$, is $M_0 \approx 6.0$.

Model ETM3AL

In contrast to ETM3M and ETM3AT, it is aspired to avoid the reflection of the lip shock by means of a convex corner in the ramp contour. With shock cancellation, the maximum allowable flow deflection at the cowl lip to avoid shock induced boundary layer separation on the ramp surface is about twice the value in comparison with shock reflection [3,4]. Shock impingement downstream of the convex corner seems to be more favourable than upstream [5]. The ramp angles $\delta_1 = 10°$ and $\delta_2 = 11.8°$ are constant; the adaptation of the throat height is achieved by a longitudinal translation of the 2nd ramp. With increasing Mach number $M_0$ and the shallower lip shock the impingement on the ramp is further downstream; simultaneosly, the 2nd ramp is moved downstream to achieve the required smaller throat height. Therefore, the requirements for lip shock position and for throat area are corresponding with changing Mach number. Model ETM3AL is shown in Fig. 4.

Model Instrumentation

The diameter of the measuring plane (AIP) of each model is 73.64 mm. During the tests in the wind tunnel SVS-2, the instrumentation of each model consisted of a rake behind the subsonic diffuser with 40 pitot probes on 8 rake arms, 8 static pressure probes and 3 dynamic probes for the static wall pressure. For the simulation of the center body with model ETM3M, this rake was replaced by 24 pitot probes on 6 radial rake arms in the annular duct between centerbody and external wall. In addition, there are 7 steady-state and 3 dynamic static pressure probes. During the tests with ETM3M at DLR, a rake with 56 pitot probes on 8 rake arms and 8 steady-state and 3 dynamic static wall pressure probes was used. In addition to the rakes, the models featured 56 - 68 steady-state and 3 dynamic pressure probes on the ramps, on the cowl lip and within the subsonic diffuser (depending on the model configuration). In the wind tunnel SVS-2, mass flow throttling is achieved by a variable throttle; mass flow is measured by a metering nozzle. In the TMK, throttling and mass flow measurement is made by means of a variable sonic exit. For each model, two different sets of glass windows are available. With these windows, Schlieren pictures at different positions of the intake duct (lip shock, duct behind throat) can be made.

Wind Tunnels SVS-2 and TMK

Tests with the models ETM3M, ETM3AT, and ETM3AL have been performed in the wind tunnel SVS-2 over a Mach number range $M_0 = 3.5 - 5.9$. SVS-2 is a "blow down" facility which, due to the high capacity of the pressure reservoir, can practically be operated continuosly. For $M_0 \leq 5$, a 2-D variable nozzle ($60 * 60\ cm^2$) is used, whereas for $M_0 = 5.5$ and 5.9 axisymmetric fixed nozzles are available. The freestream total pressure is about 5 - 7 bar. At $M_0 \leq 5.0$, the freestream total temperature is 280 K. At higher Mach numbers the air is heated up to 400 K to avoid condensation. Thus, the Reynolds number is in the range $Re/L = 0.8 \ldots 3.7 * 10^7\ m^{-1}$. Based on the intake capture height $h_C$, the Reynolds numbers during these intake tests are 60% - 80% of the Reynolds numbers for the full scale intake at flight conditions. After the tests at TsAGI, ETM3M was tested in TMK at DLR ($M_0 \leq 5.7$). This wind tunnel is operating at significantly higher total pressures and Reynolds numbers, respectively. For comparison, test points with the same Mach numbers $M_0$ and ramp angles $\delta_2$ have been repeated. The Mach number range has been extended to $M_0 = 2.5$ and 3.0, and additional $\delta_2$ have been investigated.

## TEST PROGRAM AND TEST PROCEDURE

For each Mach number and intake configuration (ramp angles $\delta$, side walls), the throat heights $h_{thr}$ for start and unstart of the internal compression have been measured. With the throttle fully open, the internal compression was started by increasing the throat height. Consequently, unstart was induced by a slow reduction of throat area. After the restart of the internal compression, throttle characteristics $\eta = f(A_0/A_C)$ have been measured at different throat heights. For model ETM3M, the investigated Mach number range was $M_0 = 3.5 \ldots 5.9$ at TsAGI (SVS-2) and $M_0 = 2.5 \ldots 5.7$ at DLR (TMK). Tests with the ramp angles $\delta_2 = f(M_0)$ according to the design ramp schedule and off-design ramp angles $\delta_2$ have been investigated. With the center body, only the design ramp angles $\delta_2$ have been tested. The test program for ETM3AT included different combinations of Mach number $M_0$ and ramp angles between $\delta_2 = 6.0° \ldots 8.5°$ and $\delta_3 = 10.5° \ldots 13.3°$. The Mach number range was $M_0 = 3.5 \ldots 5.9$. Due to the fixed ramp angles $\delta_1$ und $\delta_2$ of model ETM3AL, the test program only includes the verification of throat heights $h_{thr}$ for start and unstart and the measurement of the characteristics at different Mach numbers $M_0 = 3.5 \ldots 5.9$ for a variation of the throat heights $h_{thr}$.

## TEST RESULTS

ETM3M

In Fig. 5, the theoretical and measured area ratios $A_{thr}/A_{lip}$ for start and unstart are compared. The experimental data for different $M_0$ and $\delta_2$ are represented by single points. The theoretical area ratio for unstart holds for isentropic flow between cowl lip and throat; in addition, curves for different total pressure ratios $p_{t,thr}/p_{t2} < 1$ are shown. The total pressure losses between lip and throat are mainly due viscosity as the shock losses downstream of the cowl lip are rather small. The area ratio $A_{thr}/A_{lip}$ and, thus, the total pressure ratio $p_{t,thr}/p_{t,lip}$ is in good agreement with the data used for the design of the area distribution. Therefore, the correction of the internal contour by the internal ramp and the resulting angle within the ramp contour is negligible. In Fig. 6, the measured characteristics $\eta = f(A_0/A_C)$ for $M_0 = 3.5 - 5.9$ and the corresponding ramp angles $\delta_2$ (design values) are shown. For $M_0 = 3.5$, data for $\delta_2 = 11.2°$ (off-design) are included. Total Pressure recovery $\eta$ is equivalent to a kinetic efficiency $\eta_{kin} \approx 0.91$ in the whole Mach number range $M_0 = 2.5 ... 5.9$. Even at $2.5 \leq M_0 < 3.5$, $\eta_{kin} = 0.90$ is exceeded.

Influence of Ramp Angle $\delta_2$ and Throat Height on Intake Performance

In Fig. 7, the mass flow ratio $A_0/A_C$ for different ramp angles $\delta_2$ is plotted vs. $M_0$. Due to increasing spillage, $A_0/A_C$ decreases significantly with increasing ramp angle $\delta_2$. The influence of the ramp angle $\delta_2$ on total pressure recovery $\eta$, however, is rather small. Fig. 8 shows the variation of total pressure recovery $\eta$ with throat height $h_{thr}$ for different Mach numbers $M_0$ and ramp angles $\delta_2$. Theoretically, it is expected that the maximum total pressure recovery is obtained with the minimum throat height (minimum shock losses). In general, this model is confirmed by Fig. 8. For low supersonic Mach numbers ahead of the terminal shock, the Mach number behind the shock (i. e. the Mach number at the entry of the subsonic diffuser) is close to unity which can result in higher diffuser losses, partly compensating the lower shock losses.

Influence of the Center Body

The center body for the simulation of the ram burner has no influence on the "start" and "unstart" behaviour (throat heights $h_{thr}$) of the intake and on total pressure recovery $\eta$.

Comparison of SVS-2 and TMK Data, Variation of Reynolds Number

In Fig. 9, the characteristics at $M_0 = 4.0$, $\delta_2 = 10.4°$ measured at TsAGI and at DLR are compared. The $M_0 = 4.0$ test condition was selected because the maximum freestream total pressure $p_{t0}$ in the SVS-2 wind tunnel is close to the minimum $p_{t0}$ in the TMK, thus giving a direct comparison of wind tunnels SVS-2 and TMK at similar Reynolds numbers. In addition, a $p_{t0}$ (or Reynolds number) variation was made in the TMK. The results show that

- the maximum total pressure recovery in the SVS-2 and TMK wind tunnels is practically the same (the critical point of characteristic (CN) 33 was not recorded during the tests; i.e. the maximum total pressure recovery is not shown in Fig. 9)
- the variation of $p_{t0}$ in the TMK does not affect total pressure recovery

However, the measured mass flow ratio $A_0/A_C$ increases with increasing $p_{t0}$ due to the smaller mass flow reduction caused by the ramp boundary layer (as it was expected and predicted by theoretical assessment). The difference in $A_0/A_C$ between SVS-2 and TMK data at comparable $p_{t0}$ is assumed to be introduced by the different method of mass flow measurement at TsAGI and DLR.

Flow Uniformity

The influence of throat height $h_{thr}$ on the turbulence $T_{RMS}$ in the rake plane (aerodynamic interface plane; AIP) is negligible. If $T_{RMS}$ is plotted vs. $\eta$ for different $h_{thr}$, the curves coincide. Only the typical increase of $T_{RMS}$ at "unstart" is affected as the critical point (maximum $\eta$) changes with $h_{thr}$. Total pressure distortion DC60 shows a similar trend. The results from a Navier-Stokes calculation [12] show that the separation of the ramp boundary layer caused by the lip shock is very small. In the Schlieren pictures it can be seen that the boundary layer is thickened by the pressure rise due to the lip shock only. The subsonic part of the boundary layer becomes thinner again downstream of the throat.

**ETM3AT**

In Figs. 10 a) and b), total pressure recovery and mass flow ratio for different ramp angle combinations are plotted vs. Mach number $M_0$. Total pressure recovery $\eta$ increases slightly with increasing external compression. Maximum $\eta$ is achieved with $\delta_2 = 8.5°$, $\delta_3 = 10.8°$, but at the expense of a severe reduction of mass flow ratio $A_0/A_C$. To a smaller degree, this effect also shows for $\delta_2 = 6.0°$, $\delta_3 = 13.3°$. In both cases, the external deflection is 29°. The separation caused by the lip shock was also predicted by Navier-Stokes calculations performed at Dasa-LM [12]; Fig. 11. The Schlieren pictures made during the tests confirm that massive boundary layer separation is triggered by the reflection of the lip shock (Fig. 12). In some test cases, the separation zone occupies nearly half of the duct height and extends far downstream. This is due to the extremely high flow deflection at the cowl lip $\delta_{lip}$ which is far above the limit to avoid shock induced boundary layer separation [3,4]. In Fig. 13, the area ratios for "start" and "unstart" are plotted vs. the Mach number ahead of the cowl lip $M_3$. Compared with ETM3M, the required area ratio $A_{thr}/A_{lip} = f(M_{lip})$ is higher because of the ramp boundary layer separation caused by the lip shock which reduces the effective throat area. The lines of constant total pressure ratio $p_{t,thr}/p_{t,lip}$ = const in Fig. 13 correspond to the *average* value in the throat plane; i.e. including the separation zone. The total pressure of the core flow is higher than this value; otherwise, the total pressure recovery measured in the AIP would be significantly lower. The level of turbulence $T_{RMS}$ and its behaviour along the intake characteristic (i.e. during throttling) is similar to ETM3M.

**ETM3AL**

At low Mach numbers, $A_0/A_C$ is considerably lower than that of model ETM3M due to the higher spillage caused by the fixed ramp angle $\delta_2$. Due to the similar design geometry, the mass flow at high Mach numbers is close to the ETM3M values. Total pressure recovery is equivalent to a kinetic efficiency $\eta_{kin} \approx 0.90$ with higher values at $M_0 \leq 5$ and slightly lower values at $M_0 > 5$. The Navier-Stokes calculations predicted a separation on the ramp side due to the lip shock if this shock impinges *upstream* of the convex corner of the ramp; Fig. 14. The Schlieren pictures of model ETM3AL show that the lip shock impinges at or (at larger $h_{thr}$) downstream of the convex corner of the ramp at all Mach numbers $M_0$; Fig. 15. However, this shock is not cancelled completely. If the lip shock impinges *close to or at* the corner, the resulting separation propagates upstream, producing an additional shock ahead of the corner. With shock impingement *downstream* of the corner, separation is completely avoided or the resulting small separation is confined to the downstream flow regime. For shock impingement farther downstream of the corner, however, an increased throat height is required. Therefore, total pressure recovery is not improved despite the avoidance of the ramp boundary layer separation.

## COMPARISONS BETWEEN THE DIFFERENT INTAKE CONFIGURATIONS

In Figs. 16 a) and b), total pressure recovery $\eta$ and mass flow ratio $A_0/A_C$ of ETM3M, ETM3AT and ETM3AL are compared. Total pressure recoveries for each model are close to each other. The highest $\eta$ is obtained by ETM3AT with ETM3M data slightly below. The lowest pressure recovery was measured with ETM3AL. As already mentioned above, ETM3AT reaches the highest total pressure recovery with high external compression, connected with high spillage and significantly reduced mass flow. The design geometry (shock on lip condition) of ETM3M and ETM3AT is different: ETM3M reaches "shock on lip" at $M_0 = 6.2$ with the *geometric* ramp angle $\delta_1 = 10°$. For ETM3AT, the corresponding Mach number is $M_0 = 6.0$ ($M_0 = 6.2$ with an *effective* ramp angle $\delta_1$, including the ramp boundary layer), resulting in an intake geometry with shorter external ramps and, thus, lower spillage $\Delta(A_0/A_C)$ and reduced total pressure losses (in comparison with a similar intake geometry with $M_{DES} = 6.2$). Due to the similar design geometries ($M_{DES} = 6.2$), the mass flow ratio of ETM3M and ETM3AL is nearly the same at high Mach numbers $M_0$. At lower Mach numbers, however, the mass flow ratio of ETM3AL is lower due to the fixed ramp angle $\delta_2$ (higher spillage).

## CONCLUSIONS

1. With each model configuration, starting of the internal compression and stable operation has been achieved over the whole Mach number range $M_0 = 3.5 \ldots 5.9$ without boundary layer bleed.
2. The design procedure and the assumptions during the design of model ETM3M have been confirmed.
3. The total pressure recovery of model ETM3M is equivalent to a kinetic efficiency $\eta_{kin} \approx 0.91$ in the Mach number range $M_0 = 3.5 \ldots 5.9$ (design Mach number range). At lower Mach numbers ($M_0 = 2.5, 3.0$; off design), $\eta_{kin} \approx 0.90$ is reached. The influence of the ramp angle $\delta_2$ on total pressure recovery is small. Compared with model ETM3AT, the high total pressure recovery of ETM3M is especially noteworthy because of the relatively simple design with only one moveable external ramp. The center body (ram burner simulation) has no effect on "start" and "unstart" limits and on total pressure recovery.
4. In model ETM3AT, the ramp boundary layer is separated by the lip shock (and its reflection) due to the high flow deflection at the cowl lip at all test conditions. Due to the separation, the required throat areas $A_{thr}/A_{lip}$ are larger than for ETM3M. Despite this separation, the measured total pressure recovery corresponds to a kinetic efficiency $\eta_{kin} = 0.90 \ldots 0.92$. However, the high total pressure recovery is obtained with high external flow deflection and low mass flow ratio $A_0/A_C$. ETM3AT is favoured by the lower design Mach number $M_{DES} = 6.0$ instead of $M_{DES} = 6.2$ (ETM3M, ETM3AL). With $M_{DES} = 6.2$, pressure recovery is expected to approach values of intake ETM3M but with a further reduced mass flow ratio.
5. Due to the constant ramp angle $\delta_2$ and increasing spillage, the mass flow ratio $A_0/A_C$ of model ETM3AL is below the value of ETM3M at low Mach numbers $M_0$. With a very similar design geometry, nearly the same mass flow ratio was measured at high Mach numbers. The cancellation of the lip shock is not achieved completely. Therefore, in some test cases the ramp boundary layer is separated by the lip shock. Total pressure recovery is equivalent to $\eta_{kin} \approx 0.90$.

## REFERENCES

1. N. C. Bissinger, D. M. Schmitz: " Design and Wind Tunnel Testing of Intakes for Hypersonic Vehicles", AIAA 93-5042, Munich, 1993.
2. G. E. Gadd: "Interactions Between Wholly Laminar or Wholly Turbulent Boundary Layers and Shock Waves Strong Enough to Cause Separation", Journal of the Aeronautical Sciences, Nov. 1953.
3. P. A. Waltrup: "Liquid Fueled Supersonic Combustion Ramjets: A Research Perspective of the Past, Present, and Future", AIAA-86-0158.
4. E. S. Love: "Pressure Rise Associated with Shock-Induced Boundary Layer Separation", NACA TN 3601, 1955.
5. K. Chung, F. Lu: "Hypersonic Turbulent Expansion-Corner with Shock Impingement", AIAA-92-5101, 1992.

6. D. M. Schmitz:
"Theoretisch-empirische Abschätzung des Druckrückgewinns von Einläufen mit gemischter Verdichtung", DASA-LME12-TN-HYPAC-443, 01.09.1995.

7. D. M. Schmitz:
"Auswertung der Messungen mit dem Hyperschall-Einlaufmodell ETM3M im Windkanal SVS-2 (TsAGI)", DASA-LME12-TN-HYPAC-451, 08.12.1995.

8. D. M. Schmitz:
"Auswertung der Messungen mit den Hyperschall-Einlaufmodellen ETM3AFL und ETM3AL im Windkanal SVS-2 (TsAGI)", DASA-LME12-TN-HYPAC-459, 22.12.1995.

9. Intake Technology (Final Report)
Contract No. 805/600005982 between IABG (Germany) and TsAGI; TsAGI, Zhukovsky, 1995.

10. B. Sander: "Analyse der Messungen mit dem Hyperschall-Einlaufmodell ETM3M im TMK", DASA-LME12-TN-HYPAC-465, 1995.

11. T. Berens: "Throat Heights and Ramp Angles for the Intake Hypersonic Technology Wind Tunnel Model ETM3M at Mach Numbers 2.5, 3.0, 4.0, 4.3, 5.5, 5.7, and 6.0", DASA-LME12-TN-HYPAC-455-95, 14.12.1995.

12. N. C. Bissinger: "Hyperschalltechnologie (Phase 1c): Einlauf; Abschlußbericht", Dasa-LMLE3-HYPAC-STY-0018-A, 24.10.1996.

## ACKNOWLEDGEMENTS

The work described here has been supported by the Ministry of Research and Technology (BMBF) within the German Hypersonic Program. The authors would like to thank P. W. Sacher, project manager at Dasa-LM within this program, for his support. They would also like to acknowledge the useful discussions and highly appreciated assistance by the colleagues in the intake, numeric, and experimental groups of Dasa-LM. The close cooperation and the professional support of our colleagues at TsAGI, especially E. V. Pavlyukov, V. G. Gurylev †, and E. V. Piotrovich and the wind tunnel teams at TsAGI and DLR Cologne made it possible to conduct very successful tests.

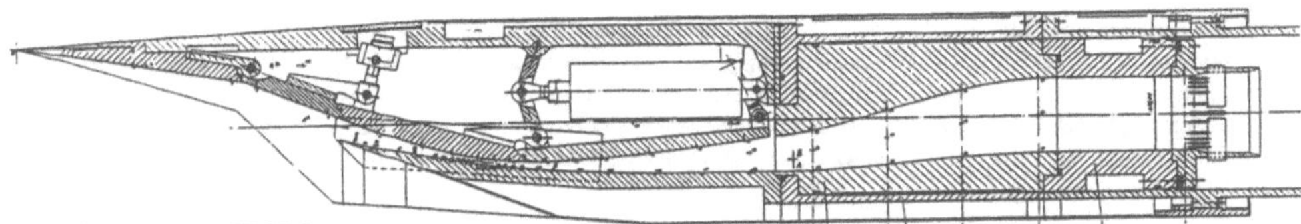

Fig. 1 Model ETM3M

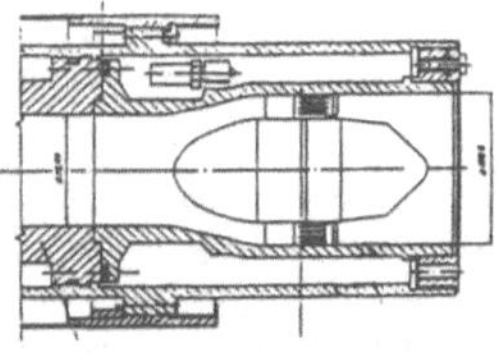

**Fig. 2 Model ETM3M Center Body**

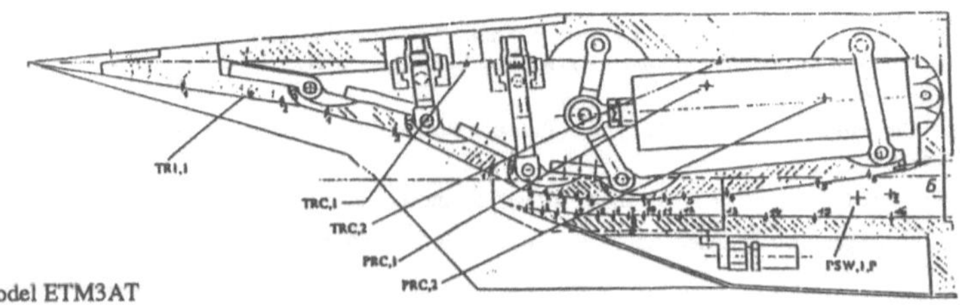


Fig. 3 Model ETM3AT

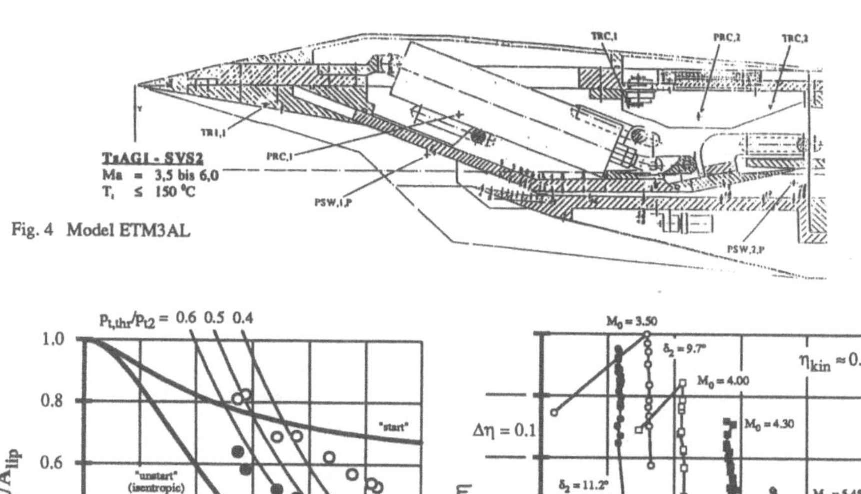


Fig. 4 Model ETM3AL

**Fig. 5 ETM3M: Contraction Ratios $A_{thr}/A_{lip}$ for "start" and "unstart"**

**Fig. 6 ETM3M: Intake Characteristics for $M_0$ = 3.5 - 5.9**

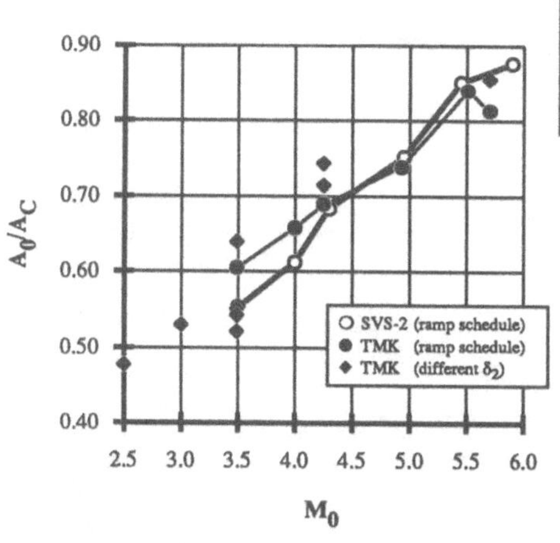


Fig. 7 ETM3M: Mass Flow Ratio (TsAGI and DLR)

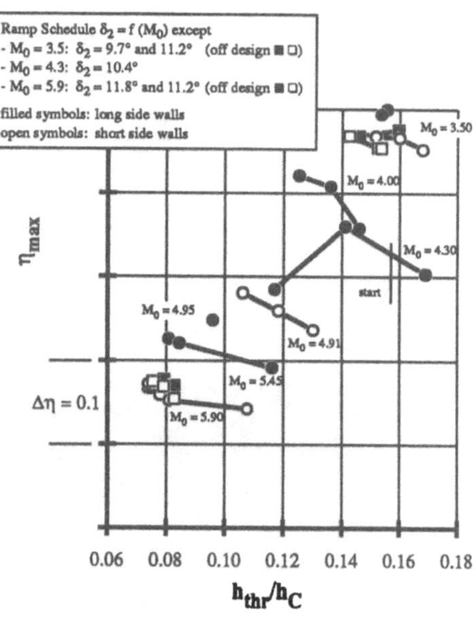


**Fig. 8 ETM3M: Influence of Throat Height on Total Pressure Recovery**

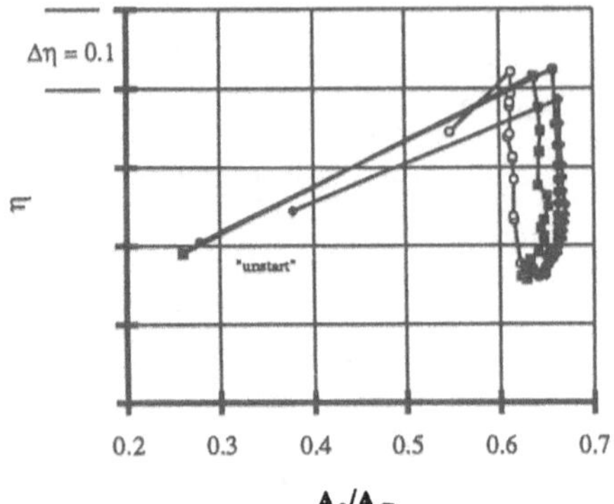


| Wind Tunnel | Characteristic | $p_{t0}$ [bar] | Re/L $[m^{-1}]$ | Symbol |
|---|---|---|---|---|
| SVS-2 | 20 | 7.36 | $3.70 * 10^7$ | ○ |
| TMK | 44 | 4.0 | $2.12 * 10^7$ | ■ |
| TMK | 30 | 7.5 | $3.95 * 10^7$ | ● |
| TMK | 33 | 13.5 | $6.91 * 10^7$ | ◆ |

**Fig. 9 Comparison of TsAGI and DLR Test Conditions $M_0 = 4.0$, $\delta_2 = 10.4$**

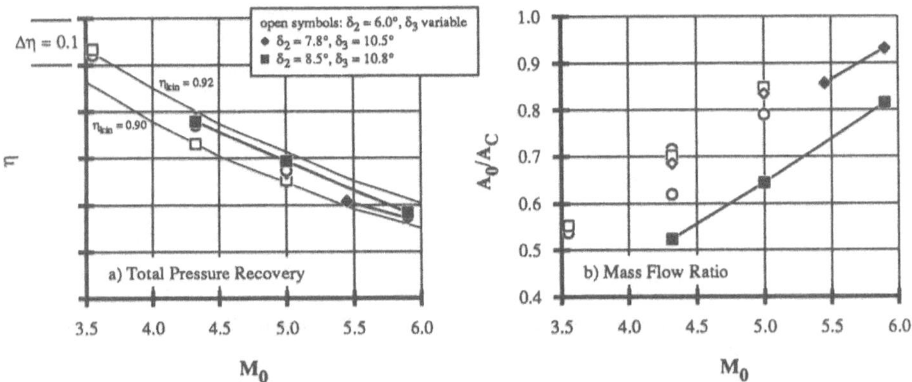


**Fig. 10 ETM3AT: Total Pressure Recovery and Mass Flow Ratio vs. Mach Number**

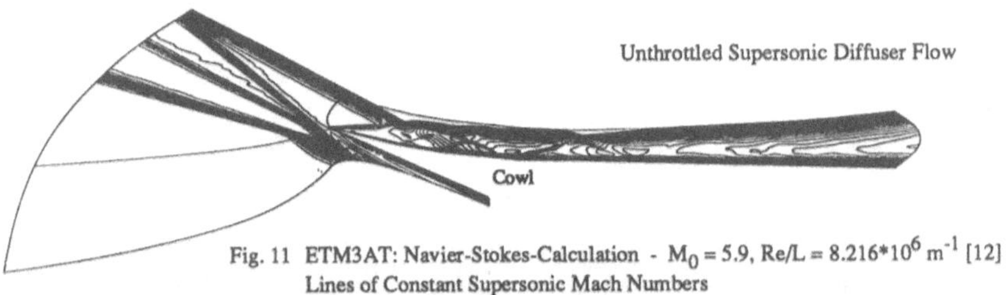


Fig. 11 ETM3AT: Navier-Stokes-Calculation - $M_0 = 5.9$, Re/L = $8.216*10^6$ $m^{-1}$ [12]
Lines of Constant Supersonic Mach Numbers

**Fig. 12 ETM3AT: Schlieren Picture - $M_0 = 5.90$ (Characteristic No. 183, Test Point 2129)**

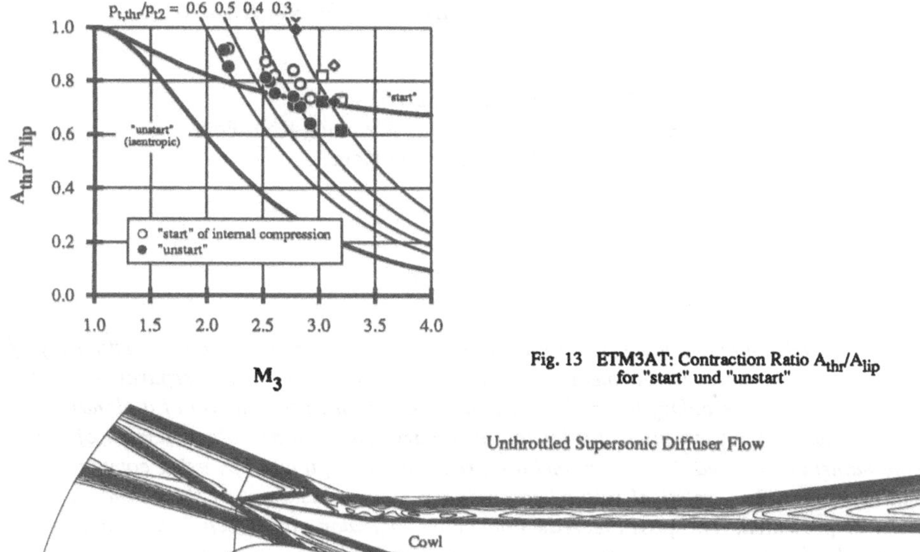


Fig. 13 ETM3AT: Contraction Ratio $A_{thr}/A_{lip}$ for "start" und "unstart"



Fig. 14 ETM3AL: Navier-Stokes-Calculation - $M_0 = 6.0$, $Re/L = 9.475 \cdot 10^6\ m^{-1}$ [12]
Lines of Constant Supersonic Mach Numbers

Fig. 15 ETM3AL: Schlieren Picture - $M_0 = 5.90$ (Characteristic No. 129, Test Point 1477; maximum total pressure recovery)

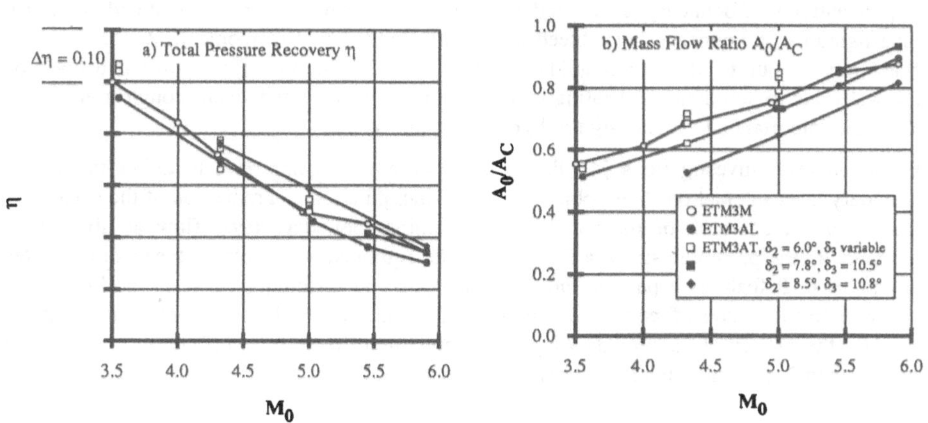


Fig. 16 Comparison of Models ETM3M, ETM3AL and ETM3AT

# BOUNDARY LAYER BLEED IN HYPERSONIC INLETS

D. Schulte, A. Henckels, I. Schell
DLR, Hauptabteilung Windkanäle, Abteilung Köln-Porz
Linder Höhe, D-51147 Köln

## SUMMARY

*Shock-induced boundary layer separation has an important influence on the efficiency of hypersonic inlet systems. A method to reduce the boundary layer separation is the implementation of boundary layer bleed. Results are presented for the use of a planar bleed slot at a generic flat plate model for laminar boundary layer flow at a Mach number of 6. The wind tunnel models and the wind tunnel test procedure are optimized by using computational fluid dynamics. An analytical method to predict the key parameters of the boundary layer bleed is presented. The possible reduction of the boundary layer separation and of the thermal loads on the walls by using bleed is demonstrated by results of wind tunnel tests and favorable design parameters for bleed slots are determined.*

## INTRODUCTION

The design of the inlet for air-breathing propulsion systems is of great importance for the performance of future hypersonic space transportation planes. A key issue of the hypersonic inlet flow is the interaction of the oblique ramp shocks with the boundary layer developing at the inlet walls. Shock impingement on the wall may result in separation of the boundary layer causing losses in total pressure recovery and a reduction of inlet efficiency.

A possibility to reduce those disadvantages of shock-induced boundary layer separation is the implementation of boundary layer bleed at the inlet walls. Known experimental and numerical investigations of boundary layer bleed are mainly limited to supersonic flow conditions at Mach numbers up to $Ma_\infty = 3$ (e.g. [1], [2], [3]). Furthermore, the effect of thermal loads on the walls due to aerodynamic heating is often completely neglected, even though it is of considerable importance for the design of hypersonic inlet systems.

Based on prior investigations [4], this paper documents an approach to affect the shock/ boundary layer interaction. Main objective of the manipulation is a reduction of the separation bubble and a decrease of thermal loads for laminar boundary layer flow at high Mach numbers. In the presented study, a flat plate model equipped with a planar bleed slot is tested at $Ma_\infty = 6$. It features a passive bleed, i.e. the pressure difference between the high static pressure on the plate surface and the low static pressure behind the rear edge of the model is used for the suction of the boundary layer. The pressure difference is sufficiently high (always greater than 2) resulting in a flow through the bleed slot that is always critical.

The application of computational fluid dynamics for the design of wind tunnel tests allows to efficiently plan the necessary test runs and it permits a validation of the model setup. Furthermore, the numerical simulation yields a very detailed view of the fluid flow. Especially at locations where by standard methods no measurements can be obtained such an insight may be valuable (e.g. inside the bleed slot). Thus the computational fluid dynamics contributes to a better understanding of the fluid flow.

In this study, a 2D finite element method is used to solve the Navier-Stokes equations on unstructured, adaptive grids [5]. Unstructured grids offer (in particular when triangular elements are employed) an increased geometric flexibility and a detailed resolution of regions with high gradients of the flow variables. The applied CFD method solves the Navier-Stokes equations by using an explicit Taylor-Galerkin scheme in a weighted residual form.

Figure 1a shows the computed flow field of the flat plate case without bleed (designated as 'reference case'). The design of the numerical problem corresponds to the experimental setup: in the upper left corner a wedge at an angle of attack of 10° induces a planar shock that impinges on the flat plate model. The boundary layer separates from the surface due to the interaction with the impinging shock and a separation bubble is formed. The separation bubble induces a shock which can be seen on top of the bubble. Downstream of the separation bubble, the reattaching boundary layer causes a reattachment shock.

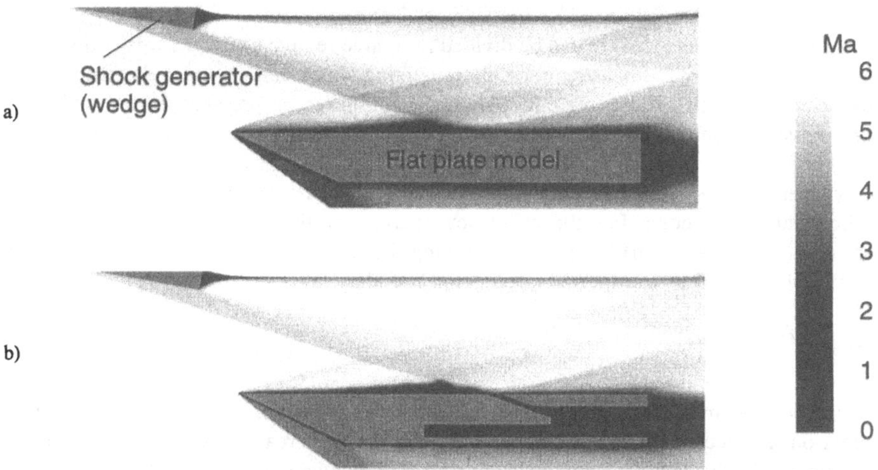


Fig. 1: Mach number distribution for the test setup computed by the finite element flow solver for the reference case (a) and a case with optimized boundary layer bleed configuration (b).

A variety of setups for the bleed configuration were designed and validated by numerical flow simulations. In this variation the geometric bleed setup regarding the critical flow through the slot was optimized. Figure 1b shows the finally selected model setup, which achieves a reduction of the maximum boundary layer thickness by more than 30% compared to the reference case.

## ANALYTICAL METHOD

In order to predict the dominating parameters of the boundary layer bleed flow an analytical method has been developed. A sketch of the model in the flowfield containing the relevant parameters is shown in Fig. 2.

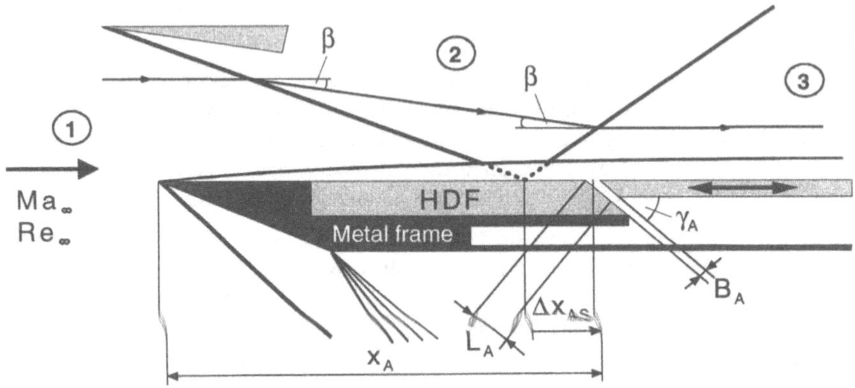


Fig. 2: Sketch of the model ($330 \cdot 180$ mm$^2$) in the flowfield containing the relevant parameters of bleed flow.

The flow field on the plate surface can be divided into three regions: *region 1* upstream of the impinging shock, *region 2* downstream of the impinging shock and *region 3* downstream of the reattachment shock. These regions also represent different pressure levels and thus different pressure gradients for the bleed flow. Depending on the pressure at the respective position, the bleed mass flow is calculated as the critical flow through the bleed slot. The boundary layer mass flow results from the displacement thickness and the velocity and density at the boundary layer edge. For the efficiency analysis of the bleed mechanism the key parameter $\Phi$ (mass flow ratio) is defined describing the relation of the mass flow within the bleed slot to the mass flow transported through the boundary layer [6]:

$$\Phi = \frac{\dot{m}_A}{\dot{m}_G} = \frac{\text{bleed mass flow}}{\text{boundary layer mass flow}} \quad (1)$$

An analysis of the mass flow ratio yields the variables influencing the bleed flow. $\Phi$ is dependent on the freestream Mach number $Ma_\infty$ and the freestream Reynolds number $Re_\infty$, furthermore, on the distance $x_A$ from the leading edge of the plate model to the bleed slot (thus influencing the boundary layer thickness at the bleed slot location). The mass flow ratio further depends on the geometric setup of the bleed slot, that is on the slot width $B_A$, the slot length $L_A$, and the slot angle $\gamma_A$. Finally, the bleed flow is dependent on the position of the bleed slot relative to the shock impingement point $\Delta x_{AS}$ (*region 1, 2* or *3*).

In this study, a variation of the bleed geometry and the position of the bleed slot relative to the shock impingement point will be carried out, while the freestream conditions $Ma_\infty$ and $Re_\infty$ and the distance $x_A$ ($x_A = 210$ mm) are kept constant. Figure 3 shows a variation of $\Phi$ with the slot width $B_A$, the slot angle $\gamma_A$, and the position of the bleed slot relative to the impingement point $\Delta x_{AS}$ as a result of the analytical method.

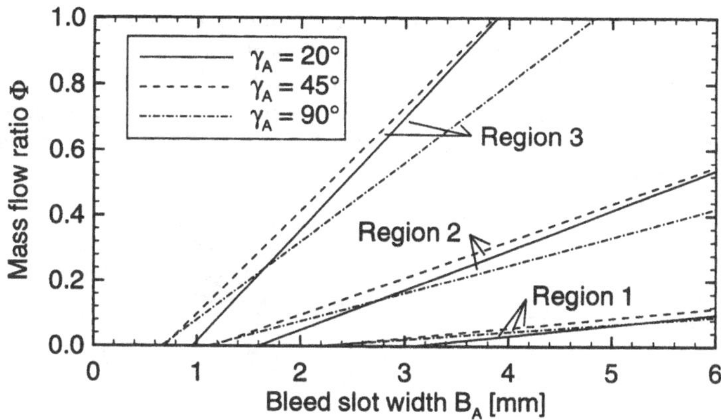


Fig. 3: Mass flow ratio $\Phi$ for $Ma_\infty = 6$, $Re_\infty = 4{\cdot}10^6$/m.

It can be seen that the bleed slot is working most efficiently when positioned downstream of the reattachment shock (*region 3*) where the static pressure on the model surface has its highest value. A steep slot angle of $\gamma_A = 90°$ is of less efficiency compared to the flatter angles. Because the thickness of the rear plate of the model is constant (see Fig. 2), a flat slot angle $\gamma_A$ leads to a greater slot length $L_A$ with an increasing boundary layer thickness inside the bleed slot. Therefore, a slot angle of $\gamma_A = 45°$ appears to be the best compromise between a smooth turning of the flow and a small boundary layer thickness in the slot. With this configuration the smallest slot width $B_A$ can be obtained, i.e. $B_A$ for $\Phi = 1$.

## EXPERIMENTAL INVESTIGATIONS

The wind tunnel tests were performed at the DLR H2K hypersonic blow down wind tunnel in Cologne at a Mach number of 6 and a Reynolds number $4{\cdot}10^6$/m. As diagnostics the facility offers a Schlieren visualization and an advanced Focal Plane Array infrared technique for the measurement of temperatures on model surfaces.

A sketch of the used wind tunnel model has already been shown in Fig. 2. The high density fiberboards (HDF), embedded in a metal frame, are changeable to allow the use of boards with different slot angles or a model with no bleed at all. The HDF boards have a low thermal conductivity which ensures a good resolution of high local heat fluxes. For the setting of different slot widths the rear board is moveable. The complete model can be moved in the wind tunnel to vary the bleed slot position relative to the shock impingement location $\Delta x_{AS}$.

The effects of the boundary layer bleed can be visualized using Schlieren pictures. Figure 4a shows a Schlieren photography for the reference case. The impinging shock (IS), the shock from the leading edge of the model (LS), the reattachment shock (RS), and the separation shock (SS) induced by the separation bubble (SB) can be observed clearly. Furthermore, an area is visible indicating the existence of vortices (V) downstream of the bubble. Prior studies at DLR [4] already demonstrated the occurrence of Görtler vortices downstream of a shock-induced boundary layer separation due to the concave curvature of the separation bubble.

In Fig. 4b, a bleed slot with a slot width of $B_A$ = 5 mm and a slot angle of $\gamma_A$ = 45° is positioned directly downstream of the shock impingement point. It is obvious that the separation bubble thickness has considerably decreased and that the separation shock has completely vanished. Also, the area indicating vortices is no longer visible. By positioning the bleed slot far downstream or upstream of the shock impingement point no such decrease of the separation bubble thickness can be achieved.

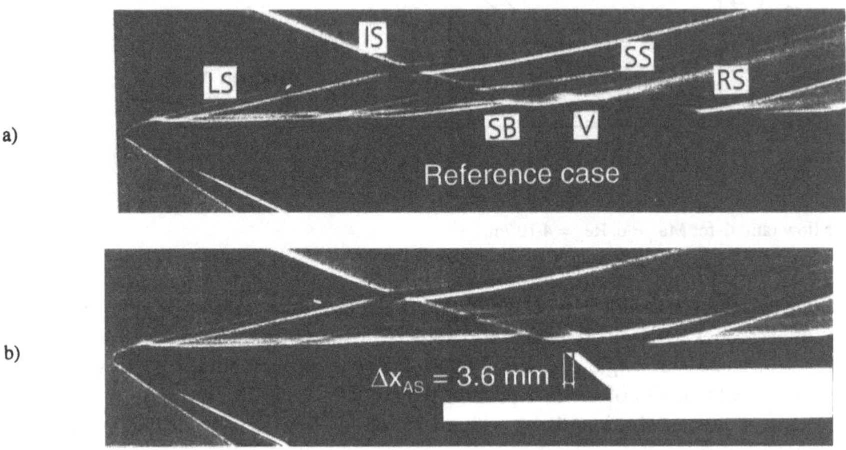


Fig. 4: Hypersonic flowfield without and with bleed for the test setup presented as Schlieren photographs.

In further wind tunnel tests, variations of several bleed flow parameters were performed. These variables are the position of the bleed slot relative to the shock impingement point $\Delta x_{AS}$, the slot angle $\gamma_A$, and the slot width $B_A$. To evaluate the results of the experiments, some key parameters are defined according to the measurement techniques. The relative boundary layer thickness $\delta_{rel}$ represents the ratio of the maximum boundary layer thickness at the shock impingement location for the case with bleed to the reference case. $\delta$ is obtained as a value proportional to the separation bubble thickness from the density gradients in the Schlieren photographs:

$$\delta_{rel} = \frac{\delta}{\delta_{ref}} = \frac{\text{maximum boundary layer thickness for the case with bleed}}{\text{maximum boundary layer thickness for the reference case}} . \tag{2}$$

The temperature rise on the plate surface is measured using the infrared thermography and represents the difference between the wall temperature $T_W$ after a certain test time and the reference temperature $T_a$ before the test. $T_W$ and $T_a$ are measured in a fixed distance to the shock impingement point (>> $\Delta x_{AS,max}$) constant for all cases. The relative temperature rise $\theta$ is defined as the relation of the temperature rise for the case with bleed to the reference case:

$$\theta_{rel} = \frac{T_W - T_a}{\left(T_W - T_a\right)_{ref}} = \frac{\text{surface temperature rise for the case with bleed}}{\text{surface temperature rise for the reference case}} . \tag{3}$$

Figure 5 shows the variation of $\delta_{rel}$ and $\theta_{rel}$ with the key parameters for the boundary layer bleed $\Delta x_{AS}$, $\gamma_A$, and $B_A$.

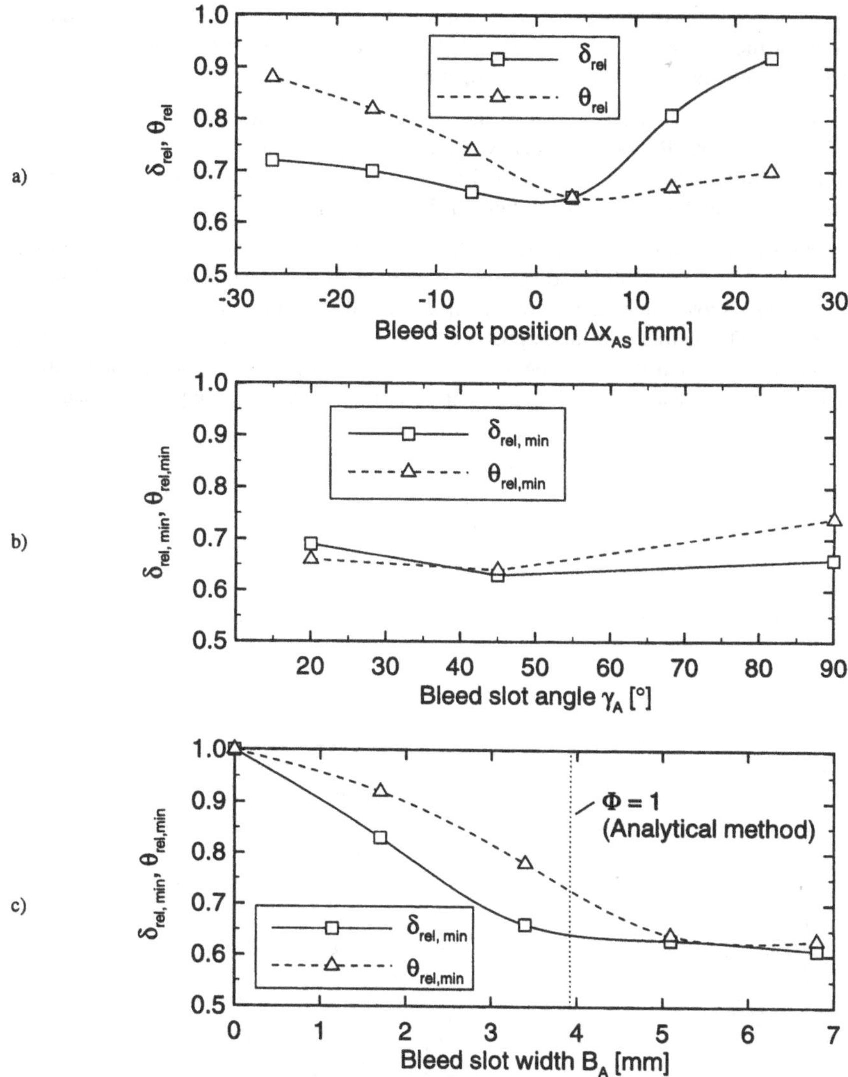


**Fig. 5: Influence of the bleed slot position relative to the shock impingement point $\Delta x_{AS}$, the bleed slot angle $\gamma_A$, and the bleed slot width $B_A$ on the key parameters.**

Figure 5a demonstrates the effect of the bleed slot position relative to the shock impingement point $\Delta x_{AS}$. In any case the boundary layer bleed causes a reduction of the separation bubble thickness $\delta_{rel}$ as well as of the temperature rise on the model surface $\theta_{rel}$. The maximum decrease for both key parameters, $\delta_{rel,min}$ and $\theta_{rel,min}$, is obtained for a position of the bleed slot directly downstream of the shock impingement point.

In Fig. 5b, the dependence of $\delta_{rel,min}$ and $\Theta_{rel,min}$ of the slot angle $\gamma_A$ is presented. As predicted by the analytical method, a slot angle of $\gamma_A = 45°$ yields the maximum decrease of both the separation bubble thickness and the surface temperature rise.

The development of $\delta_{rel,min}$ and $\Theta_{rel,min}$ with the slot width $B_A$ is shown in Fig. 5c. First an increase of the slot width $B_A$ causes a steady decrease of $\delta_{rel,min}$ and $\Theta_{rel,min}$. Above a certain slot width $B_A$, no further decrease of $\delta_{rel,min}$ and $\Theta_{rel,min}$ can be achieved. This slot width can be approximately predicted by the analytical method ($B_A$ for $\Phi = 1$, dotted line in Fig. 5c).

From the surface temperatures measured by IR-thermography heat fluxes are calculated for the reference case and for cases with bleed at certain positions of the bleed slot $\Delta x_{AS}$. The upper half in Fig. 6a and b shows the heat flux on the model as a distribution of Stanton numbers for the reference case. Downstream of the impingement point, an area with high Stanton numbers can be observed. This again indicates the above mentioned counterwise rotating longitudinal vortices that cause high local thermal loads forming a striated pattern on the plate surface.

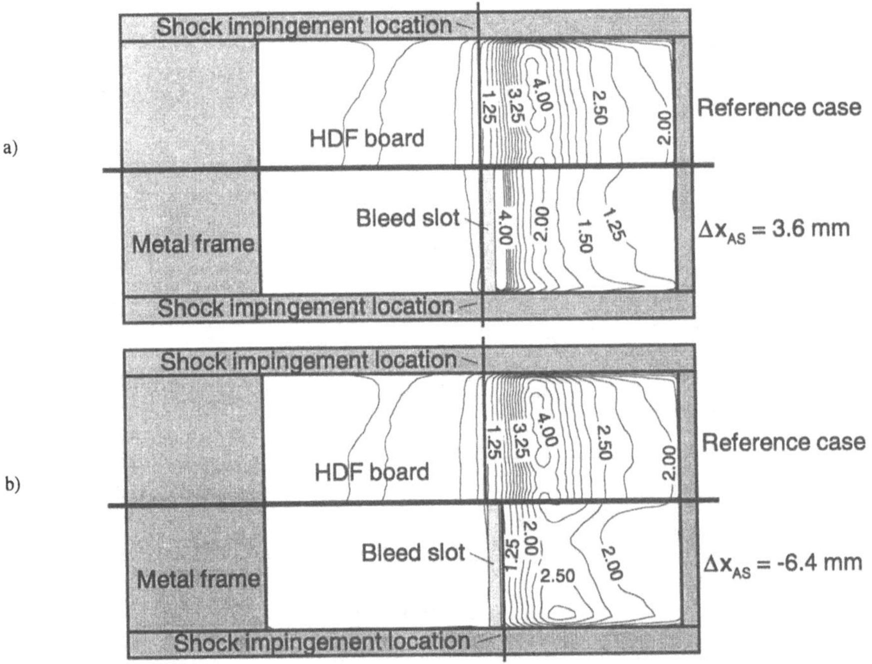


Fig. 6: Thermal loads on the HDF plate surface presented as Stanton number distribution ($St \cdot 10^3$).

The high local thermal loads on the plate surface are shifted and concentrated over a short distance if a bleed slot is implemented and located directly downstream of the impingement point (lower half of Fig. 6a). Even though the rear edge of the bleed slot is exposed to a high thermal load, most of the plate shows smaller thermal loads compared to the reference case. However, the slot rear edge represents a stagnation point for the boundary layer flow because of the flow turning into the bleed slot. Thus the temperature resistant setup of the bleed slot proves to be one of the major problems for the implementation of bleed in hypersonic inlets.

A possible solution of this problem is to move the position of the slot slightly upstream. In this case (lower half of Fig. 6b, the bleed slot is located upstream of the shock impingement point), the thermal load is also small at the rear edge of the bleed slot. The heat flux on the plate surface shows only little increase compared to a bleed slot position directly downstream of the impingement point and the decrease of $\delta_{rel}$ and $\theta_{rel}$ is similar (see Fig. 5a). Regarding technical feasibility of boundary layer bleed in hypersonic inlets the bleed slot is best positioned at or directly upstream of the shock impingement point.

## CONCLUSION

In this paper the capability of boundary layer bleed to influence the shock/laminar boundary layer interactions was demonstrated in a hypersonic wind tunnel at Mach 6 and a Reynolds number of $4 \cdot 10^6$/m using a flat plate model. Numerical, analytical and experimental methods for the analysis of the bleed efficiency were used:

- A finite element numerical flow simulation was used for the setup of wind tunnel tests and the design of an optimized wind tunnel model was examined.
- Results of an analytical method for the prediction of the key parameters according to the implementation of bleed slots were presented and confirmed by experiments.
- In wind tunnel tests, a reduction of the boundary layer separation (i.e. the thickness of the separation bubble) and of thermal loads on the wall was found. Favorable design parameters for the bleed slot geometry and the preferable location of the bleed slot relative to the shock impingement point were determined. Furthermore, considerations in regard to the technical feasibility of boundary layer bleed in hypersonic inlets were made.

Thus, the results of the presented study are the basis for further investigations and may offer support for the implementation of boundary layer bleed in hypersonic inlets.

This study is part of the DFG Sonderforschungsbereich 253 at the University of Aachen. The authors gratefully acknowledge the support by DFG.

## REFERENCES

[1] Harloff, G. J.; Smith, G. E.: *Supersonic-Inlet Boundary-Layer Bleed Flow.* AIAA Journal, Vol. 34, No. 4 (1995), pp. 778 - 785.

[2] Willis, B. P.; Davis, D. O.; Hingst, W. R.: *Flow Coefficient Behavior for Boundary-Layer Bleed Holes and Slots.* NASA Technical Memorandum 106846, AIAA-95-0031 (1995).

[3] Chyu, W. J.; Rimlinger, M. J.; Shih, T. I.-P.: *Control of Shock-Wave/Boundary-Layer Interactions by Bleed.* AIAA Journal, Vol. 33, No. 7 (1995), pp. 1239 - 1247.

[4] Henckels, A.; Kreins, A. F.; Maurer, F.: *Experimental Investigation of Hypersonic Shock/Boundary-Layer Interaction.* Zeitschrift für Flugwissenschaften und Weltraumforschung, Vol. 17, No. 2 (1993), pp. 116 - 124.

[5] Rick, W.: *Adaptive Galerkin Finite Elemente Verfahren zur numerischen Strömungssimulation auf unstrukturierten Netzen.* Shaker, Aachen (1994).

[6] Schell, I.: *Experimentelle Untersuchungen zur Grenzschichtabsaugung an einem Prinzipmodell im Hyperschallwindkanal.* Master Thesis at Hauptabteilung Windkanäle, DLR / Lehr- und Forschungsgebiet Betriebsverhalten der Strahlantriebe, RWTH Aachen (1996).

# High Resolution Measurement of Turbulence in Axial Compressors

A. Sentker and W. Riess
Institut für Strömungsmaschinen,
Universität Hannover, Appelstr.9, 30167 Hannover, Germany

## SUMMARY

In a low speed axial compressor the instationary flow field behind a rotor at 50% blade height has been measured with hot-film probes in split-film technique. Continuous measurement over many rotor revolutions and evaluation with the 'Ensemble Averaging Method' and high temporal and spatial resolution renders average values and momentary fluctuations for the flow velocity in each point. Description of velocities and fluctuations in a mainflow/crossflow system seems more appropriate for a correct physical understanding than the axial/circumferential coordinate system familiar in turbomachines.

Since the measuring technique applied cannot render a continuous velocity signal for a certain point, but is rather a sampling at a relatively low time rate, the fluctuations measured are not necessarily resulting from turbulence, but could have also other causes. Statistical evaluations are used to investigate their character in more detail.

## INTRODUCTION

The wider use of stationary and instationary Navier-Stokes-Codes for the calculation of the flow in turbomachines necessitates a validation of such programs with experimental data. The special flow conditions in turbomachines such as

- periodical instationary velocity fluctuations in space and time due to wake disturbances of the preceding blade row
- periodical instationary velocity fluctuations resulting from the wake flow of the blade row after the upstream one, which are converted by the following relatively moving blade row
- the resulting anisotropic turbulence field

have to be considered. As the turbulence models in Navier-Stokes Codes are of such importance for their efficiency a validation and modification with macroscopical flow data is insufficient. Therefore instationary velocity fields and turbulence data have to be acquired by experiments.

## EXPERIMENTAL FACILITY AND TECHNIQUE

The two stage low speed axial compressor of the Institute for Turbomachinery (figure (1)) is suitable for the experimental recording of the instationary flow field because of its large dimensions. This allows an easy instrumentation with different kinds of probes without excerting too much influence on the measuring result, which is the main problem in industrial compressors. A detailed description of the compressor is presented in [1]. Some stage parameters of the compressor in the design point are described in figure (1).

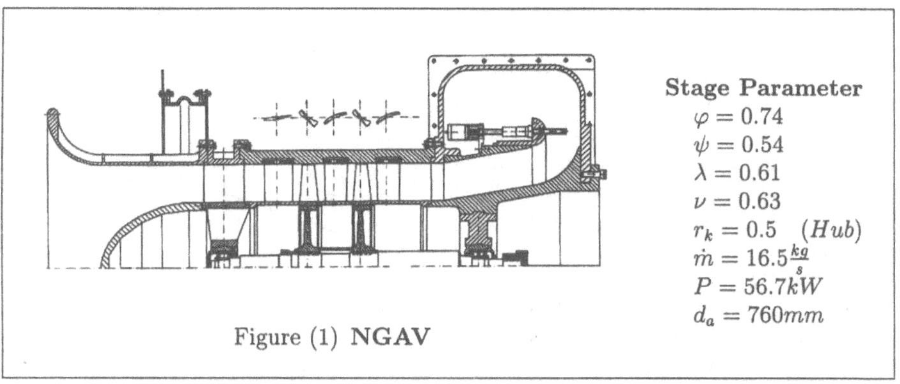


Figure (1) NGAV

For this investigation two different kinds of split film probes were used. They consist both of two $3\mu m$ Nickel films deposited on a $200\mu m$ diameter quartz fibre with an active length of $1.25mm$. With an overheat ratio of 1.7 and a mean film temperature of $250^oC$ the operation of the two probes in the constant temperature (CT) mode guarantees a high velocity sensitivity [2]. The data are recorded by a 12bit analogue/digital converter and stored on a personal computer. With this equipment sample rates from $2Hz$ to $2MHz$ and voltages from $0.05V$ to $50V$ can be acquired. For further informations see [3].

With the radially (R57) and tangentially (R56) oriented split film probes the instationary velocity components in three directions are acquired behind the first rotor. In this investigations only results from measurements at 50% blade height near the the design point of the compressor are reported.

## EVALUATION METHOD

The datasets are recorded continously for 2.62 seconds together with a triggering pulse from the motorshaft. This leads to $2^{19}$ single data which are collected and stored into 4200 time windows per revolution or 140 time windows per pitch. Each time window has a size of $4.92\mu s$ and ensures a very detailed covering of the flow field over the circumference of the rotor. Using the 'Ensemble Averaging Method' [4] the mean instationary flow components for each window are evaluated [3]. Another important point in the evaluation of the instationary flow field in turbomachines is the choice of the reference system. Usually the flow field in turbomachines is described in an axial-circumferential plane. This method did not lead to satisfactory results. Mainly in the wake region the differences concerning the fluctuation components were significant and did not describe the character of the instationary flow field in a correct way. Instead an evaluation in a mainflow-crossflow reference system seemed more appropriate.

In figure (2) the resulting velocity fluctuations are presented. $c'_{mf}$ and $c'_{cf}$ are the instationary velocity components for the absolute system gained simultanously by the probe type with the radial sensor fibre (R57). The resulting instationary velocity components in the relative system are designated $w'_{mf}$ and $w'_{cf}$.

With the other probe type (R56) the radial velocity component $c'_r$ and the mainflow component $c'_{mf*}$ can be measured simultanously as is indicated in the velocity diagram at the bottom of figure (2). Therefore the probe has to be placed perpendicular to the mean flow direction. Because of the 1.25mm extension in circumferential direction of probe type R56 the fluctuations in mainstream direction $c'_{mf}$ are leveled off somewhat mainly in the wake region and do not reach the same range as the fluctuations measured with probe R57 [5].

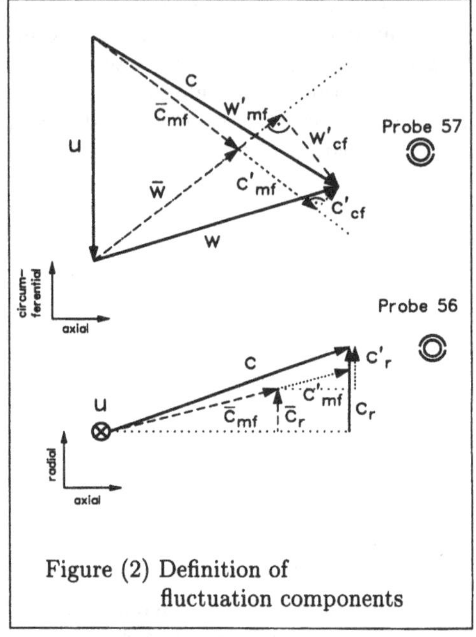


Figure (2) Definition of fluctuation components

As a consequence the turbulence intensity is calculated here considering $\overline{c'^2_{mf}}$ and $\overline{c'^2_{cf}}$ for the absolute system and $\overline{w'^2_{mf}}$ and $\overline{w'^2_{cf}}$ for the relative system only:

$$Tu_c = \frac{1}{\bar{c}}\sqrt{\frac{1}{2}(\overline{c'^2_{mf}} + \overline{c'^2_{cf}})} \qquad (1) \qquad\qquad Tu_w = \frac{1}{\bar{w}}\sqrt{\frac{1}{2}(\overline{w'^2_{mf}} + \overline{w'^2_{cf}})}. \qquad (2)$$

## EXPERIMENTAL RESULTS

The absolute and relative velocity components measured with probe type R57 over half a rotor revolution are presented in figure (3). With $80m/s$ the relative velocity is higher than the absolute velocity with $65m/s$ for this measuring conditions behind the first rotor. Both velocity components decrease in the rotor wake region, the relative velocity by about $10m/s$, the absolute velocity by about $5m/s$. The higher decrease of the relative velocity is mainly the result of the strong decrease of the absolute flow angle in the wake region.

In figure (4) the mean square fluctuations in the absolute system are presented over one pitch. Between the wakes both fluctuations reach values of $2m^2/s^2$. In the wake the mean square fluctuations in main stream direction $\overline{c'_{mf}}^2$ increase towards $22m^2/s^2$. The mean square fluctuations in crossflow direction $\overline{c'_{cf}}^2$ reach values of $18m^2/s^2$ in the wake region. Characteristical for the fluctuations in crossflow direction $\overline{c'_{cf}}^2$ is the double peak in the wake region. The peak related to the suction side of the blade is distinctly higher. The double peak is the result of the high velocity gradients in the flanks of the rotor wake in crossflow direction.

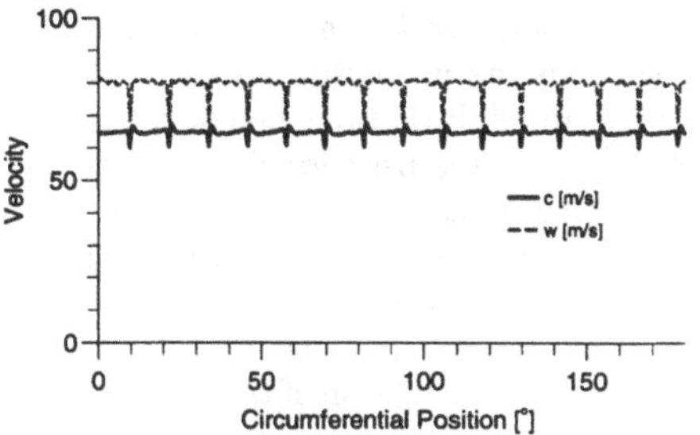


Figure (3) Absolute and relative velocity

Considering the same data evaluated in the relative system, the mean square fluctuations between the wakes have a value of $2m^2/s^2$, too. In the wake region both fluctuations reach values of $20m^2/s^2$ (see figure (5)). In the relative system, however, the mean square fluctuations in mainflow direction $\overline{w'_{mf}}^2$ exhibit the double peak in the wake region. The mean square fluctuations in crossflow direction $\overline{w'_{cf}}^2$ have a single maximum at $20m^2/s^2$, which is slightly lower than the mean square fluctuations in mainflow direction $\overline{c'_{mf}}^2$ in the absolute system.

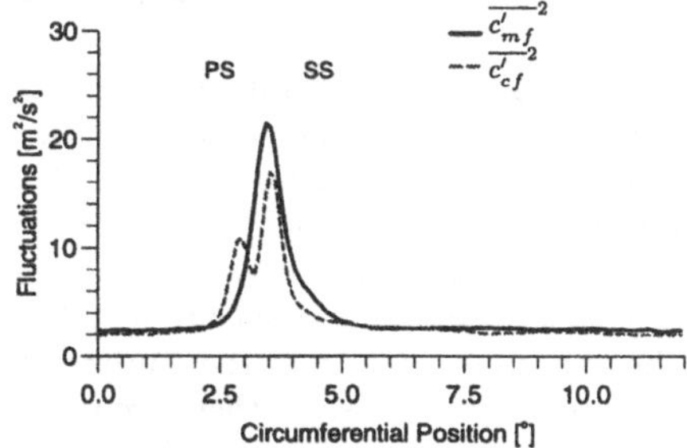


Figure (4) Mean square fluctuations in absolute system

This result is the consequence of the velocity distribution in the velocity diagram in figure (2). A fluctuation in mainflow direction in the absolute system results in fluctuations in crossflow direction in the relative system and vice versa. A slight difference exists only in the absolute values.

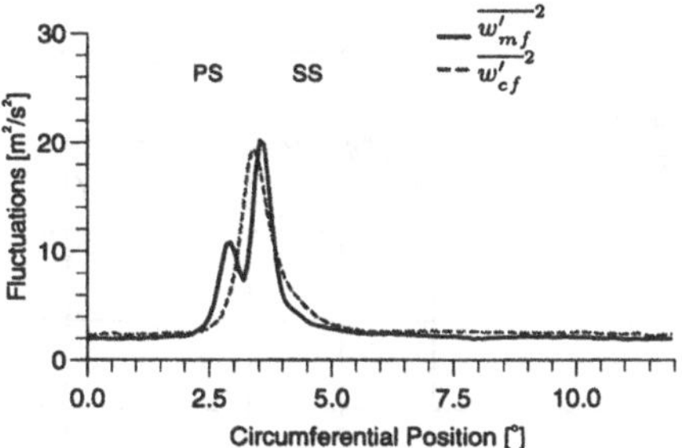


Figure (5) Mean square fluctuations in relative system

The results presented above show a nearly isotropic turbulence field behind the first rotor in the region between the wakes. In the wake region the instationary flow field has a complex character, but the chosen reference system allows the examination of the real physical behaviour of the flow.

In addition to the mean square velocity fluctuations the turbulence intensity for the absolute and the relative system over one pitch is presented in figure (6). Between the wakes the turbulence intensity of the relative system $Tu_w$ reaches values of 2%, that of the absolute system 2.5%. In the wake the turbulence intensity increases sharply towards

6.5% and 7% respectively. Figure (2) shows, that a momentary velocity fluctuation - simultaneously influencing value and angle - results in different fluctuation components in the absolute and the relative system. In addition the value of the mean velocity, which is used for the calculation of turbulence intensity, is different in both systems. These influences might result in different values of turbulence intensity in the absolute and the relative system. A systematic investigation of these effects has yet to be made.

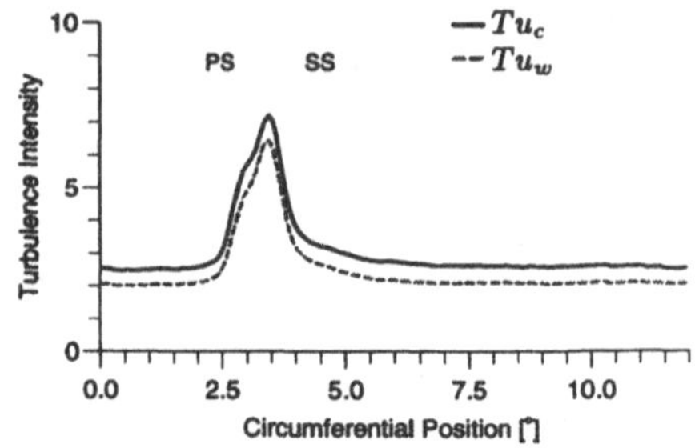


Figure (6) Turbulence intensity after first rotor

## STATISTICAL INTERPRETATION

The measuring and evaluation method applied cannot render a continuous velocity signal for a certain point of the flow field, the temporal fluctuations of which would be a direct measure for turbulence. The data acquired by ensemble averaging the velocity signal and calculating the difference between momentary and average velocity for each time window, as it passes by the probe, are rather velocity samples at this point in the flow field at a sampling rate considerably lower than characteristic turbulence frequencies to be expected. A test of the 'turbulent' character of the velocity fluctuations by the usual frequency analysis is therefore impossible.

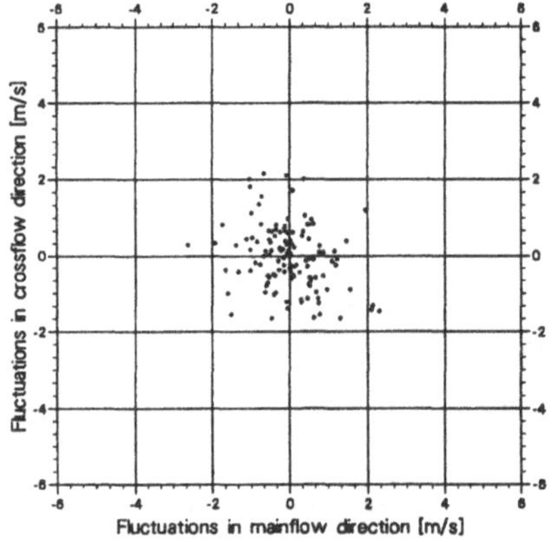


Figure (7) Dispersion of the fluctuations in the undisturbed flow field

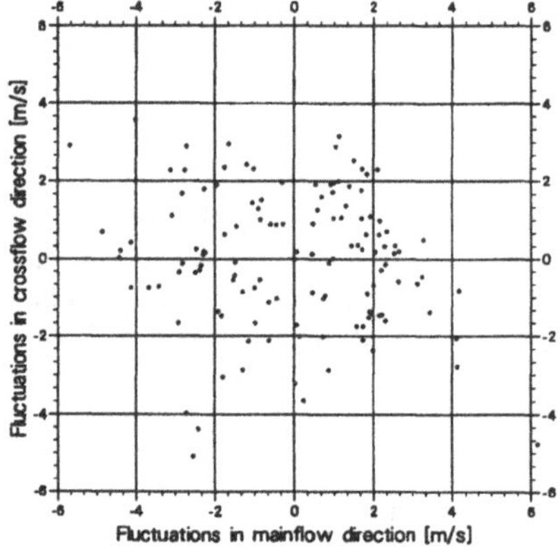


Figure (8) Dispersion of the fluctuations in the rotor wake region

It is necessary therefore to utilize other characteristic attributes of turbulence for a quality check of the measured fluctuations. Because of continuity reasons longitudinal fluctuations in turbulence must be compensated by simultaneous lateral fluctuations. In figure (7) and (8) this test is made for simultaneous fluctuations in mainflow and crossflow direction in the undisturbed flow field and in the wake region. Evidently compensating fluctuations could occur also in radial direction, but these cannot be measured simultaneously. The investigations have shown, that their values are lower in the wake region. Figures (7) and (8) exhibit a rather random distribution of the points, which supports rather the turbulent character of the fluctuations, since large scale fluctuations of the whole flow field should rather produce very specific patterns of distribution of the points.

While a frequency analysis of the measured fluctuations in a time window is not possible, an investigation of the amplitude distribution might be suitable to show their random character, which would be some evidence for their turbulent origin. Figures (9) and (10) show the distribution of the absolute values of the mainflow fluctuations in the undisturbed flow and in the wake region. While figure (9) exhibits a nearly Gaussian distribution with a peak at about 1.65m/s figure (10) has a peak in the same region but in addition nearly constant lower percentages up to 5m/s.

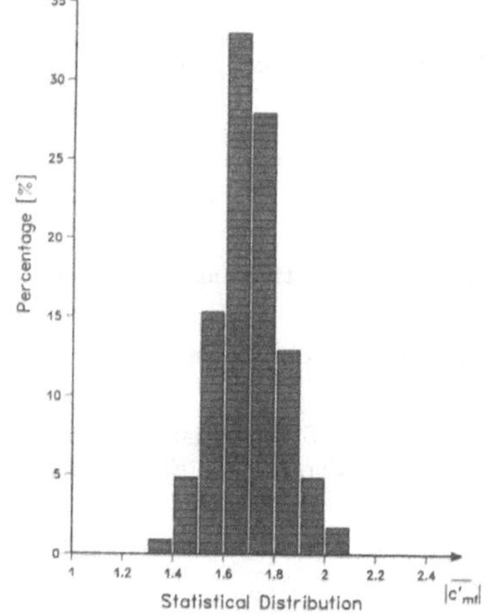


Figure (9) Statistical distribution of $\overline{|c'_{mf}|}$ in the undisturbed flow field

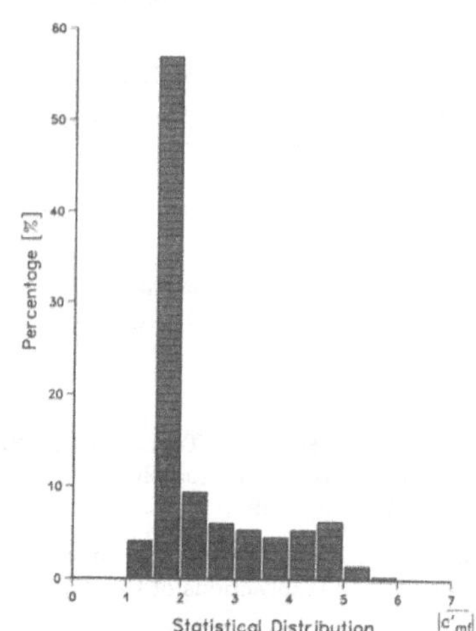


Figure (10) Statistical distribution of $\overline{|c'_{mf}|}$ in the rotor wake region

The results of the statistical evaluations will have to be investigated more in depth with the aid of turbulence theory. For the moment they seem not to contradict the assumption of the turbulent character of the measured fluctuations.

## CONCLUSIONS

The measurement of the instationary flow field behind the first rotor with high temporal and spatial resolution reveals details such as a nearly isotropic turbulence in the region between the wakes. In the wake region the fluctuations increase sharply and show a characteristic double peak of the absolute fluctuations in crossflow direction in the flanks of the wakes.

The evaluation in a mainflow/crossflow system adapted to the fluid flow enhances the interpretation of the measurements considerably.

The investigation of the coincidence of longitudinal and lateral fluctuations renders a rather random correlation, which points more to turbulent than a systematic character of the fluctuations. Similarly their amplitude distributions have a more or less statistical character, so that the assumption of the turbulent origin of the fluctuations is not improbable. More and specialized measurements and comparison with the results of turbulence theory will be necessary.

The investigations try to supply data, which might be usefull to understand the complex instationary flow field in multistage turbomachines. Data of this kind will be indispensable in near future for the validation of instationary Navier-Stokes Codes and their turbulence models for flow calculations in turbomachines.

## ACKNOWLEDGEMENT

The measurements reported here were supported by the DFG in the project Ri375/12-1. The authors want to express their sincere thanks.

# References

[1] Traulsen, D.: "Axialverdichterprüfstand zur Untersuchung von Randzonenströmungen", Fortschritt-Berichte VDI, Reihe 7, Nr.161 (1989).

[2] Bruun, H.H.: "Hot-Wire Anemometry: Principles and signal Analysis", Oxford Science Publications, 1995.

[3] Sentker, A., Riess, W.: "Turbulence Intensity Measurements in a Low Speed Axial Compressor", Engineering Turbulence Modelling and Experiments 3, Eds. Rodi, W., Bergeles, G., Elsevier Science B.V., pp. 773-783, 1996.

[4] Camp, T.R., Shin, H.-W.: "Turbulence Intensity and Length Scale Measurements in Multistage Compressors", Transactions of the ASME, Vol.117, Jan. 1995.

[5] Riess, W., Sentker, A., Walbaum, M.: "Experimental Investigation of Unsteady Flow in Axial Compressors", I Mech E Automobile Division Conference, Rugby, 1996.

# The **Pr**eliminary Aircraft **D**esign and **O**ptimization Program for **Sup**ersonic Commercial Transport Aircraft **PrADO-Sup**

Rainer Seubert
DLR, Institute of Design Aerodynamics
Lilienthalplatz 7, D-38108 Braunschweig, Germany

**SUMMARY**

This paper presents the multi-disciplinary design code PrADO-Sup for supersonic commercial transport aircraft which is a strongly modified and extended version of PrADO, the Preliminary Aircraft Design and Optimization Code developed at the Institute of Aircraft Design and Structural Mechanics of the Technical University of Braunschweig. The modular structure of PrADO-Sup is shown and all topics within the iteration and optimization loops are addressed. Detailed information is given on the integration of an advanced aerodynamic model. Here the higher order panel code HISSS, developed by DASA-LM, was implemented to provide data for a wide range of both subsonic and supersonic onflow conditions.

## INTRODUCTION

Continuously rising air traffic and extended routes will lead to a decision of the aerospace industry, aircraft operators and governments from different countries for the development of future high capacity aircraft within the next few years. On the one hand this will be a conventional ultra high capacity aircraft with 600+ passengers cruising at high subsonic speed and on the other hand a new supersonic transport aircraft with increased capacity of 250+ passengers and increased range of 5000+ nm compared to the Concorde, the one and only commercially operated supersonic transport aircraft in the world.

This new supersonic aircraft has to fulfill current and future environmental requirements such as pollution or take-off noise and must be economically competitive against new subsonic aircraft.

Due to the complexity of a totally revised new aircraft and due to the sensitivity of a supersonic transport configuration with respect to aerodynamics, structural weight, engine performance and environmental constraints a multidisciplinary design and optimization tool has to be used to get the most economic result fitting within all the requirements and constraints.

That is why DLR, Institute of Design Aerodynamics, is currently adapting a preliminary aircraft design and optimization program for supersonic commercial transport aircraft, called PrADO-Sup. This code is a strongly modified and extended version of PrADO [1], a code developed at the Technical University of Braunschweig, Institute of Aircraft Design and Structural Mechanics.

Besides the redesign of the existing supersonic transport configuration, the Concorde, for the validation of PrADO-Sup the code will be run for the design and optimization of advanced configurations, parameter and sensitivity studies within the design and the analysis of future unconventional configurations like blended wing bodies or canard configurations.

This paper shows the modular structure of PrADO-Sup. All the topics within the iteration and optimization loops are addressed and detailed information is given on the integration of an advanced aerodynamic model. Here the higher order panel code HISSS [2], developed by DASA-LM, was implemented to provide valid data in a wide range of both subsonic and supersonic flow.

## TASKS AND OBJECTIVES OF PrADO-Sup

The task of PrADO-Sup is the design and optimization of an aircraft for a specified mission. All disciplines within a multidisciplinary design loop have to be addressed including design layout, aerodynamics, flight mechanics, structure and weight, engine evaluation and integration. Finally a desired objective function, e.g. take-off weight or direct operating costs(D.O.C.) will be minimized. Sensitivity studies concerning future technologies like carbon structures, modern engine designs, e.g. mid tandem fan or a laminar wing, have to be performable to evaluate the potential in weight and drag reduction. Also the integration of high lift devices has to be considered to increase take-off and landing performance and to reduce take-off noise. Finally PrADO-Sup should be able to analyze future unconventional aircraft designs such as blended wing body configurations within a feasibility study to obtain possible improvements.
Starting point of every new configuration is the definition of specific requirements such as the demands for payload, range, aircraft performance or environmental viability. As it is not only sufficient to find one possible configuration which fits into the requirements for given constraints but also to find that configuration which fits best into the required objective function e.g. minimized take-off weight, minimized D.O.C. or minimized air pollution and noise, all the possible derivatives have to be addressed, analyzed and optimized within the preliminary design. That means multiple loops for the design strategy and emphasizes the utmost importance of a fast and very efficient code. The objective of that tool is maximum accuracy and maximum reliability achieved at minimum run time and costs.
To ensure compatibility to higher sophisticated tools for more detailed analysis of preliminary designed configurations user defined interfaces must be available to initiate further investigations easily e.g. using Euler or Navier Stokes codes for detailed aerodynamic optimization, finite element codes for structural design or both for flutter simulation.

## STRUCTURE OF PrADO-Sup

A code for preliminary design has to meat the following requirements: flexiblity and modularity, with well defined interfaces and a fast and reliable databank management. Flexibility is needed to adapt the code to new design tasks, modularity is needed for an easy installation of new design features and to switch to different data bases. Well arranged interfaces make the code easy to handle and allow the user to run other tools using the same data. A highly sophisticated data bank management allows flexible and reliable data transfer between the different code segments.
Considering these requirements and following the different tasks within the design loop PrADO-Sup is divided into four levels: data in-/output, optimization loop, multidisciplinary design loop and the design program libraries (see Fig. 1). Finally a complex data management system D.M.S. handles the data transfer between the different levels.

<u>1. Level - Data Input/Output:</u> This level includes routines for the data input of the user defined baseline design, including mission requirements, constraints and objective functions. A complete internal data base is configured with missing parameters calculated from statistics and/or external data bases. Finally the data base of each converged configuration is stored and open for further investigations or a graphic postprocessor.
<u>2. Level - Optimization Loop:</u> A given objective function e.g. take-off weight, fuel consumption or D.O.C. is minimized varying user defined parameters such as aspect ratio or wing thickness and checking the constraints defined in the input file.

3. Level - Multidisciplinary Design Process: On this level all the modules necessary for the interdisciplinary design process (geometry definition, aerodynamics, propulsion system, flight mechanics, weight and structure, D.O.C.) are run, the aircraft and the powerplant system are sized within several internal iteration loops until all the requirements are satisfied and no constraints are violated anymore. If this cannot be reached with user defined accuracy the program stops printing out the parameters which did not converge.
4. Level - Design Program Libraries: The program libraries include all the subroutines for the different disciplines called by Level 3. These subroutines represent the physical background and form the kernel for the internal design process. New modules developed by different disciplines can be easily integrated and switched on by the user to enhance performance and accuracy of the code.
Databank Management System - D.M.S.: The D.M.S. guarantees multiple data exchange within or between the different levels 1-4, all values which are necessary for the interdisciplinary design process are stored [1], values changing within an iteration or optimization are automatically replaced. The final dataset is stored and can also be used as a starting design for further investigations.

## DESIGN PROCESS

The first step within the design process is the initial geometry definition of the aircraft. The fuselage is defined providing the required planform area for the integration of the whole equipment, e.g. floors, seats, cockpit. The wing is initially defined for a given wing loading and is designed by its planform, e.g. an ogee wing and the airfoil sections. The nacelles are added and a vertical fin is defined to ensure directional stability (no horizontal tail at the moment).
Within the next step the aerodynamics are calculated. As the original PrADO only considered subsonic flow running a simple linearized method (lifting line theory only for the wing) for aerodynamic calculations a completely different aerodynamic model had to be installed into PrADO-Sup. This model running the panel code HISSS [2] currently handles a combined wing-fuselage with flaps and is described in detail in the following chapter.
With the aerodynamic data achieved, the engines can be defined to fit into the FAR requirements, so take-off distance, climb angle, single engine climb and transonic transmission are checked. Engine performance, e.g. thrust, fuel consumption, $NO_x$ emission is either calculated within a thermodynamic cycle for adaptable engines or simply read from a data file (fixed engine).
With that initial virtual aircraft a typical mission for a supersonic transport configuration is simulated including take-off, climb, cruise, decent, approach and landing. The equations of motion are solved for the aircraft at each point of the flight envelope with the mass of the aircraft initially estimated by statistics. The aircraft is trimmed by setting the flaps within an iterative process. At the moment the static longitudinal stability is set constant. To achieve the appropriate center of gravity as much fuel as necessary is pumped to the trim tanks. The difference between the calculated maximum weight of available fuel and the fuel burned within that mission will be later used to rescale the aircraft.
The next step is a more detailed mass calculation for the different aircraft components according to the maximum critical load factors defined in the FAR. For the calculation of the fuselage mass a semi-empirical method developed at the University of Braunschweig [1] is used. For the wing the user can choose between several empirical methods [1], [3] and a simple finite element code [1]. This finite element code will be improved soon and extended to wing-body configurations to incorporate a flexible aircraft (slender body, thin wing) into the design

process. Also the masses of the empennage, the landing gear and operational equipment are calculated and added to get the total empty weight. For the calculation of the take-off weight the mass of passengers, payload and fuel is added.
The result is now used to rerun the code iteratively for a periodical updating of the results meeting finally all user defined values, e.g. take-off weight.
This converged result represents an aircraft which fits into the user defined requirements and restrictions and can now be used as a starting design for optimization to minimize the defined objective function, e.g. D.O.C. or take-off weight.
The initial design may simply start with only a few user defined requirements such as payload, range, cruising speed and some airplane specific input, e.g. wing planform, type of airfoil, kind of powerplant. All the missing data, which are absolutely necessary for the design code, are provided by a statistical module. Unfortunately this module includes only data derived from Concorde publications.

## THE AERODYNAMIC MODEL OF PrADO-Sup

For PrADO-Sup a new aerodynamic model had to be installed. PrADO, formerly developed for the design of subsonic transport aircraft could only handle subsonic flow running a simple linearized method (lifting line theory only for a wing). For supersonic flow a more expensive model has to be used to consider planform-, thickness-, chamber-, twist-effects or the influence of the fuselage, empennage and nacelles. Furthermore, requirements for a code to run within a preliminary design process are robustness, high accuracy and minimum run time.Considering all these topics the panel code HISSS [2] was chosen in conjunction with a DLR grid generator [4] to provide valid data in a wide range of both subsonic and supersonic flow.

## ACCURACY OF THE PANEL CODE HISSS

Several test runs have been performed to show reliability and accuracy of the panel code HISSS. Fig. 2 shows the results for a missile-type configuration from Ma=0.5 to Ma=2.0 (lift, drag, pitch moment) compared to wind tunnel results measured in the transonic windtunnel at DLR Göttingen TWG [5]. As HISSS does not calculate viscous drag, this was added by calculating the viscous drag (turbulent flat plate theory) for the wing and body. It can be seen that HISSS generates reliable results in both subsonic and supersonic flow with sufficient accuracy for the integration into a preliminary design code.

## INTEGRATION OF THE PANEL CODE HISSS

HISSS is not directly incorporated into PrADO-Sup but it is started by a "System Call" to ensure a maximum of flexibility (see Fig. 3). Currently the grid generator handles a wing body combination and writes the input files for HISSS automatically to a user defined directory. Geometry files are created for the wing body combination with and without flap deflection. The integration of the flap is shown in Fig. 4. Defining a transition region between the fuselage and the flap allows the same network structure with the wake identically fitted to the fuselage for all flap settings.
HISSS is run for 6 different Mach numbers (3 for sub/supersonic) with flap settings of $0^{o}$ and $10^{o}$. For each Mach number the lift curve slope, pitch moment, wave and induced drag, and the corresponding flap increments are temporarily stored within PrADO-Sup (see Fig. 5 for typical results for a Mach number of 0.7). To simulate the aircraft's whole flight envelope it is not sufficient to have isolated points only for a few Mach numbers. On the other hand it is too time

consuming to run HISSS for more Mach numbers, so polynomial interpolations are used for curve fits through the isolated points within the subsonic and the supersonic regime. PrADO-Sup stores only the polynomial coefficients, so results for every Mach-Number can be rapidly recalculated. The transonic regime is added empirically. Fig. 6 shows the interpolation and the transonic modeling for the lift curve slope (the same procedure is used for the pitching moment). The vortex lift is added empirically using a polynomial function for the lift curve slope with an additional amount of 50% of the linear lift at $20^{\circ}$ angle of attack as derived from Concorde measurements [6]. See Fig. 7 for Ma=0.3 for the redesigned Concorde with PrADO-Sup (symmetrical, uncambered, untwisted wing).
Fig. 8 shows the approximation of the drag polars integrated into PrADO-Sup. Each drag polar is approximated by a polynomial starting with the zero lift drag. The zero lift drag and the k-factors are again interpolated by polynomials and only their coefficients are stored by the D.M.S. So the drag can be recalculated for every Mach number very efficiently. To achieve realistic drag values the viscous drag of the wing, fuselage, fin and nacelles, and the wave drag of the stream tube of the engines are added. Fig. 8 shows the Ma=2.0 polar of the redesigned Concorde calculated from PrADO-Sup with the design lift coefficient predicted correctly.

## CONCLUSIONS

The general function of PrADO-Sup and the integration of the new aerodynamic model is shown. First aerodynamic results for the redesign of the Concorde aircraft are presented. It is shown that the aerodynamic characteristics of the Concorde, both high-speed and low-speed, are predicted accurately by PrADO-Sup. With some minor improvements speeding up the panel code HISSS the program system PrADO-Sup runs fast and reliable and can be efficiently used for the design of future supersonic aircraft. Next topic is the validation of the complete PrADO-Sup program system. The validated PrADO-Sup will then be used for calculations to evaluate the advantage of modern powerplant systems, e.g. the mid tandem fan or laminarization of the wing for preliminary designed or redesigned configurations. Finally PrADO-Sup will be run for the preliminary design of unconventional configurations, e.g. a blended wing body.

## REFERENCES


[1] W. Heinze: Ein Beitrag zur quantitativen Analyse der technischen und wirtschaftlichen Auslegungesgrenzen verschiedener Flugzeugkonzepte für den Transport großer Nutzlasten, Dissertation, ZLR-Forschungsbericht 94-01, 1994.

[2] L. Fornasier: HISSS - A Higher - Order Subsonic/Supersonic Singularity Method for Calculating Linearized Potential Flow, AIAA-Paper 84-1646, 1984.

[3] M. Kaufman, V. Balabanov, S.L. Burgee, A.A. Giunta, B. Grossman, W.H. Mason, L.T. Watson, R.T. Haftka: Variable-Complexity Response Surface Approximations for Wing Structural Weight in HSCT Design, AIAA-Paper 96-0089, 1996.

[4] M. Orlowski, U. Herrmann: Aerodynamic Optimization of Supersonic Transport Configurations, In: Proc. 20th. Congress of the International Council of Aeronautical Sciences, Sorrent, Italien, 8.-13.9.1996, pp. 924-930.

[5] W. Schneider: Experimentelle Bestimmung von Flügel-Rumpf-Interferenzen im Machzahlbereich Ma=0,5 bis Ma=2,0, AVA-Bericht 66 A 31, 1966.

[6] Jean Rech and Clive S. Leyman: A Case Study by Aerospatiale and British Aerospace on the Concorde, AIAA Professional Study Series, 1980.

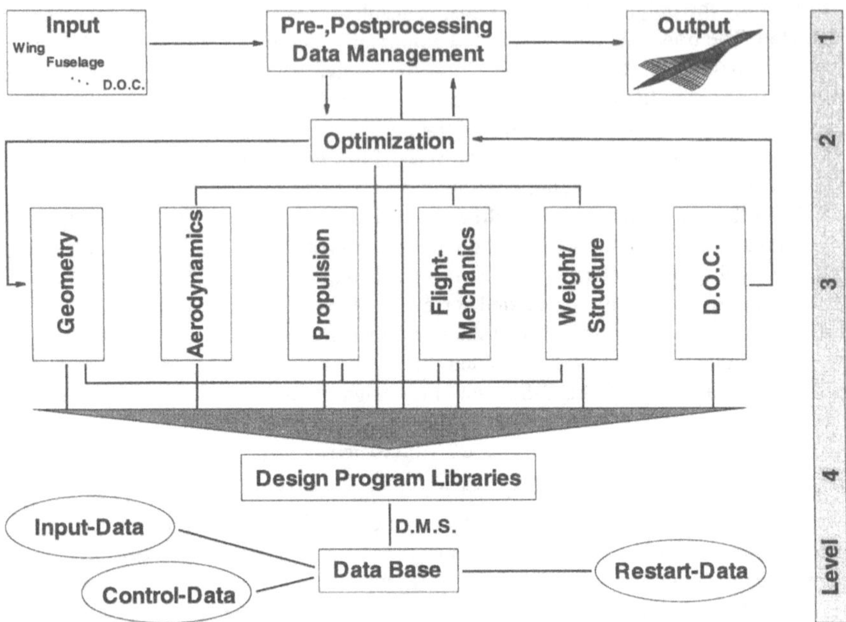


**Fig. 1:** The Structure of Prado-Sup

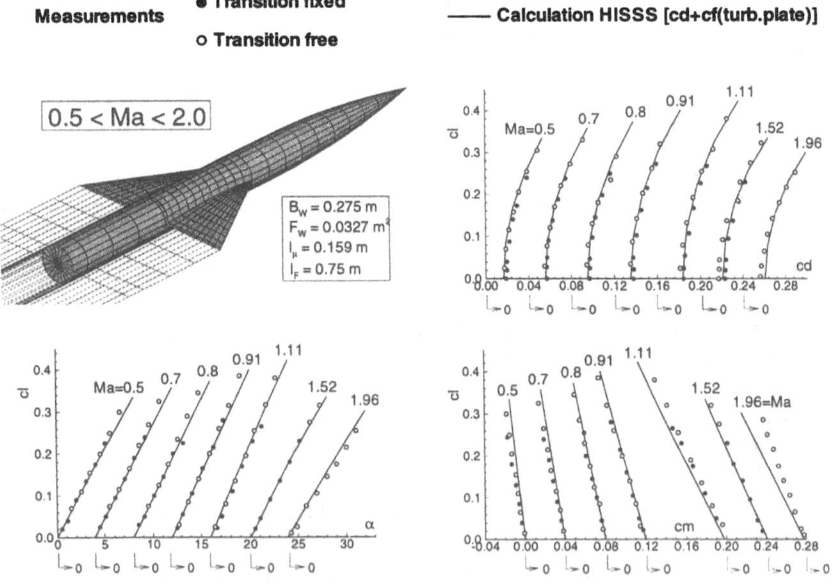


**Fig. 2:** Validation of the Panel Code HISSS for the Present Purpose

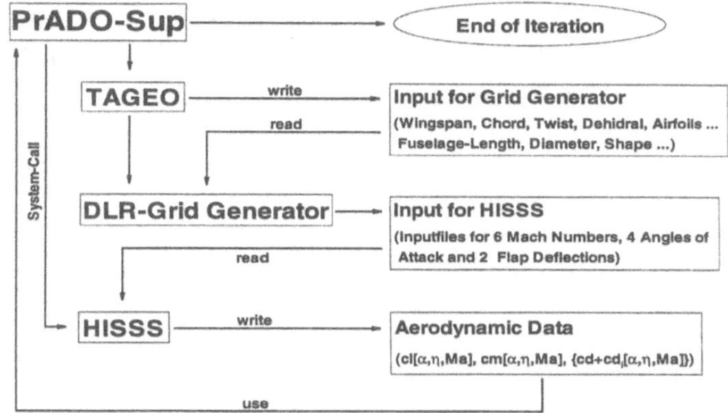


**Fig. 3:** Integration of the Panel Code HISSS

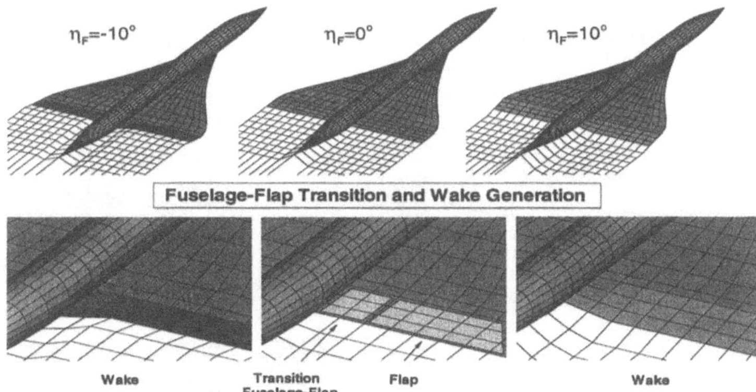


**Fig. 4:** Flap Integration for Panel Calculation

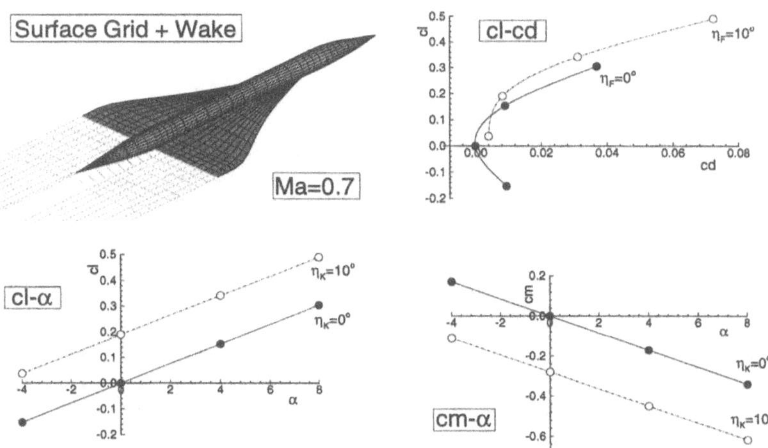


**Fig. 5:** Evaluation of the HISSS Results

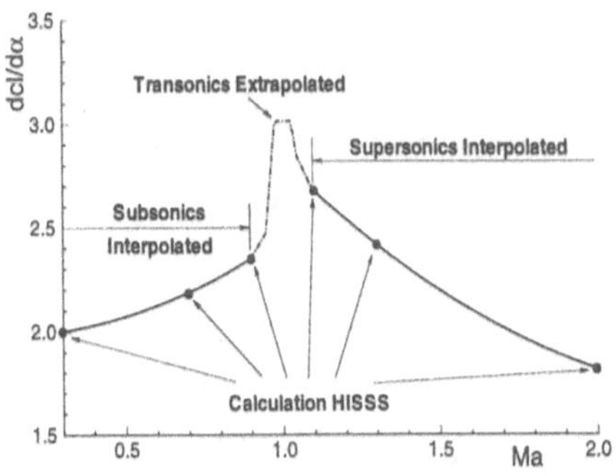


**Fig. 6:** Interpolation of the Lift Curve Slope

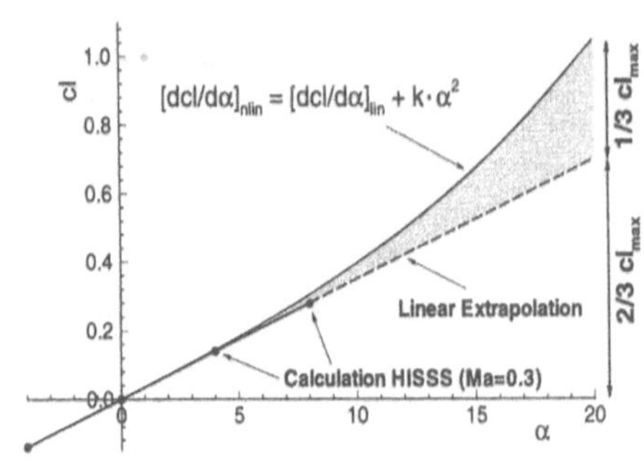


**Fig. 7:** Approximation of Vortex Lift

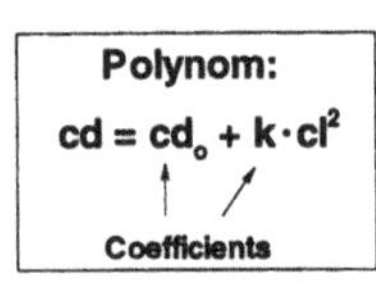


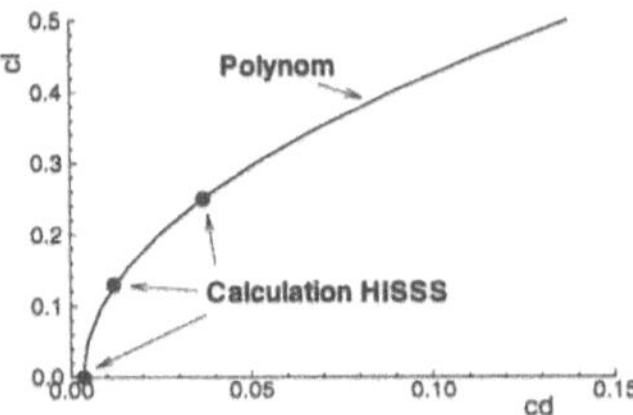


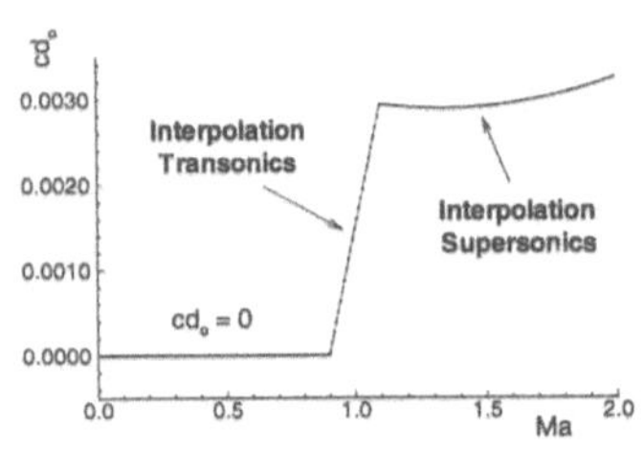


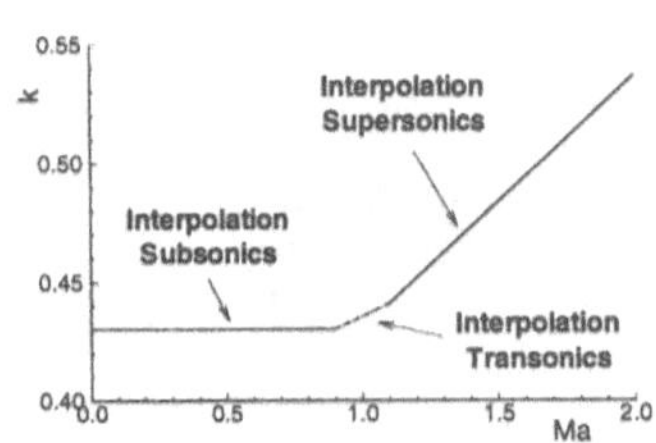


**Fig. 8:** Approximation of the Drag Polars

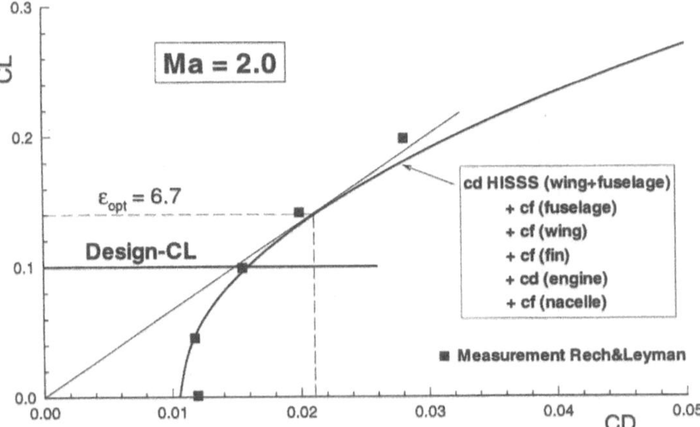


**Fig. 9:** Drag Polar of Redesigned Concorde

# Design of a laminar wing for a commuter airplane using a 3D-design method

Th. Streit, W. Bartelheimer, H. Bleecke, C.-H. Rohardt
DLR, Institute of Design Aerodynamics, Braunschweig, Germany
W. Wohlrath
Dornier Luftfahrt GmbH, Oberpfaffenhofen, Germany

## Summary

Using conventional design methods a laminar wing was designed for the DO 328 commuter airplane in a preceding work. However, in the vicinity of the body and the nacelle, strong interference effects diminished the extent of laminar flow. In this work a 3D design method based on the solution of the Euler equations was used to enlarge the extent of laminar flow for the wing region placed nearby nacelle and body. In order to solve numerically the Euler equations, a structured 3D computational mesh was generated around the wing, nacelle and body of the DO 328 configuration. For the design task it was necessary to elaborate a design procedure which retained the original wing twist distribution and wing thickness in the spar region. In addition to the geometrical constraints, the lift values were prescribed at the design points being the maximum cruise and climb flight conditions. Using the laminar wing of the preceding work as starting condition, a new wing was designed with improved stability results of the laminar boundary layer for the aforementioned design points. Improvements were obtained especially in the regions near to the nacelle and the body.

## Introduction

While laminar technology has become conventional for gliders and is used successfully for small general aviation aircrafts its application to large commercial aircraft is being tested through extensive research programs [1] [2]. In order to obtain laminar flow for transonic flight conditions, all the possibilities of available laminar technology have to be exhausted and further research will be required before a transonic laminar wing becomes operational. A different situation is encountered for medium sized subsonic commercial airplanes. Here the laminar flow can be obtained through the use of natural laminar flow technology, i.e. the natural growth of instabilities in the laminar boundary layer is limited through favourable pressure gradients by using adequate contour shape. For a subsonic commuter airplane by careful design laminar boundary layers can be obtained not only at cruise flight conditions but also at climb flight condition. Throughout a flight test program conducted on a DO 228 laminar wing glove, the efficiency of laminar technology at commuter aircraft flight condition could be proved [3]. In addition it has been shown that in the wing region influenced by the propeller the flow becomes turbulent only for the time it takes that the turbulent slipstream of a propeller blade needs to pass downstream over the wing; also it proved the efficiency of de-icing and insect protection systems placed at the leading edge [4]. Therefore laminar flow technology has reached an operational stage for subsonic commuter airplanes, and consequently a laminar wing predesign was performed for the DO 328 [5].
The wing designed in [5] showed satisfactory extent of laminar flow at design conditions of the DO 328, except for regions placed near to nacelle and body. As shown in Fig. 1, at the inner wing (wing region between nacelle and body) destabilising suction peaks appeared near to the

nacelle at the lower wing side for the max. cruise design point (Fig. 1a), while for max. range cruise (Fig. 1b) and climb flight condition (Fig. 1c) weak suction peaks appeared on the upper wing side near to the body. For the outer wing the interference of the nacelle leads to similar losses of laminar flow. In the present work, a 3D design method based on the solution of the Euler equations was used in order to extent the laminar flow region. In this study the influence of the propeller was not taken into account, and the study was restricted to operating points with laminar flow regime. These are the max. cruise design point, which represents the lower edge of the laminar bucket, and the climb flight condition which represents the upper edge of the laminar bucket. For the DO 328 the max. range cruise is also placed at the upper edge of the laminar bucket.

## Design Procedure

### Design-algorithm and its implementation

A design method for transonic airfoils and wings based on the solution of the Euler/Navier-Stokes equations has been developed at the DLR [6] The design strategy applied is an inverse design method based on the work of Takanashi [7]. The difference between the computed pressure distribution of a given geometry and the prescribed target pressure distribution is iteratively reduced by solution of an inverse formulated transonic small perturbation equation (TSP-equation). Takanashi's method was modified in order to ensure the convergence of the design in transonic flow. In addition, a smoothing algorithm based on Bézier curves is used to obtain a smooth surface. The design method is implemented as a component of the DLR-FLOWer code, which also provides the analysis component in order to solve the Euler/Navier-Stokes equations. In the first step of the design cycle, the analysis module is used to compute the pressure distribution at the wall. This module is a multiblock multigrid code to solve the 3D Euler/Navier-Stokes equations for structured meshes. A finite volume cell-vertex discretization and a 5 stage Runge Kutta time stepping scheme are applied. By using local time stepping, implicit residual smoothing and a multigrid method, the solution is accelerated to steady state. If, after having checked the convergence, the design requirements are not satisfied, the design module is used. Otherwise, the designed geometry data and the calculated flow field solution are returned.

In the design routine, the difference between the prescribed target pressure and the calculated wall pressure is computed. From the solution of the inverse formulated TSP-equation the geometry correction is derived. Before adding the correction to the current geometry, a smoothing technique has to be applied in order to avoid geometry oscillations. A smoothing technique based on Bézier polynomials is introduced which carefully smoothes the surface.

When the new geometry is computed, the grid module is called and a design cycle is completed. The generation of a mesh for a wing is cpu-time consuming and difficult to be handled automatically. Therefore, an efficient algebraic technique is used to move the nodes of the grid according to the computed geometry correction. In addition this technique is only locally applied in regions subject to geometry corrections.

A detailed description of the underlying theory and of the algorithms for the analysis and the design modules can be found in [8].

### Design procedure for a laminar wing with subsonic onflow conditions

As described in the last subsection, the design method used computes geometry corrections as a function of the difference between the prescribed target pressure and the calculated wall pressure of the actual geometry. Therefore, the required design properties and constraints in this procedure have to be formulated consistently with the target pressure. In the design case of a complete wing with several constraints of aerodynamic and geometric origin, the formulation of an adequate target surface pressure field is itself an iterative process. In order to avoid ill-posed target pressure prescriptions, target pressure prescriptions were obtained introducing small modifications into calculated pressure distributions of actual geometries. These modifications successively took account of the required constraints. In the present work, the following requirements and constraints had to be satisfied:

1. Laminar boundary layer: The laminar boundary layer has to remain stable up to a prescribed extent and shall be fulfilled on nearly the whole wing at the two main design points which define the lower and upper border of the laminar bucket.

2. Twist distribution: The twist distribution of the predesigned wing had to be retained. This means that in order to obtain a wing with the prescribed target pressure distribution, only geometry corrections are allowed which do not involve rotations of the cord line of the wing sections.

3. Wing thickness: Since the geometry corrections lead to wing sections with different shape, it was required that at the position of the spar the upper and lower part of the wing is placed at the predesign location.

4. Lift coefficient: The lift coefficient had to be retained for both main design points. Furthermore, alterations of the wing lift distribution should not increase the induced drag.

In the present work, also the intersections of the wing with the nacelle and the body were left unchanged. This is due to technical reasons: a modification of these intersections implies that the mesh generation of the new geometry had to be extended to the body and the nacelle. This was not intended within the frame of this work. The design of the wing was performed up to a wing section placed very close to the nacelle or to the fuselage. The obtained geometry corrections for this section were then linearly blended in spanwise direction up to the corresponding intersection.

The wing designed in [5] was used as a start configuration, with exception of the inner wing where the NLF7 airfoil was replaced by the NLF9 airfoil. Compared to the NLF7 airfoil, the NLF9 one elaborates better off-design properties, especially regarding maximum lift [9]. Laminar stability properties for the two airfoils are similar and the Euler solution for the DO 328 with the NLF9 airfoil for the inner wing showed similar problems in the body and the nacelle regions as for the NLF7 airfoil.

Keeping in mind the restrictions mentioned above, the following design procedure was chosen:

1. Design at the upper edge of the laminar drag bucket: Concerning stability the critical region for this design point is the upper wing side. An analysis and design method for low Reynolds number airfoils [10] is used in order to establish the required pressure gradient to obtain stability up to the pressure minimum, representative for a section placed in the middle of the inner wing. Then pressure targets were constructed for each wing section of the Euler computational mesh. These pressure targets were constructed in such a way that for each section the predesign pressure distribution was conserved at the nose, at the lower side of the wing and at the trailing edge, but at the upper side of the wing modifications were introduced in order to obtain the required pressure up to the pressure minimum (see Fig. 2a). By adjusting the gradient on the upper side care had to be taken to obtain the predesign local lift for each of the prescribed target sections. In order to keep the predesign twist constant, the design method was used with the no-twist option. With this option after each complete geometry update, each wing section has to be rotated around the trailing edge in a way that the leading edge is placed at the predesign

position. Using a target constructed as described above, the design method was used to produce a new wing geometry. This first new wing will in general not satisfy all required constraints. Especially due to camber changes at the upper wing side, in some sections the wing thickness at the spar location may have changed. Since the wing thickness is determined by the level of the pressure distribution, it was adjusted by constructing new targets by shifting the pressure distribution at the lower and upper side parallel up to the point given by the pressure minimum (see Fig. 2a). With this new target a second new wing was obtained. This process was iterated until a target was prescribed which lead to a wing geometry satisfying the constraints. With this procedure the resultant wing has sections with the predesign wing thickness at the spar location, but compared to the corresponding predesign sections a displaced position of upper and lower surface position may have resulted (see Fig. 2b). Therefore, the final step for the wing desig for this operation point was to displace each airfoil by such an amount that the upper and lower side position coincided with the predesigned wing, i.e. to allow a straight spar.

2. Design at the lower edge of the laminar drag bucket: Concerning boundary layer stability, the critical region for this design point is placed at the lower wing pressure distribution. The wing designed for the upper edge of the laminar bucket is now used as starting configuration. The pressure targets are constructed by using the pressure distribution obtained for the designed wing for the upper edge of the laminar bucket and introducing modifications on the lower wing side. It is important that the pressure distributions are not altered on the upper side in order to keep the already achieved stability situation on the upper wing side at the upper edge of the laminar bucket. Then the lower surface target pressure distributions are modified in the same manner as described above in order to fullfill the boundary layer stability requirements at the lower edge of the laminar bucket.

## Results

The surface mesh for the DO 328 is shown in Fig. 3. The 3D Euler mesh was divided into 36 blocks with a total of $3.13 \cdot 10^6$ points. It was generated with the DOGRID mesh generation program [11].
Pressure distribution results for $M_\infty = 0.28, \alpha = 2.6^o$ are given in Fig. 4. Inner wing results of the predesigned wing (Fig. 4a using the NLF9 airfoil for the inner wing) are compared to the new wing design (Fig. 4b). Note that results in Fig. 4b show an improved gradient on the upper wing side, especially for the sections placed close to the nacelle and the fuselage. Results for the outer wing are given in Fig. 4c and Fig. 4d. Especially, for sections lying outboard, a more favourable pressure gradient was obtained (Fig. 4d). For the maximum cruise design point, $M_\infty = 0.54, \alpha = -0.5^o$ the resulting pressure distributions are shown in Fig. 5. Note that compared to the predesign results, the new pressure distributions do not show suction peaks on the lower wing side in the region close to the nacelle. Fig. 6 compares the inner wing contour of the NLF9 airfoil with selected sections of the new wing. It shows that large geometry changes were required in order to obtain the prescribed target pressure distributions for sections placed close to the fuselage (Fig. 5a) and the nacelle (Fig. 5b).
The position of the transition line is shown in Fig. 7 for $M_\infty = 0.28$ and in Fig. 8 for $M_\infty = 0.54$. It was obtained by analysis of the pressure distribution of selected wing sections using the 2D panel method of Drela [10]. Note that the noncritical wing sides (lower side at $M_\infty = 0.28$ and upper wing side $M_\infty = 0.54$), show already a satisfactory extent of laminar flow for the predesign. For the critical sides the laminar flow region is enlarged, especially for regions placed close to nacelle and body. In addition, the width of the laminar bucket was increased. This is illustrated for selected wing sections in Fig. 9 showing the local lift value as function of transition line position. The lift values shown in Fig. 9 correspond to the flight conditions of the two design

points; and two additional points were obtained by increasing (decreasing) the angle of attack by $0.5^o(-0.5^o)$ for $M_\infty = 0.28(M_\infty = 0.54)$.

## Conclusions

A laminar wing was designed for a commuter aircraft using a 3D design method. The extent of laminar flow boundary layers for maximum cruise and climb flight conditions had to be maximized. In addition, the lift value was prescribed for the two design points (which represent the lower and upper edge of the laminar bucket). Furthermore, the wing twist distribution and the wing box thickness were prescribed. A design procedure was elaborated which took account of these requirements. Using the laminar wing of a previous design effort as starting solution, a new wing was designed with improved stability results of the laminar boundary layer at the aforementioned design points. Improvements were obtained especially for the region close to the nacelle and the body. In comparison with the previous design, the width of the laminar bucket was increased.

### Aknowledgement

The authors would like to thank K.H. Horstmann and A. Seitz for their engagement through helpful discussions for this work.

## References

[1] Redeker G.,Wichmann G. Die Technik der Laminarhaltung an ungepfeilten und gepfeilten Flügeln. Luftfahrttechnisches Handbuch, Vol. Aerodynamik 1995.

[2] Barnwell R. W., Hussaini M.Y. Natural Laminar Flow and Laminar Flow Control. ICASE/NASA LaRC Series Springer-Verlag 1992.

[3] Horstmann, K.H., Müller R., Dick P., Wohlrath W. Flugversuche am Laminarhandschuh der DO 228 DGLR-Bericht 92-07 p. 91-99, 1992.

[4] Müller R., Engemann S.,Horstmann, K.H.,Welte D., Moeken B. Flugversuche in simulierten Vereisungsbedingungen mit Laminarhandschuh an der DLR DO 228 und einem fluidischen Enteisungssystem. Deutscher Luft- und Raumfahrtkongreß 1996, Dresden 24-27.9.1996 to be published in DGLR-Jahrbuch 1996.

[5] Wohlrath W. Vorentwurf eines Laminarflügels für die DO 328 als Technologieträger (Stand 5.12.94). Dornier Internal Report AV EV31-006/95 1995.

[6] Bartelheimer W. An improved Integral Equation Method for the Design of Transonic Airfoils and Wings. AIAA Computational Fluid Dynamics Conference, San Diego, CA, 19.-22.6.1995, Paper No. AIAA-95-1688, p. 463-473, 1995.

[7] Takanashi S. An Iterative Procedure for Three-Dimensional Transonic Wing Design by the Integral Equation Method. AIAA Paper 84-2155, 1984.

[8] Bartelheimer W. Ein Entwurfsverfahren für Tragflügel in transsonischer Strömung. Doctoral thesis TU Braunschweig to appear as DLR-FB 1996.

[9] Wohlrath W. Überarbeitung des Vorentwurfes hinsichtlich maximal Auftrieb. Dornier Internal Report AV EV31-007/95 1995.

[10] Drela M. XFOIL: An Analysis and Design System for Low Reynolds Number Airfoils. Low Reynolds Number Aerodynamis, Proceedings of the Conference Notre Dame, Indiana, USA, 5-7 June 1889, T.J. Mueller (Editor) Lectures Notes in Engineering No. 54, p. 1-12, No.54, Springer-Verlag 1989.

[11] Seibert, W. Interaktive Netzgenerierung für Strömungsberechnung (System DOGRID). Anwendung und Weiterentwicklungs. Dornier-Bericht BF 15/91 B, December 1991.

## Figures

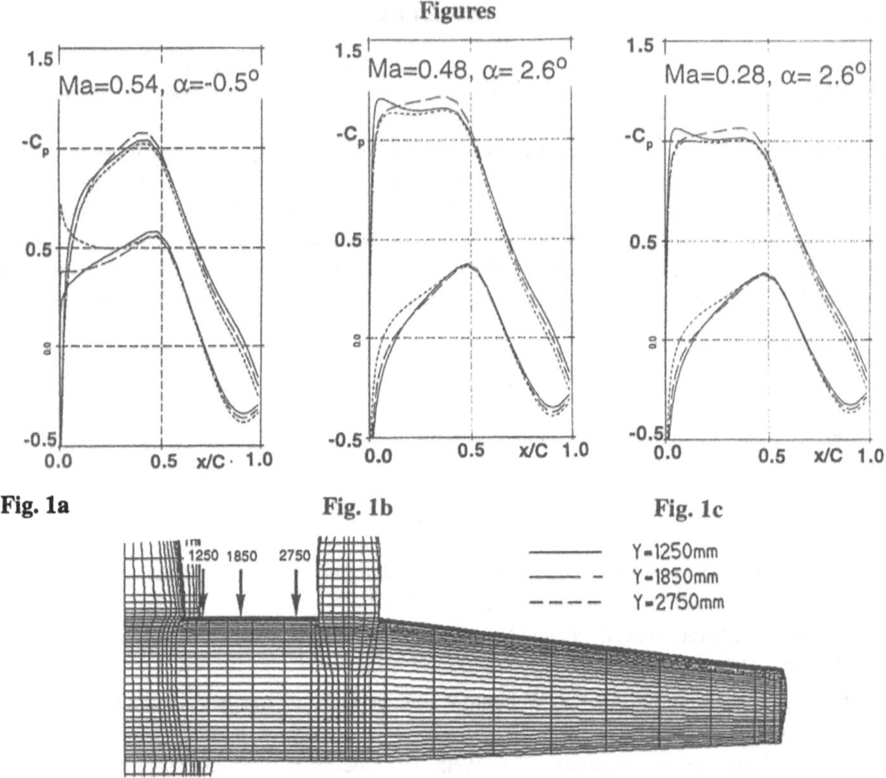


Fig. 1 Pressure distributions for a preceding laminar wing design [5] for the DO 328. For inner wing sections the NLF7 airfoil is used. Fig. 1a corresponds to max. cruise, Fig. 1b to max. range cruise and Fig. 1c to climb flight condition. These predesign results are taken from ref. [5].

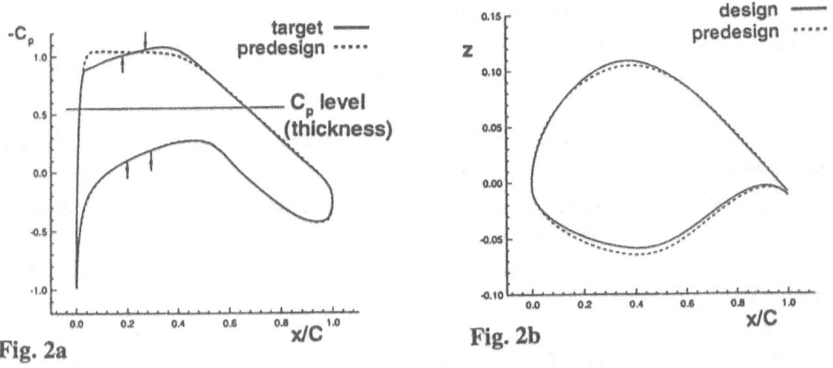


Fig. 2. Design procedure for upper edge of laminar bucket. Fig. 2a shows predesign and prescribed target pressure distributions. Fig. 2b compares predesign and resulting geometry.

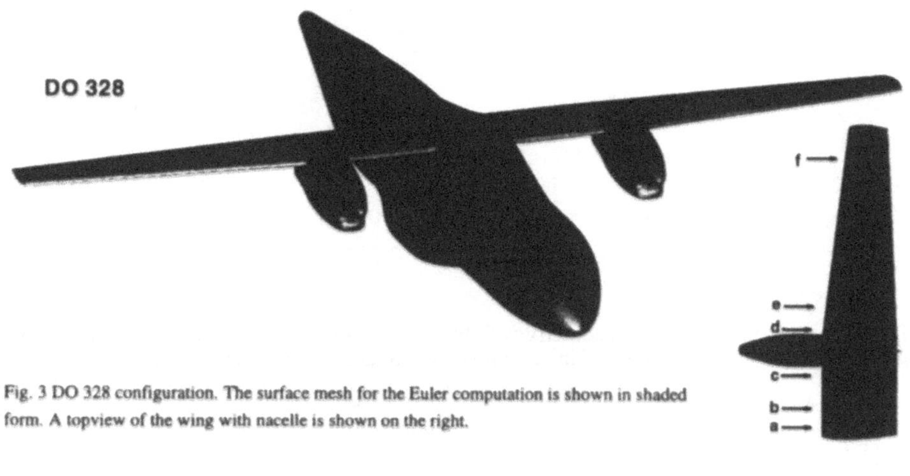


Fig. 3 DO 328 configuration. The surface mesh for the Euler computation is shown in shaded form. A topview of the wing with nacelle is shown on the right.

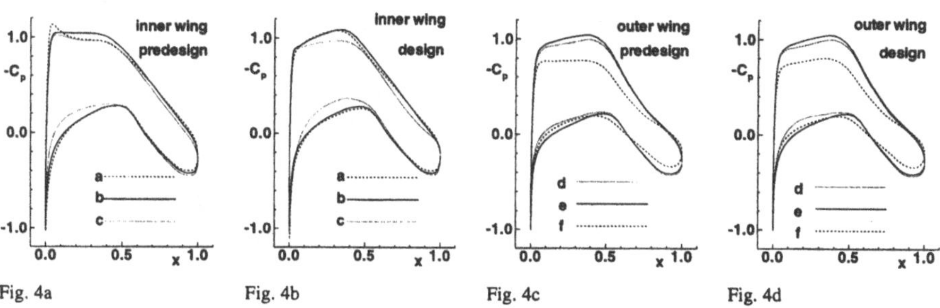


Fig. 4a Fig. 4b Fig. 4c Fig. 4d

Fig. 4 Pressure distributions for climb flight condition ($M_\infty$=0.28, $\alpha$=2.6°) at selected wing sections. Comparison of predesign results with results obtained with the new design.

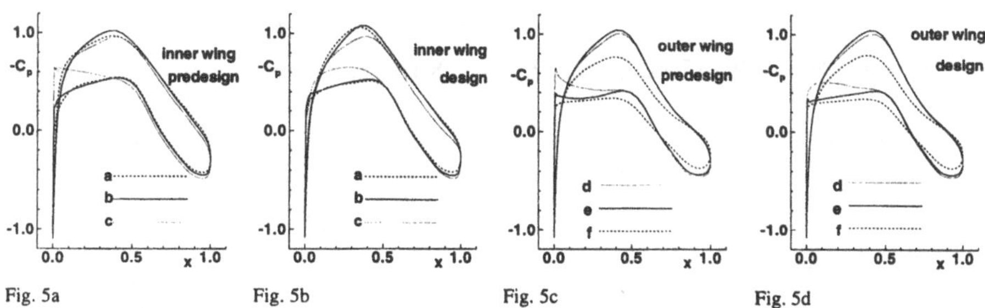


Fig. 5a Fig. 5b Fig. 5c Fig. 5d

Fig. 5 Pressure distributions for maximum cruise flight condition ($M_\infty$=0.54, $\alpha$=-0.5°) at selected wing sections. Comparison of predesign results with results obtained with the new design.

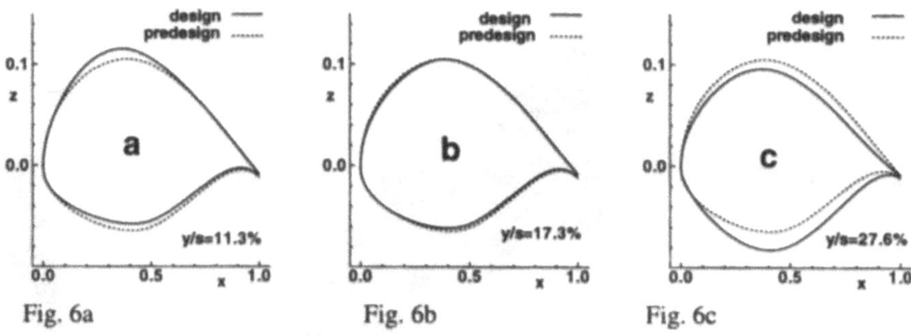

Fig. 6a Fig. 6b Fig. 6c

Fig. 6 Comparison of designed geometry with the (NLF9) airfoil for a section placed close to the body (Fig. 6a), in the middle of inner wing (Fig. 6b) and close to the nacelle (Fig. 6c).

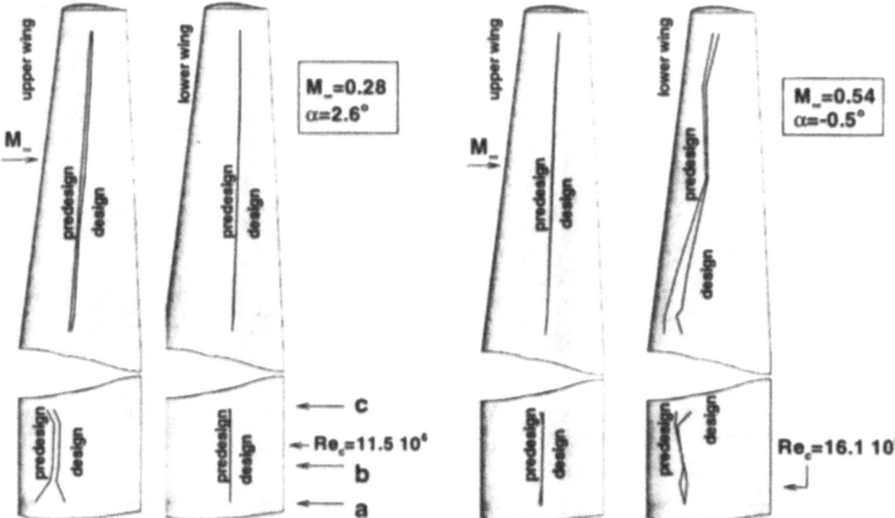

Fig. 7 Transition line position for climb flight. Fig. 8 Transition line position for max. cruise.

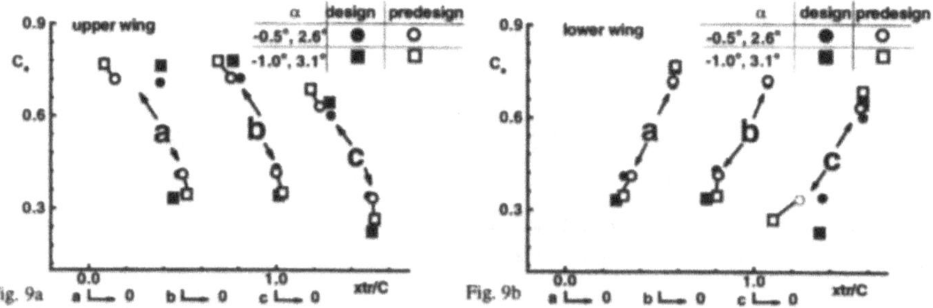

Fig. 9a Fig. 9b

Fig. 9 Local lift values as function of transition position for selected wing sections. Upper (lower) $C_a$ values correspond to $M_\infty=0.28$ ($M_\infty=0.54$), uppermost (lowermost) $C_a$ value corresponds to an design point $\alpha$ changed by $0.5^0$, $(-0.5^o)$.

# Impact of Planform Geometry on Waverider Aerodynamics

D. Strohmeyer, Th. Eggers
DLR, Institute of Design Aerodynamics, Lilienthalplatz 7
D-38108 Braunschweig, Germany

## Summary

In the present paper the influence of the planform shape of Osculating Cones Waveriders on the aerodynamic behavior with respect to the longitudinal motion is discussed. The inviscid flowfield around the configurations is simulated along the complete trajectory of an airbreathing two stage to orbit system using the DLR Euler-/ Navier-Stokes code CEVCATS.

It is found, that the modification of a gothic planform towards combined forebody - delta wing planforms allows a significant improvement of the aerodynamic efficiency L/D in sub- and transonic flow. In addition the longitudinal stability is increased without compromising the favourable high speed qualities. These benefits are partly compensated by an increasing neutral point shift along the trajectory.

## Introduction

The payload transport into the lower earth orbit during the last 3 decades shows a rapid increase [1]. In order to reduce the specific launch costs, to obtain a higher mission flexibility and mission safety, as well as for environmental protection, work on the next generation of space transportation systems is necessary. The investigated projects include single stage to orbit (SSTO) as well as two stage to orbit transportation systems (TSTO). Compared to SSTO vehicles the most important advantage of a TSTO system is a higher payload ratio (percent payload of total launch mass) leading to a lower sensitivity concerning design and mission parameters. Since integrated predesign studies of airbreathing TSTO systems show a high sensitivity with respect to the aerodynamic efficiency L/D (or $C_L/C_D$ respectively) of the first stage, waverider configurations with a high $C_L/C_D$ in the design point as well as under off-design conditions seem to be a realistic alternative to conventional blended body and wing body configurations respectively [2].

Based on integrated predesign studies the DLR-F8 waverider configuration was designed as a promising waverider shape. For the aerodynamic investigation of this configuration detailed numerical simulations along a complete TSTO trajectory [3] as well as wind tunnel experiments were performed [4]. During more detailed predesign studies it was found that the aerodynamic potential of DLR-F8 is not put to full use. For this reason two new waverider configurations with modified planforms and reduced planform area have been designed and aerodynamically investigated.

## Configurations

The design of waverider geometries at DLR Braunschweig is based on the Osculating Cones Concept [5], using the program WIPAR [6]. In order to fulfill the requirements coming from the integrated predesign - such as integration of tanks, engine and other systems - as well as to reduce the base area and the resulting base drag especially in sub- and transonic flow, the WIPAR geometries have to be modified. This is illustrated in Fig. 1a by means of

WR-12-G-FR (WR ... waverider; 12 ... design Mach number, $M_{Des.}$; G ... gothic planform; FR ... upper surface = freestream surface) which was realized with a CAD system and led to the configuration WR-12-G-EXP (EXP ... upper surface = expansion surface). This waverider is also called DLR-F8 since it is built as a wind tunnel model. The mean features of the modification are:

- Introduction of an expansion surface in the wing region, leading to a sharp trailing edge and to the reduction of the base area
- Introduction of straight hinge lines for the integration of aileron and elevator
- Shaping the body in spanwise direction to allow the integration of tanks and systems as well as to guarantee a low base drag

A side effect of the introduced expansion surface is a rigging angle of incidence between wing and body ridge line of approximately $\alpha_{inc.} \approx 5°$, leading to a zero-lift angle of about $\alpha_0 \approx -5°$. As the geometry was also used for a wind tunnel model, the wing tip was cut. Furthermore, inlet, propulsion box and nozzle are not modeled and the propulsive jet is represented by a solid sting with constant cross-section in the aerodynamic analyses. In order to get data for global design work the effect of the propulsion system (inlet, propulsion box and nozzle) on the aerodynamic behavior should be taken into account by a propulsion bookkeeping method.

The detailed integrated predesign studies of DLR-F8 showed that the configuration as the lower stage of a TSTO system fulfills the mission with flight $C_L$ below the $C_L$ for optimum $C_L/C_D$. To improve the mission $C_L/C_D$ two waveriders with about 20% reduced planform area, $S_{Ref.}$, and different aspect ratio, A, were designed and modified to WR-12-GD-EXP-A and WR-25-GD-EXP-A in order to fulfill the boundary conditions coming from the integrated predesign, Fig. 1b.

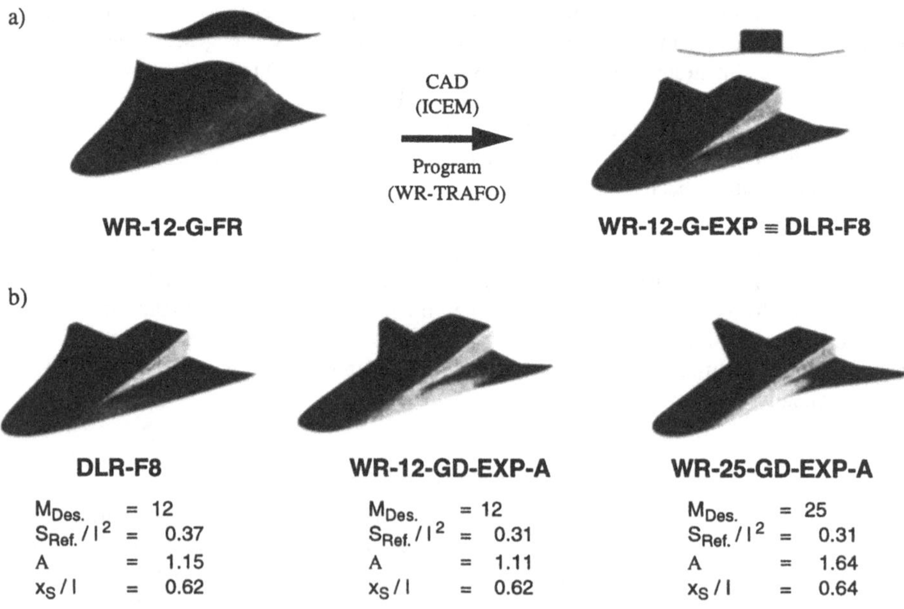


Fig. 1 Waverider modification (a) and aerodynamically investigated waverider configurations (b)

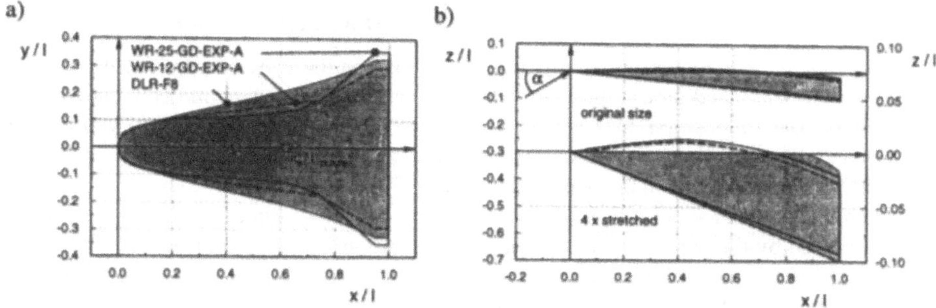


Fig. 2 Comparison of planform (a) and symmetry plane (b) of the investigated configurations

The area reduction was obtained by changing the gothic planform of DLR-F8 into a combined planform, coupling a gothic forebody with a delta wing (-GD- ... combined planform: gothic forebody - delta wing), which resembles the planform of typical wing body configurations. For the geometric modification of both waveriders a numerical program was used instead of the CAD system. This procedure is significantly less costly but leads also to less even surfaces.

In Fig. 2a the three planforms are illustrated. A comparison of WR-12-GD-EXP-A and DLR-F8 shows the reduced planform area, whereas the centre of gravity of the planform area remained the same. Due to the smaller span WR-12-GD-EXP-A has almost the same aspect ratio as DLR-F8, see Fig. 1b. WR-25-GD-EXP-A is characterized by the largest span and an aspect ratio which is almost 50% larger than those of DLR-F8 and WR-12-GD-EXP-A. In addition the center of gravity of its planform area is shifted 2% of the body length closer to the trailing edge. This leads to a greater longitudinal stability in hypersonic flow since under high speed conditions the neutral point coincides approximately with the center of gravity of the planform area.

For an additional increase of the longitudinal stability the usable body volume of WR-12-GD-EXP-A and WR-25-GD-EXP-A was also redistributed, as Fig. 2b illustrates. The thickness of the former body with upper freestream surface was smoothly increased in the nose region and reduced in the tail region, leaving the usable body volume unchanged (-A ... variation with modified distribution of the body volume). This procedure allows a shift of the center of gravity to the nose, increasing the longitudinal stability.

## Numerical Simulations

The numerical simulation of the waverider aerodynamics is based on the solution of the Euler equations using the DLR Euler-/Navier-Stokes code CEVCATS [7], [8]. In sub- and transonic flow a code version with central differencing and artificial dissipative terms was used. In the super- and hypersonic speed regime upwind discretization was applied. For the spatial discretization of the flowfield in sub- and transonic flow a grid with about 660000 grid points was generated, covering a domain of about 5 body lengths, l, around the configuration. A second grid was designed for the simulation of the super- and hypersonic flows in order to follow the bow shock more closely. It consists of about 330000 points. The influence of viscous effects on the aerodynamic efficiency was estimated by adding the skin friction of a flat plate in turbulent flow. The Reynolds numbers, Re, were chosen according to typical trajectories of airbreathing TSTO systems.

## Planform Effects on the Waverider Aerodynamics

Key parameters for the investigation of planform effects on the waverider aerodynamics with respect to the integrated predesign of a TSTO system are the aerodynamic efficiency $C_L/C_D$ and the neutral point position. In the discussion of both parameters for the three waverider configurations special emphasis is laid on the subsonic and lower supersonic flight regime up to $M_\infty = 1.5$ for two reasons: On the one hand the changes of the aerodynamic characteristics of a flight vehicle in higher super- and hypersonic flow are very small since the aerodynamic forces are determined by the pressure side with an almost constant pressure distribution. The reason for this behavior is the limited force maximum on the suction side (vacuum) in contrast to the unlimited maximum force on the pressure side. On the other hand the speed range up to $M_\infty = 1.5$ has a great influence on the mission performance: Mission simulations show that almost 25% of the fuel are consumed in this segment of the mission [9].

The aerodynamic efficiency as a function of the lift coefficient for different Mach numbers is given in Fig. 3a for the three waverider configurations. The corresponding drag polars shows Fig. 3b. As a result of the asymmetry of the configurations with respect to the horizontal plane, the apex of the drag polars is shifted in positive $C_L$- and $C_D$-direction. Neglecting the $C_L$-shift of the curves, which is permissible up to $M_\infty = 1.5$ since the offset is small and changes only little with increasing Mach number, the drag polars can be approximated as:

$$C_D(C_L) = C_{D,\,min.} + \frac{(C_L - C_{L,\,C_{D,\,min.}})^2}{e\;\pi\;A} \approx C_{D,\,min.} + \frac{{C_L}^2}{e\pi A}\;, \tag{1}$$

with $C_{D,min.}$ as the minimum drag coefficient and e as the Oswald factor, describing the deviation of the lift distribution in spanwise direction from the elliptical lift distribution (e = 1 for elliptical lift distribution, e < 1 for deviation from the elliptical one). Since $\pi$ and A in Eq. (1) are constants the shape of the parabolic polar is controled by the Oswald factor. The $C_L/C_D$ maximum of the $C_L/C_D$-$C_L$ curves in Fig. 3a is located at:

$$C_L\left(\left(\frac{C_L}{C_D}\right)_{max.}\right) = \sqrt{C_{D,\,min.}e\pi A}\;, \qquad \text{with} \qquad \left(\frac{C_L}{C_D}\right)_{max.} = \frac{1}{2}\sqrt{\frac{e\pi A}{C_{D,\,min.}}}\;. \tag{2}$$

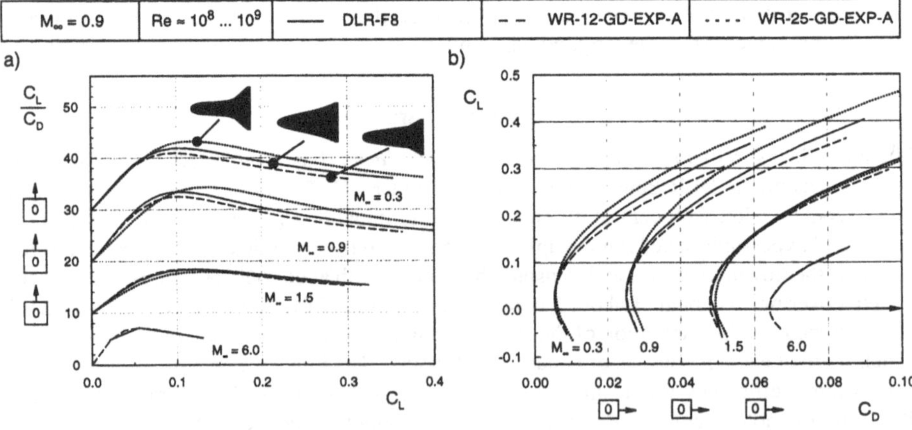


Fig. 3 Influence of the planform on the aerodynamic efficiency (a) and the drag polars (b)

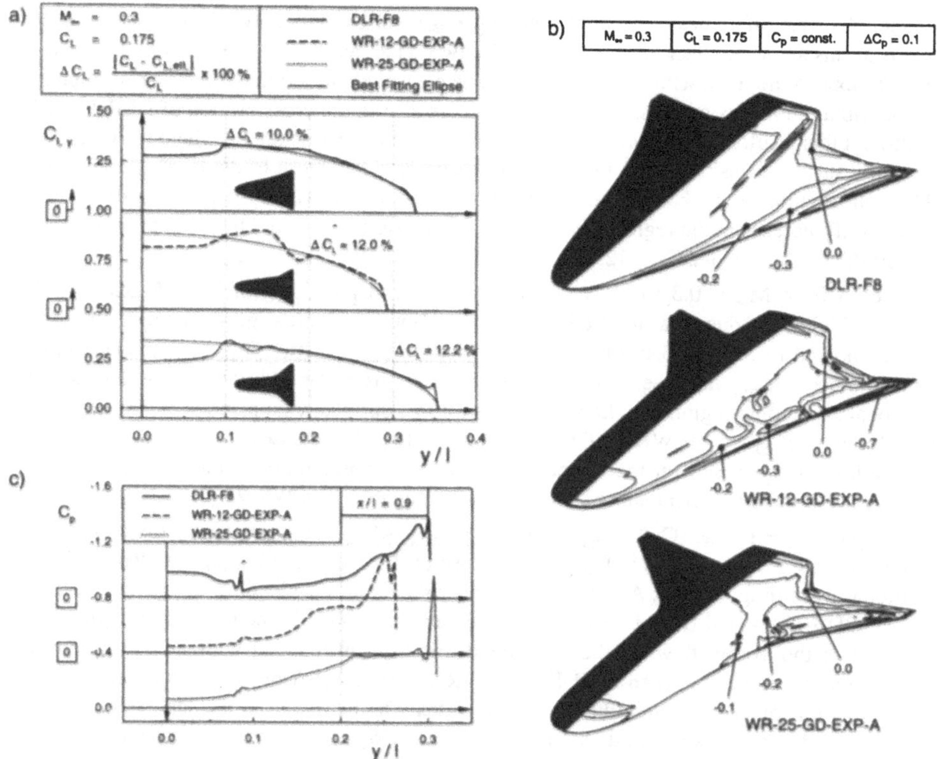


Fig. 4 Influence of the planform on the lift distribution (a) and pressure distribution on the upper surface (b) and in the cross-section x/l = 0.9 (c)

A comparison of the $C_L/C_D$-$C_L$ curves at $M_\infty = 0.3$ shows that WR-25-GD-EXP-A obtains the greatest $(C_L/C_D)_{max.}$ at the highest $C_L$ due to its large aspect ratio, Fig. 3a. Although WR-12-GD-EXP-A and DLR-F8 have almost the same aspect ratio the $(C_L/C_D)_{max.}$ of WR-12-GD-EXP-A is smaller and obtained at a slightly lower $C_L$. The reason for this effect is the less elliptical lift distribution and hence the smaller e of WR-12-GD-EXP-A.

In order to illustrate this phenomenon the lift distribution of the configurations in spanwise direction is shown in Fig. 4a for constant $C_L$, with $C_{l,y} = C_l(y/l) \cdot l(y/l) \cdot 1 / S_{Ref.}$. Here $C_l(y/l)$ is the local lift coefficient of the section y/l and l(y/l) describes the local chord length at y/l. Additionally the best fitting ellipse for each lift distribution as well as the total difference between best fitting ellipse and real lift distribution $\Delta C_L$ is given.

All lift distributions show a lift decrease close to the symmetry plane (y/l ≈ 0.0 ... 0.1) which results from the lower angle of attack of the body compared with the wing, see Fig. 1. This effect is strongest for WR-25-GD-EXP-A on account of its large body part, leading to the largest deviation from the elliptical lift distribution of $\Delta C_L = 12.2\,\%$. In comparison with DLR-F8, WR-12-GD-EXP-A shows an additional deviation from the best fitting ellipse in the region y/l ≈ 0.1 ... 0.2. The reason for this additional lift is illustrated by means of the pressure distribution on the upper surface of all configurations for $C_L = 0.175$, see Fig. 4b, and the corresponding $C_p$ distribution in the cross-section x/l = 0.9 in Fig. 4c. Compared with DLR-F8, where the pressure distribution indicates a single leading edge vortex, the flowfield of WR-12-GD-EXP-A at this angle of attack is characterized by two separated leading edge vortices: A forebody vortex and the delta wing vortex. Both can be clearly identified in the cross-section pressure distribution. A second pressure minimum occurs also in the cross-section of DLR-F8 ($y/l \approx 0.25$) but this results from the swelling of the upper surface in this region where the modified expansion surface transitions into the original waverider shape close to the leading edge. The additional suction force below the forebody vortex of WR-12-GD-EXP-A leads to the observed lift increase in the region $y/l = 0.1 ... 0.2$ (see also Fig. 2a) and hence to the less elliptical lift distribution with $\Delta C_L = 12.0\%$.

Proceeding from $M_\infty = 0.3$ to $M_\infty = 0.9$ in Fig. 3a the $(C_L/C_D)_{max.}$ of WR-25-GD-EXP-A moves to higher $C_L$ due to a more elliptical lift distribution (increasing gradient of the drag polar compared to $M_\infty = 0.3$ in Fig. 3b). However, the value of $(C_L/C_D)_{max.}$ remains almost constant since $C_{D,min.}$ increases, an effect which results from the additional wave drag based on a local supersonic region and the shock on the upper surface. Compared with WR-25-GD-EXP-A the $(C_L/C_D)_{max.}$ of WR-12-GD-EXP-A and DLR-F8 increases a little and moves to a slightly higher $C_L$, based on the increasing e. For both configurations the influence of the supersonic flow and the terminating shock is small compared with WR-25-GD-EXP-A.

Going up to $M_\infty = 1.5$ the $(C_L/C_D)_{max.}$ of WR-25-GD-EXP-A decreases drastically and moves to an even higher $C_L$, as shown in Fig. 3a. The reason is the increased $C_{D,min.}$ on account of the additional wave drag, Fig. 3b. At this Mach number the delta wing leading edge is supersonic. WR-12-GD-EXP-A and DLR-F8 show the same behavior. Since in the supersonic regime the influence of the vortex flow on the pressure distribution is reduced with increasing Mach number, the difference between DLR-F8 and WR-12-GD-EXP-A decreases. Two effects are responsible for this decreasing influence of the vortex. On the one hand the supersonic part of the leading edge grows and avoids the feeding of the downstream vortex with circulation, on the other hand with the existing bow shock new boundary conditions - the oblique shock relations - have to be fulfilled closer to the vortex in comparison with the subsonic case.

For hypersonic flight conditions, $M_\infty = 6.0$, only the drag polars and $C_L/C_D$-$C_L$ curves of DLR-F8 and WR-12-GD-EXP-A were determined. Both $C_L/C_D$-$C_L$ curves are nearly identical, Fig. 3a, the influence of the upper surface of WR-12-GD-EXP-A is almost negligible. This was expected since both configurations have the same body angle and the influence of the suction side on the aerodynamic behavior in hypersonic flow plays a minor role as discussed before.

Compared with $M_\infty = 1.5$ the polar is now symmetric to the abscissa. Every deviation from $\alpha = 0°$ leads to an additional wave drag and decreasing $C_L/C_D$: The numerical simulation shows that increasing the angle of attack leads to a worse $C_L/C_D$ of the dominating pressure side, decreasing the angle of attack improves the lower side $C_L/C_D$ which is overcompensated by the additional pressure force on the upper side. The resulting location of the drag polar leads to a further reduction of $(C_L/C_D)_{max.}$ and the shift to a significantly smaller $C_L$.

The analysis of the aerodynamic efficiency shows that both configurations WR-12-GD-EXP-A as well as WR-25-GD-EXP-A promise a mission with higher $C_L/C_D$ compared to DLR-F8. In super- and hypersonic flow where all configurations show the same $C_L/C_D$ characteristic the reduced reference area of WR-12-GD-EXP-A and WR-25-GD-EXP-A leads to a mission $C_L$ closer to $(C_L/C_D)_{max.}$ In sub- and transonic flow WR-25-GD-EXP-A is the most promising configuration due to the highest $(C_L/C_D)_{max.}$, an advantage which is slightly moderated by the shift of $(C_L/C_D)_{max.}$ to higher $C_L$. Even though the $(C_L/C_D)_{max.}$ of WR-12-GD-EXP-A is below that of DLR-F8, the $C_L$ location is almost the same and the smaller reference area overcompensates the $(C_L/C_D)_{max.}$ disadvantage.

The second aspect concerning the planform effects on the waverider aerodynamics is the influence on the neutral point position. The neutral point is defined as the point where, in any case of change of the steady state of a flight vehicle, the additional aerodynamic force attacks. In Fig. 5a the behavior of the neutral point as a function of the Mach number is illustrated for the three waverider configurations at $\alpha = 1.25°$ and compared with the neutral point regime of typical blended body and wing body configurations. All waveriders show the similar characteristic which is discussed comprehensively in [3]:

- Slight rearward movement in subsonic flow
- Rearward movement in subcritical transonic flow of approximately 6% of the body length with a distinct noseward drop close to $M_\infty = 1.0$ due to the shock influence on the upper surface pressure distribution
- Almost constant position in the high speed regime according to the increasing influence of the lower surface pressure distribution on the lift of the vehicle

But the more the planform is changed from the gothic shape (DLR-F8) to the combined type (WR-12-GD-EXP-A, WR-25-GD-EXP-A) the more rearward is the neutral point located. This effect is based on the different aspect ratios and hence different lift curve slopes of the slender forebody and the delta wing. The larger this difference is, the larger the lift curve slope of the delta wing is compared to that of the forebody, shifting the neutral point rearwards.

This behavior is illustrated in Fig. 5b for $M_\infty = 0.3$ showing the lift distribution for the two angles of attack $\alpha = 0.0°$ and $\alpha = 2.5°$ in x/l-direction, with $C_{l,x} = C_l(x/l) \cdot s(x/l) \cdot 1 / S_{Ref.}$. Here $C_l(x/l)$ is the local lift coefficient of the cross-section x/l and s(x/l) stands for the local half span at x/l. The corresponding difference lift distribution in Fig. 5c shows clearly the influence of the different aspect ratios of forebody and delta wing of WR-12-GD-EXP-A and WR-25-GD-EXP-A on the lift distribution, shifting the neutral point, which can be interpreted as the x/l-position of the center of gravity of the area below the $\Delta C_L$ curve, to the tail.

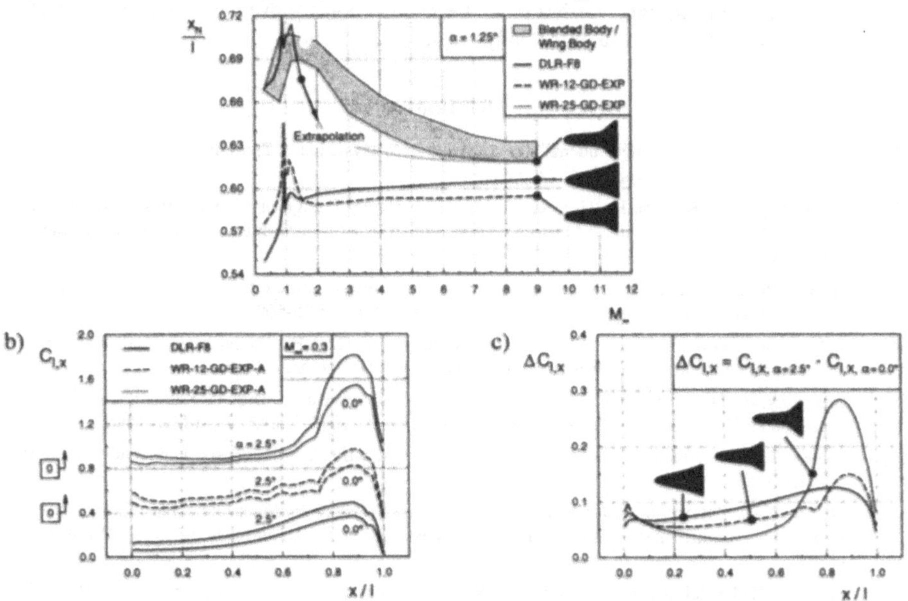


Fig. 5 Influence of the waverider planform on the neutral point position (a), lift- (b) and difference lift distribution (c) of the investigated waverider configurations

Having in mind that at hypersonic speeds the neutral point coincidences approximately with the center of gravity of the planform of a flight vehicle, the resulting movement of the neutral point from transonic to hypersonic flow increases the more the gothic planform is modified towards the combined one which may lead to higher trim losses.

In this context it should be mentioned that for all three configurations the neutral point at hypersonic speeds is located slightly upstream the center of gravity of the planform area, see also Fig. 1b. This effect is primarily due to the introduced expansion surface: In contrast to a freestream upper surface, the expansion surface, which is introduced up from $x/l \approx 0.3$, provides for small angles of attack no additional contribution to the aerodynamic forces in the case of a disturbance of the angle of attack. The vacuum remains unchanged, leading to the neutral point shift in upstream direction compared with a configuration with freestream upper surface. For WR-12-GD-EXP-A and WR-25-GD-EXP-A this effect is increased by a change of the compression flow in the nose region based on the upper surface modification in this region as illustrated in Fig. 2b.

## Conclusions

In the present study an analysis of the planform effects on the aerodynamic characteristics of Osculating Cones Waveriders was presented. In the aerodynamic analysis the DLR Euler-/Navier-Stokes code CEVCATS was used for the simulation of the aerodynamic behavior in inviscid flow with respect to the longitudinal motion along the complete trajectory of a typical TSTO mission.

The aerodynamic investigation shows that the modification of a gothic waverider planform towards a combined planform with gothic forebody and delta wing can be used for planform area reduction and neutral point shift almost without an influence on the aerodynamic performance in super- and hypersonic flow. Limiting factor of the modification is the overall stability characteristic from sub- up to hypersonic flow: The more the gothic planform is changed towards a wing body design, the larger is the neutral point movement along the trajectory, increasing from 6.5% up to 10% of the body length.

## Bibliography


[1] Koelle, D.E.; "Entwicklungstendenzen bei Raumtransportsystemen", *Zeitschrift für Flugwissenschaften und Weltraumforschung*, Vol. 16, 1992, pp. 67-76.

[2] Bardenhagen, A.; Kossira, H.; Heinze, W.; "Interdisciplinary Design of Modern Hypersonic Waveriders Using the Integrated Program PrADO-Hy", ICAS Paper 94-1.4.1, 19th International Council of the Aeronautical Sciences, 1994.

[3] Eggers, Th.; Strohmeyer, D.; Nickel, H.; Radespiel, R.; "Aerodynamic Off-Design Behavior of Integrated Waveriders from Take-Off up to Hypersonic Flight", AIAA Paper 95-6091, 1995.

[4] Seltsam, M.; Strohmeyer, D.; "Windkanalmessungen am DLR-F8 Wellenreiter-Windkanalmodell im Transsonischen Windkanal Göttingen (TWG)", DLR-IB 29112-96A05, 1996.

[5] Sobieczky, H.; Doughtery, F.C.; Jones, K.D.; "Hypersonic Waverider Design for Given Shock Waves", 1st International Hypersonic Waverider Symposium, Oct. 17-19, 1990, University of Maryland.

[6] Center, K.B.; "Interactive Waverider Design and Optimization", PhD Thesis, University of Colorado, Boulder, 1993.

[7] Rossow, C.-C.; "Berechnung von Strömungsfeldern durch Lösung der Euler-Gleichungen mit einer erweiterten Finite-Volumen Diskretisierungsmethode", Dissertation TU Braunschweig, 1988, DLR-FB 89-38, 1989.

[8] Kroll, N.; Radespiel, R.; "An Improved Flux Vector Split Discretization Scheme for Viscous Flows", DLR-FB 93-53, 1993.

[9] Kossira, H.; Bardenhagen, A.; Heinze, W.; "Investigation on the Potential of Hypersonic Waveriders with the Integrated Aircraft Design Program PrADO-HY", AIAA Paper 93-5098, 1993.

# Visualization of the Spatial-Temporal Instability Wave Development in a Laminar Boundary Layer by Means of a Heated PVDF Sensor Array

D. Sturzebecher, W. Nitsche
Technische Universität Berlin, Institut für Luft- und Raumfahrt, F2
Marchstraße 12, 10587 Berlin, Germany

## Summary

The spatial and temporal development of artificially excited instability waves in a laminar boundary layer on a flat plate is investigated by means of the PVDF foil measuring technique. Compared to a simple exploitation of the piezoelectric effect, the signal to noise ratio can be improved considerably when deliberately exploiting the pyroelectric effect by means of an integrated surface heating device. In this context, the present paper describes simultaneously recorded measurements, employing a total number of 142 wall sensors, allowing to detect traveling instability waves of small amplitude. In order to illustrate the flow phenomena investigated, the obtained results are visualized in space and time.

## 1 Nomenclature

| | | |
|---|---|---|
| RMS | [mV] | root mean square |
| x, y, z | [mm] | coordinates |
| F | [-] | frequency parameter |
| f | [Hz] | frequency |
| $\lambda$ | [mm] | wavelength |
| $U_\infty$ | [m/s] | freestream velocity |
| Re | [-] | Reynolds number |
| $Re_{\delta 1}$ | [-] | displacement thickness Reynolds number |
| u' | [%] , [m/s] | streamwise velocity fluctuation |
| Tu | [%] | turbulence intensity |
| Q | [C] | electric charge |
| $\Delta T$ | [K] | temperature gradient |
| p | [Pa] | pressure |
| $S_{Sensor}$ | [$m^2$] | surface area of single sensor |
| $d_T$ | [$C/(m^2K)$] | pyroelectric coefficient |
| $d_{33}$ | [C/N] | piezoelectric coefficient |
| $\lambda^*$ | [%] | POD eigenvalue |
| PVDF | | Polyvinylidenfluorid (polarized plastic film) |
| CTA | | constant temperature anemometer |

## 2 Introduction

Periodical pressure oscillations on vanes, location of transition, amplification rates of boundary layer instabilities as well as shock positions on airfoils have been successfully determined in the past by means of multi-sensor piezofoil arrays (e.g. [1],[2],[3]). In all these cases, the special merits of this measuring technique are that it can be easily operated and applied without interference, even on rather strongly curved surfaces. In the field of transition research in particular, PVDF sensors have to meet exceptional high standards with regard to spatial and temporal resolution (i.e. the detection of wave trains) as well as regarding the detection of small perturbations occurring within the instability region. Because of these requirements, the piezofoil technique has been further developed selectively. Apart from its piezoelectric properties, the material used for the foil (PVDF-Polyvinylidenfluorid) is distinguished by excellent pyroelectric properties as well. For this reason and by deliberately exploiting the pyroelectric effect, a modified sensor technique can be used for such measurements. Operated in this way, foil sensors measure wall temperature or heat flux fluctuations instead of pressure fluctuations. Hence, with regard to quality, the signals can be compared to those of surface hot film. In comparison with piezoelectric sensors, the signal to noise ratio is approximately doubled with regard to a difference in temperature between wall and fluid per Kelvin. This facilitates the detection even of rather small perturbations like travelling waves in laminar boundary layers to quite a considerable extent.

## 3 PVDF Foil Measuring Technique

The linear relation between mechanical deformation and electric polarization of a piezoelectrical material is called piezoelectric effect. This interaction is mainly used in the field of pressure measurements [1],[3]. Assuming both a piezo element, which is electrically and mechanically unloaded, and an isothermic or adiabatic change of state [4], this relation can be described by

$$Q = d_{33} \cdot p \cdot S_{Sensor} \, . \tag{1}$$

The electric polarization of a piezo element also depends on temperature. Thus, the following equation results when assuming that both, the electric field strength and the mechanical tension are constant, but the temperature is changed:

$$Q = d_T \cdot \Delta T \cdot S_{Sensor} \, . \tag{2}$$

The material of the foil employed in the present investigation consists of Polyvinylidenfluorid (PVDF), having a semi-crystalline structure which is distinguished by piezoelectric and pyroelectric properties. The basic set-up of a single PVDF sensor is shown in Figure 3-1. The small electric charges, actively produced by the sensor material, can be measured by means of charge amplifiers converting the electric charge to an output voltage, which optionally can be postamplified by a power amplifier.

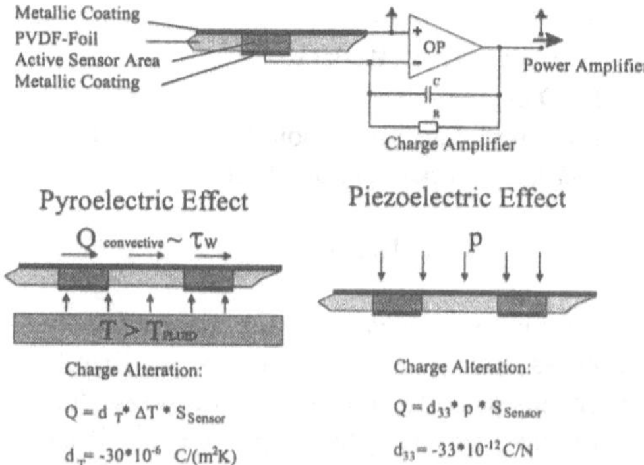


*Figure 3-1: PVDF sensor and measuring principle*

A preceding investigation [5] proved that piezo foil sensors show an improved signal to noise ratio when operated under a slight temperature gradient between flow and test body. This temperature gradient can be generated either by means of an integrated heating layer under the PVDF sensors or by slight flow heating and / or structure cooling. Alterations in charge then result mainly from the pyroelectric properties of the foil and therefore, this measuring technique can be considered as a „low temperature hot film". Within the area of excited laminar boundary layers in particular, the exploitation of the comparatively large pyroelectric effect provides signals which can more easily be correlated in a spatial way. Using only the piezoelectric properties implies that clearly smaller signal amplitudes are measure, which, among other things, can be influenced by noise and vibration. This is shown in Figure 3-2, illustrating a sensor pre-test with an artificial excitation (f = 200 Hz) in a laminar flat plate boundary layer with and without heating.

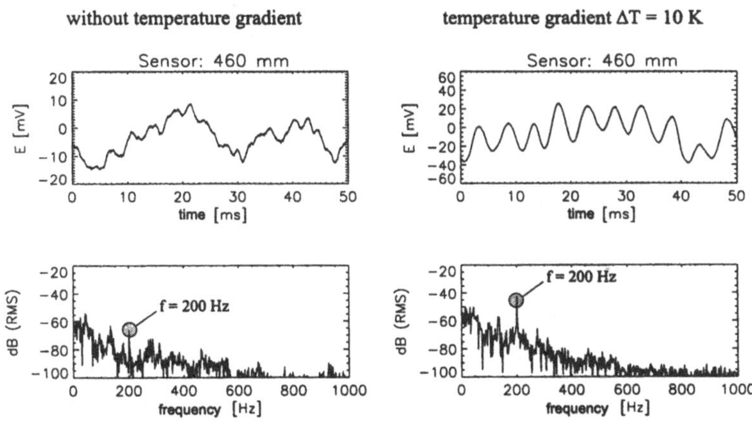


*Figure 3-2: Improving the signal to noise ratio in case of artificial excitation*

# 4 Experimental Set-Up

All measurements were carried out in the ILR boundary layer wind tunnel [6]. The test section used has a height of 0.4 m and a width of 0.6 m. The maximum velocity is 25m/s and the freestream turbulence intensity is approximately Tu = 0,1 %. For the experiments a flat plate (with a length of 1,2 m, including a trailing-edge flap of 0,155m) was used, which allowed a flush fitting of a measuring insert (of approximately 200 x 700 mm). A heating sublayer was integrated into the insert. In the present investigation a temperature gradient of ΔT = 8 K between flow and plate was chosen. On this insert, a PVDF sensor array consisting of 142 single sensors was applied. The experimental set-up and the sensor arrangement on the flat plate is shown in Figure 4-1. In order to artificially excite the laminar boundary layer, a slot source (x = 200mm) and a point source (x = 290mm) had been integrated, which allowed to induce periodical velocity fluctuations perpendicular to the wall. Figure 4-2 shows the stability diagram of the laminar boundary layer containing the parameters of the conducted experiments along with the different perturbation sources.

The data acquisition device consists of 3 separate measuring units which are synchronized and contain 48 channels each. By means of these single measuring units, 48 separate sensors can be operated on charge amplifiers respectively, which are followed by power amplifiers. A 48-channel multiplexer operated with 1 MHz successively connects each channel to a 12-bit-A/D-converter. The A/D converter is interfaced by a bi-directional serial fiber optical transmission to a PC-slotcard. This card has a RAM-buffer, which can be read by the processor e.g. to store the binary data on hard disk.

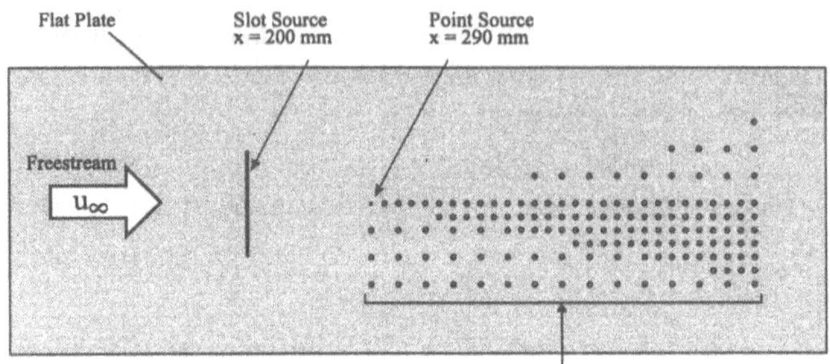


*Figure 4-1: Experimental set-up and arrangement of the sensors*

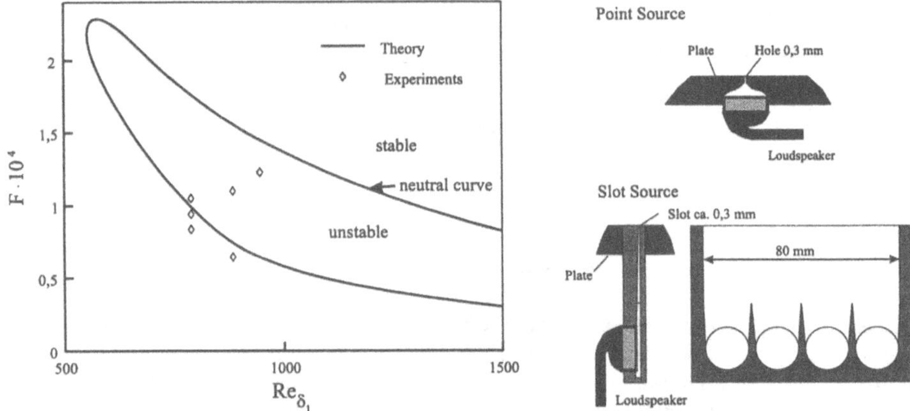


*Figure 4-2: Diagram of stability and set-up of the point and slot sources*

# 5 Results

For an excitation frequency of f = 225 Hz and flow velocity of $U_\infty$ = 15,1 m/s some selected time traces, as well as their corresponding spectra, which are measured downstream of the slot source, are show in Figure 5-1. The downstream travelling waves, the increase of perturbation amplitudes, the growing of spectra in a non-linear way up to the late transitional and turbulent flow state, respectively, can be observed. Two-dimensional distributions of the RMS-values (Figure 5-2) show typical transition locations (dark areas). When the laminar boundary layer is excited by the point source, a wedge-shaped structure develops downstream. Due to the slot excitation, a straight front of transition forms in the inner region. However, another wedge-shaped formation is clearly recognizable at the edges. When exciting at the most unstable frequency, transition is located in the center of the array. In order to take a better employment of the sensor array, artificial perturbations are excited with a smaller frequency, close to the lower branch of the neutral curve. Thus, the area to be measured can be used more efficiently, because transition can be shifted downstream, while greater waves lengths can be employed as well.

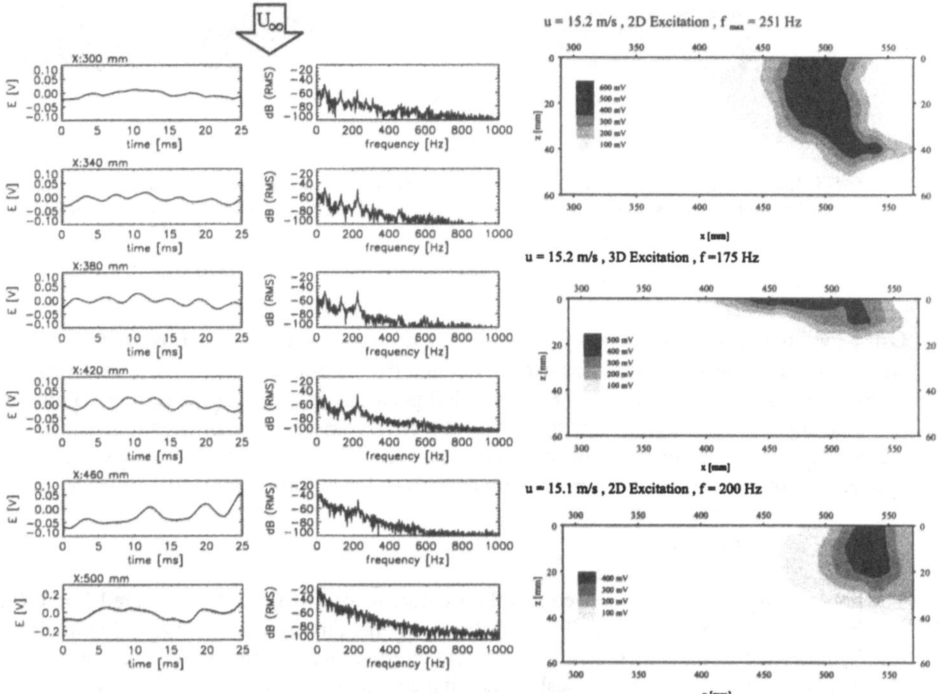


*Figure 5-1: Selected time traces and power spectra downstream a slot source*

*Figure 5-2: Selected RMS-value distribution*

The snapshots of the sensor voltage (Figure 5-3 , Figure 5-4) are presented in a time-shifted way. Both for the point and slot excitation, the snapshots show the wave development with a typical wave length. It is clearly recognizable that the waves move obliquely in the case of the excitation by the point source ($\lambda$ = 30 mm). The two-dimensional waves excited by the slot source can be clearly identified ($\lambda$ = 25 mm). Apart from that, it can be recognized that spanwise waves are occurring further downstream.

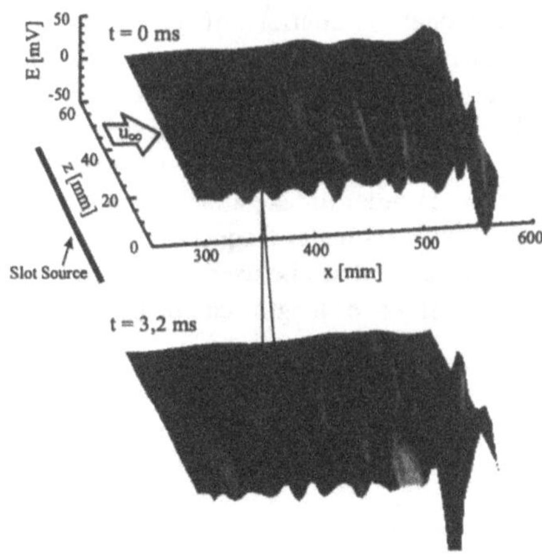


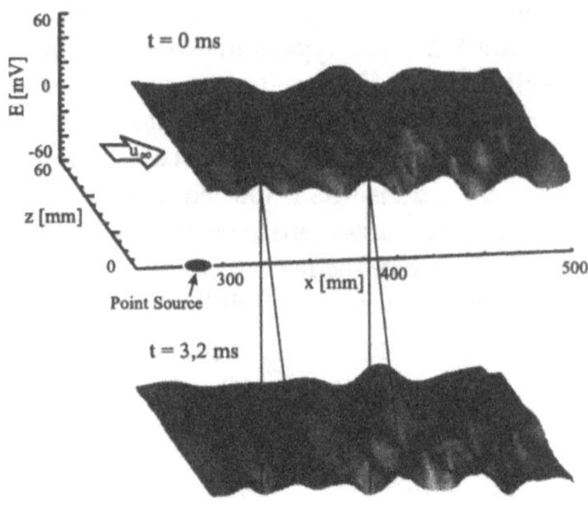


*Figure 5-3: Spatial and temporal development of a perturbation downstream a slot source, $U_\infty$=15,1m/s, f=200 Hz*

*Figure 5-4: Spatial and temporal development of a perturbation downstream a point source, $U_\infty$=15,1m/s, f=175 Hz*

Using the same test set-up, measurements were carried out by means of a traversed single hot-wire probe. The measurements was carried out at a freestream velocity of $U_\infty$ = 15,1 m/s with an excitation frequency of f = 250 Hz. For 2300 positions in a certain measuring volume the hot-wire signals were digitized, triggered by the excitation sine signal, i.e. the measurement was started at the zero crossing of the excitation signal. The traversed volume begins 10 mm upstream of the slot, located at x = 200 mm and z = -40 to 40 mm, and covers the area:

x = 190 to 270 mm,
y = 0,2 to 4,5 mm,
z = -15 to 55 mm.

The hot-wire signals were converted in velocity fluctuations (u'), bandpass filtered (f = 120 to 600 Hz) and ensemble-averaged. The distribution of the „momentary" velocity fluctuation in the volume were evaluated both in a spatial and temporal way (animation of the „snapshots") as isosurfaces. In Figure 5-5, a snapshot showing a cross sectional view of the boundary layer downstream (z = 5mm) is depicted as well as its top view of a slice in spanwise and streamwise direction (y = 0,5 mm). It is obvious that the wave length of approximately 20 mm corresponds with the excitation frequency and the group velocity. The phase shift between the region close to and far from the wall is also recognizable.

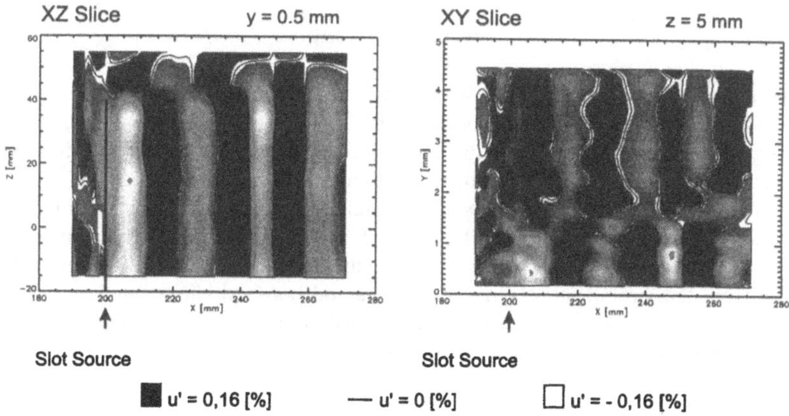


*Figure 5-5: „Snapshot" of averaged hot wire time traces downstream of the slot source, $U_\infty$=15,1m/s, f=250 Hz*

Separating the modes of the proper orthogonal decomposition (POD) [7] is a further means of evaluation. However, the POD modes provide only information on the signal coherence here, not on the Fourier modes. The resulting decomposition, which is mode-related, is likewise presented as a snapshot (Figure 5-6). It is obvious that, in the case of a slot excitation, the modes 1 and 2 describe the phenomenon of traveling waves and amplification rather well. Further modes, which accumulate less energy, mainly consist of stochastic parts (Figure 5-7).

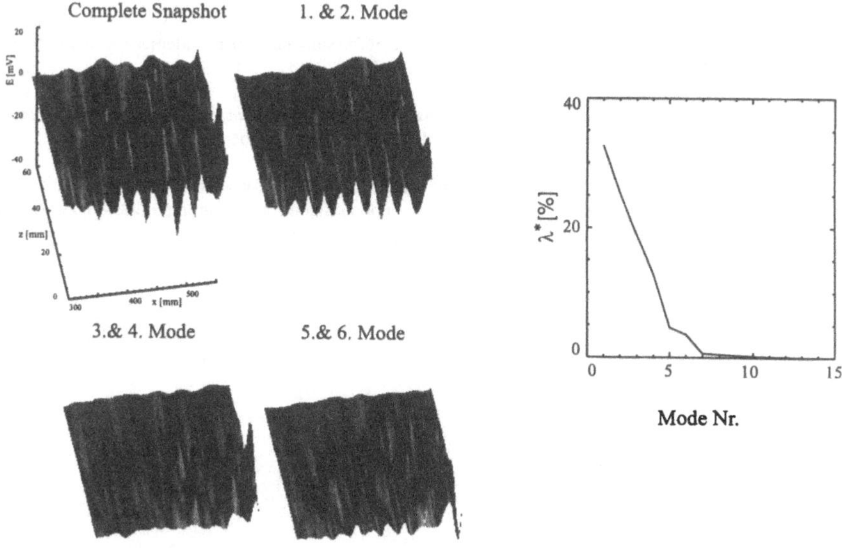


*Figure 5-6: POD snapshot*

*Figure 5-7: POD mode distribution*

# 6 Conclusions

The spatial and temporal development of artificially induced perturbations in a laminar boundary layer on a flat plate was successfully visualized by means of the PVDF foil measuring technique. Compared to the piezoelectric effect, the signal to noise ratio has considerably been improved by exploiting the pyroelectric effect by means of an integrated heating sublayer. Two-dimensional measurements, employing 142 sensors simultaneously, made it possible to measure the development of instability waves downstream of artificial disturbances (point source, slot source). Here, the structure of the travelling waves was identified from the simultaneously measured data and non-periodic as well as single events have been visualized. Advanced methods of evaluation allow to visualize large amounts of two-dimensional and spatial data and illustrate the flow phenomena.

# 7 References

[1] Nitsche, W.; Mirow, P.; Szodruch, J.: Piezo-electric Foils as a Means of Sensing Unsteady Surface Forces, Experiments in Fluids, 7 (1989), pp. 111-118.

[2] Nitsche, W.; Swoboda, M.; Mirow, P.: Shock detection by means of piezofoils, Z. Flugwiss. Weltraumforsch., 15 (1991) 223-226, Springer-Verlag 1991.

[3] Brauckhoff, R.; Nitsche, W.: Flächige Messung periodisch-instationärer Oberflächendrücke an einem hydraulischen Strömungsgitter mit Hilfe piezoelektrischer Sensorarrays, Forschung im Ingenieurwesen - Engineering Research Bd. 60 (1994) Nr. 11/12, 323-331.

[4] Tichý, J. ; Gautschi, G.: Piezoelektrische Meßtechnik, Springer-Verlag, Berlin Heidelberg New York,1980.

[5] Baumann, M.; Nitsche, W.: Aktive Grenzschichtbeeinflussung laminar-turbulenter Profilströmungen, ILR Berlin, 1995.

[6] Sturzebecher, D.: Untersuchungen zum laminar-turbulenten Strömungsumschlag an einer ebenen Platte mit Temperaturgradient zwischen Struktur und Strömung, Diplomarbeit ,TU Berlin, 1996.

[7] Hilberg, D.; Lazik, W.; Fiedler, H.E.: The Application of classical POD and Snapshot POD in a Turbulent Shear Layer with Periodic Structures, Applied Scientific Research 53: 283-290, 1994 , Kluwer Academic Publishers.

# IN-FLIGHT RESEARCH ON BOUNDARY LAYER TRANSITION - WORKS OF THE DFG - UNIVERSITY RESEARCH GROUP -

J. Suttan [1], M. Baumann [1], S. Fühling [2], S. Becker [3], H. Lienhart [3], C. Stemmer [4]

[1] Institut für Luft- und Raumfahrt, TU Berlin
Sekr. F2, Marchstraße 14, D-10587 Berlin
[2] Aerodynamisches Institut, RWTH Aachen,
Wüllnerstraße zw. 5 u. 7, D-52062 Aachen
[3] Lehrstuhl für Strömungsmechanik, Univ. Erlangen-Nürnberg,
Cauerstraße 4, D-91058 Erlangen
[4] Institut für Aerodynamik und Gasdynamik, Univ. Stuttgart,
Pfaffenwaldring 21, D-70569 Stuttgart

## Summary

The paper reports on the recent research activities of a joint universitary research group, consisting of scientists from five German universities. The main objective is the in-flight investigation of laminar-turbulent boundary layer by means of different measurement techniques as well as numerical simulations. The measurement techniques employed for the flight experiments are surface hot-film arrays, piezofoil arrays and Laser-Doppler anemometry. Additionally, direct numerical simulations concerning the spatial and temporal development of transition on the wing glove are carried out. An overwiev of the employed measuring equipment and numerical methods is given, followed by a discussion of obtained results.

## Introduction

With respect to detailed investigations concerning the mechanisms which lead to transition and finally to turbulent flow, a group of scientists from five German universities conducted a joint research work to investigate boundary layer transition downstream from a harmonic point source on an unswept laminar wing. The experiments are carried out under realistic flight conditions to obtain full scale Reynolds numbers and to avoid influences from wind tunnel flow, e.g. noise or wind tunnel turbulence. In a previous research project [1], different measurement techniques based on low-disturbant respectively disturbance free sensor types, like hot-film arrays and multisensor piezofoil arrays as well as LDA-probes were developed and measurements concerning the detection of transition location were conducted successfully. Based upon this, the main focus of the ongoing research activities is the investigation of the development of artificially induced disturbances in the laminar boundary layer downstream from a harmonic point source. The experiments are carried out on different wing gloves mounted on a powered glider Grob G109b provided by the TH Darmstadt. This aircraft is equipped with a flight attitude measurement system [2], which enables the experimentally working groups to carry out their measurements at a predefined test condition, necessary to compare the experimentally obtained results among each other as well as with the numerical results. In the following sections the results from the different working groups are presented.

## Contribution of the TU Berlin

The measurement of characteristic surface force fluctuations in the boundary layer on the wing glove is the main objective of the TU Berlin in the framework of the research project. The experiments are carried out by means of multisensor piezofoil arrays, being capable to measure dynamic forces on nearly arbitrary surfaces without influencing the flow. To allow measurements with high spatial and temporal resolution, stipulating the usage of large sensor numbers and a high sampling frequency, a compact, lightweight digital measurement system was developed. It allows a modular extension from 48 up to 192 sensors. Each single module features 48 channels with a sampling frequency of 20 kHz per channel. To make maximum use of the A/D-converter bandwidth and to avoid overmodulation of the power amplifiers, each amplifier is coupled with a programmable potentiometer, allowing computer-controlled autoranging. The complete system is mounted inside the wing glove and is connected to a standard PC in the cockpit by electronically insensitive electro-optical transmission lines.

Results from recent flight tests with a sensor array of 46 individual sensors, arranged in one row downstream from the point source with a sensor spacing of 6 mm, clearly show the amplification of artificially excited frequencies in the boundary layer. Relative amplification rates ($n_{rel.}=\ln(A_{(x)}/A_{(x)min.})$) show a linear amplification of the excitation frequency (in this case f=800 Hz) up to x/c=0.36 (Fig. 1), where a strong nonlinear development starts. This onset of nonlinear development is visible from the kink in the amplification curve and corresponds with the sharp increase in the RMS-value distribution. The spectral analysis of this test case (Fig. 2) shows the growth of the excitation frequency $f_1$=800 Hz, accompanied by the growth of a frequency $f_2 \approx 900$ Hz, which could be observed in all measurements, including the test case without artificial excitation. Since both frequencies are very close to the natural instability frequency, it is presumed that an interaction of waves with these frequencies leads to the emergence of subharmonic ($f_2-f_1=\Delta f=100$ Hz, $2\Delta f=200$ Hz, $f_1-2\Delta f=600$ Hz, $f_1-\Delta f=700$ Hz etc.) and higherharmonic frequencies ($2f_1=1600$ Hz) that are amplified and thus trigger the nonlinear development of transition and finally the turbulent breakdown of boundary layer flow [3]. A timestep diagram (Fig. 3), where the relative momentary sensor amplitudes ($A_{(x,t)}/A_{(x,t)max.}$) for f=800 Hz are depicted over x/c-position and time, shows that wave propagation in the laminar region occurs with about 30 % of the outer flow speed and accelerates in the turbulent region to about 65 % of the outer flow speed.

## Contribution of the RWTH Aachen

For the in-flight experiments in the framework of the joint universitary research project, a large multisensor hotfilm-array (1100x380 $mm^2$) with 168 discrete sensor elements was employed. Sensors are arranged asymmetrically downstream from the harmonic point source, located at a chordwise position of x/c=0.27, covering a chordwise region up to x/c=0.5 and a wedge angle in one spanwise direction of approximately 9°. To increase temporal and spatial resolution, dimensions of sensors, consisting of a 0.1 µm thick nickel film, have been reduced to 0.925x0.09 $mm^2$. Sensor spacing was chosen to 2.75 mm in spanwise and 5.5 mm in streamwise direction (Fig. 4). The hotfilm elements are operated by a completely computer-controlled Constant-Current-Anemometer (CCA) [1,4] and signals are stored on a 96-channel transient-recorder. Since not all sensors can be operated simultaneously, with regards to phase relationship between sensor signals, the operating signal of the point source was sampled, too. Data processing by means of spectral analysis [5] allows for determination of phase velocity and wavelength of traveling waves from phase shifts of sensor signals as well as for determination of phase shifts.

Excitation frequency during the flight tests was chosen to 900 Hz, while the amplitude of the point source was varied in fine steps. Natural transition as well as onset of transition using small excitation amplitudes, detected by a strong increase of RMS-values of sensor signals in the transition regime, is located at $x/c \approx 46\%$. Fig. 5 shows power spectra of centerline sensor records at different chordwise positions downstream from the harmonic point source. Directly downstream from the point source the excited frequency is visible within the power spectrum of the first sensor. Traveling downstream, this discrete wave seems to be dampened initially. However, further downstream and increasing in flow direction, an amplification of this discrete wave is visible from a marked peak at this frequency. The last sensor of this plot lies within the transition regime, where discrete waves are owerwhelmed by broadband disturbances. An interesting subject in context with the traveling waves is the spatial coherence between neighbouring sensors. Fig. 6 compares obtained coherence functions for different chordwise positions, the first closely behind the point source ($x/c=0.32$) and the second one ($x/c=0.42$) located a short distance upstream from onset of transition. At a chordwise position of $x/c=0.32$ signals are coherent only at the induced frequency (900 Hz). Boundary layer receptivity for the excited wave train is indicated by the marked peak (coherence function nearly 1) at this frequency. Further downstream, at $x/c=0.42$, coherence of sensor signals is observed for a wide frequency range. This turns out to be the frequency range of the most amplified waves travelling downstream.

## Contribution of the University of Erlangen-Nürnberg

Within the scope of the university cooperation project, specially designed LDA systems were developed, built, and tested at the Dept. of Fluid Mechanics of the University of Erlangen-Nürnberg (LSTM). In a first approach, a LDA probe using an integrated semiconductor laser diode light source and a semiconductor photodetector in direct backscatter was adopted [1,6]. The experiences gained with this system can be summarized as follows: local velocity measurements in the flow close to a flying aircraft using LDA were feasible, however the data rate in tests in free atmosphere was quite low and not sufficient for spectral analytical studies of Tollmien-Schlichting instability. Measurements of the natural scattering particles diameter distribution during the flight tests using a cascade impactor device showed that this resulted from the low concentration of aerosols with diameters exceeding $d_P \approx 1\mu m$, needed to give detectable signals. From the steep increase in concentration for particles of smaller size it was obvious that the data rate could be improved considerably if one succeeded in detecting signals from particles of diameters less than $d_P \approx 0.5\text{-}1\ \mu m$. This led to the development of an improved LDA-system with emphasis posed on maximum signal power from very small scattering particles. To this end the new system was designed featuring a laser diode pumped frequency-doubled Nd-YAG laser, which provides for higher light power with smaller wave length compared to the semiconductor laser, while having still limited electrical power consumption. The optical arrangement was completely modified to a forward scattering system. Despite the losses due to the fibre links needed, since the size of the laser no longer allows an integrated construction, this new system yields a gain in Doppler signal power of two to three orders of magnitude. Additional improvements were achieved by adopting specially designed electronics for the detection and evaluation of the LDA-signals. The severe spatial restrictions imposed by the research aircraft and the demand for a forward scattering arrangement resulted in a quite unconventional probe design which is shown in Fig. 7. It uses the narrow gap between wing and wing glove for the optical components. The beam path is diverted twice by mirrors for both the transmitting and the receiving path. Thus all the optics apart from the upper mirrors are placed underneath the wing glove surface and only the latter are protruding. Therefore the distortion of the flow induced by the measuring system is minimal. The requirements for laser beam intersection angle and sufficient receiving aperture area are met by employing optical components cut into narrow slices of 6x40 $mm^2$. The size of the measuring control volume created is about 35μm in diameter. The LDA probe is connected to the laser and the photodetector that are placed in the cockpit of the aircraft by glass fibre cables respectively and is mounted on a traversing mechanism that allows for automated measurements of boundary layer profiles. For completion of the measuring system, laser, traversing controller, photomultiplier, power supplies, and signal acquisition/processing unit were installed on the instrumentation platform in the cockpit behind the pilot seats.

In the actual flight tests, the LDA-system proved to be very reliable, stable in alignment, and insensitive to the harsh environment. The data rate achieved in clear atmosphere was in the order of 150 - 200 Hz, in hazes of growing clouds a data rate up to several kHz was observed. In Fig. 8 typical results for laminar and turbulent boundary layer velocity profiles are shown.

## Contribution of the University of Stuttgart

The investigations carried out by the University of Stuttgart are aimed at DNS-studies of the streamwise disturbance evolution in the laminar boundary layer on the upper side of the glove airfoil at flight conditions. A well-tested finite-difference/spectral method for the solution of the incompressible 3-D Navier-Stokes equations is used, being capable of handling strongly decelerated boundary layers. This code is now applied to a base flow with streamwise pressure distribution representing a section of the airfoil boundary layer at flight conditions. For a selected flight test case, the streamwise evolution of a three-dimensional wave train emanating from a harmonic point source was simulated to allow comparison with corresponding flight experiments. A finite rectangular box is selected to represent a certain region of a boundary-layer flow on a flat plate. For the calculation of the steady base flow, the external velocity distribution $u_e(x)$ imposed at the free-stream boundary is taken from the potential flow results [7]. In a second step, disturbances are introduced by timewise periodic blowing and suction through a narrow spanwise disturbance strip at the wall [8]. The spatial evolution of these disturbances is calculated by solving the unsteady three-dimensional Navier-Stokes equations in vorticity-transport formulation for the vorticity components and three Poisson equations for the three velocity components. The numerical method is discussed in detail by Kloker et al. [9,12]. The numerical method compares with results from linear (spatial) stability theory, secondary stability theory, amplification rates, eigenfunctions and experiments [10-12]. For selected flight test cases (spatial) linear stability calculations have been performed. They were carried out for 2-D and 3-D disturbances to solve the sixth-order Orr-Sommerfeld-Squire equations [13]. As an example for the results, the stability diagrams for 2-D waves are shown in Fig. 9 for two cases at $Re=2.8x10^6$ ($U_\infty=32$ m/s, c=1.3 m) for $\alpha= 4.5°$ and $\alpha=5.71°$, respectively. The results of the stability calculations serve for choosing the streamwise position of the point source for the flight experiments, as well as for comparisons with the DNS results. Some results of DNS-studies for the spatial disturbance evolution in the airfoil boundary layer at $U_\infty=32$ m/s, $\alpha=5.71°$ with a global Reynolds number of $Re=1.4x10^5$ (c'=0.065 m, so x is related by a factor of 20 to the surface coordinate s/c) are discussed. A detailed simulation was performed where the boundary layer was excited by a small-amplitude harmonic point source disturbance. Different combinations of 2-D- and pairs of 3-D waves with initial amplitudes of approximately $u'/U_\infty \approx 10^{-4}$ are generated. The streamwise amplitude growth for different fourier modes (n,k), with disturbance frequency $n_F$ and spanwise wave-number $k_y$, is plotted in Fig. 10. Waves with oblique wave angles larger than 45° (modes k≥16) are initially dampened, however, downstream of x≈8.7 all 3-D waves show strong nonlinear amplitude growth. At the onset of this nonlinear evolution, the 2-D wave has an amplitude of approximately $u'/U_\infty=0.2\%$. The instantaneous wall vorticity distribution is plotted in Fig. 11. Initially, lines of constant vorticity $\omega_z'$ at the wall visualise the well-known linear evolution of a wave train, where the curved wave fronts indicate the superposition of unstable 3-D waves with the 2-D disturbance. Development of wall vorticity further downstream shows nonlinear interactions of 2-D- and 3-D waves and requires further studies.

## Conclusions

During the course of a joint universitary research project, aimed at experimental and numerical investigations on boundary layer transition, encouraging results were obtained from all applied measuring techniques as well as from the direct numerical simulation. The presented results by the TU Berlin clearly showed the ability of the measurement system to measure the temporal as well as the spatial distribution of highly dynamic surface forces in transitional boundary layers. Data related to the shear stress fluctuations, measured by the RWTH Aachen at a large number of positions over the surface provided informations from which the growth and development of the wave train could be followed as it travels downstream. Due to the application of spectral analysis for data processing, phase velocity and wave length of the travelling wave were obtained. The University of Erlangen developed a miniaturised LDA-probe featuring a forward scattering arrangement for obtaining higher data rates under flight conditions. First experiments with the new probe have proven the ability of measuring mean velocities and turbulence intensities inside the boundary layer over the wing. The results obtained by the University of Stuttgart show that spatial DNS can be applied to investigate the evolution of disturbances in a boundary layer with varying pressure gradient corresponding to an airfoil flow. When the ratio of boundary layer thickness to surface curvature is small, as within the chord region considered here, the flat plate assumption is well justified. Results of such simulations will allow direct comparison with the experiments in flight. Further activities of the research partners will concentrate on measurements with further improved spatial resolution as well as on comparative analysis of the obtained results.


## Acknowledgements

The authors wish to acknowledge the financial support for their work by the German Research Foundation (DFG). Additional appreciation goes to the heads and members of all working groups for their input to this work and for their technical advise.


## References


[1] Ewald, B., Durst, F., Krause, E., Nitsche, W.: In-flight measuring techniques for laminar wing development, Z. Flugwiss. Weltraumforsch., vol. 17 (1993), pp. 294-310.

[2] Erb, P., Ewald, B.: Flight experiment guidance technique for research on transition with Grob G109b aircraft of the Technische Hochschule Darmstadt, Presentation held at the 10th DGLR-Fach-Symposium „Strömungen mit Ablösung", Braunschweig, November 1996, to be published in „Notes on Numerical Methods", spring 1997.

[3] Kachanov, Yu. S.: On the resonant nature of the breakdown of a laminar boundary layer, J. Fluid Mech. (1987), vol. 184, pp. 43-74.

[4] Kornberger, M.: Multisensor-Heißfilmtechnik zur Transitionserkennung im Windkanal - und Flugversuch. Dissertation, RWTH Aachen, Verlag Shaker, 1992.

[5] Romano, G.P.:Analysis of two-point velocity measurements in near-wall flows, Experiments in Fluids, vol. 20 (1995), pp. 68-83, 1995.

[6] Durst, F., Lienhart, H., Müller, R.: Application of a semiconductor LDA for inflight measurements, Proc. 6th Int. Symp. on Appl. Laser-Anemometrie to Fluid Mech., Lisbon, 1992.

[7] Eppler, R., Somers, D.: Computer Program for Design and Analysis of Low Speed Airfoils, NASA T. M. 80210, (1980).

[8] Konzelmann, U., Fasel, H.: Numerical Simulation of a Three-dimensional Wave Packet in a Growing Flat-Plate Boundary Layer, in: Proc. Conf. on Boundary Layer Transition and Control, April 8-12, 1991, Peterhouse College, Cambridge, UK.

[9] Kloker, M., Konzelmann, U., Fasel, H.: Outflow boundary conditions for Spatial Navier-Stokes Simulations of Transition Boundary Layers, AIAA J., vol. 31 (1993), pp. 620-628.

[10] Fasel, H., Rist, U., Konzelmann, U.: Numerical Investigation of the Three-Dimensional Development in Boundary-Layer Transition, AIAA J., vol. 28 (1990), pp. 29-37.

[11] Rist, U., Fasel, H.: Direct Numerical Simulation of Controlled Transition in a Flat-Plate Boundary, J. Fluid Mech. (1995), vol. 298 , pp. 211-248.

[12] Kloker, M.: Direkte numerische Simulation des laminar-turbulenten Strömungsumschlags in einer stark verzögerten Grenzschicht, Dissertation, Univ. Stuttgart, (1991).

[13] Cebeci, T., Bradshaw, P.: Momentum Transfer in Boundary Layers, McGraw-Hill (1977).

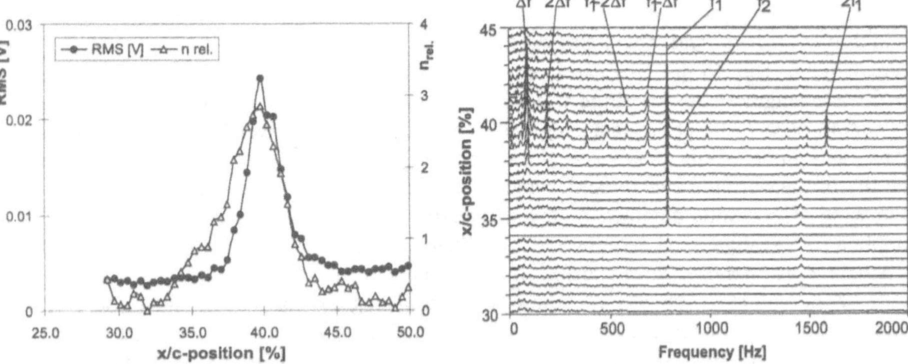


Fig. 1: RMS-values/relative amplification rates for f=800 Hz, piezoarray flight test

Fig. 2: Power spectra from sensor signals, artificial excitation with f=800 Hz, piezoarray flight test

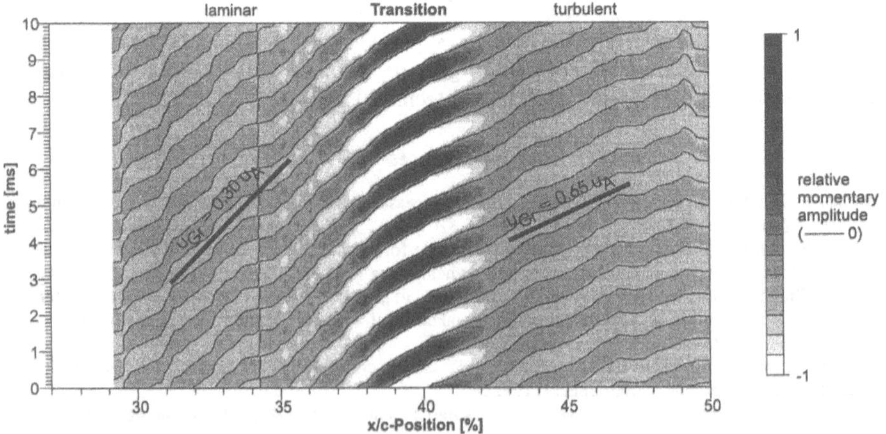


Fig. 3: Timestep diagram for excited frequency f=800 Hz, piezoarray flight test

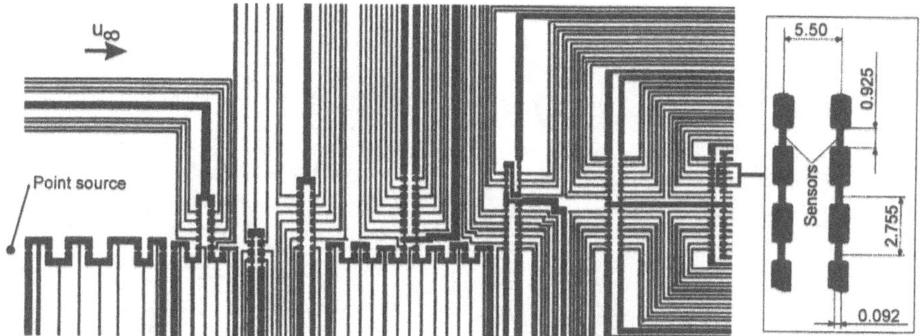

Fig. 4: Layout of multisensor hotfilm array for flight test

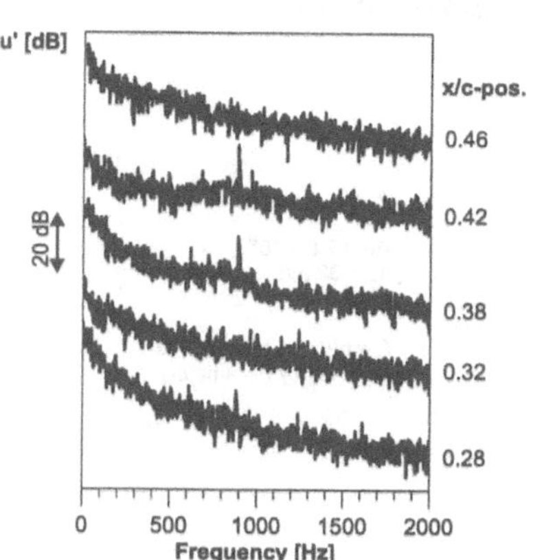

Fig. 5: Power spectra of centerline hotfilm sensors

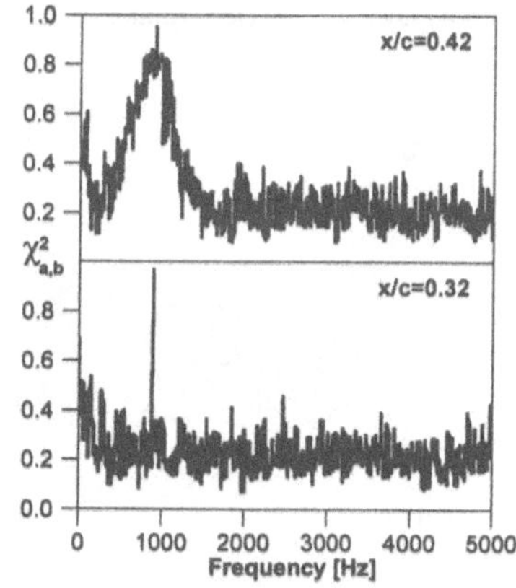

Fig. 6: Coherence of hotfilm sensor signals

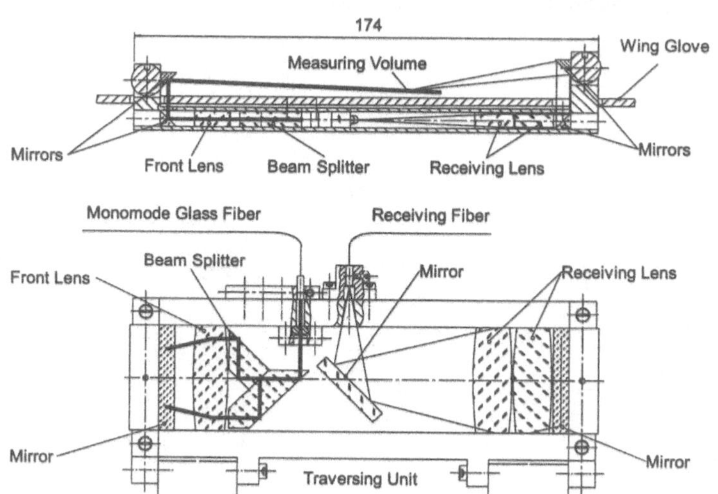

Fig. 7: Cross sectional view of Nd-YAG LDA-probe for in-flight measurements

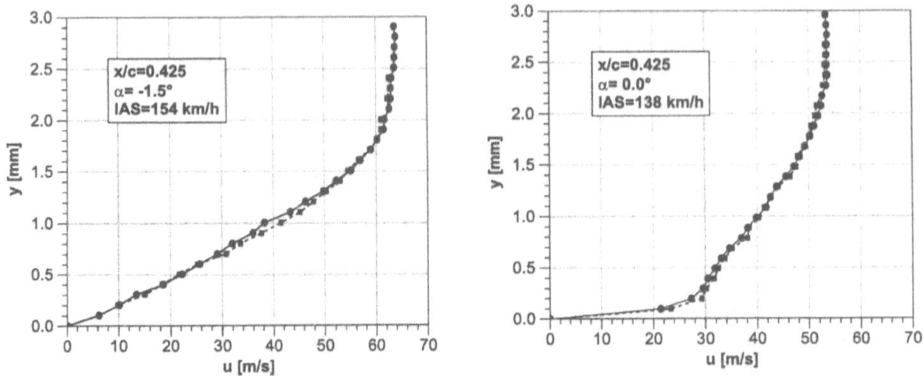


Fig. 8: Boundary layer profiles (left: laminar, right: turbulent) from in-flight LDA-measurements

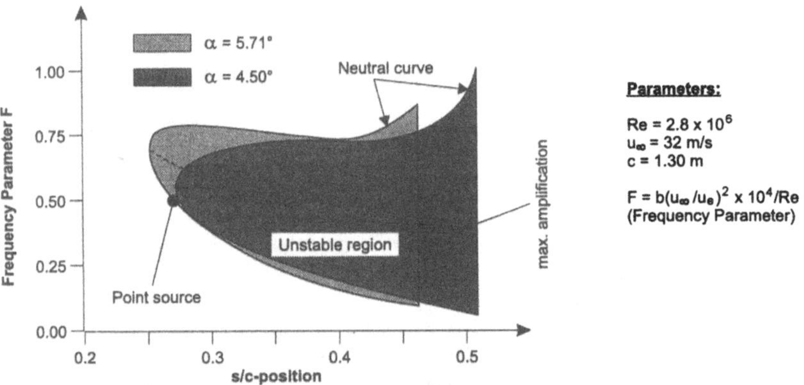


Fig. 9: Stability diagram for two different flight test conditions

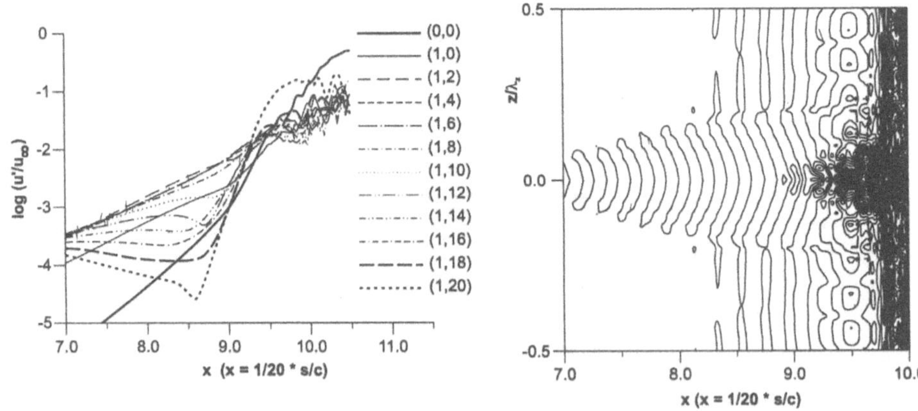


Fig. 10: Amplitude growth of fourier modes (n,k) for harmonic point source excitation

Fig. 11: Instantaneous wall vorticity distribution $\omega'_z$ for harmonic point source excitation

# On subcritical breakdown to turbulence in swept wing attachment line boundary layer flows

V. Theofilis & U. Dallmann

DLR, Institute for Fluid Mechanics, Bunsenstraße 10, 37073 Göttingen, Germany

## SUMMARY

A spectral collocation algorithm is used to study the secondary instability of the boundary layer formed on the windward surface of swept wings. Numerical validation of our results is provided by reference to the well-understood secondary stability of the classic Blasius boundary layer; spectral convergence is exhibited. The secondary eigenspectrum of the flow in question is found to comprise one real (phase-locked with the primary wave) mode which can go unstable depending on the value of the Reynolds number. Additionally, pairs of complex conjugate eigenvalues are delivered by the algorithm that correspond to decaying waves and, as such, are less interesting from a stability analysis point of view.
In the region where subcritical turbulence has been observed in either experiments or direct numerical simulations our model predicts that small-amplitude transiently decaying primary perturbations suffice to act as catalyst for strongly growing secondary instabilities. Consistent with the assumptions of the Floquet theory utilised, we predict very good agreement between the experimental result on the breakdown Reynolds number and that calculated using the secondary instability model proposed herein.

## INTRODUCTION

Flow on the windward face of swept wings is of great technological interest, not least with respect to maintenance of natural laminar flow on the wing surface [1,20]. At the leading edge region an attachment line is formed, schematically depicted in figure 1, separating flow on either side of the wing. Maintenance of laminar flow in this boundary layer is a necessary condition for laminar flow to be established further downstream on the wing surface. Therefore we address the existing unacceptable difference between experimental observation and available theoretical predictions of transition onset by a new approach.

One of the best investigated scenaria of transition to turbulence follows the growth of infinitesimal (linear) instability waves in laminar flow; when these primary waves grow to levels that can interact with the base flow secondary instability sets in and breakdown to turbulence occurs on a fast convective time-scale. While this scenario is fairly well understood both experimentally [17] and theoretically [2,5,13] in the classic Blasius boundary layer, little is known on the respective attachment line problem.

## WHAT IS KNOWN AND THE ISSUE CONSIDERED PRESENTLY

Active practitioners in the field of attachment line instability often quote the early work of Gray [9] as the origin of interest in this problem. Actually experimental studies of instability in this boundary layer may be traced back to the Horten brothers' efforts to build a 'tailless-plane' [14]. Much of their knowledge was obtained under not entirely scientific conditions; nevertheless it was clear from the early works that an instability mechanism, distinct from that of cross-flow instability, is plaguing maintenance of laminar flow on a swept wing. The experiments of Gaster [7] in Europe and Pfenninger [18] and co-workers in the US clearly indicated the role of a small-amplitude wave disturbance that was naturally linked with the well understood Tollmien-Schlichting instability of the flat-plate boundary layer.

Poll [19] was the first to undertake theoretical investigation into this problem. Although his critical Reynolds number result was updated by subsequent theoretical work, the principle of a linear instability wavefront propagating along the attachment line has been central to all linear inverstigations of this flow that followed, such as the receptivity work of Floryan and Dallmann [6]. Better agreement with experiment regarding the linear critical Reynolds number value was obtained by Hall et al. [11] who utilised the ansatz originally proposed by Görtler[8] and Hämmerlin [10] in order to arrive rationally at a system governing small perturbations in the neighbourhood of the leading edge. Theofilis [22] extended the approach of Hall et al. by considering the initial-boundary value problem (IBVP) for small-amplitude disturbances in its linear form. While there is little experimental evidence on the relevance of the GH model in the nonlinear regime, Hall and Malik [12] proceeded to study the nonlinear version of the IBVP and predicted that nonlinear equilibrium solutions can be subcritically unstable starting for Reynolds number values larger than Re=535. This numerical result was considered an improvement in the agreement of the GH linear theory with experiment; however, Spalart [21], Jiménez and co-workers [4,15], and Theofilis [23] have been unable to verify the existence of subcritically unstable nonlinear equilibria, in contrast to Joslin [16] who also reported their existence.

This remaining open question is of fairly academic interest compared with the crucial issue of turbulence at the attachment line boundary layer at Reynolds numbers as low as Re=235. The latter phenomenon is clearly beyond the reach of linear primary theory or weakly nonlinear analysis. Interestingly, subcritical turbulence was observed already in the first direct numerical simulation (DNS) of this flow due to Spalart[21], as well as the subsequent attachment line DNS of Corral and Jiménez[4]. Spalart's DNS is important on a number of counts. First, it bridged the gap between the ad-hoc use of the GH model for primary parturbations and the small-amplitude disturbance structure observed in experiment or DNS. Second, it showed that traditional methods of control of linear instability in the boundary layer, such as suction, may have little impact on the growth of disturbances once the latter have been permitted to reach nonlinear levels. Interestingly, Spalart treated the chordwise direction in a way which permits capturing the physics of the flow in this direction, introducing what is now commonly referred to in DNS terminology as a buffer-zone. Joslin[16], on the other hand, used periodic boundary conditions in his DNS of the attachment-line boundary layer and reported that his three-dimensional disturbance results differ from those of the (two-dimensional) GH eigenproblem by only a few percent. In line with the earlier two-dimensional IBVP results of Theofilis and Jiménez et al., this latter finding suggests that the GH perturbations determine the linear instability behaviour, and this result sets the scene for the approach we have taken in what follows.

The novelty in the present approach lies in the physical modelling of the secondary instability. While the chordwise dependence of the basic flow and the primary perturbations may suggest that secondary disturbances should be seeked that are also linearly dependent on the chordwise coordinate, it is immediately clear that introduction of such an ansatz into the governing equations prohibits the admittance of wave-like disturbances propagating in the chordwise direction with wavefronts parallel to the attachment line. Such waves have been clearly identified in experiments at subcritical conditions during the systematic studies of Görtler vortices in this problem by Bippes [3]. If, on the other hand, one restricts the secondary study on the plane defined by the attachment line and the wall normal as far as the basic flow and the primary perturbations are concerned there is no conceptual or mathematical restriction to introduce chordwise periodic disturbances as a model for secondary instability. The response of this basic flow model, which consists of small-amplitude Görtler-Hämmerlin disturbances superimposed upon the generalised Hiemenz boundary layer may then be monitored using classic Floquet theory along the lines developed for the Blasius boundary layer [13].

It might be argued that this isolation of the attachment line plane from its surroundings is only introduced for mathematical convenience. However, the recent three-dimensional DNS of Joslin [16] has utilised periodic boundary conditions in the chordwise direction in order for the available resources to be devoted to a spatial treatment of the spanwise direction. The results of such a model were in very good agreement with those of the previous two-dimensional investigations, indicating that the treatment of the chordwise direction may not be as significant as that along the span; in other words, the behaviour of the mathematically elliptic leading-edge region may be dictated by the disturbance flowfield at the attachment line itself. This being the case, the first candidate to examine from the point of view of secondary instability is indeed the chordwise periodic waves considered herein.

From a numerical point of view, the building blocks of the algorithm are well-established for the study of hydrodynamic stability problems and are described in detail in [25] where results on the high-Reynolds number regime were presented; we chose to monitor this flow regime first in view of the analytical merging of the full eigenvalue spectrum of the attachment line boundary layer into that of the Blasius flow [24] and the ample theoretical results available on the secondary instability of the latter problem. By contrast, here we focus on the behaviour of secondary perturbations at the physically easier attainable (linearly) neutral and subcritical conditions.

We monitor the full spectrum of eigenvalues delivered by a global solution of the temporal secondary eigenproblem. The reason for selecting the resource-intensive global approach is that when one starts the study of a new problem an estimate for the seeked eigenvalue(s) is not available. This results in the likelihood of failure of a local solver to converge. Even if, fortuitously, an eigenvalue is found and subsequently followed by the local search procedure a local search offers no guarantee that no other unstable eigenmode exists. The boundary conditions in the far-field are straightforward for this incompressible problem, namely vanishing of perturbations, such that the disturbance field merges smoothly into the undisturbed flow. Should one attempt the compressible problem, however, care must be taken in designing the far-field boundary conditions so that they permit the appearance of the analoga of Mack modes in this flow as well.

## NUMERICAL RESULTS

Figure 2 depicts the full spectrum of secondary eigenvalues in the subcritical region. At the highest value of Reynolds number considered, Re=500, linear theory predicts that the least stable mode is damped at a damping rate which increases as the Reynolds number decreases. The presence of the primary mode at a small amplitude, on the other hand, suffices to trigger off secondary waves as shown in figure 2. Of particular interest is the real mode, singled out and indicated by a circle. This is the most unstable of all modes in the spectrum at all Reynolds numbers and chordwise wavenumbers examined. It is to be seen that at Re=500 this mode is unstable ($\sigma>0$) while its growth rate decreases progressively as the Reynolds number decreases.

Special mention should be made of the role of the magnitude of the primary amplitude wave in the present instability. In all calculations performed this was typically taken between A=1% and A=2% , never exceeding A=3% for secondary growth to be obtained. This is of particular significance since the IBVP numerical solutions of Theofilis [22], amongst others, have demonstrated (at supercritical, neutral or subcritical conditions) that white noise initially imposed on this flow at arbitrary amplitudes eventually develops into the GH primary disturbance. It is thus physically plausible that the low primary disturbance amplitudes presently utilised can easily be attained if a general disturbance of arbitrary size enters the attachment line boundary layer. Interestingly, all experimental investigations to-date have clearly indicated that an increasingly large amplitude is necessary for subcritical turbulence to be observed as the Reynolds number decreases.

Of prime importance is also the effect of the chordwise wavenumber value on the secondary growth rate result. Should the present model have delivered very small values (compared with the unstable primary wavenumber) to be the most unstable secondary waves then little credibility should be attached to a model based on chordwise periodic waves of long wavelengths to be responsible for instability, in view of the strong pressure gradient in the chordwise direction. It turns out, however, that the most unstable secondary wavenumber range is of the same order of magnitude as that of the (spanwise-developing) primary wavenumber. As a matter of fact secondary growth extends to quite high secondary wavenumber values, much as the case is in the Blasius boundary layer, implying that a whole range of short wavelength chordwise periodic waves can become unstable due to the mechanism proposed herein. This point is graphically depicted in figure 3.

For the sake of comparison with experiment, when available, we present in figure 4 the spatial structure of secondary eigenfunctions at primary neutral conditions. It should be noted, however, that no experiment is known to the authors to have been performed with the aim of recovering even the primary perturbations. Finally, in figure 5 we consider the maximally amplified mode of secondary instability which is found to be the subharmonic mechanism. Secondary growth rates at the maximally amplified value of chordwise wavenumber as a function of the primary wave amplitude A are plotted and it may be seen that the highest value of A considered, which is consistent with the assumptions of the theory utilised, delivers a (secondary) critical Re=234, which is very close to the observed experimental value for subcritical turbulence, Re=240. Higher amplitudes have also been examined; however, amplitudes as high as A=10%, which clearly violate the assumptions of linearity inherent in our model, have resulted in finite critical Reynolds number values of the order Re=200. More important than the precise value of the secondary critical Reynolds number in our view is the fact that the entire linearly subcritical domain may be destabilised by secondary mechanisms triggered by small-amplitude primary disturbances. This we propose to be one of the reasons accounting for the experimentally observed subcritical turbulence.

## CONCLUSIONS

At neutral primary conditions and values of the superimposed primary disturbance amplitude A of O(1%) the swept attachment line boundary layer is demonstrated to support instability waves that grow according to secondary theory. At subcritical Reynolds numbers the present model predicts that transiently decaying primary instabilities suffice to act as catalyst for strongly growing subcritical secondary waves. In line with the assumptions of the linear model utilised it is demonstrated that the amplitude of the primary wave need not exceed A=2% for growing secondary waves to be obtained. At conditions maximising secondary growth a secondary critical Reynolds number Re=234 is calculated for the first time, in very good agreement with the experimental value of attachment line transitional Re=240.

## REFERENCES


1. Arnal, D. 1993 Boundary layer transition: Predictions based on linear theory. In *AGARD Rep*. **793**, pp. 2-1 - 2-63.
2. Bertolotti, F. P. 1991. Linear and Nonlinear Stability of Boundary Layers with Streamwise Varying Properties. Ph. D. Thesis, The Ohio State University, Columbus, Ohio.
3. Bippes, H. 1989. Instability features appearing on swept wing configurations. *IUTAM Laminar-Turbulent Transition Symposium*, D. Arnal and R. Michel (eds.), Toulouse, pp. 419 - 430.
4. Corral, R. & Jiménez, J. 1994 Direct numerical determination of the minimum bypass Reynolds number in boundary layers. *AGARD CP*-**551** pp. 19-1 - 19-10.
5. Fischer, T. M. & Dallmann, U. 1991. Primary and secondary stability analysis of a three-dimensional boundary-layer flow. *Phys. Fluids* A **3**, pp. 2378 - 2391.
6. Floryan, J. M. & Dallmann, U. 1990 Flow over a leading edge with distributed roughness *J.Fluid Mech*., **216**, pp. 629 - 656.
7. Gaster, M. 1965 A simple device for preventing turbulent contamination on swept leading edges. *J. R. Aero Soc*. **69**, pp. 788 - 789.
8. Görtler, H. 1955 Dreidimensionale Instabilitaet der ebenen Staupunktstroemung gegenueber wirbelartigen Stoerungen. In *50 Jahre Grenzschichtforschung* (eds. H. Goertler & W. Tollmien) pp. 304 - 314. Vieweg und Sohn.
9. Gray, W. E. 1952 The effect of wing sweep on laminar flow. RAE TM Aero 255.
10. Hämmerlin, G. 1955. Zur Instabilitaetstheorie der ebenen Staupunktstroemung. In *50 Jahre Grenzschichtforschung* (eds. Goertler & W. Tollmien). pp. 315 - 327. Vieweg und Sohn.
11. Hall, P., Malik, M. R. & Poll, D. I. A. 1984. On the Stability of an Infinite Swept Attachment-Line Boundary Layer. *Proc. R. Soc. Lond*. A **395**, pp. 229 - 245.
12. Hall, P. & Malik, M. R. 1986 On the instability of a three-dimensional attachment-line boundary layer: weakly nonlinear theory and a numerical approach. *J. Fluid Mech*. **163**, pp. 257 - 282.
13. Herbert, Th. 1988. Secondary instability of boundary Layers. *Ann.Rev.Fluid Mech*. **20**, pp. 487 - 526.
14. Horten, R. & Selinger, P. F. 1983 Nurfluegel - Die Geschichte der Horten-Flugzeuge 1933-1960. H. Weishaupt.
15. Jiménez, J., Martel, C., Agui, J. C. & Zufiria, J. A. 1990 Direct numerical simulation of transition in the incompressible leading edge boundary layer. Techn. Note ETSIA MF-903.
16. Joslin, R. D. 1995 Direct simulation of evolution and control of three-dimensional instabilities in attachment-line boundary layers. *J. Fluid Mech*. **291**, pp. 369 - 392.
17. Kachanov, Yu. S. & Levchenko, V. Ya. 1984. The resonant interaction of disturbances at laminar-turbulent transition in a boundary layer. *J. Fluid Mech*. **138**, pp. 209 - 247.
18. Pfenninger, W. & Bacon, J. W. 1969 Amplified laminar boundary layer oscillations and transition at the front attachment line of a 45 degree swept flat-nosed wing with and without suction. In *Viscous Drag Reduction* (ed.\C. S. Wells), pp. 85 - 105. Plenum Press.
19. Poll, D. I. A. 1979 Transition in the infinite swept attachment line boundary layer *Aero.Q*. **30**, pp. 607 - 629.
20. Second European Forum on Laminar Flow Technology 1996. Bordeaux, France.
21. Spalart, P. R. 1988 Direct numerical study of leading-edge contamination. *AGARD CP*-**438**, pp. 5-1 - 5-13.
22. Theofilis, V. 1993 Numerical experiments on the stability of leading edge boundary layer flow: A two-dimensional study. *Intl J. Numer. Meth. Fluids* **16**, pp. 153 - 170.
23. Theofilis, V. 1994 On subcritical instability of the attachment-line boundary layer. *AGARD CP*-**551** pp. 31-1 - 31-8.
24. Theofilis, V. 1995 Spatial stability of incompressible attachment-line flow. *Theor. Comput Fluid Dyn*. **7**, pp. 159 - 171.
25. Theofilis, V. 1996 On the transition along the leading edge of swept wings: High Reynolds number secondary instability in leading edge boundary layer flow *DLR IB 223-96* A 26.

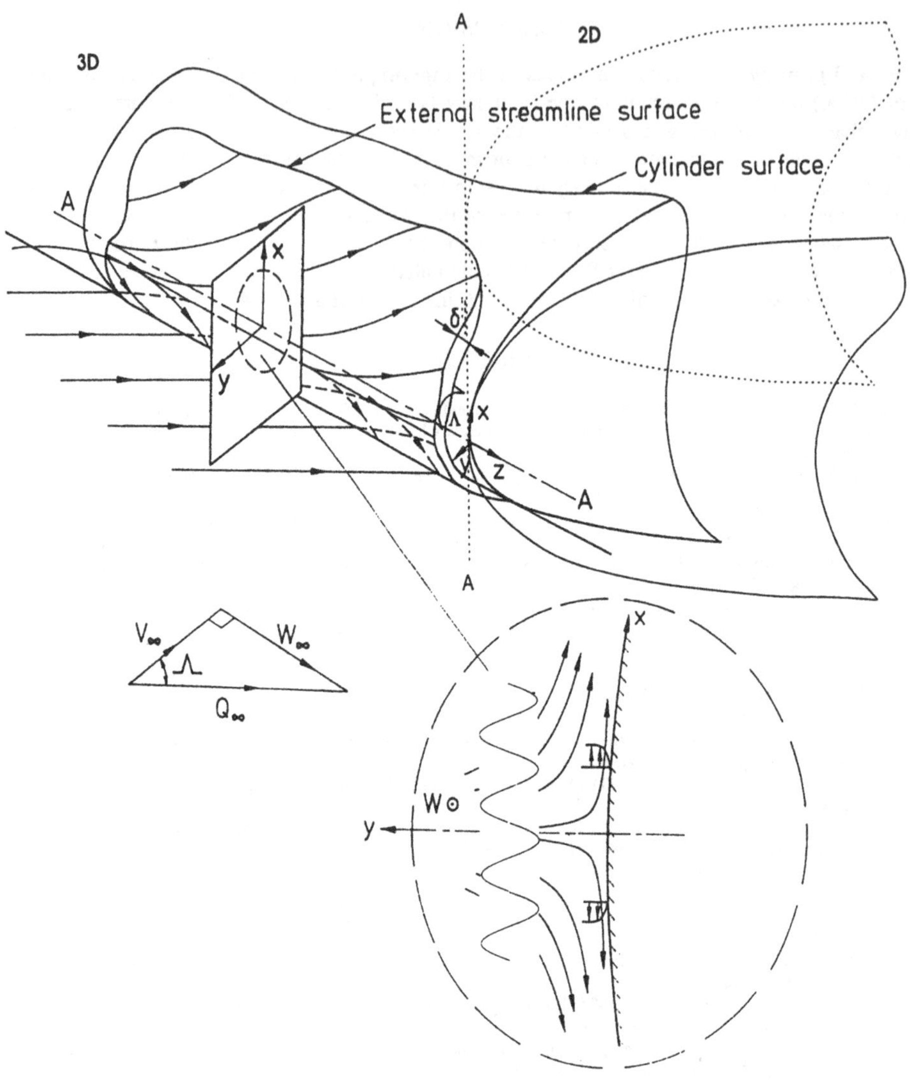


**Figure 1**. Schematic representation of flow in the stagnation region of a swept wing.

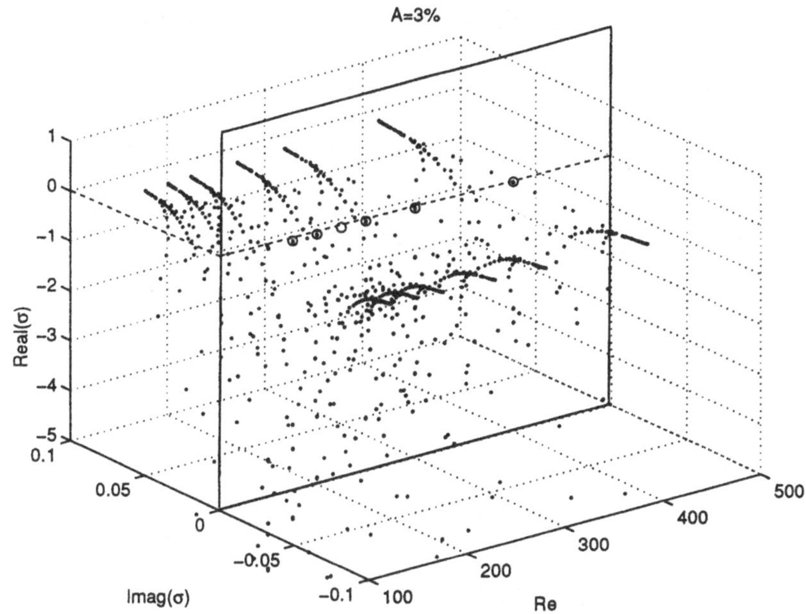


**Figure 2.** The unstable real secondary mode and the stable full secondary spectrum at subcritical Reynolds numbers.

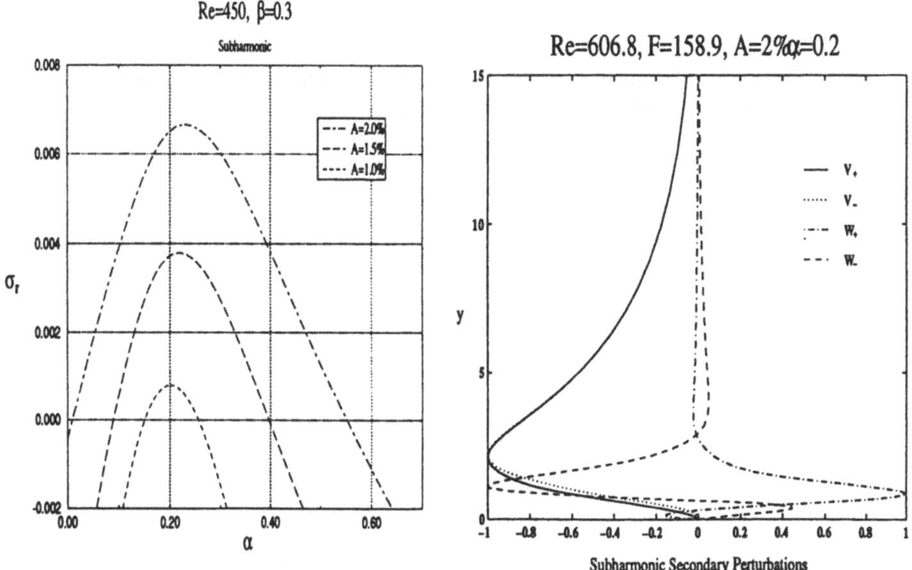


**Figure 3.** Chordwise wavenumbers subcritically excited by secondary mechanisms.

**Figure 4.** The structure of subharmonic secondary perturbations at neutral primary conditions.

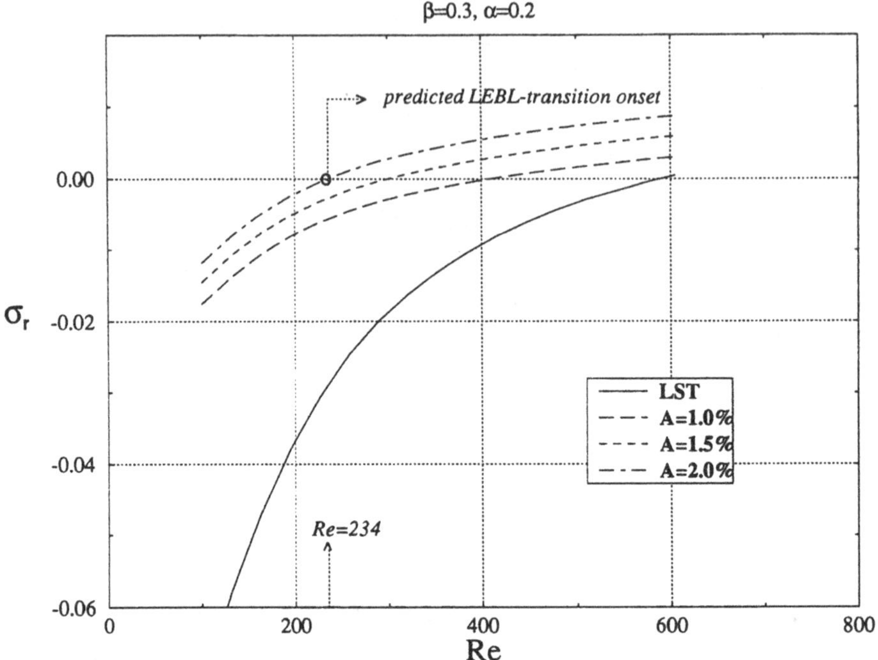


**Figure 5**. Subcritical growth of subharmonic secondary perturbations at maximum growth conditions

# Numerical simulation of the vortex sheet roll-up behind wings with different lift distributions

**L. Türk** and **D. Coors**
Institute for Aeronautical and Aerospace Engineering
RWTH Aachen, Wüllnerstraße 7, D-52062 Aachen, Germany

## Summary

Numerical simulations using the two-dimensional vorticity-stream function formulation were performed to investigate the correlation between the spanwise lift distribution of a wing and the forming vortex structures. The intent was to identify the potential for the reduction in vortex strength, which can be achieved by the use of winglets or flaps. The structures of the forming vortices were analyzed and compared. As a measure for the danger of the vortex systems the maximum induced rolling moments on a following wing were calculated.

## List of Symbols

| | | |
|---|---|---|
| $R$ | = | induced rolling moment |
| $T$ | = | second moment of the 2D vorticity distribution |
| $U_\infty$ | = | flight speed |
| $C_R$ | = | rolling moment coefficient |
| $r$ | = | distance from vortex center |
| $r_c$ | = | core radius |
| $s$ | = | half wing span |
| $s_{ref}$ | = | reference half wing span |
| $t$ | = | time |
| $v_y, v_z$ | = | horizontal and vertical velocity |
| $v_t$ | = | tangential velocity |
| $x,y,z$ | = | cartesian coordinates |
| $y_s, z_s$ | = | position of the center of gravity of the 2D vorticity distribution |
| $\Delta$ | = | grid spacing |
| $\Gamma_0$ | = | initial circulation |
| $\Gamma_{0.15}$ | = | circulation within 0.15 $s_{ref}$ of the vortex center |
| $\Psi$ | = | stream function |
| $\nu$ | = | kinematic viscosity |
| $\omega$ | = | vorticity |

## I. Introduction

The wake vortex systems behind large aircraft pose a considerable danger for following smaller aircraft. Depending on the flight pass of the following aircraft relative to the large aircraft's vortex system the following plane might experience strong rolling moments, a severe loss in altitude or rate of climb and huge load factors. In order to reduce these threats a minimum separation distance between aircraft has to be maintained. Since this separation distance limits the capacity of airports and flight routes it is necessary to investigate ways to reduce the effects of the vortex systems so that separation distances can be reduced.

In the present study the correlation between the spanload distribution of a wing and the resulting vortex system was investigated. Theoretical spanloads, which in reality might be achieved by the use of winglets or flaps, were assumed and their initial vorticity distribution

$\omega(x=0,y,z)$ was calculated. The roll-up of the vortex sheets was simulated by the 2D-vorticity transport equation in combination with the stream function. The forming vortices were compared by their vorticity profiles, core radii $r_c$, circulation $\Gamma_{0.15}$, maximum tangential velocities $v_t$ and the rolling moment R they induce on a following wing.

## II. Computational Method

Assuming vortex stretching effects to be small and a dominating vorticity component $\omega_x$ ($\rightarrow \omega \approx \omega_x$), the roll-up of a given vorticity distribution $\omega(x=0,y,z)$ can be calculated by the 2D-vorticity transport equation [Fell, Ref. 1]

$$\frac{\partial \omega}{\partial t} + v_y \frac{\partial \omega}{\partial y} + v_z \frac{\partial \omega}{\partial z} = \nu \cdot \left( \frac{\partial^2 \omega}{\partial y^2} + \frac{\partial^2 \omega}{\partial z^2} \right) \tag{1}$$

with $t = x/U_\infty$. The velocities $v_y$ and $v_z$ are obtained by solving the Poisson equation

$$\frac{\partial^2 \Psi}{\partial y^2} + \frac{\partial^2 \Psi}{\partial z^2} = -\omega \tag{2}$$

for the stream function $\Psi$ and

$$v_y = +\frac{\partial \Psi}{\partial z} \; ; \; v_z = -\frac{\partial \Psi}{\partial y} . \tag{3}$$

The boundary conditions for equation (2) [Weston & Liu, Ref. 2] are

$$\Psi(y,z,t) = -\frac{1}{2\pi} \int_{-\infty}^{+\infty} \int_{-\infty}^{+\infty} \omega(y',z',t) \cdot \ln\left[ \sqrt{(y-y')^2 + (z-z')^2} \right] dy'dz' . \tag{4}$$

The transformation of a given spanwise distribution of circulation into the initial vorticity distribution $\omega(x=0,y,z)$ is shown in Fig. 1. The spanload distribution is divided into a great number (ns) of small steps and the circulation $\Delta\Gamma_i$, i.e. the change in circulation over each step, is distributed over a circular area by the 2D Gauss function

$$\omega_i(y,z) = \frac{\Delta\Gamma_i}{2 \cdot \pi \cdot \sigma^2} e^{-\frac{(y-y_i)^2 + (z-z_i)^2}{2 \cdot \sigma^2}} \tag{5}$$

thus forming one „vorticity pile" per step. The coordinates $y_i$ and $z_i$ represent the center of each „vorticity pile" and are given by the form of the bound vortex in the y,z-plane. $\sigma$ equals the standard deviation in the Gauss function and influences the thickness of the vortex sheet as can be seen in Fig. 1. The final vorticity distribution is achieved by the summation over all „vorticity piles" according to

$$\omega(y,z) = \sum_{i=1}^{ns} \omega_i(y,z) . \tag{6}$$

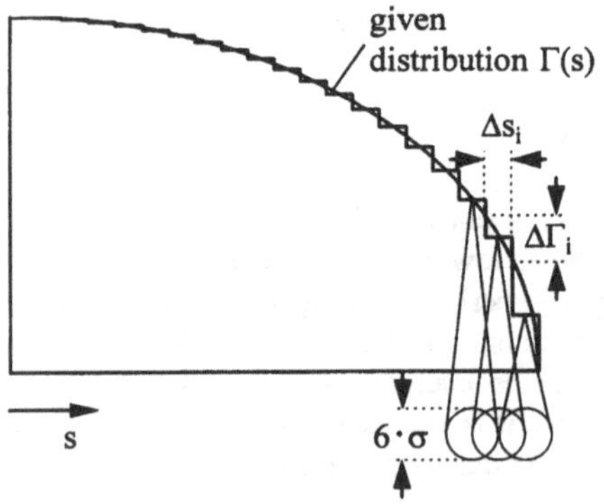

**Fig. 1 Transformation of spanwise circulation distribution into vorticity distribution $\omega(x=0,y,z)$.**

## III. Studied Cases

Eleven spanload distributions were studied, representing cases forming one, two or three vortex pairs. An elliptical distribution with $\Gamma_{0,ref}$ = 0.1 $m^2/s$ and $s_{ref}$ = 1 m was chosen as reference. In analogy to water tunnel experiments the freestream velocity was set to $U_\infty$ = 0.5 m/s and the density to $\rho$ = 1000 $kg/m^3$. With $\nu = 1.01 \cdot 10^{-6}$ $m^2/s$ the Reynolds number was calculated to $Re_{ref} = 4.95 \cdot 10^5$. Besides the elliptical reference case distributions with one or more winglets at various angles, a distribution with less load on the outer part of the wing and distributions creating several vortex pairs were studied. For all cases the circulation $\Gamma_0$ was adapted to produce the same lift as the elliptical reference case. Fig. 2 shows the spanload distributions for selected cases and reveals that the case with one horizontal winglet is equal to an elliptical distribution with s = 1.1 $s_{ref}$. All other cases with winglets were generated by changing the inclination of the outermost 0.1 $s_{ref}$ of this distribution to the desired angle and adapting $\Gamma_0$. All cases with winglets showed a reduction of $\Gamma_0$ relative to $\Gamma_{0,ref}$ whereas in all other cases $\Gamma_0$ had to be increased, a result supported by fundamental theory [Schlichting & Truckenbrodt, Ref. 3].

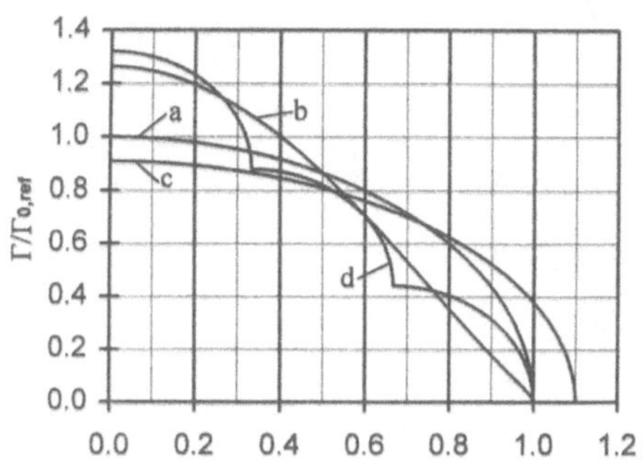


**Fig. 2 Spanload distribution for selected cases; a → elliptical spanload; b → unloaded wingtip; c → winglet 0°; d → 3 equal vortices.**

The spanloads lead to initial vorticity distributions $\omega(x=0,y,z)$, which are schematically shown in Fig. 3. Filled areas represent vorticity extremes and black lines show vorticity sheets. The width of the vorticity sheets was set by choosing $\sigma$ in equation (5) to $\sigma/s_{ref}$ = 0.06, a value observed in water tunnel experiments. The cases in Fig. 3 are classified by the number of

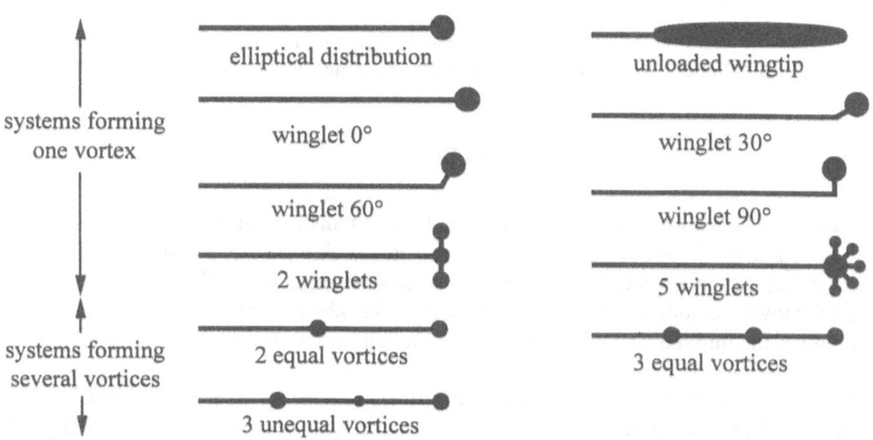


**Fig. 3 Schematic initial vorticity distributions $\omega(x=0,y,z)$.**

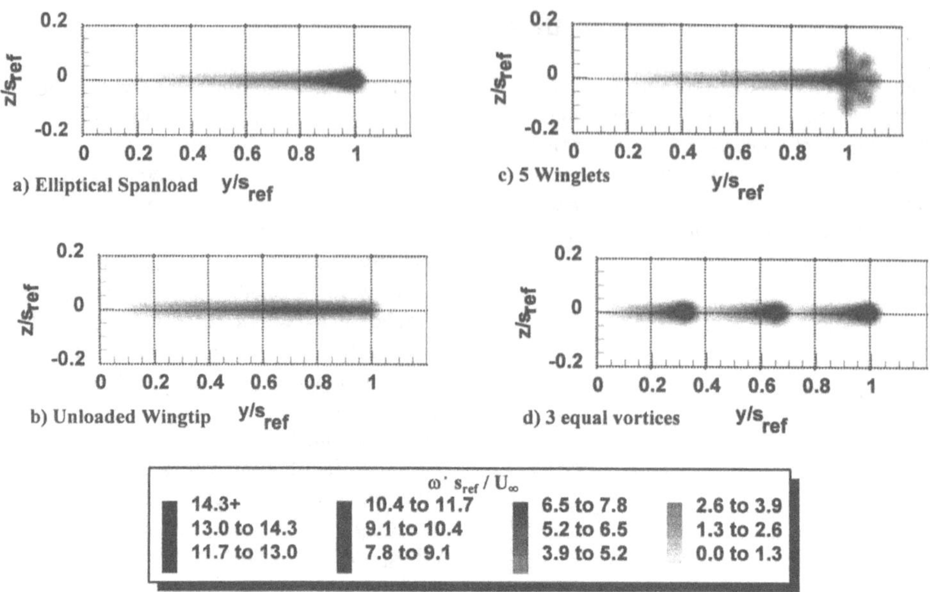


**Fig. 4 Initial vorticity distributions ω(x=0,y,z) for selected cases.**

vortices forming in the y,z-semiplane within one wingspan behind the initial plane. Fig. 4 shows the initial vorticity distributions ω(x=0,y,z) for selected cases and the strong reduction of the maximum vorticity for the „unloaded wingtip" (-65%) and the wing with five winglets (-62%) compared to the elliptical spanload. For the case with three equal vortices the reduction in $\omega_{max}$ is much less (-25%). The case with two winglets shows a 55% reduction in $\omega_{max}$ whereas all cases with one winglet and the cases with two equal or three unequal vortices produce only minor reductions.

For all cases the wake vortex flow was calculated up to $x/s_{ref} = 30$, using an equidistant grid with $\Delta/s_{ref} = 0.004$ and a second order Runge-Kutta scheme.

## IV. Results

The vortex structures were analyzed for different x-positions behind the wing up to $x/s_{ref} = 30$. In almost all cases the roll-up process was terminated within $x/s_{ref} = 2$ and no major changes in the vortex structures were observed further downstream. Only the case with unloaded wingtips showed a longer roll-up process due to the absence of a distinguished vorticity maximum in the initial distribution. In this case the roll-up was finished within $x/s_{ref} = 4$.

Fig. 5 shows the vorticity profiles through the vortices of selected cases at $x/s_{ref} = 8$. The cases with unloaded tips and with five winglets show considerably broadened and flattened profiles whereas in all other cases only minor changes can be observed. The broadened and flattened

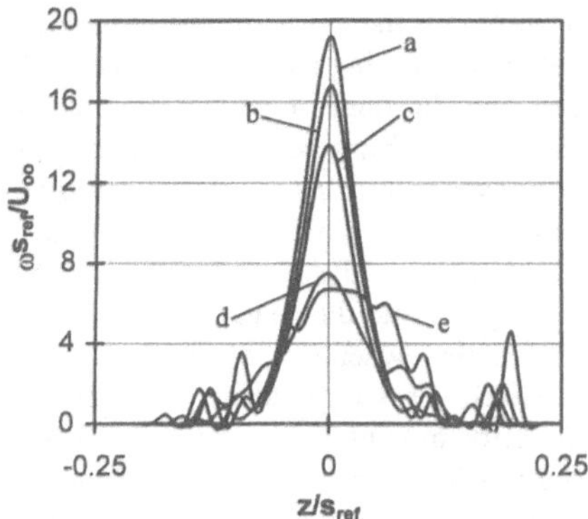


**Fig. 5 Vorticity profiles for selected vortices at $x/s_{ref} = 8$; a → elliptical spanload; b → winglet 0°; c → 3 equal vortices; d → 5 winglets; e → unloaded wingtip.**

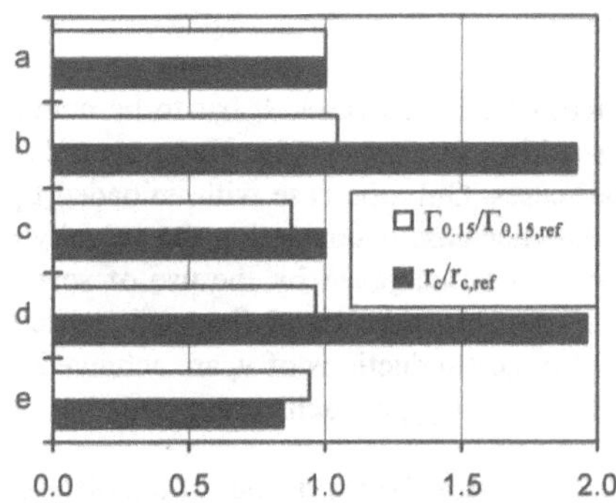


**Fig. 6 Core radius $r_c$ and circulation $\Gamma_{0.15}$ for selected cases at $x/s_{ref} = 8$; a → elliptical spanload; b → unloaded wingtip; c → winglet 0°; d → 5 winglets; e → 3 equal vortices.**

profiles of the two named cases are a result of the more evenly spread vorticity in the initial vorticity distributions (see Fig 4 a-c).

The broadened vorticity profiles of the case with unloaded tips and the case with five winglets result in a considerably increased core radius $r_c$, as can be seen in Fig. 6. For all other cases the changes in $r_c$ are minor with an increase between 0 and 2% for the cases with one winglet and an increase of 20% for the case with two winglets. All cases with more than one vortex pair show a reduction of $r_c$ by up to 15%. Therefore the only way to effectively increase the core radius $r_c$ is to form systems with only one vortex pair and evenly spread vorticity in the initial distribution $\omega(x=0,y,z)$.

The circulation $\Gamma_{0.15}$ within 0.15 $s_{ref}$ of the vortex center is also shown in Fig. 6. Only the distribution with unloaded wingtips leads to an increased circulation $\Gamma_{0.15}$ (+5%) whereas all other cases show a reduction relative to the elliptical reference case. The reductions of the cases with winglets are due to a reduced $\Gamma_0$ whereas the reductions for the cases with several vortex pairs are a result of the division of $\Gamma_0$ among the vortices. For the cases with several vortex pairs the achieved reductions are however less than the division would predict. One cause is the increased $\Gamma_0$ for these cases. Another reason is a higher concentration of circulation within the vortices. Fig. 7 displays the concentration of circulation $\Gamma_{0.15}/\Gamma_0$ for selected cases and shows for cases with only one vortex pair a maximum concentration of less than 70% whereas for the case with three equal vortices almost 100% of $\Gamma_0$ are contained within the three vortices (33% each). The high concentration was observed for all cases forming more than one vortex pair. The reason for the different levels of concentration between cases with one and several vortex pairs was found to be the second moment

$$T = \int_0^{+\infty}\int_{-\infty}^{+\infty}\left[\left(y-y_s\right)^2+\left(z-z_s\right)^2\right]\cdot\omega\,dz\,dy \approx \text{const.} \tag{7}$$

T describes the structure of the 2D-vorticity distribution within the y,z-semiplane and limits the maximum concentration of circulation for cases with one vortex pair. The limitation is

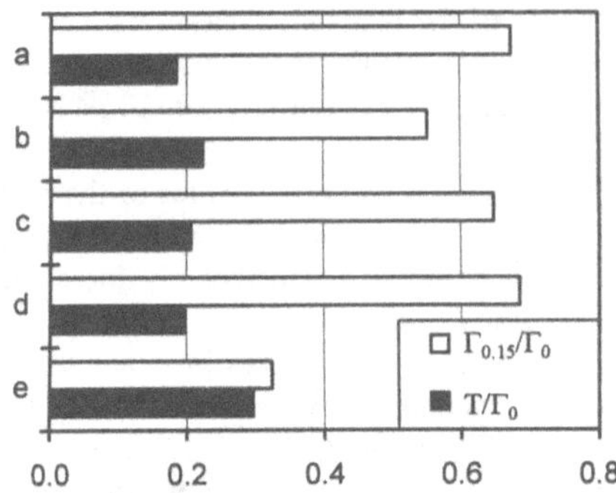


**Fig. 7 Concentration of circulation $\Gamma_{0.15}/\Gamma_0$ and second moment T for selected cases at $x/s_{ref} = 8$; a → elliptical spanload; b → unloaded wingtip; c → winglet 0°; d → 5 winglets; e → 3 equal vortices.**

a result of the necessity to maintain the value of T during the roll-up process which can only be done if parts of the vorticity move away from the center of gravity $y_s, z_s$ whenever other parts move towards it. Since for cases with one vortex in the y,z-semiplane the center of gravity is usually close to the center of the vortex this means that whenever vorticity is moving towards the vortex there also has to be vorticity moving away from it. Therefore the maximum concentration is limited for cases with one vortex pair. For systems with several vortices per y,z-semiplane there is no limitation since T can be maintained for all levels of concentration by different distances between the vortices and the center of gravity. Therefore the second moment T is the reason for the different levels of concentration $\Gamma_{0.15}/\Gamma_0$ and thus for the small reductions of $\Gamma_{0.15}$ for cases with several vortex pairs.

Fig. 8 shows the tangential velocity $v_t$ for the vortices of selected cases. It has to be noticed that the displayed curves represent the velocity induced by one vortex only. The curves show reductions of $v_t$ within the vortex core for almost all cases. Only the case with unloaded tips has in certain areas a $v_t$ greater than the $v_t$ of the reference case. Considering the velocity in the inner part of the vortex the strongest reductions can be achieved by the use of several winglets or unloaded tips, a phenomenon caused by the broadened and flattened vorticity profiles of these cases. Outside the vortex core the strongest reductions of $v_t$ are achieved by cases with several vortex pairs due to the reduced circulation $\Gamma_{0.15}$ of each vortex.

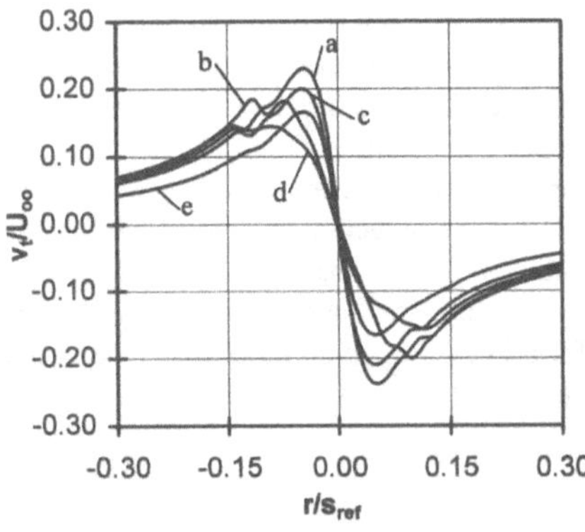


**Fig. 8 Tangential velocity $v_t$ for the vortices of selected cases at $x/s_{ref} = 8$; a → elliptical spanload; b → unloaded wingtip; c → winglet 0°; d → 5 winglets; e → 3 equal vortices.**

As a measure for the danger of the different vortex systems, the maximum induced rolling moment on a following rectangular wing was calculated. The span of the following wing was set to $b_f = 0.15\ b_{ref}$, a value used by Rossow [Ref. 4] for experimental wake vortex studies behind a B747 model. The velocity $U_{\infty,f}$ was set equal to the velocity $U_\infty$, and the induced rolling moment was calculated by

$$R = \frac{\rho \cdot U_\infty^2}{2} \cdot C_{L\alpha} \cdot l \int_{-b_f/2}^{+b_f/2} \frac{v_n}{U_\infty}(s^*) \cdot s^* \cdot ds^* \quad (8)$$

with $v_n$ naming the velocity perpendicular to the following wing, $C_{L\alpha}$ naming the derivative of the lift coefficient and l naming the wing chord of the following wing. To

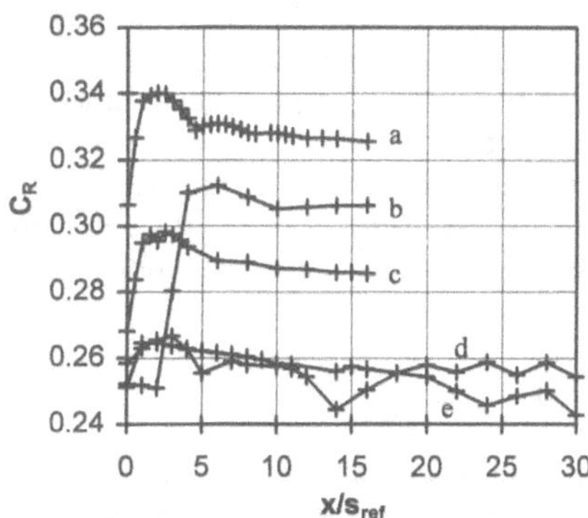


**Fig. 9 Maximum induced rolling moment on a following wing for selected cases; a → elliptical spanload; b → unloaded wingtip; c → winglet 0°; d → 5 winglets; e → 3 equal vortices.**

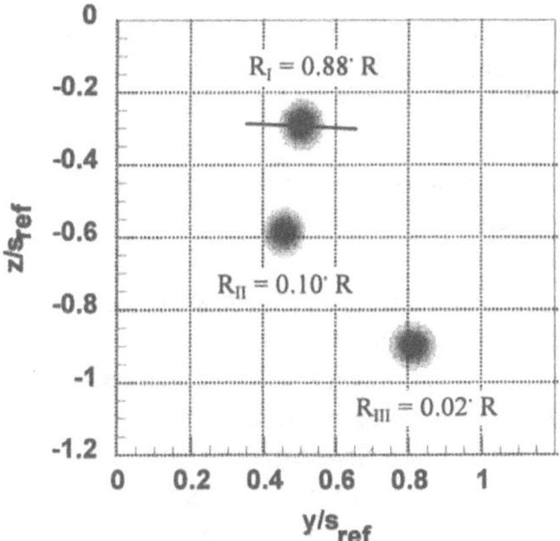


**Fig. 10 Superposition of the induced rolling moments of each vortex for the case with 3 equal vortices at $x/s_{ref} = 18$.**

achieve realistic proportions, $C_{L\alpha}$ was set to 1.8 and l to 0.1 $s_{ref}$. In order to find the maximum induced rolling moment the position and the bank angle of the following wing - both influencing the velocity $v_n$ - were varied and optimized. The calculated maximum rolling moments for selected cases are displayed in Fig. 9. The curves show a reduction of R relative to the reference case for all cases. The strongest reductions are achieved by the case with three equal vortices and the case with five winglets. In the case with three equal vortices the strong reduction is a result of the reduced circulation of each vortex, leading to a reduced tangential velocity $v_t$. In the case with five winglets the strong reduction is caused by the increased core radius $r_c$ and the resulting reduction of $v_t$ in the inner part of the vortex. The phenomenon of the case with three equal vortices and the case with five winglets showing almost identical rolling moments is caused by the chosen ratio $b_f/b_{ref}$. For a smaller ratio $b_f/b_{ref}$ the case with five winglets shows the smaller moments due to the lower velocity $v_t$ in the inner part of the vortex whereas a greater ratio $b_f/b_{ref}$ leads to an advantage for the case with three equal vortices. The cases with one winglet show smaller reductions of R than the cases with several winglets. A bigger inclination of the winglet leads to a slightly increased R due to an increased $\Gamma_{0.15}$. Due to the higher circulation $\Gamma_{0.15}$ and the in certain areas increased tangential velocity $v_t$ (see Fig. 8) the case with unloaded wingtips shows the smallest reduction in R.

Overall the most effective way to reduce the rolling moment R is to produce systems with several vortex pairs even though the reductions are smaller than the division of the circulation among the vortices would predict. One reason is the higher concentration of circulation within the vortices. Another reason is the adding up of the rolling moments induced by each vortex. This effect is shown in Fig. 10. The maximum rolling moment R on the following wing is a result of the moments $R_I$, $R_{II}$ and $R_{III}$ induced by the vortices I, II and III. Therefore, depending on the position of each vortex within the vortex system, the maximum rolling moment can be bigger or smaller than the moment induced by just one vortex. This effect causes the rough curve for the rolling moment of the case with three equal vortices in Fig. 9.

An effect not observed in this study except during the roll-up process of the cases with several winglets is the merging of vortices. Whenever several vortex pairs exist there is the possibility of some vortices merging, thus forming stronger vortices. None of the studied cases with several vortex pairs showed merging of vortices up to $x/s_{ref} = 30$ but merging further downstream cannot be excluded. Therefore the vortex structures calculated for the cases with several vortex pairs are „best cases". If in any of these cases vortices merge the new vortex will show an increased circulation and core radius and will induce a stronger rolling moment.

## V. Conclusion

A 2D-vorticity-stream function formulation was used to investigate the correlation between the spanload over a wing and the resulting vortex structures. Spanloads producing a constant lift and resulting in wake vortex systems with one or several vortex pairs were studied.

The study showed the strongest reductions of the circulation $\Gamma_{0.15}$ within a single vortex for systems forming several vortex pairs, even though the reductions were smaller (-36% max.) than the division of the circulation $\Gamma_0$ among the vortices would predict. One reason for this smaller than predicted reduction is the increased circulation $\Gamma_0$ relative to the elliptical reference case. Another reason is a higher concentration of circulation within the vortices for cases with several vortex pairs. The cause for the different levels of concentration for cases with one and several vortex pairs was found to be the second moment T of the two-dimensional vorticity distribution. The cases with winglets showed only small reductions of $\Gamma_{0.15}$ with the biggest reductions for cases with several winglets. For distributions with only one winglet the case with no inclination of the winglet was the most advantageous.

The biggest increase in the core radius $r_c$ was achieved by the use of several winglets or a spanload with unloaded tips while the vortices of systems with several vortex pairs showed a smaller core radius than the elliptical reference case. The tangential velocity $v_t$ was reduced in almost all studied cases. Only the case with unloaded tips showed a slight increase of $v_t$ in a small area. In analogy to $v_t$ the rolling moment R induced on a following wing was reduced in all treated cases with the biggest reductions for distributions with several vortex pairs and cases with several winglets. For small ratios of $b_f/b_{ref}$ the cases with several winglets are advantageous, for bigger ratios $b_f/b_{ref}$ the cases with several vortex pairs show the greater reductions of R even though the observed reductions are smaller than the structure of each single vortex would predict. This effect is a result of the superposition of the rolling moments induced by each vortex within the vortex system.

## References


1. S. Fell, „Formierung und Struktur von Randwirbeln verschiedener Flügelkonfigurationen", VDI Fortschrittberichte, Reihe 7, Nr. 279, VDI Verlag, Düsseldorf 1995.

2. R. P. Weston & C. H. Liu, „Approximate Boundary Condition Procedure for the two-dimensional numerical solution of vortex wakes", AIAA Paper No. 82-0951, 1982.

3. H. Schlichting & E. Truckenbrodt, „Aerodynamik des Flugzeuges", Springer Verlag, 1962.

4. V. J. Rossow, „Effect of Wing Fins on Lift Generated Wakes", Journal of Aircraft, Vol. 15, No. 3, 1978.

# Turbulent flow in idealized Czochralski crystal growth configurations

C. Wagner and R. Friedrich
Lehrstuhl für Fluidmechanik
Technische Universität München
Boltzmannstr.15
D-85747 Garching

## Summary

Turbulent convection of a Si-melt in an idealized Czochralski crystal growth configuration was simulated with a three-dimensional, time-dependent Navier-Stokes solver. The analysis of the flow data focuses on the influence that thermocapillar forces and rotation of crystal and crucible have on the flow structures, the heat transport and the development of velocity and temperature fluctuations. Thermocapillarity is found to be insignificant for the bulk flow structure and overall heat transfer. It is important though if intensities of the velocity field or the fluctuating flow field are of interest. Introducing rotation of the crystal and crucible creates a complex flow with three recirculation zones, the dynamics of which is controlled by centrifugal forces counteracting buoyancy and surface tension effects. Temperature and velocity fluctuations are enhanced by up to one order of magnitude due to rotation. The maximum values are located in a thin layer underneath the edge of the crystal within the cristalization zone.

## 1 Introduction

In semiconductor industry there is a demand for a high amount of pure crystals, which are commonly grown in the Czochralski process. In this configuration the melt of a semiconductor is held in a cylindrical crucible which is heated from the side walls. The cylindrical crystal is pulled from the free surface at rates between a few millimeters to centimeters per hour. The heating of the melt gives rise to buoyant convection. Crucible and crystal are commonly rotated in opposite directions, so that centrifugal forces counteract free convection. At the free surface, open to the ambient air, Marangoni convection develops due to the radial temperature gradient between the sidewalls and the comparably cold crystal (kept below melting temperature). In addition thermal conduction and heat radiation at the free surface induce losses of heat to the surrounding.

The fluid flow and heat transfer processes in the melt of a Czochralski growth system are extremely complex. The experimental investigation of these low Prandtl number melts faces lots of difficulties since in large scale growth systems the flow is mostly turbulent, non-transparent and because of the high temperatures involved, unsuitable for tracer particles.

In an experiment with a transparent fluid of Prandtl number O(1) Jones [1] investigated the convection forced by rotation of the crystal in a non-rotating heated crucible. In the vincinity of the crystal he observed temperature fluctuations of high amplitude. Temperature measurements by Kuroda et al. [2] showed that temperature fluctuations of large amplitude are responsible for an increased concentration of micro-defects in the crystal.

For a Si-melt Elwell et al. [3] performed temperature measurements in the case of a rotating crucible with a non-rotating crystal. They found that the temperature fluctuations strongly vary in space and with the rotation rate of the crucible.

There have been a number of time-dependent three-dimensional numerical simulations of Czochralski Si-melt flow for low Grashof numbers. Mihelcic et al. [4-5] were among the first to attempt three-dimensional simulations studing the bulk flow strukture and transition from a two to three-dimensional flow state. Bottaro and Zebib [6] simulated the buoyant flow in a cylindrical confinement, without modeling rotation of the crystal or crucible. Numerous numerical simulations (Miyahara et al. [7], Dupret et al. [8], Ryckmans et al. [9], Kobayashi [10], Fontaine et al. [11] and others) have been performed applying different numerical approaches in order to investigate various physical processes in the low Prandtl number melts such as global heat transfer, radiation, magnetic fields and encapsulation of the melt flow.

The more recently reported publications on three-dimensional numerical simulations concentrate on the flow of fluids with higher Prandtl numbers $Pr = O(10)$.

Concerning buoyancy and surface tension driven flow in a cylindrical container heated from below Wagner et al. [12] presented three-dimensional numerical simulations of laminar flow using a time-dependent finite volume code.

Xiao and Derby [13] apply a theoretical bulk-flow model to compute the flow with a Galerkin finite element method on a grid with 266987 nodes. The Czochralski configuration was idealized in the sense that surface tension effects and heat transport across the flat free surface were neglected. For a Grashof number $Gr = 2.54 \cdot 10^5$, a Prandtl number $Pr = 8$ and a Reynolds number based on the crystal rotation rate of $Re = 2000$ (the crucible is assumed to be stationary) they consider the steady axisymmetric flow and the time-dependent nonaxisymmetric state. They find that this transition strongly affects the temperature distribution and heat transfer through the melt.

Using a Galerkin finite element method and a bulk flow model as well, but imposing symmetry conditions on the axis, Fontaine et al. [14] compute the flow for a Grashof number regime of 75000 to 100000 and Reynolds numbers between 100 to 600. The flat free surface is assumed to be adiabatic and the pulling rate of the crystal is neglected. Most of the simulations were performed without taking thermocapillary forces into account. By comparing axisymmetric flows for increasing thermocapillary forces the authors conclude that surface tension does not have a significant effect either on the heat and mass transfer or on the bulk flow structure. For increasing rotation rates and a fixed

Grashof number they observe a deminishing oscillatory frequency.

It is the aim of this work to investigate the influence that thermocapillary forces as well as rotation of crucible and crystal have on the growth of temperature fluctuations in the melt at high Grashof numbers where the flow seems to be turbulent. This is done by means of Direct Numerical Simulation. In this study the Czochralski configuration is idealized in the sense that a flat free surface and a flat crystal/melt interface are assumed. Finally heat radiation, conduction at the free surface and the pulling of the crystal are neglected and the crucible bottom wall is taken to be adiabatic.

## 2 Numerical method

The incompressible Navier-Stokes momentum equations in Boussinesq approximation, including a heat conduction equation, are integrated applying Schumann's volume balance procedure [15] in a cylindrical coordinate system.

The momentum equations are evaluated on staggered grids. Integration in time is performed with a second order semi-implicit time step. All convection and diffusion terms of the momentum equations containing derivatives in circumferential direction as well as all diffusive terms of the heat conduction equation are integrated implicitly by a Crank-Nicholson time step. The remaining terms are treated explicitly with a Leapfrog time step, which is restricted by a linear stability argument.

The coupling between pressure and velocity fields is provided by a fractional step approach. This leads to a three dimensional Poisson equation for the pressure, which has to be solved at each time step. The same is true for the implicit treatment of the temperature. The direct solutions of the Poisson equations for pressure and temperature are obtained using FFT's in $\varphi$-direction and cyclic reduction algorithms for the remaining 2D Helmholtz problems. The 1D Poisson problems associated with the implicit treatment of the $\varphi$-derivatives are solved by a tridiagonal matrix algorithm.

## 3 Initial/boundary conditions and geometrical outline

The discrete Navier-Stokes equations are solved in the cylindrical domain shown in Fig.1. The ratio between crystal radius $R_c$ and crucible radius $R_t$ is assumed to be one half ($\frac{R_c}{R_t} = 0.5$) and the aspect ratio of the system $\zeta = \frac{H}{R_t} = 1.0$, where $H$ denotes the height of the melt. The temperature of the Si-crystal is $1685K$ and that of the crucible wall $T_t = 1772K$.

Simulations were performed using dimensionless quantities, for which the crucible radius $R_t$ has been used as a reference length, the arithmetic average of crucible temperature $T_t$ and crystal temperature $T_c$ as reference temperature and $u_{ref} = (\chi g R_t(T_t - T_c))^{1/2}$ as reference velocity, $\chi$ being the thermal expansion coefficient and $g$ the gravitational acceleration. The dimensionless value of the crystal temperature $T_c = -0.5$ and the temperature at the crucible sidewall $T_t = 0.5$. For the simulation

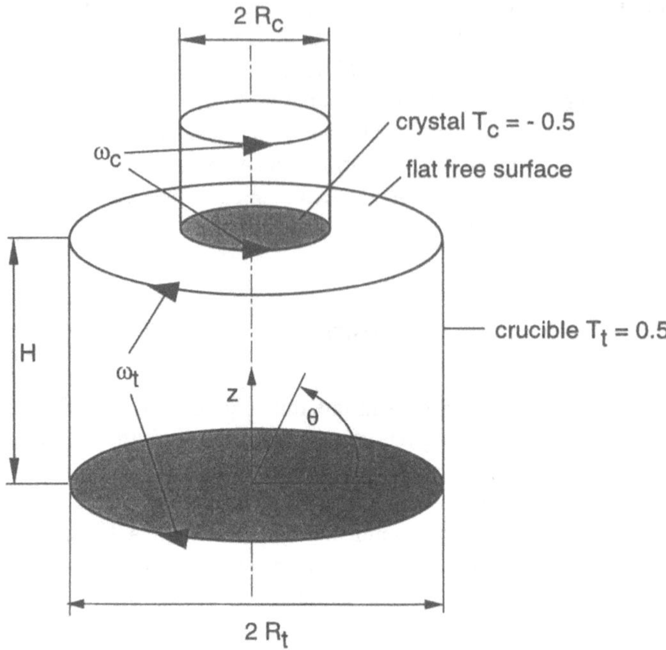


Figure 1: Geometrical Model of the Czochralski crystal growth configuration

with rotation the crucible rotates counterclockwise with the frequency $\omega_t = -1.3195$, while the crystal rotates clockwise with $\omega_c = 1.885$. This leads to a crystal rotation Reynolds number $Re = \frac{2\pi R_c^2 \omega_c}{\nu} = 1380$, with the kinematic viscosity $\nu$, and a rotation ratio $\frac{\omega_t}{\omega_c} = -0.7$. The Grashof number was chosen to be $Gr = \frac{\chi g R_t^3 (T_t - T_c)}{\nu^2} = 10^8$ and the Prandtl number $Pr = \frac{\nu}{\gamma} = 0.0175$ (which is the value of a Si-melt), with $\gamma$ the thermal diffusivity.

At time $t = 0$ the simulations were started with

$$u_r = u_\varphi = u_z = 0, \qquad T = T_c + T_{random} \tag{3.1}$$

where $T_{random}$ stands for numerically created random numbers between $-0.1$ and $0.1$.

Boundary conditions at rigid walls (i.e the side and bottom walls of the crucible and the interface to the crystal) are the impermeability condition for the wall normal components and the no slip conditions for the tangential velocity components. Because of the rotation of the crucible and the crystal the circumferential velocity component describes the rotation of these boundaries. Temperature boundary conditions are of Dirichlet type at the heated/cooled boundaries, i.e. at the side wall of the crucible and the crystal/melt interface. At the bottom of the crucible a Neumann boundary condition for the temperature namely vanishing temperature gradient in axial direction is assumed. Finally, at the free surface, strain boundary conditions (3.5) - (3.7) were imposed.

Crucible side wall:

$$u_z = 0, \quad u_\varphi = \omega_t, \quad u_r = 0, \quad T = T_t \tag{3.2}$$

Crucible bottom:

$$u_z = 0, \quad u_\varphi = r \cdot \omega_t, \quad u_r = 0, \quad \delta_z T = 0 \tag{3.3}$$

Crystal/melt interface:

$$u_z = 0, \quad u_\varphi = r \cdot \omega_c, \quad u_r = 0, \quad T = T_c \tag{3.4}$$

Flat free surface:

$$\delta_z u_r = -\frac{Ma}{Pr\sqrt{Gr}} \delta_r T \tag{3.5}$$

$$\delta_z u_\varphi = -\frac{Ma}{Pr\sqrt{Gr}} \delta_\varphi T \tag{3.6}$$

$$u_z = 0, \quad \delta_z T = 0 \tag{3.7}$$

where the Marangoni number is $Ma = \frac{c_{tc} R_t (T_t - T_c)}{\gamma} = 3.6034 \cdot 10^4$ and contains the capillary coefficient $c_{tc}$. The discrete operator $\delta_\alpha$, $\alpha = (z,\varphi,r)$ in equations 3.3 and 3.5 to 3.7 symbolizes a central difference in $\alpha$-direction.

# 4 Results

## 4.1 Thermocapillarity

In order to investigate the effect that thermocapillarity has on the turbulent flow, we performed two simulations, one for a vanishing Marangoni number $Ma = 0$ and a second one for the Marangoni number mentioned above. In both flow cases the crystal and crucible were kept stationary.

A mesh with 64 × 128 × 64 equidistantly spaced grid points in $(z,\varphi,r)$-direction was used. For the following comparison statistical averages were calculated from 500 realizations of the turbulent flow with a time lag of 200 time steps and based on spatial averging over 128 grid points in $\varphi$-direction.

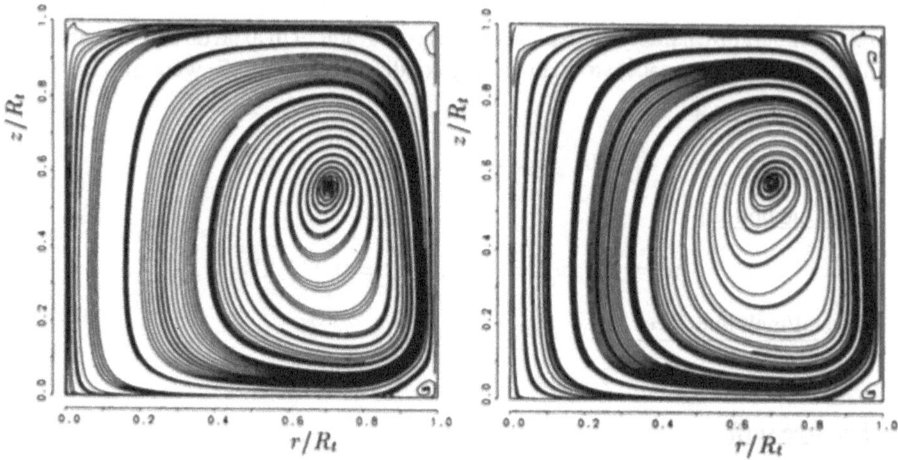


Figure 2: Streamlines of the mean velocity fields projected onto $(z,r)$-planes. Without thermocapillarity $Ma = 0$ (right), with thermocapillarity $Ma = 3.6034 \cdot 10^4$ (left).

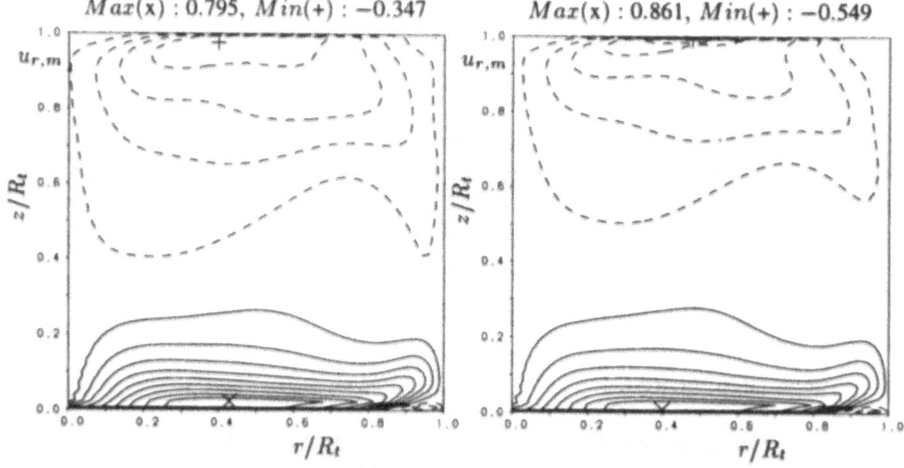


Figure 3: Contours of mean radial velocity component. Right: $Ma = 0$, Left: $Ma = 3.6034 \cdot 10^4$. Solid/dashed lines represent positive/negative values

The streamlines in Fig.2 representing the two cases show almost identical recirculation zones. Buoyancy forces cause the fluid to rise at the heated crucible wall turning inward towards the axis below the free surface. Below the crystal the flow starts decending to the crucible bottom forming one mean recirculation. Only the small recirculation at the meniscus is slightly intensified by surface tension effects. Consequently the mean isotherms in Fig. 4 are of similar shape, which supports Fontaine et al's [8] findings, that thermocapillarity has an insignificant effect on the bulk flow structure and global

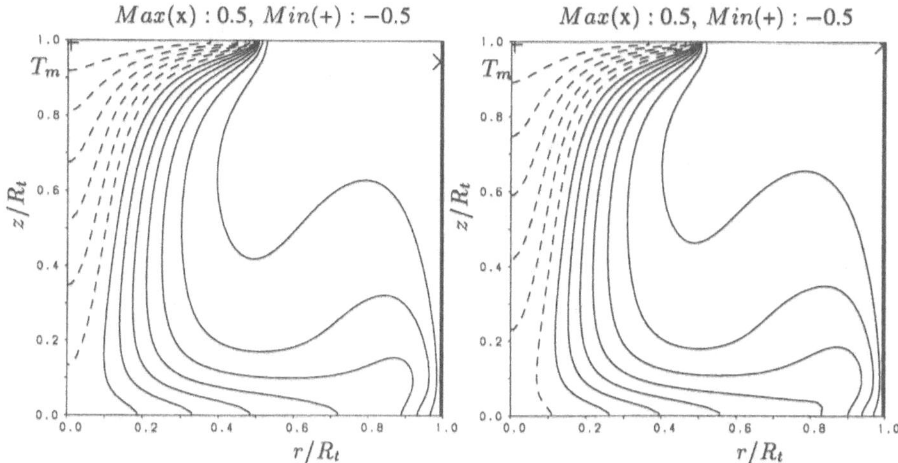


Figure 4: Contours of mean temperature. Right: $Ma = 0$, Left: $Ma = 3.6034 \cdot 10^4$. Lines as in Fig.3.

heat and mass transfer.

In order to highlight how surface tension affects the flow we compare the minimum values of the mean radial velocity component depicted in Fig. 3. Accounting for thermocapillarity increases the negative value by 58% and shifts its location towards the crystal edge where the radial temperature gradient is maximum (see Fig.4). Since surface tension supports the buoyant recirculation it accelerates the flow below the free surface.

A significant change is also observed in the fluctuating flow fields. The maximum value of the axial rms velocity fluctuations shown in Fig.5 is decreased from 0.194 in the case of vanishing surface tension to 0.151, when termocapillarity is taken into account. Isothermes of rms temperature fluctuations depicted in Fig. 6 reveal an increase in the maximum value by nearly 20% together with a remarkable shift of its location towards the crucible bottom by more than half a melt height.

Although thermocapillarity has little influence on the bulk flow structure, there is a significant change in the intensities of the mean and fluctuating flow fields. For the following flow case thermocapillarity is therefore taken into account.

## 4.2 Rotation effects

For the simulation with rotation a grid with $96 \times 128$ points (equidistant) and 96 cells in radial direction was chosen. Grid points were clustered in radial direction below the crystal edge and in the vincinity of the crucible wall.

In a perspective view contours of the instantaneous velocity components and temperature are presented in Fig.7,8,9 and 10. The axial and radial velocity components in Fig. 7 and 9 reveal spiral-like structures just below the free surface, reflecting the

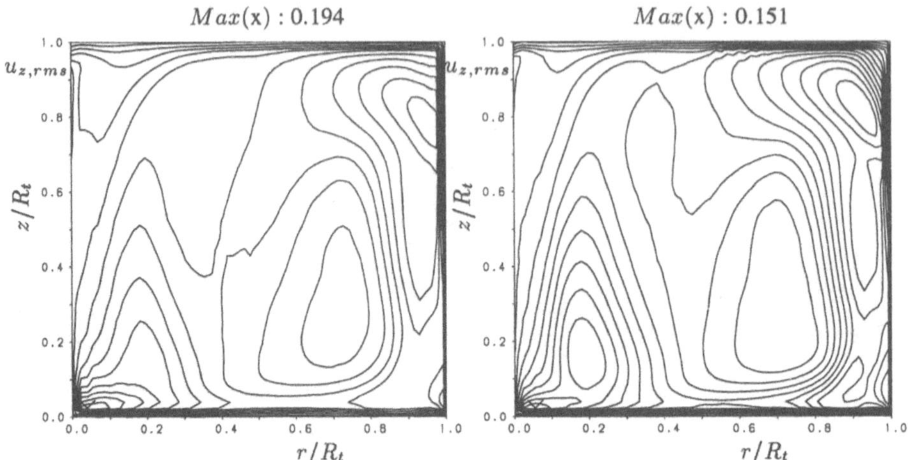


Figure 5: Contours of axial rms velocity fluctuations. Left: $Ma = 0$, Right: $Ma = 3.6034 \cdot 10^4$. Lines as in Fig.3.

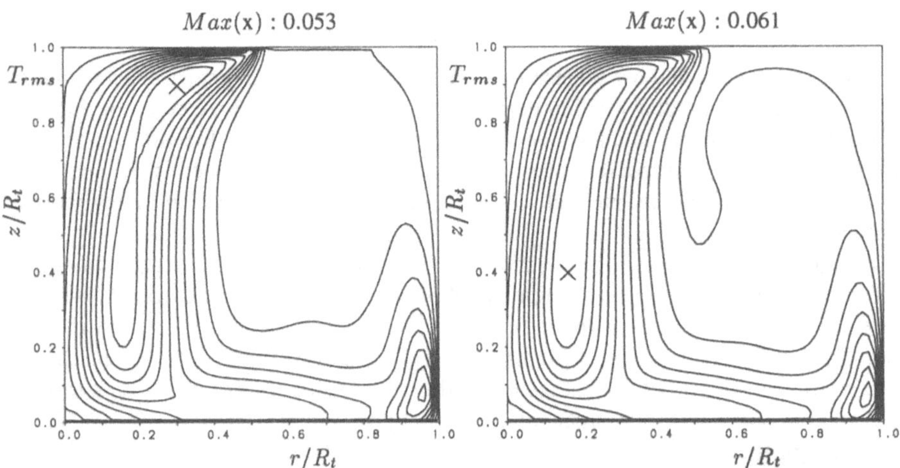


Figure 6: Contours of rms temperature fluctuations. Right: $Ma = 0$, Left: $Ma = 3.6034 \cdot 10^4$. Lines as in Fig.3.

transport of cold fluid (see the isothermes in Fig. 10) from below the crystal towards the crucible wall and the free surface. The circumferential velocity component is dominated by the crucible rotation as indicated by the predominantly negative values in Fig. 8. The highest turbulent activity is obviously found in a thin layer below the free surface and underneath the crystal.

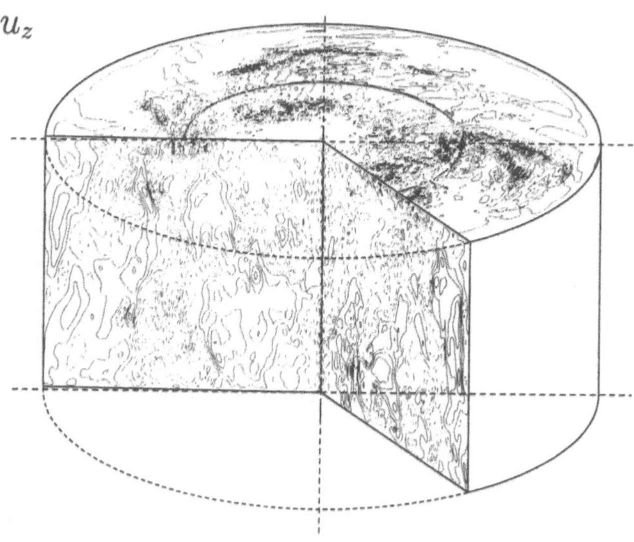


Figure 7: Snapshot of the instantaneous axial velocity component $u_z$. Solid/dashed contour lines represent positive/negative values. $(u_z)_{max} = 0.371$, $(u_z)_{min} = -0.653$.

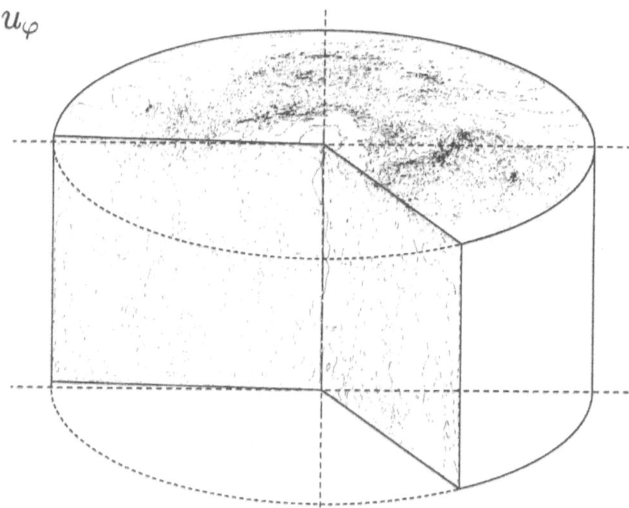


Figure 8: Snapshot of the instantaneous circumferential velocity component $u_\varphi$. Lines as in Fig. 7.

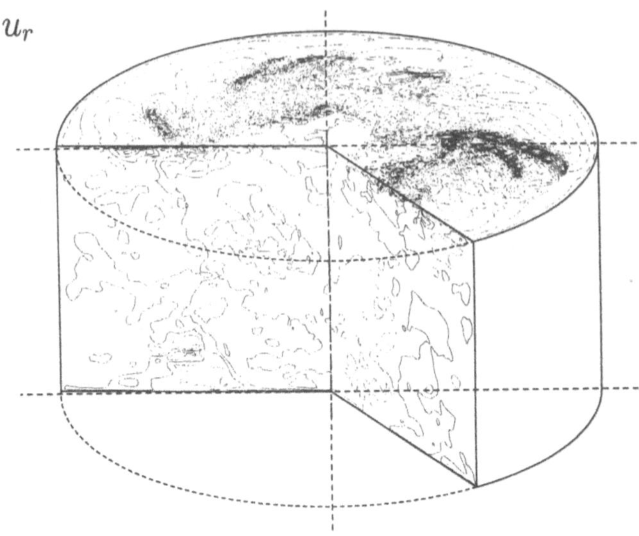


Figure 9: Snapshot of the instantaneous radial velocity component $u_r$. Lines as in Fig. 7. $(u_r)_{max} = 1.497$, $(u_r)_{min} = -1.581$.

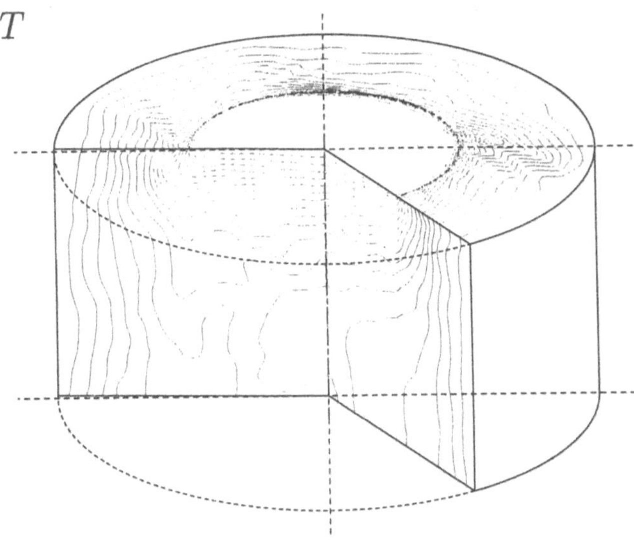


Figure 10: Snapshot of the instantaneous temperature $T$. Lines as in Fig. 7.

In order to obtain stable statistical values approximatly 1000 realizations with a time lag of 400 time steps were averaged in circumferential direction and in time. The global bulk flow structure and heat transport is visualized by streamlines of the mean velocity field in Fig.11 and mean isotherms in Fig. 12.

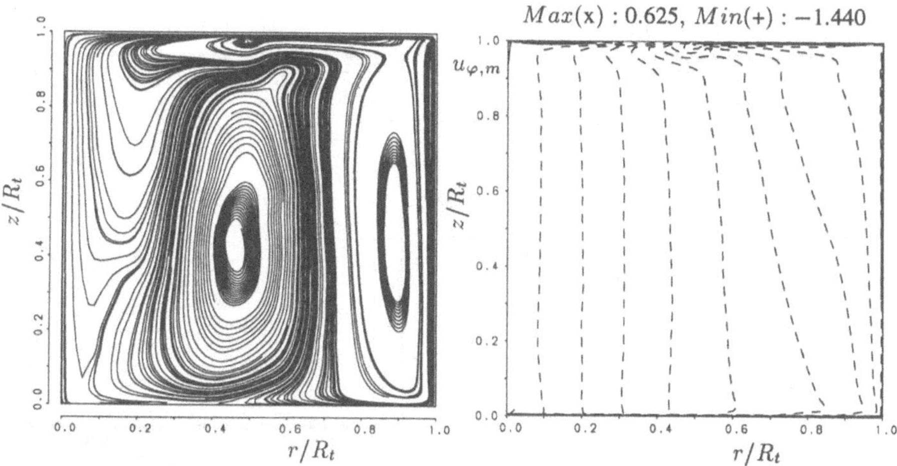


Figure 11: Left: Streamlines of the mean velocity field. Right: Contours of the mean circumferential velocity component. Lines as in Fig.3.

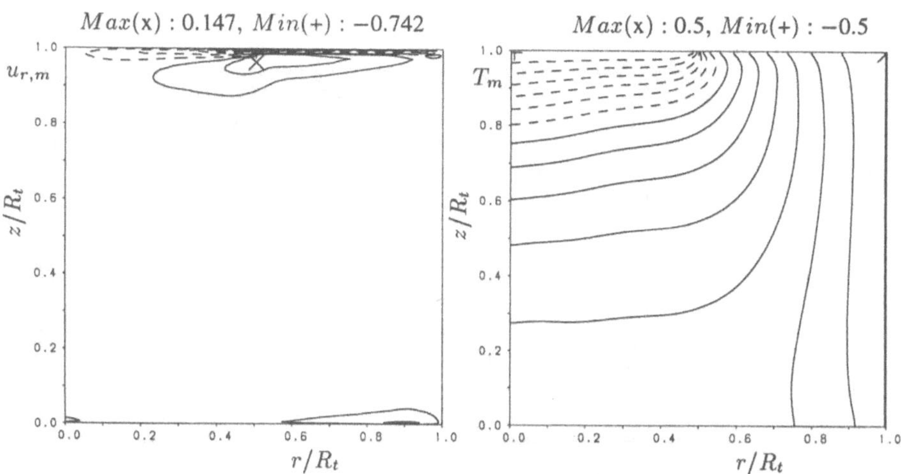


Figure 12: Left: Contours of mean radial velocity component, Right: Contours of the mean temperature. Lines as in Fig.3.

Buoyancy drives the melt upward along the heated crucible side wall as indicated by the wall parallel isotherms in this region (see Fig. 12). This flow turns inward at the meniscus and moves along the free surface creating a flow towards the cylinder axis close to the free surface. In the vicinity of the crystal the flow encounters centrifugal forces due to crystal rotation. Contours of the mean radial velocity component in Fig. 12 reveal a thin recirculation below the free surface, with a centrifugally forced flow towards the crucible wall below a layer of hot fluid.

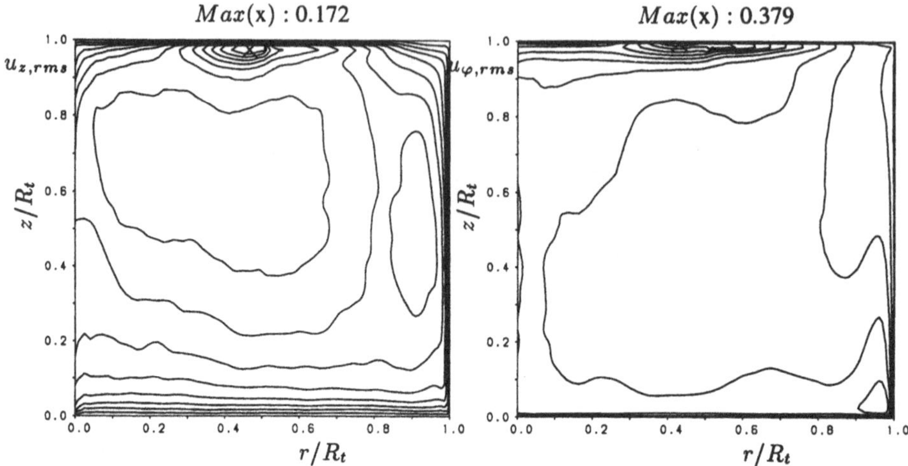


Figure 13: Left: Contours of axial rms velocity fluctuations, Right: Contours of circumferential rms velocity fluctuations. Lines as in Fig.3.

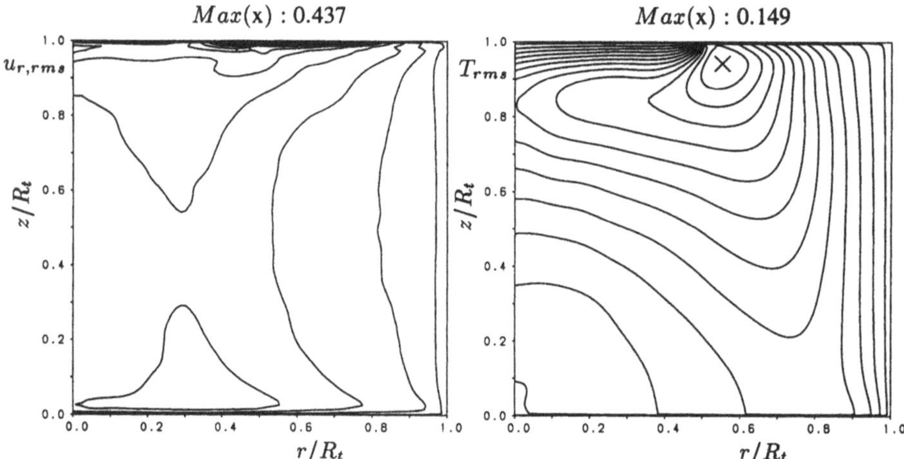


Figure 14: Left: Contours of radial rms velocity fluctuations, Right: Contours of rms temperature fluctuations. Lines as in Fig.3.

These two counteracting mechanisms lead to local ejections (as observed in Fig. 10) of cold fluid into the buoyancy driven hot recirculation zone creating large amplitude rms temperature fluctuations with positions close to the crystal edge in Fig. 12. The maximum rms temperature fluctations reach values of $13K$. Instantaneous temperature fluctuations even reach values of $28K$. It is most probable that these fluctuations create the inhomogeneities in the crystal.

The mean circumferential velocity component (Fig. 11) reveals a dominant counter-clockwise flow forced by the crucible rotation which encapsulates the zone of cold fluid underneath the crystal as shown in Fig. 12. The rotation of crystal and crucible forces the mean isotherms to be aligned with the crystal/melt interface, which is a necessary condition for a uniform growth of the crystal.

The locations of maximum rms velocity fluctuations in Fig. 13 to 14 are observed in a thin layer below the free surface and underneath the crystal, as already discussed in terms of the instantaneous flow fields.

## 5 Conclusions

We have performed Direct Numerical Simulátions of the turbulent flow in an idealized Czochralski crystal growth configuration.

Comparing the results of the pure buoyancy governed flow with crystal and crucible assumed to be stationary and vanishing thermocapillarity ($Ma = 0$) to the flow field with $Ma = 3.6034 \cdot 10^4$ we observed a nearly equal bulk flow structure and global heat transfer. But intensities of the mean flow and the temperature fluctuations were increased significantly. Since temperature fluctuations in the Si-melt seem to be responsible for inhomogeneities in the crystal, we feel that surface tension effects have to be modeled in simulations of the turbulent flow.

Introducing rotation of the crystal and the crucible at rates comparable to realistic growth conditions changes the flow structure and heat transfer in the melt tremendously. The crucible rotation forces the flow on spiral-like paths. Two recirculation zones of the bulk flow are observed. One is driven by buoyancy at the crucible wall and another driven by crystal rotation is located below the crystal. The cold fluid is encapsulated underneath the crystal with isotherms being aligned with the crystal/melt interface, a temperature distribution which allows for a uniform growth of the crystal and avoids undercooling of the melt. Just below the free surface centrifugally forced flow by the crystal rotation counteracts buoyancy and surface tension convection forming a third recirculation. In this situation cold fluid from underneath the crystal is ejected locally by centrifugal effects into the comparably warm buoyant recirculation zone and vice versa. This leads to large temperature fluctuations encountered in the measurements by Jones [1] and Kuroda et al. [2]. It is therefore recommended to decrease the rates of rotation so that fluctuations are minimized but still encapsulation of the cold fluid is achieved.

## References

[1] Jones, A.D.W (1989): "Flow in a Model Czochralski Oxide Melt.", J. of Crystal Growth 94, pp. 421-432.

[2] Kuroda, E., Kozuka, H., Takana, Y. (1982): "Influence of Growth Conditions on Melt Interface Temperature Oscillations in Silicon Czochralski Growth.", J. of Crystal Growth 68, p. 613.

[3] Elwell, D.,Andersen, E., Dils, R.R. (1989): "Temperature Oscillations in Silicon Melts.", J. of Crystal Growth 98, pp. 667-678.

[4] Mihelcic, M., Wingerath, K., Pirron, Chr. (1984): "Three-Dimensional Simulations of the Czochralski Bulk Flow.", J. of Crystal Growth 69, pp. 473-488.

[5] Mihelcic, M., Wingerath, K. (1989): "Instability of the Bouyancy Driven Convection in Si Melts During Czochralski Crystal Growth.", J. of Crystal Growth 97, pp. 42-49.

[6] Bottaro, A. and Zebib, A. (1989): "Three-Dimensional Thermal Convection in Czochralski Melt.", J. of Crystal Growth 97, pp. 50-58.

[7] Miyahara, S. Kobayashi, S., Fujiwara, T., Kubo, T., Fujiwara H. (1990): "Global heat transfer model for Czochralski crystal growth based in diffusive-gray radiation.", J. of Crystal Growth 109, pp. 696-701.

[8] Dupret, F., Nicodeme P.,Ryckmans, Y., Wouters, P., Crochet, M.J. (1990): "Global modeling of heat transfer in crystal growth furnace.", Int. J. Heat Mass Transfer 33 (9), pp. 1849-1871.

[9] Ryckmans, Y., Nicodeme, P., Dupret, F. (1990): "Numerical simulation of crystal growth: influence of melt convection in global heat transfer and interface shape.", J. Crystal Growth, 99, pp. 702-706.

[10] Kobayashi, N. (1990): "Oxygen transport under axial magnetic field in Czochralski silicon growth.", J. of Crystal Growth 108, pp. 240-246.

[11] Fontaine, J.P., Randriamampianina, A., Bontoux, P. (1991): "Numerical simulation of flow structures and instabilities occuring in liquid-encapsulated Czochralski process.", Phys. of Fluids A, 3 (10), pp. 2310-2331.

[12] Wagner, C., Friedrich, R., Narayanan, R. (1994): "Comments on the numerical investigation of Rayleigh and Marangoni convection in a vertical circular cylinder.", Phys. of Fluids, 6 (4), pp. 1425-1433.

[13] Xiao, Q. and Derby, J.J. (1995): "Three-Dimensional melt flows in Czochralsky oxide-growth: high-resolution, massively parallel, finite element computations.", J. of Crystal Growth 152, pp. 169-181.

[14] Fontaine, J.P., Bountoux, P., Quazzani, J., Extremet, G.P., Raspo I., Chevrier, V. and Launay, J.C. (1996): "Oscillatory convection and control by rotation in a large Prandlt number melt during Czochralski process.", Eur. J. Mech., B/Fluids, 15, pp. 665-694.

[15] Schumann, U. (1973): "Ein Verfahren zur direkten numerischen Simulation turbulenter Strömungen in Platten- und Ringspaltkanälen und über seine Anwendung zur Untersuchung von Turbulenzmodellen.", Dissertation, Univ. Karlsruhe.

# Euler calculations for flows around longitudinally accelerated profiles and profiles in shear flow

C. Weishäupl

Lehrstuhl für Fluidmechanik, Technische Universität München
Boltzmannstraße 15, 85748 Garching, Germany

## Summary

The present paper is concerned with steady and unsteady flows around wing profiles, in particular with longitudinally accelerated flows in the transonic flow regime. Unsteadiness can be caused actively by an unsteady motion of the considered body itself, like a Mach number oscillation or passively by an unsteady or inhomogeneous incoming flowfield. Here investigations are performed with the Euler method of Yee, Roe and Davis for a NACA 0012 airfoil, discretizing the evaluation region by a structured multiblock grid. First a short overview of the importance of longitudinally accelerated flows is given. The used Euler method and the grid generation algorithm are explained in their basics. Results are presented for two flow types, namely for a horizontal oscillation at two different reduced frequencies and for the steady flow in an inhomogeneous incoming flowfield. The behaviour of the aerodynamic coefficients is analyzed with special regard to the influence of shock motion.

## Nomenclature

| | |
|---|---|
| $c_d$, $c_l$, $c_m$, $c_p$ | drag, lift, moment, pressure coefficient |
| $\Delta h$ | height of the shear wind field |
| $k$ | reduced frequency |
| $l$ | wing chord |
| $M$ | Mach number |
| $\mathbf{R}$ | matrix of the right eigenvectors of the Jacobian |
| $x, y, z$ | cartesian coordinates |
| $\alpha$ | angle of attack |
| $\varphi$ | phase shift |
| $\tau$ | dimensionless time |
| $\xi, \eta, \zeta$ | curvilinear coordinates |

## Introduction

Unsteady flows occuring at an aircraft have significant influence on the dynamics of the aircraft and its components. Consequently for the development of an aircraft effects induced by unsteady aerodynamics are of great importance, in particular to ensure the safety of the airplane. Unsteady flow conditions can be created in an active or passive way, namely by an unsteady motion of the aircraft itself like oscillations or any maneouver or by unsteady or inhomogeneous incoming flowfields.

Hitherto existing investigations of unsteady flows, using Euler methods, mainly deal with vertically accelerated flows of active origin, like pitching oscillations of an airfoil [1,2], that means a variation in angle of attack. But also longitudinal accelerations, corresponding to an oscillation of Mach number, affect the flowfield around the airfoil considerably. This can be seen for example from upper side wing pressure distributions in the transonic velocity regime. Both, shock strength and position are influenced significantly already by small Mach number variations in comparison to changes in angle of attack [3]. Longitudinally accelerated flows are in comparison to vertically accelerated ones less investigated, applying Euler codes. Horizontal oscillations are studied e.g. by Grünspahn [4], using an Euler code based on a bicharacteristic method and by Habibie [5]. For inhomogeneous incoming flowfields some experimental results exist [6].

A further reason for detailed investigations is the complexity of longitudinal accelerations, exemplarily obvious from the wave propagation of a source, accelerated uniformly from rest to supersonic speed: At uniform velocities any field point is reached by one disturbance in the subsonic case and by two in the supersonic. But for uniformly accelerated flows up to four disturbances emitted at different times may reach one point at the same time, that means that a complex flowfield occurs [3].

Therefore in the present paper, longitudinally accelerated flows, which occur in many practical cases, are investigated. As active causes Mach number changes corresponding to changes in the horizontal airfoil velocity are to be mentioned. They occur e.g. in connection with yawing motions of a wing or together with in-plane oscillations at variable swept wing configurations. Incoming unsteady flows come up in shear fields or at rotor blades, when the free flow is not aligned with the rotor axis.

In this paper a Mach number oscillation as active cause and an inhomogeneous incoming flowfield as passive cause are selected from the wide field of longitudinally accelerated flows. Here the transonic regime is of particular interest, as already small variations of Mach number lead to remarkable changes in aerodynamic loads [5].

The performed calculations for a NACA 0012 airfoil deal with unsteady inviscid flow and are carried out with the Euler method of Yee, Roe and Davis. The evaluation region is discretized by a structured multiblock grid, generated with a Poisson algorithm.

## Euler code

The performed investigations are based on inviscid compressible flow. For this case the flow variables are described with the Euler equations in curvilinear coordinates:

$$\frac{\partial \mathbf{Q}}{\partial \tau}+\frac{\partial \mathbf{F}}{\partial \xi}+\frac{\partial \mathbf{G}}{\partial \eta}+\frac{\partial \mathbf{H}}{\partial \zeta}=0 \tag{1}$$

with the solution vector $\mathbf{Q}$ and the fluxes $\mathbf{F}$, $\mathbf{G}$, $\mathbf{H}$ in the three spatial directions $\xi$, $\eta$, $\zeta$ [1]. To determine the flow variables at the new time level $(\tau+\Delta\tau)$ from the old time level $\tau$ the Euler equations (1) are discretized as

$$\mathbf{Q}^{\tau+\Delta\tau}=\mathbf{Q}^{\tau}-\frac{\Delta\tau}{\Delta\xi}(\tilde{\mathbf{F}}^{\tau}_{j+1/2}-\tilde{\mathbf{F}}^{\tau}_{j-1/2})-\frac{\Delta\tau}{\Delta\eta}(\tilde{\mathbf{G}}^{\tau}_{k+1/2}-\tilde{\mathbf{G}}^{\tau}_{k-1/2})-\frac{\Delta\tau}{\Delta\zeta}(\tilde{\mathbf{H}}^{\tau}_{l+1/2}-\tilde{\mathbf{H}}^{\tau}_{l-1/2}) \tag{2}$$

with $j, k, l$ as cell indices. For the Euler code of Yee, Roe and Davis the numerical flux is formulated as

$$\tilde{\mathbf{F}}_{j+1/2} = \frac{1}{2}\left(\mathbf{F}_{j+1} + \mathbf{F}_j + \mathbf{R}_{j+1/2}\mathbf{\Phi}_{j+1/2}\right) \tag{3}$$

with
$$\Phi^l_{j+1/2} = -\Psi(a^l_{j+1/2})\left(\alpha^l_{j+1/2} - Q^l_{j+1/2}\right)$$

for the $l$-th component of $\mathbf{\Phi}$, where $\Psi$ denotes the entropy correction, $a^l$ the eigenvalue of the jacobian, $\alpha^l_{j+1/2}$ the wave strength and $Q^l_{j+1/2}$ the TVD–minmod–Limiter

$$\Psi(z) = \begin{cases} |z| & |z| \geq \delta \\ \frac{(z^2+\delta^2)}{2\delta} & |z| < \delta \end{cases}$$

$$\alpha^l_{j+1/2} = \mathbf{R}^{-1}_{j+1/2}\Delta\mathbf{Q}_{j+1/2}$$

$$Q^l_{j+1/2} = minmod\left(2\,\alpha^l_{j-1/2}, 2\,\alpha^l_{j+1/2}, 2\,\alpha^l_{j+3/2}, \tfrac{1}{2}\left(\alpha^l_{j-1/2} + \alpha^l_{j+3/2}\right)\right).$$

The used Euler code is an upwind finite–volume shock–capturing TVD code. The integration in time is performed explicitly. Concerning unsteady flows, the code considers the unsteady metric terms and reveals a robust behaviour, tested in many applications [1,7,8]. As boundary conditions characteristic and non–reflecting boundary conditions are used at the farfield and characteristic and kinematic boundary conditions are implemented at the body contour. Details concerning the code are given in [1].

## Grid generation

In the following the grid generation method is described in its fundamental relations. Further informations can be obtained from [1]. At the body contour points are distributed analytically. On this basis points in space are distributed iteratively using a Poisson algorithm.

$$g^{11}\left(\frac{\partial^2\mathbf{r}}{\partial\xi^2}+P\frac{\partial\mathbf{r}}{\partial\xi}\right)+g^{22}\left(\frac{\partial^2\mathbf{r}}{\partial\eta^2}+Q\frac{\partial\mathbf{r}}{\partial\eta}\right)+g^{33}\left(\frac{\partial^2\mathbf{r}}{\partial\zeta^2}+R\frac{\partial\mathbf{r}}{\partial\zeta}\right)+2\left(g^{12}\frac{\partial^2\mathbf{r}}{\partial\xi\partial\eta}+g^{13}\frac{\partial^2\mathbf{r}}{\partial\xi\partial\zeta}+g^{23}\frac{\partial^2\mathbf{r}}{\partial\eta\partial\zeta}\right) = 0$$

with $\mathbf{r} = [x, y, z]^T$. $g^{ij}$ are the contravariant metric coefficients, whereby here the terms $g^{12}$, $g^{13}$, $g^{23}$ are set to zero to realize considerable orthogonalism and so the last term disappears. The remaining coefficients $g^{ii}$ can be evaluated as

$$g^{11} = \frac{g_{22}g_{33} - g_{23}g_{32}}{det(\mathbf{G}_{ij})}, \qquad g^{22} = \frac{g_{33}g_{11} - g_{31}g_{13}}{det(\mathbf{G}_{ij})}, \qquad g^{33} = \frac{g_{11}g_{22} - g_{12}g_{21}}{det(\mathbf{G}_{ij})}$$

resulting from the covariant metric coefficients $g_{ij} = \sum_{k=1}^{3} \frac{\partial x^k}{\partial\xi^i}\frac{\partial x^k}{\partial\xi^j}$. $\mathbf{G}_{ij}$ denotes the tensor of the covariant metric coefficients. $P$, $Q$, $R$ are the source terms, controlling the distribution of points in space. For their evaluation in space a Laplace–algorithm is suitable [1].
The described algorithm enables an exact representation of the body contour and avoids cell interactions. Cells can be concentrated in regions with high gradients, continuous growth and small deformation are realized. Regarding the multiblock topology the elliptic smoothing over block boundaries is to be mentioned as the characteristic feature.

Unsteady motions of bodies require a dynamic adaptation of the grid at the respective body position, here guaranteed by the Poisson algorithm at each timestep. So the grid generation and the Euler code are coupled for unsteady cases. The used method has been validated for many steady and unsteady cases [1,7,8] and supplies good agreement with results of other Euler codes.

## Discretization

The presented investigations are performed for a NACA 0012 airfoil. The discretization of the calculation region is performed with a structured grid, consisting of 12 blocks with totally 6900 cells. The farfield boundary lies 20 wing chords away from the body. The topology of the grid shown as an exploded view in Fig. 1a) is characterized as follows. In the region near the profile, the grid has C–topology with a high resolution of the leading edge nearly without deformation. In the outer region the topology is similar to an H–topology with cells oriented normally to the undisturbed flow direction. Fig. 1b) shows the detail near the body. Each airfoil side is resolved by 60 cells, the distance of the first cell line to the body is 0.001 of the wing chord. The grid quality is confirmed by a small deformation, continuous cell growth outwards and smooth transition over block boundaries.

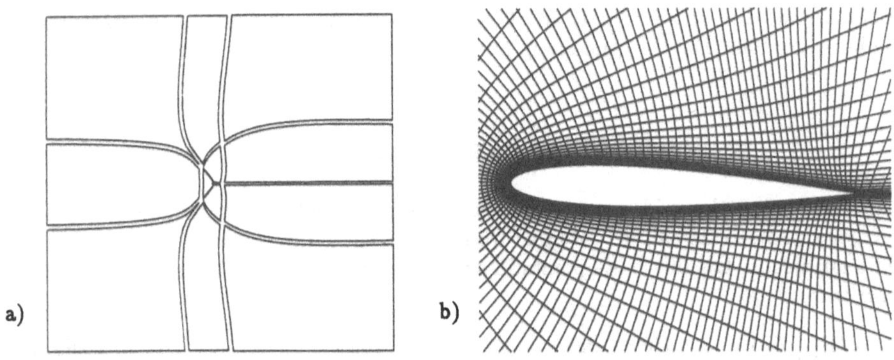


**Fig. 1** Discretization of the evaluation regime.

## Results and Discussion

Concerning longitudinally accelerated flows in this paper two cases are considered: first a motion of the body, that means an actively caused flow, namely a horizontal oscillation (case I) and secondly the steady flow in an inhomogeneous incoming flowfield (case II).

**Case I**: The motion of case I can be described by an oscillation of the Mach number $M = M_0 + \Delta M \cos(k\tau)$ with the mean value $M_0$, the amplitude $\Delta M$, the reduced frequency $k$ and the dimensionless time $\tau$, corresponding to the oscillation of the position of the airfoil $x = x_0 - \Delta x \sin(k\tau)$ and to the oscillation of the body velocity $u = u_0 - \Delta u \cos(k\tau)$. Horizontal oscillations are of major interest at flutter calculations where in–plane motions of the airfoil – occuring e. g. at airplanes with variable sweep – play an important role. At

transonic flow conditions periodic shock motions go along with the oscillations. Aerodynamic coefficients are largely affected. Therefore an oscillation in the transonic flow regime with $M_0 = 0.8$ and $\Delta M = 0.01$ is selected. The airfoil is investigated at an incidence of $\alpha = 2°$. Two different reduced frequencies $k = 0.1$ and $k = 0.5$ are investigated.
All figures concerning the horizontal oscillation show the fully developed response. The results for the lower reduced frequency $k = 0.1$ are given in Fig. 2 and 3. Fig. 2a) shows the lift coefficient $c_l$ over the Mach number $M$, Fig. 2b) presents $c_l$ over the location $x$, related to the amplitude $\Delta x$ of the profile.

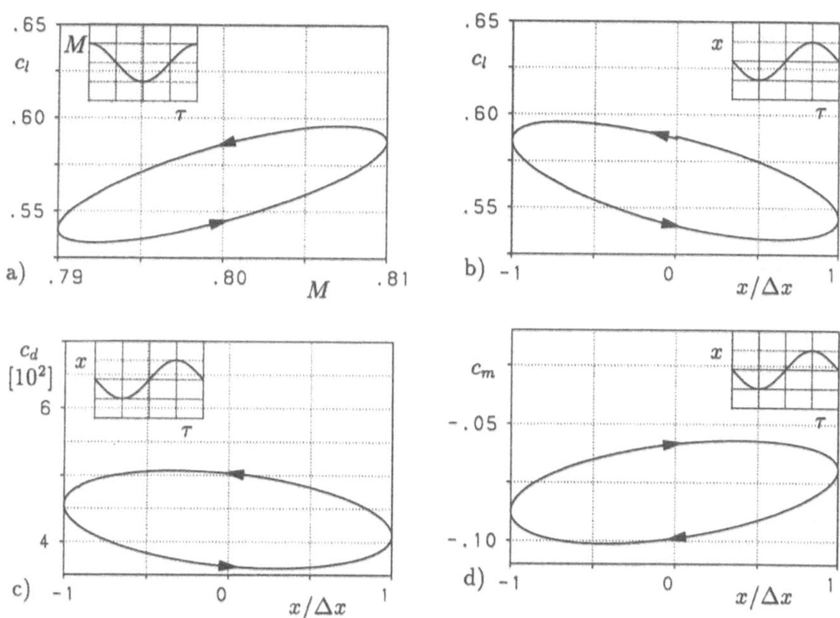


**Fig. 2** $M_0 = 0.8$, $\Delta M = 0.01$, $\alpha = 2°$, $k = 0.1$: characteristics of aerodynamic forces.

The answer of the lift coefficient $c_l$ appears with a phase shift of $\varphi_{c_l} = 41°$ after the impulse. From there first the lift coefficient decreases, corresponding to the decreasing Mach number and then $c_l$ increases again with growing Mach number. $c_l$ varies between 0.533–0.596. From Fig. 2b) it can be seen that $c_l$ increases when the profile moves forward and decreases when it moves backward. The presented results show principal agreement with the results of Habibie [5,9] for a similar case, who examined longitudinally accelerated flows, namely Mach number ramps and oscillations with the Euler code EUFLEX, developed by Eberle. Nevertheless the here presented results are smoother, showing an improvement versus Habibies solutions. Fig. 2c) and d) give the drag coefficient $c_d$ and the moment coefficient $c_m$ for the reference point $x_m = 0.25\,l$ over body location $x/\Delta x$. $c_d$ lies between 0.036 and 0.051 with $\varphi_{c_d} = 18°$. $c_m$ varies between $-0.1013$ and $-0.0571$ and is characterized by a phase shift $\varphi_{c_m}$ of 24°. This variation of the aerodynamic forces is in the regarded transonic regime caused by the shock motion and the mentioned time lag occurs due to the retarded establishing of the flowfield around the oscillating body and the slowness of the shock motion. This can be seen from the pressure distributions for different phases $\varphi = 0, \pi/2, \pi, 3\pi/2$ in Fig. 3. The distribution of the pressure coefficient on the upper side is characterized by following shock motion. With a certain time lag the shock follows the reduction of the Mach number and moves upstream, corresponding to the decreasing $c_l$ and increasing $c_m$ and similar for increasing Mach number the shock moves downstream, $c_l$ increases and $c_m$ decreases. Simultaneously the velocity on the lower side decreases and increases. That means that the shock motion is of great importance for the response of the aerodynamic coefficients, its slowness is responsible for the time lag.

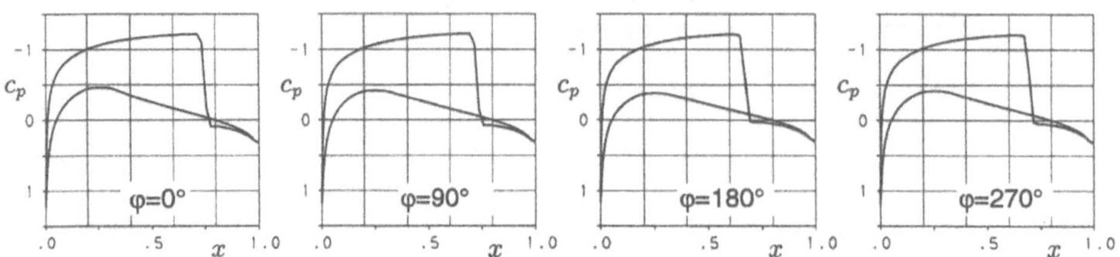


**Fig. 3** $M_0 = 0.8$, $\Delta M = 0.01$, $\alpha = 2°$, $k = 0.1$: **pressure distributions.**

The results for an increased reduced frequency $k = 0.5$ with all other parameters unchanged are given in Fig. 4 and 5:

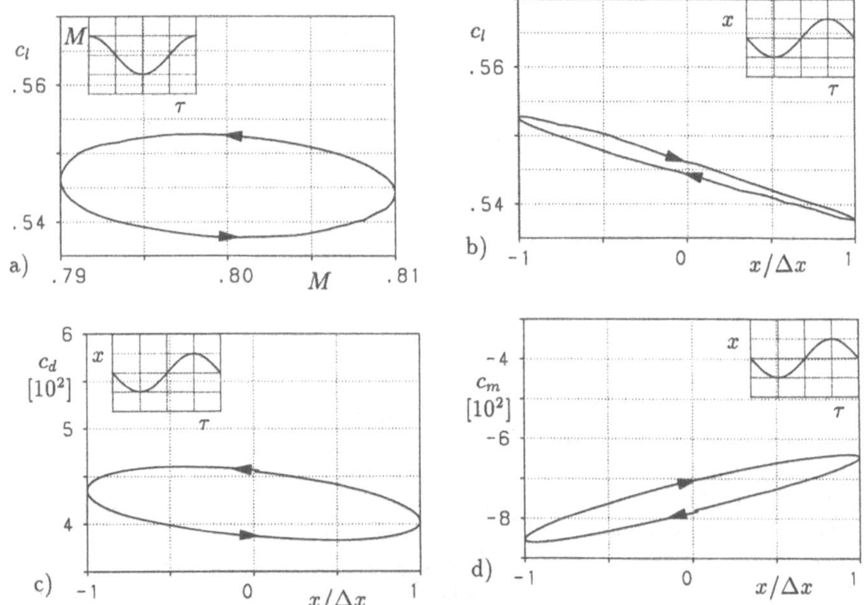


**Fig. 4** $M_0 = 0.8$, $\Delta M = 0.01$, $\alpha = 2°$, $k = 0.5$: **characteristics of aerodynamic forces.**

Fig. 4a) shows $c_l$ over the Mach number, Fig. 4b) $c_l$ over $x/\Delta x$. In Fig. 4c) $c_d$ over $x/\Delta x$ and in 4d) $c_m$ over $x/\Delta x$ is presented. All curves are characterized by a smaller amplitude in the aerodynamic forces and a larger time lag as for $k = 0.1$. The range of $c_l$ reduces to 0.538 - 0.553, for $c_d$ one obtains values between 0.038 and 0.046 and for $c_m$ between $-0.086$ and $-0.064$. The phase shifts increase considerabely to $\varphi_{c_l} = 98°$, $\varphi_{c_d} = 26°$ and $\varphi_{c_m} = 70°$. Both, the decreasing amplitude and the increasing phase shift are caused by the slowness of the shock motion. Due to the retarded development of the flowfield the shock can follow the faster motion of the body only in a restricted way and therefore the shock position does not vary as much as for the lower reduced frequency. This can also be seen from the pressure distributions in Fig. 5.

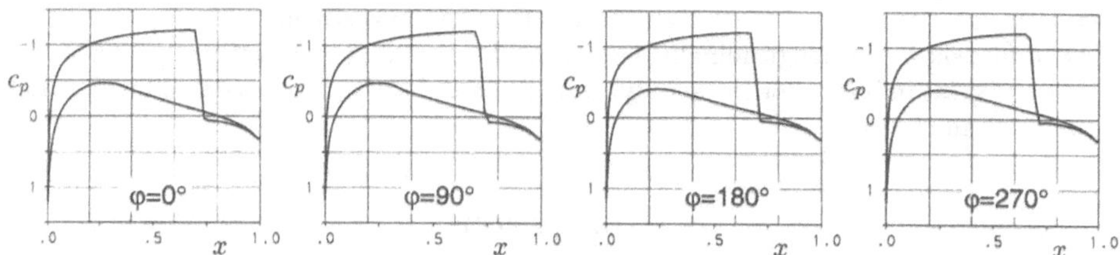


**Fig. 5** $M_0 = 0.8$, $\Delta M = 0.01$, $\alpha = 2°$, $k = 0.5$: pressure distributions.

Analogous to $k = 0.1$ after a time lag the shock moves upstream with decreasing Mach number and downstream with increasing Mach number, but the timelag is greater and the shock motion smaller as for $k = 0.1$. That means a higher reduced frequency results in a smaller amplitude and a greater phase shift of the response due to the slowness of the shock motion.

**Case II**: Case II deals with the steady flow in an inhomogeneous linear incoming flowfield with $M(z) = M_l + (M_u - M_l)(z - z_l)/\Delta h$ shown in Fig. 6a) in the same velocity regime as the horizontal oscillation. The lower Mach number $M_l$ is 0.7, the upper $M_u$ 0.9. Such flow conditions occur for a motion of an airfoil with $M = 0.8$ through a linear shearfield with positive and negative extrem values of $\Delta M = 0.1$. The airfoil is located at the half height of the shear field at an angle of attack of two degrees. Two heights for the shear field are investigated, namely $\Delta h = 5l$ and $1l$. As reference a homogeneous flow with $M = 0.8$ and $\alpha = 2°$ is regarded. Fig. 6b) shows the pressure distributions for the homogeneous and the two inhomogeneous cases with reference Mach number 0.8.

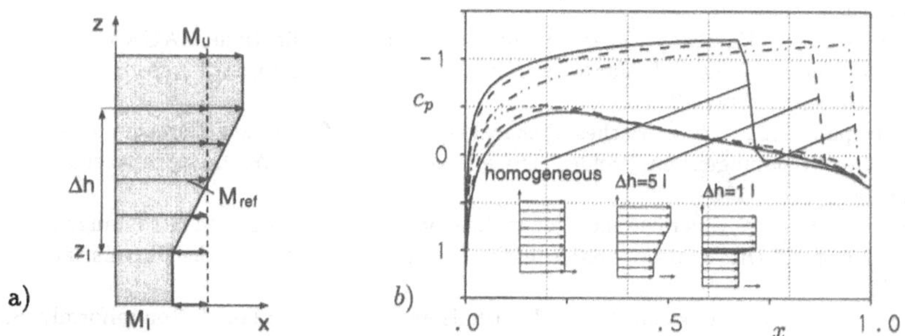


**Fig. 6** $M_l = 0.7$, $M_u = 0.9$, $\alpha = 2°$: inhomogeneous flowfield.

For the homogeneous case the shock on the upper side lies at 70% of the chord. For inhomogeneous flow with $\Delta h = 5\,l$ the velocity on the upper side becomes smaller than in the homogeneous case and therefore the $c_p$-values are higher and the shock is shifted downstream. The velocity on the lower side grows and the pressure becomes smaller, due to a motion of the stagnation point on the lower side towards the leading edge. The total

lift coefficient $c_l$ grows in comparison to the homogeneous case from 0.534 to 0.675. For a higher velocity gradient in the shear wind with $\Delta h = 1\,l$ the velocity on the lower side is increased and on the upper side is decreased further. The shock position is pushed further downstream and the stagnation point moves further upstream. Because of the approaching pressure curves of lower and upper side near the leading edge $c_l$ does not reach its value for $\Delta h = 5\,l$ although the shock moves downstream.
The results show that longitudinally accelerated flows caused in an active or passive manner are of great influence for the aerodynamic loads and are to be considered for the safety reasons of an aircraft.

## Conclusions

Longitudinally accelerated flows are of great importance in many practical cases. In this paper two types of such flows, namely a horizontal oscillation and an inhomogeneous incoming flowfield are evaluated for a NACA 0012 airfoil with the Euler method of Yee, Roe and Davis, using a structured 12 block-grid. The investigations are performed in the transonic velocity regime, because here complex flow conditions occur. For the horizontal oscillation a large influence of the shock motion and its slowness on the aerodynamic coefficients can be seen. A higher reduced frequency leads to a larger phase shift and a reduced amplitude of the aerodynamic response. Concerning incoming inhomogeneous flowfields for steady flow conditions, a positive velocity gradient results in a lower velocity on the upper side and a displacement of the shock.

## References


[1] Rochholz, H., " Eulerlösungen für den Separationsvorgang von Träger/Orbiter-Systemen im Hyperschall," Dissertation, Lehrstuhl für Fluidmechanik, Technische Universität München, 1994.

[2] Wegner, W., " Vollständige Riemannlösung der ein- und zweidimensionalen Euler-Gleichungen," Dissertation, Lehrstuhl für Fluidmechanik, Technische Universität München, 1992.

[3] Laschka, B., " Unsteady Flows – Fundamentals and Applications," AGARD CP-386, Unsteady Aerodynamics – Fundamentals and Applications to Aircraft Aerodynamics, Göttingen, Germany, May 1985.

[4] Grünspahn, K., " Numerisches Bicharakteristikenverfahren zur Berechnung stark instationärer Strömungen um Profile," Dissertation, Rheinisch-Westfälische Technische Hochschule Aachen, 1988.

[5] Habibie, I. A., " Eulerlösungen für instationär längsbeschleunigte Strömungen um Tragflügelprofile," Dissertation, Lehrstuhl für Fluidmechanik, Technische Universität München, 1994.

[6] Ludwig, G. R., Erickson, J. C., " Airfoils in Two-Dimensional Nonuniformly Sheared Slipstreams," Journal of Aircraft, Vol. 8, NO. 11, November 1971.

[7] Heller, G., Kreiselmaier, E., " Eulerluftkräfte für Klappenschwingungen," Institutsbericht Nr. 96/05, Lehrstuhl für Fluidmechanik, Technische Universität München, 1996.

[8] Weishäupl, C., " Euler-calculations for steady and unsteady flows around a NACA 0012 airfoil," Overview of Research Projects on the CRAY Y-MP at the Leibniz-Rechenzentrum München, LRZ-Bericht Nr. 9601, Aug. 1996, pp. 357-362.

[9] Habibie, I. A., Laschka, B., Weishäupl, C., " Analysis of Unsteady Flows around Wing Profiles at Longitudinal Accelerations," ICAS-94-2.7.1, 19th Congress of the International Council of the Aeronautical Sciences, Anaheim, CA, Sept. 18-23, 1994.

# Numerical Investigation of the Flow Around a Fuselage-Canard Configuration for a Long Range High Capacity Aircraft

G. Wichmann, C.-H. Rohardt

DLR, Institute of Design Aerodynamics

Lilienthalplatz 7, 38106 Braunschweig, Germany

## Summary

In the conceptual development for a long range, ultra high capacity aircraft (UHCA) at Airbus Industrie also the application of a three surface configuration including a front fuselage canard is discussed. Therefore, as a contract work for Airbus Industrie, at DLR a numerical study concerning the flow around a fuselage-canard configuration has been performed with the aim to prove its aerodynamic feasibility especially in the high subsonic Mach number range. Starting from flow investigations about the isolated fuselage a fuselage-canard configuration could be designed. The analysis showed that the prescribed design requirements at cruise and off-design conditions had been met.

## Introduction

In the discussion of a possible configuration of a future long range, ultra high capacity aircraft the concept of a three surface configuration is under consideration, too. It is expected that higher flight performances can be achieved and specific problems related to the uncommonly large weight and inertia effects can be solved. That concept includes a small wing at the front fuselage - the canard -, then the main wing with high aspect ratio and a small rear horizontal stabilizer. These three surfaces act together in order to reduce trim drag by minimizing the downward force of the horizontal stabilizer to compensate the nose-down pitching moment of the main wing. Regarding higher manoeuverability and longitudinal balancing, especially during take-off and landing, the canard concept is promising, too.

As part of a contract work for Airbus Industrie, the DLR Institute of Design Aerodynamics carried out a theoretical study with the aim to develop the critical design boundary conditions of a corresponding fuselage-canard configuration and to prove the aerodynamic feasibility of such a configuration in the high subsonic cruise Mach number range.

## Numerical method

The study has been performed using the DLR Euler methods CEVCATS and FLOWer [1]. The computational grids necessary for the application of these methods are based on two types of grid topologies. The calculations of the fuselage alone have been done using a H-O type topology and C-H type grids have been applied for fuselage-canard calculations. Smooth surfaces have been generated using special CAD-routines and algebraic procedures have been used for surface mesh generation. The corresponding three-dimensional field grids have been generated using elliptical mesh generators with source term control in wall proximity to obtain equally spaced cells near canard and fuselage surfaces.

## Design conditions

The general arrangement of the basic configuration including a possible canard arrangement is given in fig. 1. It shows a conventional aircraft geometry having an overall length of 75 m and a span of 78 m which is supplied with a canard. That aircraft is assigned to operate on long distance routes at a cruise Mach number of $M_\infty = 0.85$.

Geometrical requirements including spanwise twist and thickness distribution had been prescribed by Airbus Industrie. For functional and structural reasons as well as limitations regarding pilot's view, the canard had to be placed in low wing position. The required planform of the canard can roughly be taken from fig. 1.

The aerodynamic requirements can be drawn from fig. 2. For the design Mach number $M_\infty = 0.85$ the fuselage-canard lift at cruise condition is prescribed to be $c_{LC} = 0.25$ and the angle of attack at $c_{LC} = 0.0$ is fixed at $\alpha_{0C} = -1.3^o$. For $M_\infty = 0.85$ the lift coefficient at buffet onset should be greater than $c_{LC} = 0.5$.

## Results of the flow calculations

### Analysis of the flow around the isolated fuselage

The first part of the investigation was the analysis of the flow around the isolated fuselage without the canard. For that purpose a smooth surface mesh was generated on the fuselage contour according to its initial geometry data. Starting from that a three-dimensional H-O type computational mesh was constructed in the surrounding flow field.

The computed surface flow about the front part of the fuselage for $M_\infty = 0.85$ and $\alpha = 2^o$ is illustrated in fig. 3 using contour lines of the Mach number. The plot shows a fairly smooth distribution of the local Mach number all over the front surface, except for a small region on the lower fuselage surface. There a shock wave with a local Mach number of $M_{local} = 1.35$ in front of the shock exists. This flow behavior is unacceptable, in particular because just there would be the location of the canard root trailing edge. A shock wave on that position would lead to even worse flow conditions. Consequently the fuselage contour in this region had to be smoothed in the next step in order to reduce the supersonic Mach numbers there by reducing the surface curvature.

The effect of surface smoothing is shown in fig. 4. The undesired large flow gradients and consequently the generated shock wave could be completely removed and the velocity distribution is now smooth everywhere. But the highest velocities still remain at the position where the shock wave was located before.

### Analysis of the flow around the fuselage-canard configuration

For the analysis of the fuselage-canard configuration a three-dimensional C-H mesh [2], [3] was generated around the fuselage-canard surfaces choosing first a canard contour without twist designed by the optimization procedure [4]. The mesh structure can be seen in fig. 5.

The results of the corresponding Euler calculation at design condition ($M_\infty = 0.85$ and $\alpha = 1.6^o$) exhibit a negative value of the total lift coefficient $c_L = -0.17$ corresponding with higher Mach numbers on the canard's lower surface (fig. 6). Consequently undesired high suction peaks appear on the lower surface, finally resulting in a negative value of the integrated lift.

In order to confirm these results the flow field about the isolated fuselage has been investigated in more detail. The result is shown in fig. 7, where the calculated flow field at the fuselage section x = 0.055 (mid chord position at canard root) is presented. The normal velocity components of the resultant flow field are plotted, which in the canard region show large velocity components directed downwards. Obviously these velocity components act like an additional negative angle of incidence of the canard, leading to the observed shift of the total lift to negative values. This result shows the strong three-dimensional influence of the fuselage on the canard pressure distribution.

To meet the prescribed aerodynamic conditions, this strong fuselage influence had to be compensated mainly by two actions, i.e. firstly by introducing an additional twist of the canard relative to the fuselage and secondly by a careful modification of the canard surface contour to obtain a reasonable pressure distribution.

After a number of analysis runs with several twist angles of the canard, finally an acceptable twist distribution was found. The twist angle at root has been fixed at $\varepsilon = 8^o$ with a linear variation to $\varepsilon = 2^o$ at the tip. Analyzing the corresponding fuselage-canard configuration the prescribed design lift could be met but unacceptable strong shock waves appeared on the upper surface inducing too early buffet onset.

Consequently, in the next step a careful modification of the existing canard surface contour was made in order to achieve reasonable pressure distributions at design and off- design conditions, as well. The design has been performed in a way similar to the procedure described in [5]. After a number of intermediate steps, a final design was achieved, which is presented in fig. 8. The Mach number distribution shows at the trailing edge of the upper surface root section a high velocity peak. The local Mach number reaches here $M_{local} = 1.4$ indicating the danger of shock induced flow separation. As stated before this effect is due to fuselage curvature effects. The trailing edge of the canard root is just at the location of the fuselage with the highest velocity peaks. This effect is even enlarged by the presence of the canard. The canard prevents the downward directed fuselage flow and additionally accelerates the flow along the fuselage-canard junction up to the canard trailing edge. As the figure shows, this disturbance could be limited to the near fuselage region with no influence on the outer part of the canard. The spanwise pressure distribution is smooth everywhere without suction peaks and shock waves. The corresponding value of the total lift coefficient is $c_L = 0.26$, meeting the design requirement. The chordwise pressure distribution of the final design in the mid section of the canard is compared with the initial one without twist in fig. 9. It demonstrates that the introduction of twist and the careful design of a reasonable canard contour leads to smooth pressure distributions as shown in the left part of the figure. The resultant lift meets the design requirement.

The off-design behaviour is presented in fig. 10 to 13. A high load case ($M_\infty = 0.85$; $\alpha = 4.1^o$; $c_L = 0.5$) is shown in fig. 10 and 11 and a high speed case ($M_\infty = 0.90$; $\alpha = 1.6^o$ with a resultant lift coefficient $c_L = 0.27$) in fig. 12 and 13, respectively. In both cases the fuselage induced suction peak of the trailing edge root section remains nearly unchanged. The larger supersonic region with a strong terminating shock wave in the high load case leads to the statement that this is the more critical case concerning buffet onset, although it must be mentioned that the aft shock in the high speed case can be critical, too.

Lift slope and zero lift angle of attack as well are plotted as functions of the freestream Mach number in fig. 14. It shows the expected increase of lift slope with increasing Mach number, whereas the zero lift angle of attack is decreasing for higher Mach numbers.

## Conclusions

In a contract work for Airbus Industrie a numerical study of the flowfield around a fuselage-canard configuration has been performed applying the DLR Euler codes CEVCATS and FLOWer. The configuration is based on the geometry of a given front fuselage of an ultra high capacity aircraft model and a specified canard planform. The results can be summarized as follows:

- a canard design is possible in the presence of the three-dimensional flowfield of the front fuselage
- design and off-design conditions have been met by the final configuration
- further improvements are possible by careful designs in more detail
- root section modifications are necessary
- aerodynamic feasibility of a fuselage-canard configuration for an ultra high capacity aircraft even in the high subsonic Mach number range of cruise has been proved for an application within a three surface concept.

In a DLR project further theoretical and experimental investigations are performed on this topic with the aim to determine the performance potentials of three surface configurations for large transport aircraft. Within an integrated predesign here aspects of flight mechanics, structural mechanics and aeroelasticity are considered, too.

## References


[1] Radespiel, R.; Rossow, C.-C.; Swanson, R.C.:
*Efficient Cell-Vertex Multigrid Scheme for the Three-Dimensional Navier-Stokes Equations.*
AIAA-Journal, Vol. 28, No.8, (1990), S. 1464 - 1472.

[2] Leicher, S.; Fritz, W.; Grashof, J.; Longo, J.:
*Mesh Generation Strategies for CFD on Complex Configurations.*
Lecture Notes in Physics, Vol. 170, Eight International Conference on Numerical methods in Fluid Dynamics. Berlin/Heidelberg: Springer Verlag, 1982, S. 329 - 334.

[3] Wichmann, G.:
*Untersuchungen zur Flügel-Rumpf-Interferenz durch Anwendung eines Eulerverfahrens für kompressible Strömungen.*
ZLR-Forschungsbericht 93-06, Dissertation TU Braunschweig, (1993).

[4] Bartelheimer,W.:
*Design of Transonic Airfoils and Wings Based on the Solution of the Euler/Navier-Stokes Equations.*
DLR-IB 129-94/19, (1994).

[5] Horstmann, K.H.; Rohardt, C.-H.; Wichmann, G.; Dressler, U.; Hansen, H.; Rill, S.:
*Auslegung eines Laminarhandschuhs für die Fokker 100.*
8. DGLR-Fach-Symposium „Strömungen mit Ablösung“, 10.-12. November 1992, Köln-Porz-Wahnheide, (1992).

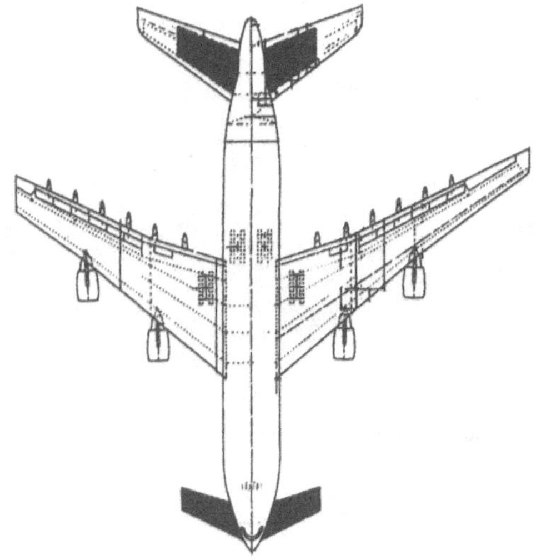

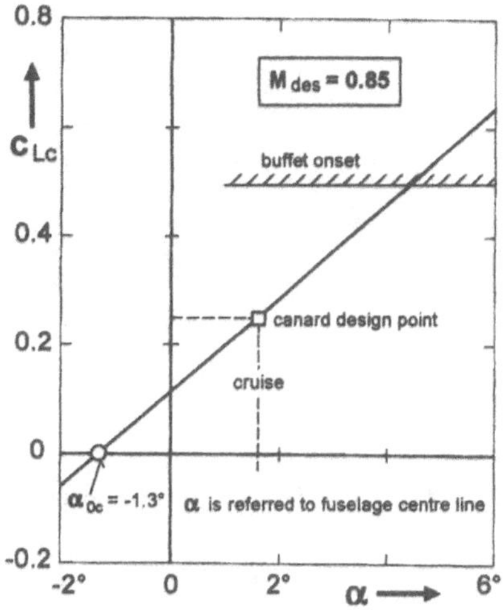

**Fig. 1** Possible canard concept on a future UHCA

**Fig. 2** Design point of canard on future UHCA

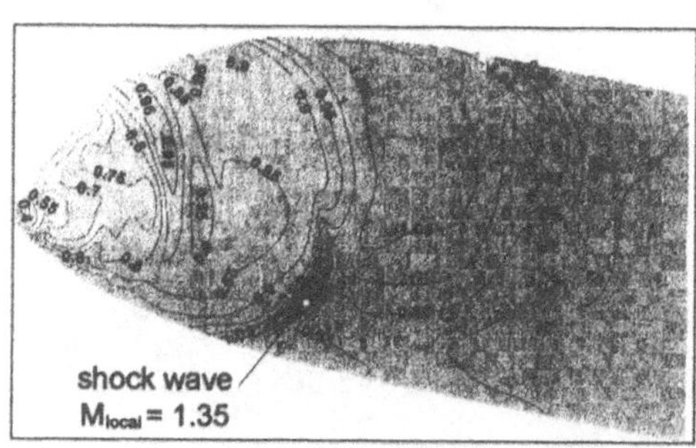


**Fig. 3** Lines of constant Mach number on front fuselage for original shape

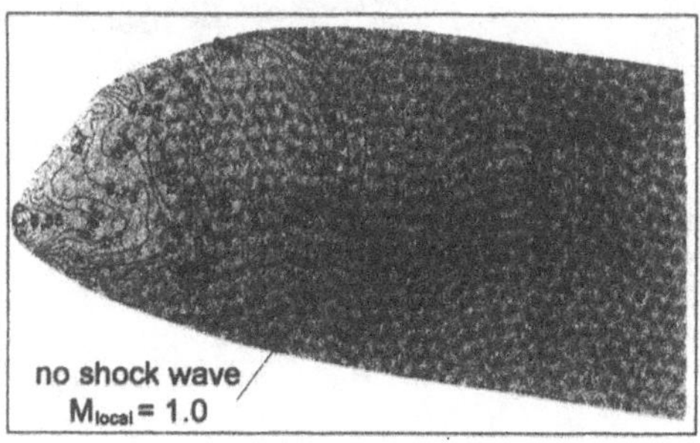


**Fig. 4** Lines of constant Mach number on front fuselage for smoothed shape

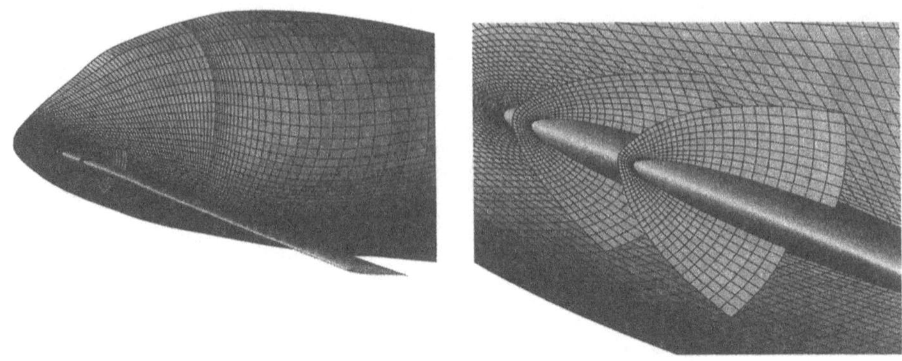

**Fig. 5** Surface grid of fuselage and C-type grid around canard wing

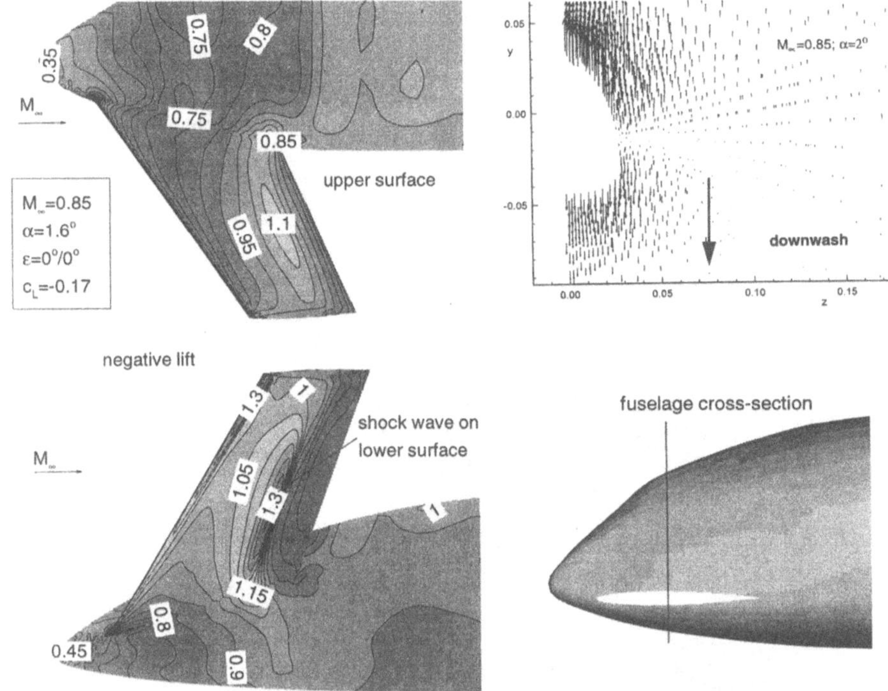


**Fig. 6** Lines of constant Mach number on canard surface of original design without twist

**Fig. 7** Analysis of flow velocities in the vicinity of the canard-fuselage junction

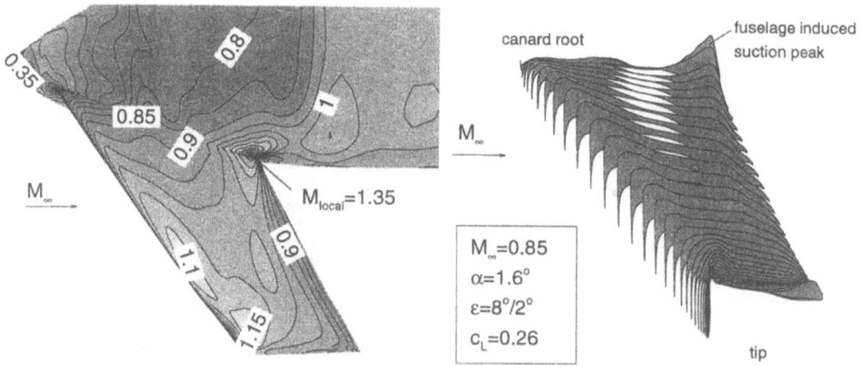


**Fig. 8** Lines of constant Mach number and surface pressure coefficients on final design canard wing

canard mid section pressure distribution

$c_p$

final design
with twist and
contour modification

$c_p^*$

design point
M=0.85 α=1.6° ε=8°/2°

canard mid section pressure distribution

initial design
without twist

$c_p^*$

negative lift

design point
M=0.85 α=1.6° ε=0°/0°

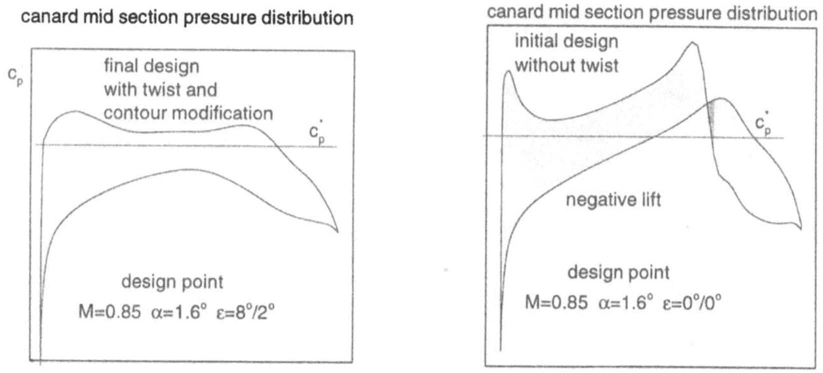

**Fig. 9** Typical section pressure distribution of final canard wing and original design canard wing

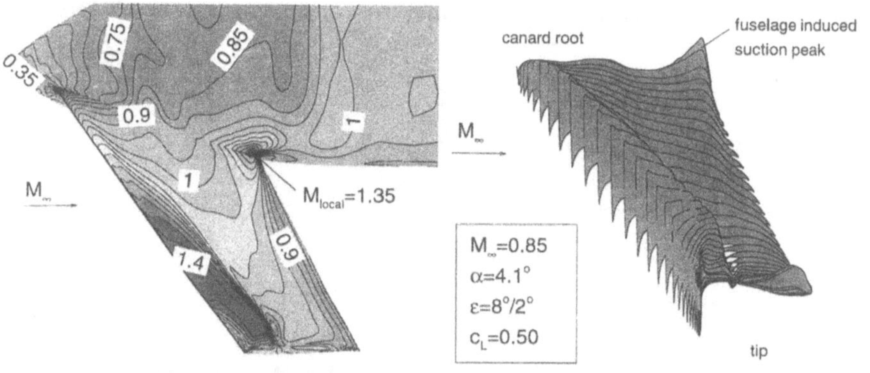


**Fig. 10** Lines of constant Mach number and surface pressure coefficients on canard wing of final design at $c_L$=0.5 (Buffet Onset)

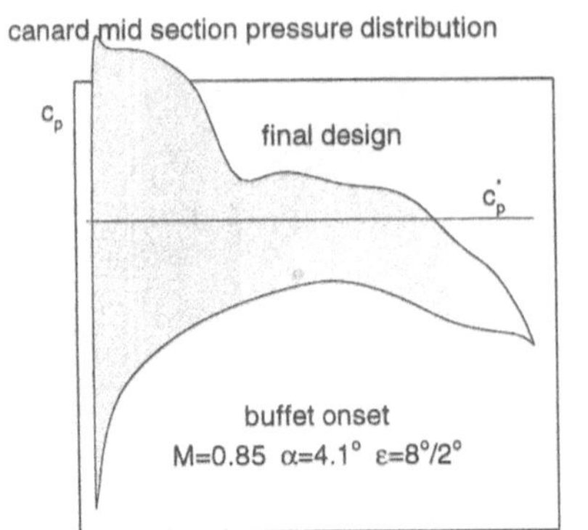


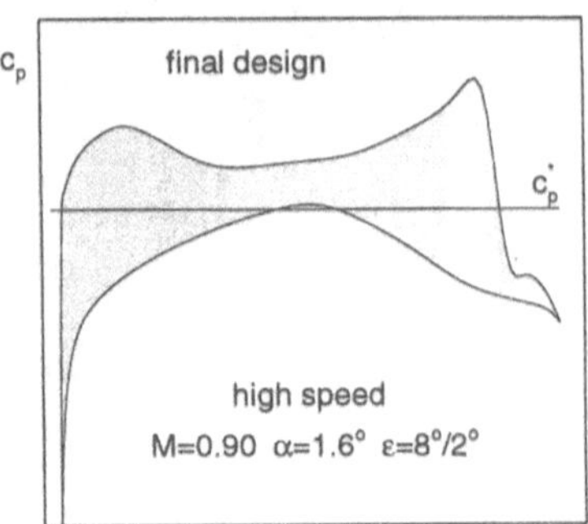


**Fig. 11** Mid section pressure distribution at $C_L$=0,5 (Buffet Onset)

**Fig. 12** Mid section pressure distribution at M=0,9 (High Speed)

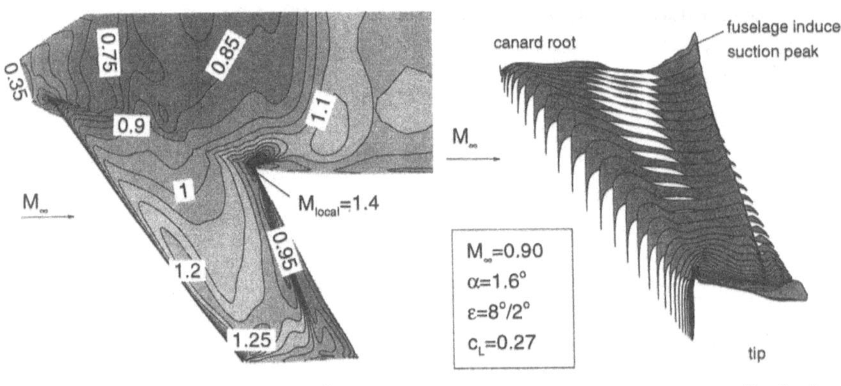


**Fig. 13** Lines of constant Mach number and surface pressure coefficients at $M_\infty$ = 0,9 (High Speed

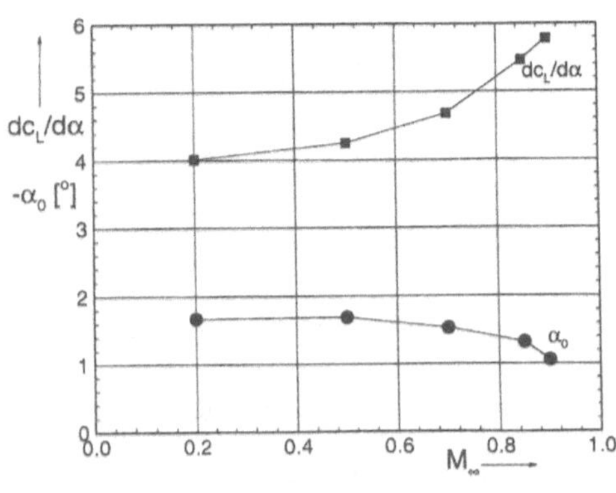


**Fig. 14** Lift slope and zero lift angle of attack as function of Mach number

# TRANSITION PROCESS OF A WAVE TRAIN IN A LAMINAR BOUNDARY LAYER

Th. Wiegand, H. Bestek, S. Wagner
Institut für Aerodynamik und Gasdynamik, Universität Stuttgart
Pfaffenwaldring 21, D-70550 Stuttgart, Germany

## SUMMARY

We report on experimental investigations of the evolution of a harmonic point-source disturbance in a laminar flat-plate boundary layer. The experiments were performed in a laminar water-channel designed for low-speed transition experiments. Small initial disturbances were introduced at a point source by periodic excitation with a single frequency. This type of artificial disturbance contains a broad spectrum of spanwise wave number modes, which are amplified or damped depending on the local stability characteristics. Detailed hot-film measurements in the linear regime of disturbance development show a wedge-like downstream evolution of the resulting wave train. A *subharmonic-type* breakdown process of the wave train with the characteristic pattern of *staggered* Λ-shaped vortices was observed by using flow visualization with hydrogen-bubbles and laser-sheet illumination.

## INTRODUCTION

Much of the present knowledge concerning possible routes of transition to turbulence has been gained from controlled experiments in flat-plate boundary layers. In such experiments, some sort of disturbance generator has been used to introduce artificial disturbances under controlled and repeatable conditions. The streamwise evolution of these disturbances up to their breakdown into turbulence was then investigated using quantitative measurements or flow visualization techniques. Most often the Tollmien-Schlichting route to transition was investigated, where a vibrating-ribbon was used to excite 2-D Tollmien-Schlichting (TS) waves [9]. When in the unstable region of the stability diagram of linear stability theory (LST) and for small disturbance amplitudes, these waves grow with downstream distance. If the amplitude of the 2-D waves reaches certain threshold values, a three-dimensional disturbance development due to a secondary instability mechanism may set in. This development leads to the formation of Λ-(lambda)-shaped vortices that appear in an aligned or staggered pattern, for fundamental or subharmonic breakdown respectively.

Another important class of experiments deals with a point source to generate localized 3-D disturbances. The early stages of natural transition do not involve single frequency disturbances, but have a broad spectrum of disturbance frequencies, resulting from rather broad-band freestream turbulence or localized initial disturbances. Therefore, point-source disturbances represent a better model of "natural" disturbance generation within the boundary layer than single frequency 2-D disturbances. When a pulse excitation with small amplitudes as in the experiments of Gaster & Grant [3] is used, a broad spectrum of modes in the frequency-spanwise wave number domain are excited. As a consequence a wave packet develops through selective amplification and superposition of the linear waves. This wave packet grows and spreads downstream and finally evolves into a turbulent spot [1].

In the literature several experiments are reported using a time-periodic point source [4, 5, 10]. In this way, a three-dimensional wave train is generated consisting of a finite spectrum of pairs of oblique waves superposed onto a 2-D wave, where all generated waves have the same frequency. However, these investigations focussed on the linear stage of wave train development, and not on the later stages of transition including the final breakdown to turbulence.

In the present paper experimental investigations of the evolution of a harmonic point-source disturbance are discussed. These experiments were carried out in a laminar water-channel at the University of Stuttgart. In this paper these experiments are discussed with particular emphasis on the linear stage of disturbance development. A comparison to linear stability theory and visualizations of the breakdown process of a wave train emanating from a harmonic point source are performed.

## EXPERIMENTAL SETUP

The experiments reported here were carried out in the Laminar-Wasserkanal (LWK) at the Institut für Aerodynamik und Gasdynamik (IAG), University of Stuttgart. This free surface, closed-loop laminar water-channel is designed and optimized for stability and transition experiments in laminar boundary layers [11, 12]. The LWK has a rectangular (1.2*0.5*10 m) test-section with bottom and side-walls, made of floatglass, allowing flow visualization observations from all directions. The "constant-temperature" design of the water-channel laboratory and the insulating effect of thick sandwich walls ensure a very stable temperature of the water: variations are less than 0.05° Celsius per day. The channel can be operated within a velocity range from 0.05 m/s up to 0.20 m/s for low-turbulence conditions which ensures a highly uniform two-dimensional mean flow. The flow quality was carefully optimized resulting in a turbulence level of 0.05 % at a freestream velocity $u_\infty = 0.145$ m/s with a frequency range of 0.10 - 10 Hz [12].

A modular, 8.6 m long, flat-plate model was mounted horizontally inside the test-section, 20 cm beneath the free surface. An elliptical (10:1) leading edge was manufactured directly onto the first Floatglass-plate. This results in a smooth local acceleration of the near wall flow, that prevents the generation of a laminar separation bubble. The coordinate system is defined with x-axis in streamwise direction (x = 0: leading edge), y normal to the wall (y = 0: surface) and z in spanwise direction (z = 0: centerline of the point source). Natural transition of the flat-plate boundary layer starts at about 6.5 m downstream of the leading edge, and a fully turbulent boundary layer is detected at x = 8.0 m.

The flow measurements were conducted using hot-film equipment based on DISA 55M10 anemometers and hot-film-probes DISA 55R15 with overheat ratio 1.05. All measurements including the artificial disturbance excitation, conditional sampling and pre-analysis of the *phase-locked* instantaneous time signals are fully computer-controlled.

In order to design a disturbance generator, that allows introducing localized disturbances of sufficient amplitudes but without the unwanted secondary effects, several types and diameters of holes have been investigated by flow visualization using dye [12]. Based on the criteria frequency and amplitude flexibility, robustness, waterproof/-resistant and available mounting space, we decided to apply a blowing/suction mechanism with a piston driven by an electric servo-actuator. This disturbance generator is fully computer controlled including an *active signal optimization* to compensate friction effects arising from O-ring sealing between piston and housing. "Smooth" disturbance excitation was realized by a disturbance input employing a "local perforation": a pattern of holes (Ø 0.5 mm) located on a circle (Type A) or a pattern of holes on two concentrical circles (Type B).

In our present experiment the point source is located on the centerline of the plate at the distance x = 1.491 m from the leading edge corresponding to $Re_{\delta 1} = 750$. The wave train development has been investigated for three disturbance frequencies: f = 0.22; 0.32; 0.46 Hz, which correspond to the dimensionless frequency parameter $F = (2\pi\nu f/u_\infty^2)*10^4 = 0.66; 0.96; 1.38$. These frequencies have been chosen to study three different paths in the linear stability diagram as shown in Figures 1, 2 and for convenient comparison with experimental results of Gilyov et al. [4, 5] and Seifert [10].

# MEAN FLOW AND LINEAR STABILITY CHARACTERISTICS

The freestream flow velocity $u_\infty$ used in the experiments was 0.145 m/s. Within $0 \leq x \leq 5$ m the variations of $u_\infty$ did not exceed 1 %, i. e. the flat plate had almost a zero pressure gradient.

Usually, "theoretical" local boundary layer profiles (e. g. from Blasius or Falkner-Skan solutions) are assumed to obtain basic velocity distributions U(y) and $\partial^2U/\partial y^2$ of adequate accuracy to perform linear stability analysis by solving the Orr-Sommerfeld equation. This may result in a significant disagreement with measured amplification rates. Therefore, we calculated a 2-D base flow by solving the incompressible Navier-Stokes equations with a well-tested numerical code (see [2] for details). The integration domain was chosen to extend from x = 1.0 m to x = 6.0 m of the flat-plate model. The measured freestream velocity distribution u(x) was prescribed at the upper boundary of the integration domain. This calculated flow is denoted as experimental mean flow.

In Figure 1 the linear stability diagram $-\alpha_i = f(\gamma, F, x)$ for 2-D waves ($\gamma = 0$) of the experimental mean flow is compared with the neutral curve for Blasius flow. Two distinctive *local* effects of the experimental mean flow are visible: a slightly accelerating flow up to x = 2.0 m (causing a higher critical Reynolds number for 2-D waves) and an increasing deceleration beyond x = 5.0 m. Further downstream, the boundary layer becomes transitional, resulting in a nonlinear distortion of the mean flow. At this location, the comparison with LST

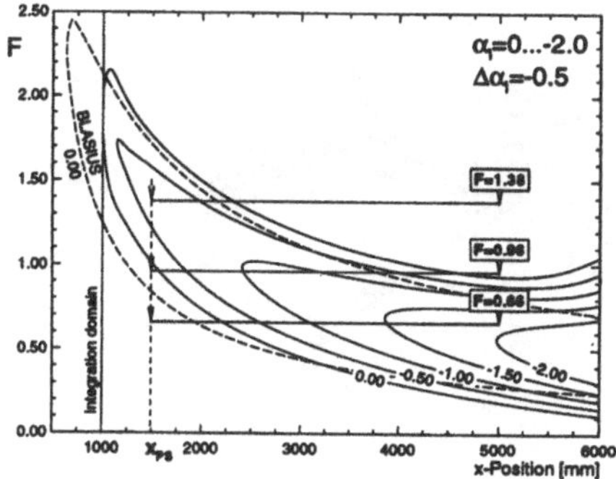


Figure 1: Linear stability diagram for the experimental mean flow and neutral curve for Blasius flow for 2-D disturbances. ($F = (2\pi\nu f/u_\infty^2)*10^4$, $u_\infty = 0.145$ m/s).

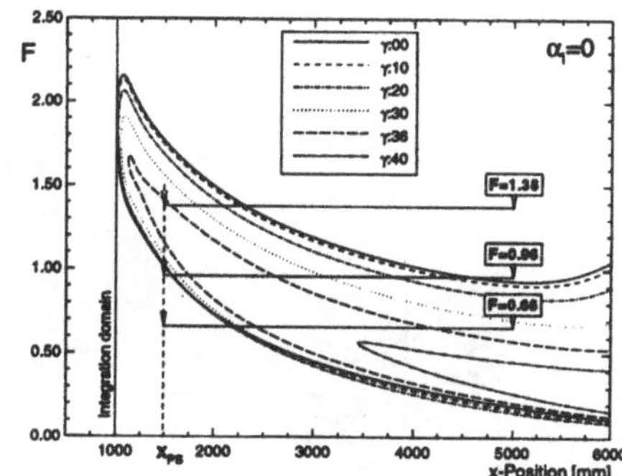


Figure 2: Linear stability diagram for the experimental mean flow for 3-D disturbances with various spanwise wave numbers γ.

is no longer useful. Within the region of the wave train development (x = 1.5 - 5.0 m) the stability characteristics are close to the Blasius case, but not identical. This arises from a slight deformation of the boundary layer profiles and shows clearly the strong sensitivity of the stability characteristics to weak local mean flow gradients $\partial U/\partial x$.

Figure 2 shows the neutral curves for 3-D waves with various spanwise wave numbers ($\gamma = 0$; 10; 20; 30; 36; 40). The stability characteristics are strongly different according to the three paths through the stability diagram which are marked by the disturbance frequencies F = 0.66; 0.96; 1.38. In case of the highest frequency investigated (F = 1.38) the 2-D wave and all 3-D wave pairs are damped downstream x = 2.5 m. Within the amplified region the 2-D portion is expected to be dominant, while the amplification of 3-D waves is weak. In contrast, the wave train development for F = 0.66 is dominated by various 3-D waves, which are amplified inside the major part of the measuring region (and further downstream).

# RESULTS

All hot-film measurements were made by employing a *phase-locking* procedure which was triggered by the "crossing zero and positive slope" position of the disturbance generator [12]. This procedure gives an exact relationship for the spatial phase $\Phi(x,y,z,t)$ of the disturbance signal relative to the temporal phase $\Phi_{PS}(x_{PS},y_{PS},z_{PS},t)$ of the forcing signal of the disturbance generator.

All amplitude and phase distributions for a single frequency $f_i$ (usually equal to the excitation frequency $f_{PS}$) were obtained from a Fast Fourier Transform (FFT) of the instantaneous time signals $u'(x,y,z,t)/u_\delta$. Spanwise wave number spectra $B_{i,k}(x_0,y_0,f_i,\gamma_k)$ were calculated by using a complex Fourier transform of the amplitude $A_i(x_0,y_0,z,f_i)$ and phase $\Phi_i(x_0,y_0,z,f_i)$ along the spanwise coordinate z. The local velocity outside the boundary layer $u_\delta$ was used to normalize all mean velocity and disturbance profiles.

Linear stability studies were performed using the computed Navier-Stokes profiles. The stability characteristics of these velocity profiles were calculated for different disturbance frequencies as well as for a number of spanwise wave numbers $\gamma$ by solving the Orr-Sommerfeld equation.

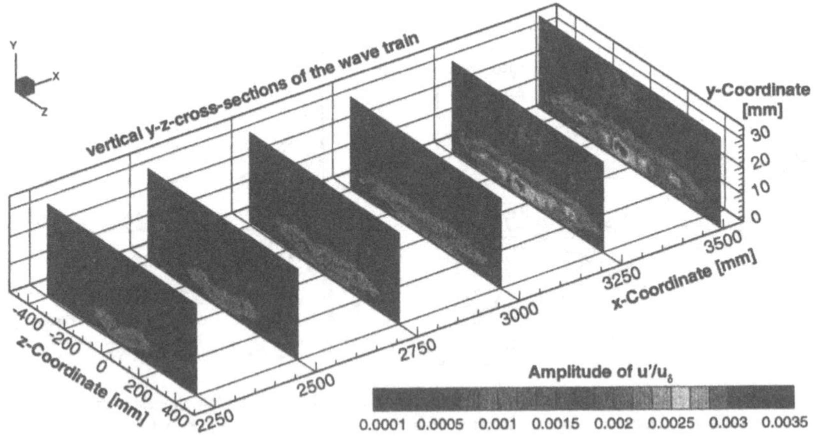


**Figure 3:** Amplitude distributions in 6 y-z-planes at x = 2.25 - 3.5 m ($Re_{\delta_1}$ = 950 - 1200) for f = 0.32 Hz.

## THE LINEAR STAGE OF DISTURBANCE DEVELOPMENT

The spanwise distributions of the mean velocity, disturbance amplitude and phase have been measured for three different frequencies at several downstream positions including the disturbance profiles along y. Figure 3 shows amplitude distributions for the frequency $f_i$ = 0.32 Hz ($F_i$ = 0.96) in six y-z-planes at streamwise positions from x = 2.25 m ($Re_{\delta 1}$ = 950) to x = 3.5 m ($Re_{\delta 1}$ = 1200). The total extensions of the measuring planes are y = 0...33 mm with the increment $\Delta y$ = 1.0 mm and z = ± 400 mm (± 480 mm at x = 3.5 m) with $\Delta z$ = 20 mm. This grid size has been chosen in order to obtain experimental data for the entire wedge of the wave train (half angle 10-12°). Thus, the frequency-wave number spectra as well as the eigenfunctions for *single frequencies* $f_i$ and *single spanwise wave numbers* $\gamma_k$ can be calculated.

The shape of the spanwise amplitude and phase distributions measured across the entire boundary layer varies strongly with distance from the wall. The lower the frequency is, the more significant the two-maxima amplitude distribution inside the boundary layer develops (Figure 4, 6: y = 6.0 mm). Outside the boundary layer the distributions are much smoother (Figure 3). This is in good qualitative agreement with the results of Gilyov et al. [4, 5] obtained at two constant nondimensional distances from the wall. It is obvious that the y-position of these maxima depends on the spanwise coordinate z which indicates an inherent phase-dependent superposition of 2-D and 3-D eigenfunctions.

The shape of the amplitude profiles (= disturbance amplitude vs. y) changes significantly along the spanwise coordinate z. Close to the centerline (z = 0), the amplitude profile is similar to the characteristic profile of 2-D TS eigenfunctions, whereas amplitude profiles near off-centerline maxima resemble 3-D eigenfunctions. Despite this similarity, the amplitude and phase profiles cannot be compared directly to eigenfunctions obtained by LST. For comparison with linear theory it is necessary to decompose the amplitude and phase distributions in a y-z-plane by complex spatial Fourier transform in order to calculate correct amplitude profiles for modes with *a single frequency* $f_i$ and *a single spanwise wave number* $\gamma_k$.

Figures 5, 7 show the spanwise wave number spectra corresponding to Figures 4, 6 using this decomposition. The 2-D component ($\gamma = 0$) almost disappears for f = 0.22 Hz (Figure 5). For f = 0.32 Hz (Figure 7), there is a strong 2-D wave component ($\gamma = 0$) as well as a pair of 3-D wave components ($\gamma = \pm 24\ m^{-1}$) with a slightly higher absolute amplitude. Symmetry requirements of point-source excitation allow 2-D wave components as well as pairs of oblique waves (with the same spanwise wave number but of opposite sign). The two maxima originating symmetrically to $\gamma = 0$ illustrate this physical effect for both f = 0.22 Hz ($\gamma = \pm 35\ m^{-1}$) and f = 0.32 Hz ($\gamma = \pm 24\ m^{-1}$). The corresponding wave angles are $\sigma = 55°$ for f = 0.22 Hz ($\alpha_r = 24.5\ m^{-1}$) and $\sigma = 33°$ for f = 0.32 Hz ($\alpha_r = 37\ m^{-1}$).

In Figures 8 and 9, the eigenfunctions of 2-D ($\gamma = 0$) and 3-D ($\gamma = \pm 15\ m^{-1}$) wave components for f = 0.32 Hz (F = 0.96) obtained from the experiment are compared with those calculated from LST. The agreement inside the boundary layer is excellent.

## VISUALIZATION OF THE WAVE TRAIN BREAKDOWN

Although several experimental as well as theoretical investigations of a wave train emanating from a harmonic point source have been published, the subsequent breakdown process and the relevant transition mechanisms are not yet understood. Most investigations focussed on the linear stage of the wave train development and show results of hot-film measurements and comparisons with LST [6, 7, 10]. The present experiments are performed in a low-turbulence/ low-speed water-channel which allows high quality flow visualizations. In order to explore the later stages of the wave train development, hydrogen-bubble-timelines and laser-sheet illumination have been used.

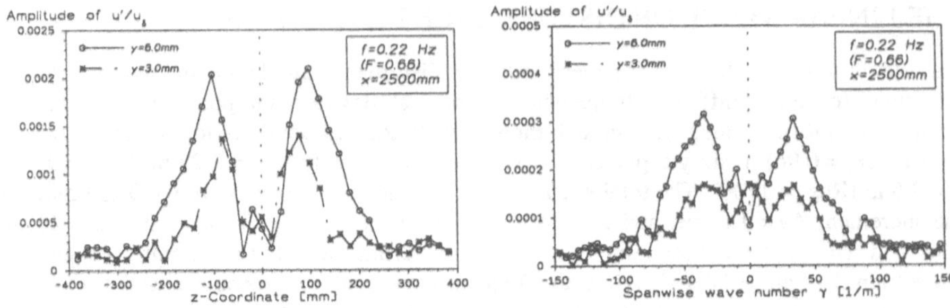


**Figure 4:** Spanwise amplitude distribution at x = 2.5 m ($Re_{\delta_1}$ = 1000) for f = 0.22 Hz.

**Figure 5:** Spectra of spanwise wave number γ at x = 2.5 m ($Re_{\delta_1}$ = 1000) for f = 0.22 Hz.

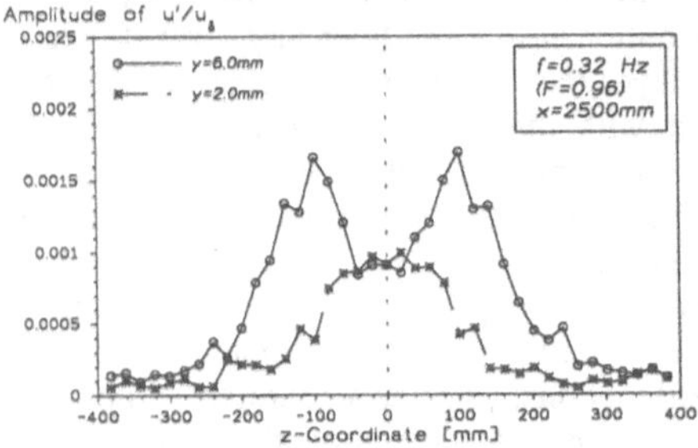


**Figure 6:** Spanwise amplitude distribution at x = 2.5 m ($Re_{\delta_1}$ = 1000) for f = 0.32 Hz.

Amplitude of u'/u_δ
y=6.0mm
y=2.0mm
f=0.32 Hz
(F=0.96)
x=2500mm
Spanwise wave number γ [1/m]

**Figure 7:** Spectra of spanwise wave number γ at x = 2.5 m ($Re_{\delta_1}$ = 1000) for f = 0.32 Hz.

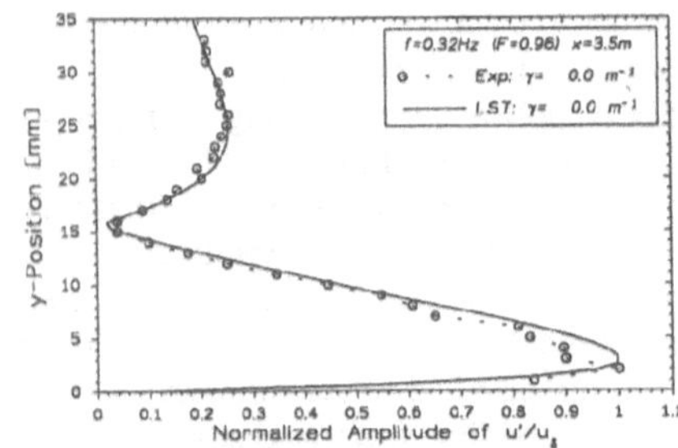


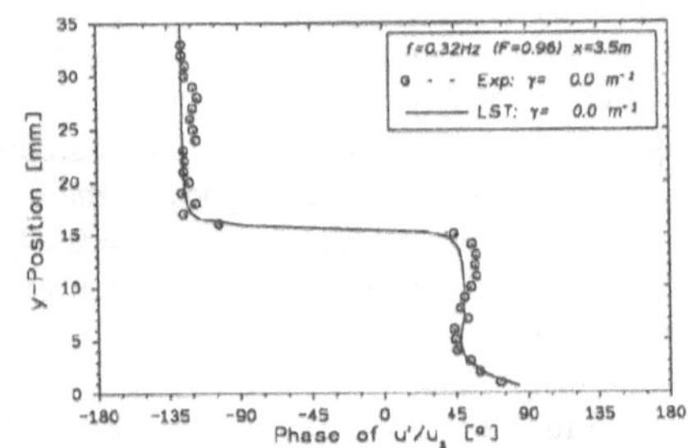


**Figure 8:** 2-D-Eigenfunctions (γ = 0) of amplitude and phase at x = 3.5 m ($Re_{\delta_1}$ = 1200) for f = 0.32 Hz. (- -o- - Harmonic point source experiment; —— Linear Stability Theory)

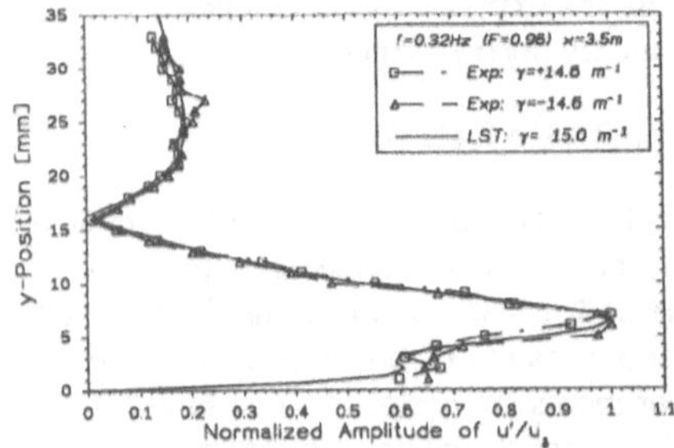


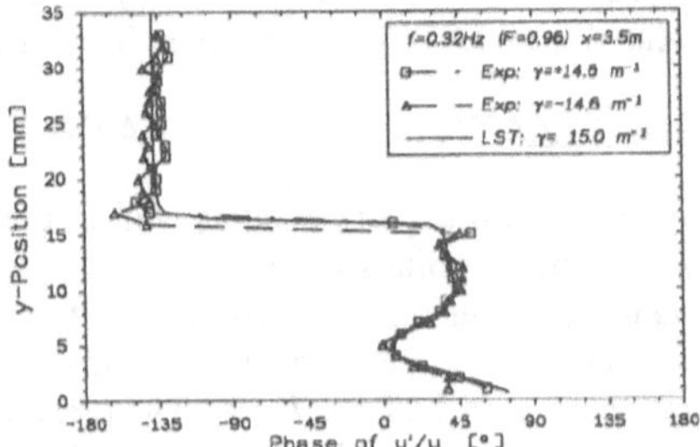


**Figure 9:** 3-D-Eigenfunctions (γ = 15) of amplitude and phase at x = 3.5 m ($Re_{\delta_1}$ = 1200) for f = 0.32 Hz. (- -□- -Δ- - Harmonic point source experiment; —— Linear Stability Theory)

For the result shown here the hydrogen-bubble wire was positioned at the distance x = 5.0 m downstream of the leading edge ($Re_{\delta_1}$ = 1500). For the case with f = 0.32 Hz (F = 0.96), the wire position is close to Branch II of the 2-D stability diagram (Figure 1). The wire was centerlined to the point source (z = 0) and it extended 550 mm in spanwise direction. It has been mounted accurately parallel to the wall at y = 10 mm, which is close to the critical layer. The hydrogen-bubbles were illuminated by a laser -sheet, which was adjusted to be exactly parallel to the flat plate. The pictures shown here were taken from a slightly off-centerline position in order to obtain a uniform distribution of scattered light.

Figure 10 shows the wave train development for an excitation frequency f = 0.32 Hz. Lambda-shaped vortices occur in the *staggered* ( = subharmonic) pattern [8]: peaks follow valleys and valleys follow peaks.

The spanwise wave length of the oblique 3-D wave can be determined approximately from Figure 10 as $\lambda_z$ = 195 mm (spanwise wave number, $\gamma = 2\pi/\lambda_z = 32\ m^{-1}$). The streamwise wave length of the 2-D fundamental wave is $\lambda_{x,2D}$ = 165 mm (streamwise wave number, $\alpha_{r,2D} = 2\pi/\lambda_{x,2D} = 38\ m^{-1}$). The wave angle for the 3-D subharmonic wave is approximately $\sigma = \arctan(\gamma/0.5*\alpha_{r,2D}) = \pm 60°$, which is close to the H-type of subharmonic breakdown [8]. The disturbance level is around $u'/u_\delta \approx 0.005$ for f = 0.32 Hz.

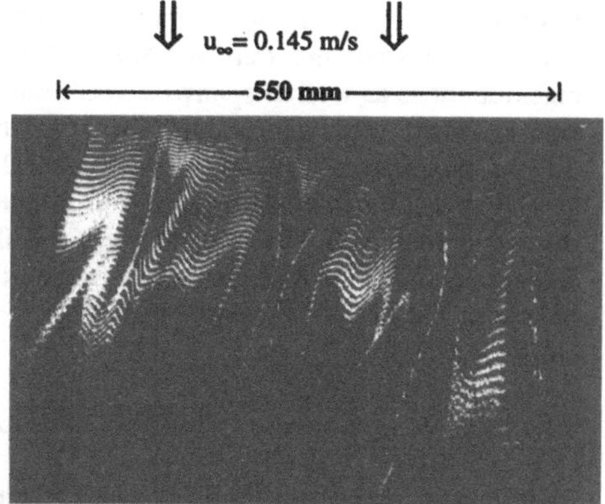


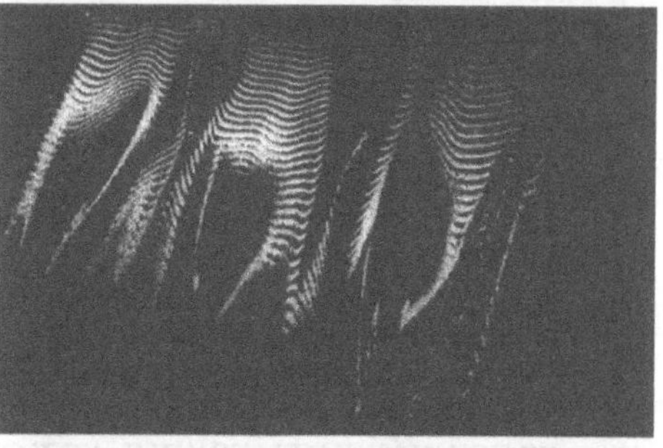

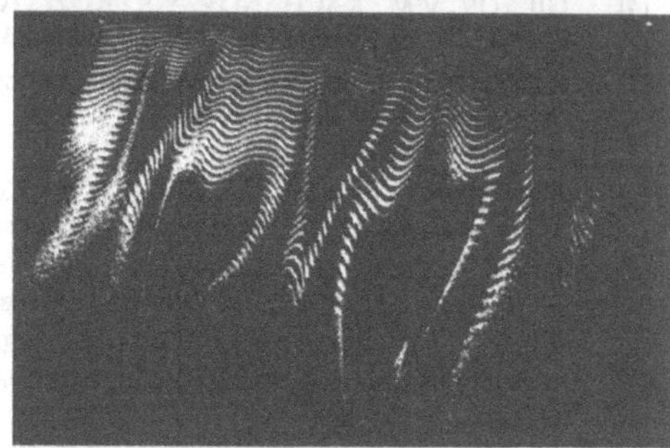

**Figure 10:** Subharmonic breakdown of a wave train emanating from a harmonic point source for f = 0.32 Hz (F = 0.96). Flow visualization of one time-period by hydrogen-bubble-timelines and laser-sheet illumination (the hydrogen-bubble wire is at y = 10 mm, x = 5.0 m, $Re_{\delta_1}$ = 1500).
Streamwise wave number of the 2-D fundamental wave $\alpha_{r,2D} = 2\pi/\lambda_{x,2D} = 38\ m^{-1}$, spanwise wave number $\gamma = 2\pi/\lambda_z = 32\ m^{-1}$, wave angle $\sigma = \arctan(\gamma/0.5*\alpha_{r,2D}) = \pm 60°$.

# CONCLUDING REMARKS

The response of a flat-plate boundary layer to the excitation by a harmonic point source was investigated. The streamwise development of the wave train emanating from the harmonic point source was measured and documented using hot-film anemometry and flow visualization. The laminar mean flow of the used facility (Laminarwasserkanal, LWK) for the present experimental investigation was close to the Blasius boundary layer, and in excellent agreement with a numerically obtained 2-D Navier-Stokes solution. A linear stability analysis of the computed 2-D flow which included the slight local pressure gradients of the experimental mean flow was carried out.

The shape of the spanwise amplitude and phase distributions measured through the entire boundary layer varies strongly with distance from the wall. The lower the frequency, the more significant is the two-maxima amplitude distribution inside the boundary layer while being much smoother outside. *Single frequency* $f_i$ and *single spanwise wave number* $\gamma_k$ eigenfunctions as well as spanwise wave number spectra were obtained by a complex Fourier analysis along the z-coordinate. These spectra show a frequency-dependent combination of 3-D wave pairs with a 2-D wave: the higher the frequency is, the stronger is the 2-D mode.

For the dimensionless excitation frequency $F = 0.96$ a *subharmonic-type* breakdown of the wave train was observed at $Re_{\delta_1} \approx 1500$. The amplitude level $u'/u_\delta \approx 0.005$ and the *staggered* pattern of Λ-shaped vortices resulting from a 2-D wave and a pair of ± 60° waves indicate a *H-type subharmonic* breakdown.

# ACKNOWLEDGEMENTS

The financial support by the Deutsche Forschungsgemeinschaft under contract Wa 424/13-1,2 is gratefully acknowledged.

# REFERENCES

[1] COHEN, J. (1994): The initial evolution of a wave packet in a laminar boundary layer. Phys. Fluids 6 (3), pp. 1133-1143.

[2] FASEL, H.F.; RIST, U.; KONZELMANN, U. (1990): Numerical investigation of the three-dimensional development in boundary-layer transition. AIAA Journal 28 (1), pp. 29-37.

[3] GASTER, M.; GRANT, I. (1975): An experimental investigation of the formation and development of a wave packet in a laminar boundary layer. Proc. R. Soc. London Ser. A 347, pp. 253-269.

[4] GILYOV, V.M.; KACHANOV, Y.S.; KOZLOV, V.V. (1981): Development of a spatial wave packet in a boundary layer. Preprint No. 34-81, Novosibirsk, ITPM SO AN SSSR (in Russian).

[5] KACHANOV, Y.S. (1985): Development of spatial wave packets in boundary layer. In: Kozlov, V.V. (ed.), Laminar-turbulent transiton, 2nd IUTAM Symposium, Novosibirsk, UdSSR, 1984, Springer-Verlag, Berlin, pp. 115-123.

[6] KACHANOV, Y.S.; MICHALKE, A. (1994): Three-dimensional instability of flat-plate boundary layers: Theory and experiment. Eur. J. Mech., B/Fluids, vol. 13, no. 4, pp. 401-422.

[7] MACK, L.M.; HERBERT, T. (1995): Linear wave motion from concentrated harmonic sources in a Blasius flow. AIAA-95-0774, 33th AIAA Aerospace Sciences Meeting, Reno, Nevada.

[8] SARIC, W.S.; THOMAS, A.S.W. (1984): Experiments on the subharmonic route to turbulence in boundary layers. In: Tatsumi, T. (ed.), Turbulence and Chaotic Phenomena in Fluids, North-Holland, pp. 117-122.

[9] SCHUBAUER, G.B.; SKRAMSTAD, H.K. (1948): Laminar boundary layer oscillations and transition on a flat plate. NACA report 909.

[10] SEIFERT, A.; WYGNANSKI, I. (1991).: On the interaction of wave trains emanating from point sources in a blasius boundary layer. Proc. Conf. on Boundary Layer Transition and Control, April 8-12, 1991, Cambridge, U.K., The Royal Aeronautical Society, pp. 7.1-7.13.

[11] STRUNZ, M.; SPETH, J.F. (1987): A new laminar water tunnel to study the transition process in a Blasius layer and in a separation bubble and a new tool for industrial aerodynamics and hydrodynamic research. In: AGARD CP-413, pp. 25-1 - 25-5.

[12] WIEGAND, TH.; BESTEK, H.; WAGNER, S.; FASEL, H.F. (1995): Experiments on a wave train emanating from a point source in a laminar boundary layer. AIAA-95-2255, 26th AIAA Fluid Dynamics Conference, San Diego, California, USA.

# Wind Tunnel and Water Tunnel Studies of Vertical Fin Buffeting

S. Woppowa, F.-R. Grosche
DLR Institut für Strömungsmechanik
Bunsenstraße 10, D-37073 Göttingen, Germany

## Summary

Many advanced combat aircrafts operating at high angles of attack experience unsteady vortex flows with strong pressure fluctuations which cause problems such as vertical fin buffeting. This paper presents selected results of comprehensive wind tunnel and water tunnel tests. Sharply peaked spectra of the surface pressure fluctuations on each side of the fin are observed at high angles of attack. Using the dominant frequency in the vortical flow field a characteristic value of a reduced frequency parameter could be obtained for the considered model configurations. Employing Digital Particle Image Velocimetry in the water tunnel the instantaneous flow field over the model was studied. A nearly periodical oscillation of the velocity vectors has been observed in front of the leading edge of the fin. The oscillation is mainly driven by the vortex shedding from the leading edge of the main wings and induces a periodic side force on the fin.

## Introduction

The agility of a combat aircraft is dependent on its capability of flying in the high angle of attack regime. Under such flow conditions aircraft structures experience aerodynamic loads which are caused by pressure fluctuations due to flow separation and the impact of vortical flows on the structure. An example of such an aerodynamic-structural interaction is the phenomenon of vertical fin buffeting. Buffeting is defined as the structural response of the airplane or its components to strong pressure fluctuations provided by separated flows [1]. Under certain flight conditions and geometric configurations the magnitude of the response may be large. This may reduce the flight envelope and is detrimental to the fatigue life of the fin.

A number of studies on high angle of attack aerodynamics and the effect of the vortical flow on vertical fin buffeting have appeared in recent years. Bean and Wood [2] studied fin buffeting on a delta wing-vertical fin-configuration and used tangential leading edge blowing to modify the flow field influencing the fin. Breitsamter and Laschka [3] investigated the flow field of a delta-canard-configuration at high angles of attack. The power spectral density of the velocity fluctuations shows significant frequency peaks which indicate a strong periodicity in the flow. The peaks are related to the concentration of kinetic turbulent energy in the flow of the wing/canard vortex sheet. Many investigations focus on the twin fin buffeting problem of the fighter aircrafts F/A-18, F-15. On the F/A-18, the source of the tail buffeting is primarily due to bursting of vortices shed off the leading edge extension [4]. Water tunnel studies are also used to investigate the interaction between the leading edge vortices and the fin [5]. Because of the typical thin and highly swept wings of fighter aircrafts the flow separates quite close to the leading edge regardless of Reynolds number. Mayori and Rockwell [6] studied the interaction of a vortex, generated by a delta wing, with a thin plate. The analysis of the instantaneous flow field revealed processes of vortex distortion and splitting. They anticipated that these features will dictate the pressure loading of the plate.

The investigations mentioned above mainly focus on delta wing configurations where fin buffeting is strongly related to vortex breakdown. It is well documented that this burst vortical flow exhibits distinct periodic oscillations which can excite the aircraft structure [7]. The present experimental study investigates the flow field over a combat aircraft configuration with variable sweep angle $\Lambda$ of the wings at high angles of attack. Strong fin buffeting was observed in flight tests for this kind of configurations, but the unsteady flow motions and their interaction with the vertical fin are not well understood. Comprehensive wind tunnel and water tunnel tests have been carried out to study the flow field over the model. The first part of the paper will report selected results obtained from wind tunnel tests and the second part of the paper presents additional tests performed in a closed circuit water tunnel with a small scale model to study the instantaneous flow field patterns. Digital Particle Image Velocimetry (DPIV) has been used to determine the velocity fields and vorticity distributions.

## Wind Tunnel Tests

Tests have been conducted in the low speed wind tunnel facility of the DLR Göttingen. The tunnel has an open test section with a 3m x 3m cross section. The flow field over a 1/6 scale combat aircraft configuration with variable sweep angle $\Lambda$ of the wings - as illustrated in Figure 1 - was investigated at high angles of attack. The model half span is 1.17 m for $\Lambda = 25°$ and 0.72 m for the case $\Lambda = 68°$. The vertical fin is instrumented with six pressure transducers (P1-P6) to sense unsteady pressure at the fin surface, three transducers are mounted flush with the fin surface on each side of the fin. An accelerometer (A) is used to detect the vibration of the fin. Figure 1 shows the instrumentation positions on the fin. Typical test Reynolds numbers were about $1 \cdot 10^6$ based on the wing mean chord.

### Results and Discussion

Flow visualization pictures with the laser light sheet technique are presented in Figure 2 for the configuration with low sweep angle ($\Lambda = 25°$) and high sweep angle ($\Lambda = 68°$). The left margin of the pictures is identical with the vertical plane of symmetry of the model (position of the vertical fin). The position of the laser light sheet and the view direction of the photographs are sketched in Figure 1. For the low sweep angle ($\Lambda = 25°$) configuration a highly turbulent flow field can be observed with strong oscillations in spanwise direction which is indicated by the relatively low smoke concentration in the presented laser light sheet picture. The unsteady flow field over the model is expanded to the vertical plane of symmetry of the model and interacts with the vertical fin. The quantitative measurements indicate a strong response of the fin. In the case of the high sweep angle ($\Lambda = 68°$) configuration a concentrated vortex separating from the leading edge of the main wing is clearly visible in the light plane. At $\alpha = 28°$, the vortex dominated flow field passes the fin well outboard and does not interact with the fin. In correspondence to this observation, the quantitative measurements show a low vibration level of the fin for this angle of attack.

The unsteady pressure measurements were conducted in the angle of attack range $20° \leq \alpha \leq 36°$. Figure 3 presents the rms values of the pressure fluctuations $p_{rms}$, normalized by the free stream dynamic pressure $q_\infty$, at the transducer locations 4 and 5 for the low sweep ($\Lambda = 25°$) and the high sweep ($\Lambda = 68°$) configurations. In the case of low sweep angle, the pressure fluctuations increase rapidly for $\alpha > 22°$ and reach a maximum value of more than 20 % of the free stream dynamic pressure. For the high sweep wing configuration, the rms values of the pressure fluctuations are much lower and remain fairly constant for $\alpha < 28°$ due to the strong leading edge vortices shown in Figure 2. The characteristics of the rms acceleration values of the fin response compare well with the rms values of the pressure fluctuations. Sharply peaked spectra of the surface pressure fluctuations on each side of the fin appear at high angles of attack. With increasing angle of attack the maximum is shifted to lower frequencies and increases in amplitude. Using the dominant frequency $f_d$ in the vortical flow field, a characteristic value of a reduced frequency parameter, $n_d = (f_d\, c \sin \alpha) / U_\infty$ based on the projection of the wing mean chord normal to the free stream direction, could be obtained for the considered model configurations (Figure 4). A comparison between water tunnel, wind tunnel and flight tests is possible by using the characteristic values for $n_d$. The value for the low sweep ($\Lambda = 25°$) configuration, $n_d \approx 0.13$, is similar to the value measured in the wake of a two dimensional airfoil at high angles of attack [8]. This indicates that the periodic vortex shedding from the leading edge of the main wings could be an important factor for the excitation of the vertical fin. More detailed information about the flow field over the model has been obtained performing water tunnel tests at low Reynolds number.

## Water Tunnel Tests

Experiments were carried out in a closed circuit water tunnel of the DLR Göttingen. The horizontal test section has a cross sectional area of 0.25 m x 0.33 m and a length of 1.25 m. The models used are 1/72 scale plastic kits with fixed main wings. Two models were fabricated representing the low sweep ($\Lambda = 25°$) and the high sweep ($\Lambda = 68°$) configurations. The model half span is 98 mm for $\Lambda = 25°$ and 60 mm for the case $\Lambda = 68°$. Typical test Reynolds numbers were between $1 \cdot 10^3$ and $1 \cdot 10^4$ based on the wing mean chord. All measurements presented here have been performed on the models with the fin removed.

For the comparison between the wind tunnel and the water tunnel tests, frequency measurements of the velocity fluctuations have been made using a hot film probe, in order to determine scale effects and to confirm that Reynolds number effects are not important. The probe was located at the position of the tip of the fin. The free stream velocity tested was in the range 46 mm/s $\leq U_\infty \leq$ 500 mm/s. The spectra show a well defined peak which allows to calculate the reduced frequency parameter $n_d$. Figure 5 presents the values of $n_d$ calculated from wind tunnel and water tunnel tests for the low sweep configuration ($\Lambda = 25°$) at $\alpha = 32°$. The good agreement suggests that the dominant flow motions generating the strong periodicity in the wind tunnel and water tunnel are similar despite the difference in Reynolds number.

### Experimental Setup for Digital Particle Image Velocimetry (DPIV)

The flow was seeded with neutrally buoyant polyamid particles having a diameter of about 70 microns. The illumination of the particles was provided by a continuous 6W Argon-ion laser with a mono-mode fiber optical cable generating a 1.5 mm thick light sheet in the test section. The total image aquisition system consisted of a usual black and white CCD-camera, shuttered at 1/250 s to minimize streaking effects of faster moving particles, and a personal computer with a commercially available frame grabber card. The image system aquires the frames with the standard 25 Hz CCIR video frame rate, that means, 25 frames per second are digitized (740 x 572 Pixels) and stored in the RAM memory (64 MB) of the computer. The image acquisition system is capable of recording 125 full frames (5 seconds) continously. Due to the interlace effect each captured frame consists of two fields, the first field displays only the odd lines and the second field displays only the even lines of the image. There is a time delay of 20 ms between the two fields. By cross-correlating spatial subsamples (windows) of two successive fields (the missing lines are linearly interpolated from the exposed adjacent lines), the average displacement vector of the particles contained in the correlation window is calculated. The DPIV analysis software used is developed by Willert and Gharib [9].

The instantaneous velocity vector fields shown in this paper are measured in an *xy*-plane over the model which has an area of 140 mm x 60 mm (Figure 6). The *z*-position is indicated by the dashed line in Figure 6 and identical with the pressure transducer position 1/4 of the wind tunnel model. The results presented in the next section were obtained at a free stream velocity of about $U_\infty = 46$ mm/s and an angle of attack $\alpha = 32°$.

**Results and Discussion**

For the averaged velocity magnitude contours given in Figure 7, fifty instantaneous images were required within a time intervall which is an order of magnitude longer than the predominant frequency of the unsteady flow field. The measurement section over the model is also sketched in Figure 7. The averaged velocity magnitude contours for the two considered model configurations ($\Lambda = 25°$, 68°) are nearly symmetrical to the centerline ($y = 0$ mm) of the model. In both cases the flow in the midsection ($x < 60$ mm) is strongly accelerated, the velocity magnitude is about 20% higher than the free stream velocity. This acceleration of the flow in the midsection is due to the wake of the main wings indicated by the low velocity areas in Figure 7. With increasing angle of attack the streamwise vortices originating from the front part of the model grow in their radial extension. The two wakes of the main wings move inboard towards the centerline of the model and generate a strong convergence effect in the plane of symmetry as mentioned in [3]. Further downstream the two wakes merge resulting in a deceleration of the flow. The merging of the two wakes occurs in the case of the low sweep angle ($\Lambda = 25°$) configuration upstream of the leading edge of the vertical fin and induces a strong flow unsteadiness. For the high sweep angle ($\Lambda = 68°$) configuration, the interaction between the two wakes is much weaker which generates only small velocity fluctuations upstream of the vertical fin [10]. This becomes more obvious by analysing time series of the instantaneous velocity vector fields.

Instantaneous velocity vector fields are presented in Figure 8 and 9 for the two considered model configurations at two different instants of time. The measurement section is shown in Figure 6. For the low sweep angle configuration ($\Lambda = 25°$), a strong flow angularity is clearly visible in front of the leading edge of the fin (Figure 8). The oscillating velocity vectors in front of the fin produce the sharp peaks in the pressure spectra and a 180° phase difference between the pressure fluctuations measured by the transducers at opposite sides of the fin. At time $t = t_0$, the angle $\beta$ between the fin and the velocity vectors in front of the fin is positive (as seen in Figure 8). The starboard side of the fin ($y > 0$) experiences an instantaneous suction, the port side ($y < 0$) experiences an instantaneous over pressure, and vice versa for a negative angle β, which is the case at time $t = t_0 + 1.24$ s. This induces a side force and a fin deflection if the fin is flexible. The phase lag of 180° between two pressure transducers positioned at opposite sides of the fin was also measured in the wind tunnel tests using cross spectral analysis [10]. Measurements of the instantaneous velocity vector fields over the high sweep configuration ($\Lambda = 68°$) are given in Figure 9. Due to the weak interaction between the two wakes there are only small velocity fluctuations in the midsection upstream of the fin. Further downstream the unsteady wake flow of the wings influence the flow in the region of the vertical fin, but there seems to be no correlation between the velocity fluctuations generated by the wake flow of the right and the left wing.

The oscillation of the velocity vectors is mainly driven by the vortex shedding from the leading edge of the main wings [10]. Due to the induced velocity field of the streamwise vortices separating from the fuselage, the shed off vortices arrange asymmetrically and interact with the vertical fin which results in a periodic force on the fin.

## Acknowledgment

This work was supported by Daimler Benz Aerospace AG (DASA Ottobrunn). The authors would like to express their gratitude to Dr. J. Kompenhans and Dr. C.E. Willert for the supply of the DPIV analysis code.

## References

[1] MABEY, D.G.: Beyond the Buffet Boundary. Aeronautical Journal Vol.11, No.148, 1973, S. 201-214.

[2] BEAN, D.E., WOOD, N.J.: A Vortex Control Technique for the Attenuation of Fin Buffet. ICAS-92-4.10.3, 1992.

[3] BREITSAMTER, C., LASCHKA, B.: Turbulent Flow Structure Associated with Vortex-Induced Fin Buffeting. Journal of Aircraft, Vol.31, No.4, July-Aug. 1994, S. 773-781.

[4] LEE, B.H.K., BROWN, D., ZGELA, M., POIRELL, D.: Wind Tunnel Investigation and Flight Test of Tail Buffet on CF-18 aircraft. AGARD CP-483, 1990.

[5] WENTZ,W.H.: Vortex-Fin Interaction on a Fighter Aircraft. AIAA Paper 87-2474, Aug. 1987.

[6] MAYORI, A., ROCKWELL, D.: Interaction of a Streamwise Vortex with a Thin Plate: a Source of Turbulent Buffeting. AIAA Journal 32, S. 2022-2029, 1994.

[7] GURSUL, I.: Unsteady flow phenomena over delta wings at high angles of attack. AIAA Journal, Vol. 32, No. 4, S. 225-231, 1994.

[8] KRZYWOBLOCKI,M.: Investigation of the Wing-Wake Frequency with Application of the Strouhal Number. Journal of the Aeronautical Sciences, Vol.12, Jan. 1945.

[9] WILLERT, C.E., GHARIB, M.: Digital Particle Image Velocimetry. Experiments in Fluids 10, S. 181-193, 1991.

[10] WOPPOWA, S.: Instationäre Wirbelströmungen hinter Flügeln kleiner Streckung bei hohen Anstellwinkeln, Dissertation, Universität Göttingen, 1996.

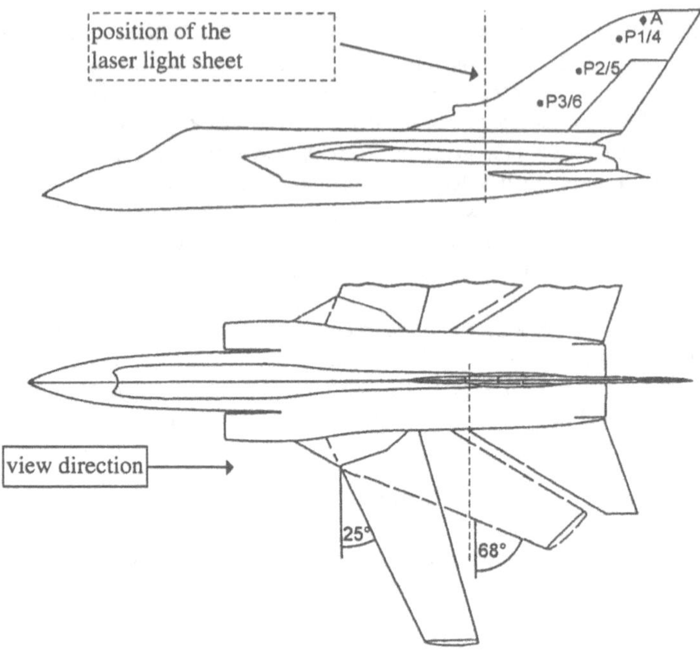


Figure 1: Model configurations, Λ = 25°, 68° , with instrumentation of the fin. Laser light sheet position and view direction of the photographs given in Figure 2.

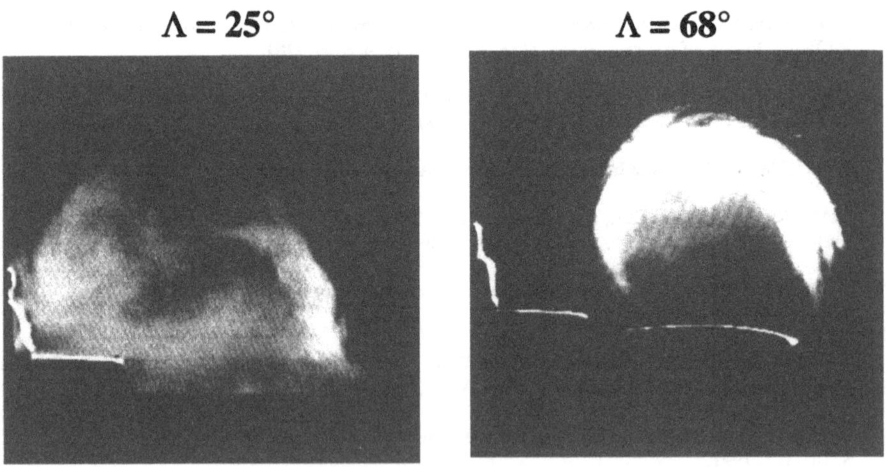


Figure 2: Laser light sheet photographs for the configuration with low sweep angle (Λ = 25°) and high sweep angle (Λ = 68°); α = 28°.

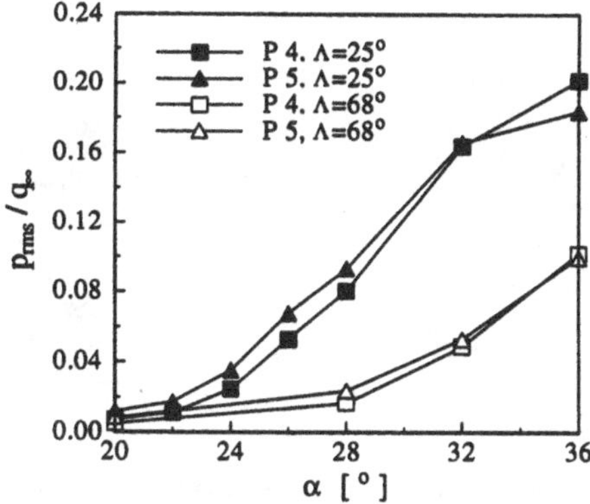


Figure 3: RMS pressure values versus angle of attack for the low sweep angle ($\Lambda = 25°$) and high sweep angle ($\Lambda = 68°$) configurations.

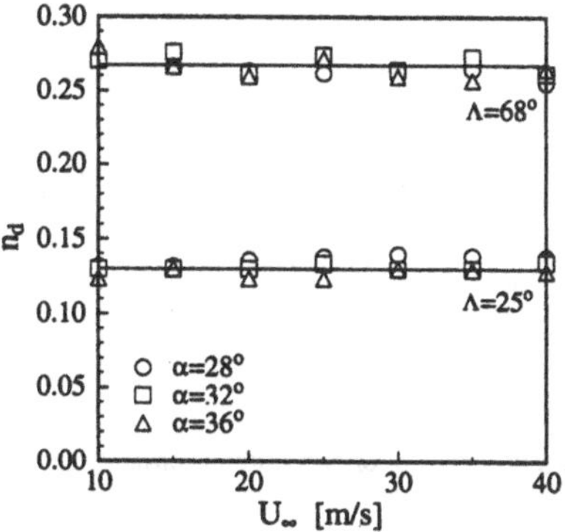


Figure 4: Reduced frequency parameter $n_d$ versus free stream velocity for the considered configurations ($\Lambda = 25°, 68°$) at different angles of attack.

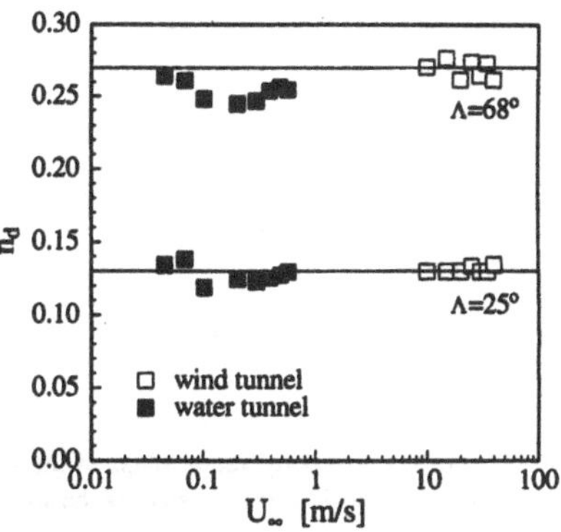


Figure 5: Values of $n_d$ calculated from wind tunnel and water tunnel tests for the two considered configurations at $\alpha = 32°$.

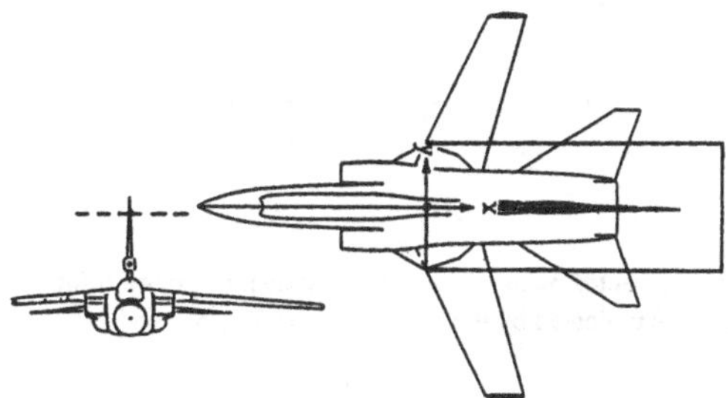

Figure 6: Measurement section in an $xy$-plane over the model; the $z$-position is indicated by the dashed line (identical with the position of pressure transducers 1/4 of the wind tunnel model).

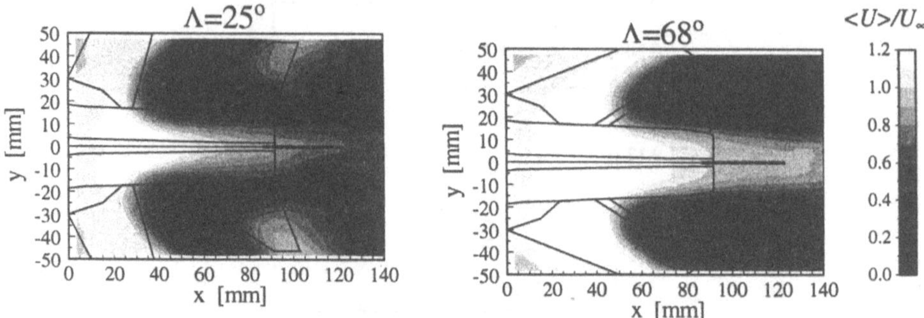


Figure 7: Averaged velocity magnitude contours for the two considered model configurations ($\Lambda = 25°, 68°$); $\alpha = 32°$, $U_\infty = 46$ mm/s. The $z$-position of the measurement section is identical with the pressure transducer position 1/4 of the wind tunnel model.

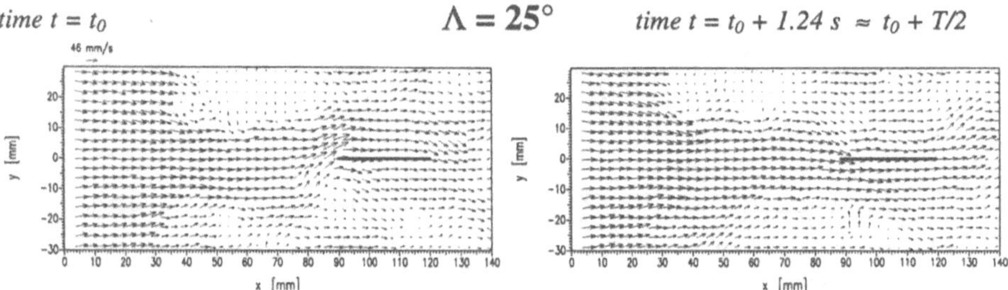


Figure 8: Instantaneous velocity vector maps of the section shown in Figure 6 for two different instants of time. The position of the vertical fin is indicated by the solid line; $\alpha = 32°$.

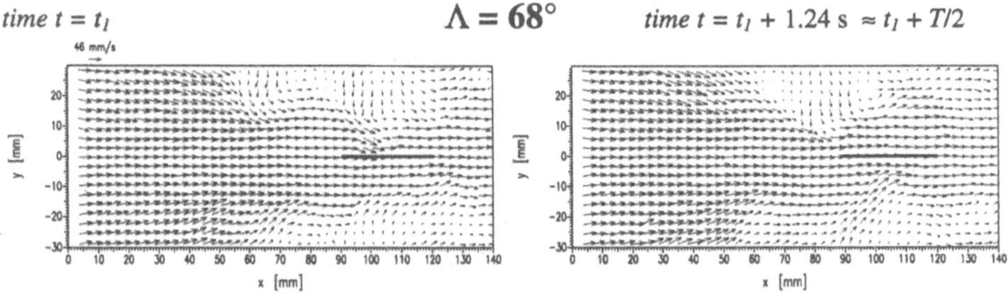


Figure 9: Instantaneous velocity vector maps of the section shown in Figure 6 for two different instants of time. The position of the vertical fin is indicated by the solid line; $\alpha = 32°$, $\Lambda = 68°$.

# Experimental Investigations of Transition Development in Attached Boundary Layers and Laminar Separation Bubbles

W.Würz, S.Wagner
Institut für Aerodynamik und Gasdynamik
Pfaffenwaldring 21, 70550 Stuttgart

## Summary

Hot-wire measurements of transitional boundary layer quantities were performed in the Laminar Windtunnel of the IAG under „natural" conditions on a 2-d airfoil section at a Reynolds number of $Re=1.2*10^6$. The measured amplification of Tollmien-Schlichting waves is compared to linear stability theory. A careful experimental setup allows the measurement of very small velocity fluctuations resulting in a large measurable amplitude ratio of $A_{max}/A_{min} \cong 1800$. The determined „onset" of transition is compared to the transition prediction with $e^n$-methods. Some remarks are made on the accuracy of simplified envelope-methods.

## Introduction

In the design of airfoils, i.e. for sailplane applications, advantage is gained from long regions of laminar flow. This influences directly the profile drag by means of the lower laminar skin friction in comparison to the turbulent one and indirectly by the possibility of a steeper pressure recovery which can be overcome by a thinner turbulent boundary layer. Transition should occur just before the beginning of the pressure rise to avoid laminar separation bubbles, which may increase the drag significantly. Therefore the resulting performance of a new airfoil depends strongly on the reliability of the method used for transition prediction. Since the fifties various local and non-local empirical methods have been used with more or less success, and today semi-empirical $e^n$-methods are state of the art [1,2].

Fig.1 shows the lift to drag polar of the SM701 airfoil, originally designed by Somers and Maughmer [3] with the Eppler code [4] which uses a local empirical transition criterion. A re-calculation was made with the airfoil design code XFOIL [5]. This code takes the displacement thickness of the boundary layer into account and predicts the transition by a simplified $e^n$-method (envelope-method). A large difference to the measured polar [6] can be observed. An attempt was made to adjust the n-factor to the experimental data but resulted also only in poor agreement (and in a very low n-factor, n≅5). From free-flight experiments performed by Horstman et al. [7]

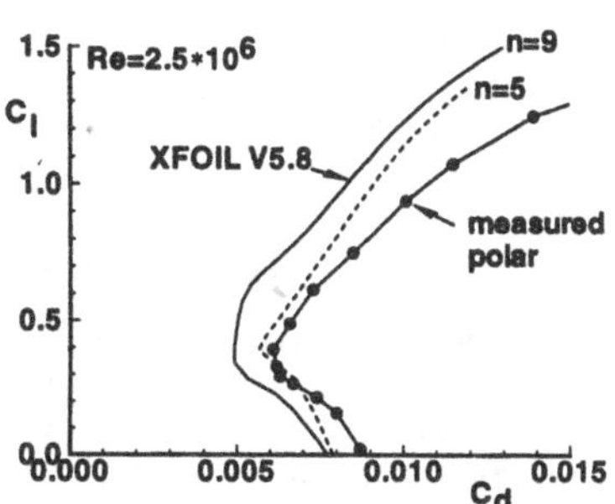


**Fig.1:** Polar of the SM701 airfoil

n-factors (from linear stability theory calculations) in the range of n≅12-14 are reported, which agree quite well with the corresponding windtunnel data.
To get more insight into this problem, boundary layer measurements were carried out under „natural" conditions, that means without additional artificial disturbances except the normal windtunnel turbulence. Two cases important for practical applications were examined [8]: transition in an attached boundary-layer under adverse pressure gradient and the transition development in a laminar separation bubble.

## Windtunnel and Turbulence level

The Laminar Windtunnel is built as an open return tunnel of the Eiffel design [9]. The rectangular test section measures $0.73*2.73m^2$ and is 3.15m long. The two-dimensional airfoil models span the short distance of the test section. The high contraction ratio of 100:1 and 5 screens and filters result in a very low turbulence level below $2*10^{-4}$.
The turbulence level was measured with a single hot-wire probe DISA55P11 centered in the middle of the test section. The probe was connected to a DISA-55M10 constant temperature anemometer.
The AC part of the signal was cut off by an 20Hz high-pass filter with a roll off rate of 12dB/octave and then amplified with a Preston amplifier. The power spectrum (fig.2) shows the frequency-distribution for a free-stream speed of 30m/s together with the noise-level of the measurement equipment. Above 200Hz the velocity fluctuations can not be distinguished from the electronic noise. Typical Tollmien-Schlichting (TS-) frequencies in the related experiments are in the order of 1kHz. Because of the required high amplification of the signal, the contamination by frequencies from the power consumption (50Hz/100Hz) makes a considerable contribution to the RMS of the signal. These frequencies are digitally filtered out. Then the turbulence level is calculated from the velocity fluctuations in streamwise direction on the assumption of an isotropic turbulence distribution.

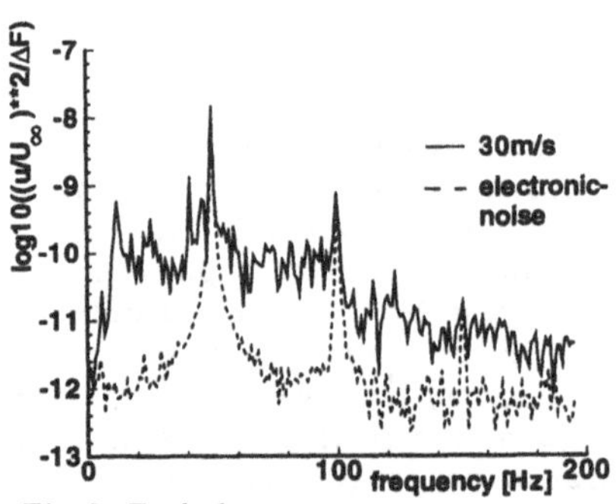

**Fig.2:** Turbulence spectrum

## Instrumentation and Procedure

For the boundary layer measurements a symmetrical airfoil (XIS40MOD) was designed which, depending on the angle of attack, allows the testing of special cases with large differences in boundary layer development and also meets all requirements for the traversing mechanism described below. One side of the airfoil was equipped with 47 pressure orifices (0.3mm diameter), and the pressure distribution was obtained by a scanivalve and a single HBM-PD1 pressure transducer. To avoid any disturbances the other side was smooth and used only for the boundary layer measurements.

A single wire boundary layer probe (DISA55P15) was used for the boundary layer surveys together with a small static pressure probe of 1mm diameter. This probe served as a velocity reference for the hot-wire at the boundary layer edge. The probes are mounted to a small traversing mechanism (fig.3). A thin support resting on the airfoil surface defines its position in relation to the wall. A small rubber between this sting and the airfoil surface prevents the coupling of mechanical vibrations. By means of a high precision rack- and pinion drive together with an optical encoder, a resolution of 5µm in wall distance is achieved. To start traversing a boundary layer, the hot-wire probe is moved towards the wall until its prongs touch a thin graphite coating, thus closing an electric circuit with high impedance which stops the motor. The direction of traverse is reversed and the probe moves until its contact with the wall breaks. By this means, eventual backlash and bending effects are removed. This position is taken as the zero wall distance.

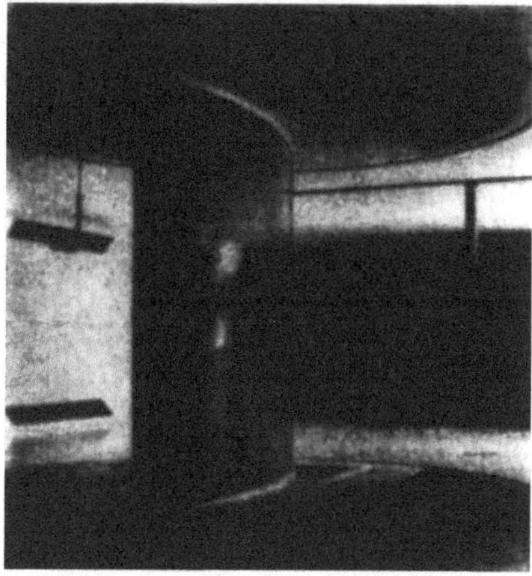

**Fig.3: Test section and probe support**

Thirty to sixty points are acquired in each boundary layer profile. The DC-output of the hot-wire anemometer is integrated by a low-pass filter at 0.3Hz. The AC-output is high-pass filtered with a special low-noise filter with 570Hz cut-off frequency and a low damping of only 4,3dB/octave. This allows high amplification of the signal in the range of the TS-frequencies by a programmable amplifier and anti-aliasing filter which are remotely controlled by the PC used for data-acquisition. The sampling rate is normally set to 10kHz and the data (4096 time signals) are collected with a 12bit AD-converter. After the FFT analysis the signal is corrected in amplitude to consider the influence of the filters and monitored online.
The calibration of the hot-wire is made according to Kings-Law and is briefly described in [10]. A comparison between the pressure distribution measured with the static probe and those measured with the pressure-orifices shows very good agreement, consequently the influence of the traversing mechanism and probe support on the mean velocity distribution must be small. Only at the rear part of a laminar separation bubble the pressure distributions are slightly diverging, which is probably due to strong curvature of the streamlines which may lead to a misalignment of the static probe.

## Boundary layer measurements

Boundary layer measurements were performed for the XIS40MOD airfoil at 1 degree angle of attack and a Reynolds number of $1.2*10^6$ based on the arc-length $s_{max}$=0.615m measured from the leading edge. The velocity distribution (fig.4) is in good agreement with the distribution calculated with XFOIL. Based on this distribution, the boundary layer parameters are evaluated with a finite difference scheme [11]. The shape factor $H_{12}$ is nearly constant

($H_{12} \cong 2.8$) from $\bar{s}=s/s_{max}=0.25$ to 0.6 and the „onset" of transition can be clearly seen as the calculated laminar shape factor diverges ($\bar{s}=0.585$) from the measured $H_{12}$, because the mean velocity profiles are influenced by the increasing turbulence production. Fig.5 shows the measured profiles in comparison to the calculated ones. At each point, a FFT is performed so that eigenfunctions for the TS-frequency can be derived, as plotted in fig.6. These eigenfunctions show a stronger second maximum inside the boundary layer in comparison to 2-d linear stability calculations, also the TS-amplitude ($0.1\%U_\delta$) is distinctly below the critical level for secondary instability effects. Under some assump-tions for the superposition of 2-d and a small amount of oblique travelling waves these eigenfunctions agree satisfactorily with the measured ones.

Under „natural" conditions, i.e. without a dominating artificial disturbance at one single frequency, it is not clear how to define a TS-amplitude and also a TS-frequency. As can be seen in fig.7, there is a broad band of frequencies which are amplified or damped according to linear theory (if they are small enough). In this study the TS-frequency will be defined as the frequency with the highest measurable amplitude (in the range of amplified frequencies) at the onset of transition. The corresponding TS-amplitude is defined as the maximum amplitude of the eigenfunction (fig.6) for a frequency-range of ±10% of the above mentioned TS-frequency. Normally this covers more than 80% of the energy of the whole amplified frequencies.

In fig.8 the development of the measured TS-amplitudes in the three maxima of the eigenfunction (fig.6) is compared to linear stability theory. A good agreement can be seen, except for $\bar{s} \leq 0.35$. In this region the hot-wire signal is dominated by electronic noise and the influence of probe vibrations. Based on this lowest measurable amplitude $A_{min}$ and the amplitude at the onset of transition $A_{tr}$ ($\bar{s}=0.585$), a high amplitude ratio could be measured, corresponding to an n-factor of n=7.5. The calculation of the n-factor with linear stability theory beginning from the instability point yields n=10.3. The value of this n-factor depends on the definition of the transition point which can be different according to the method used for determining transition. In the present experiment for example the total RMS-maximum of the fluctuation amplitudes is reached at $\bar{s}=0.65$. At the same station the measured wall shear stress increases significantly. For this position the calculated n-factor reaches a value of n=11.5.

A second boundary layer experiment was performed under the same conditions but with an angle of attack of -3 degrees. In this case a laminar separation bubble occurs at $\bar{s}=0.717$ with turbulent reattachment at $\bar{s}=0.82$. The velocity distribution and the development of the integral boundary layer parameters can be seen in fig.9. An additional velocity distribution was measured with a small trip in front of the separation point which causes transition and prevents the formation of the bubble. The influence of the bubble on the velocity distribution is clearly visible. The instability point is at $\bar{s}=0.27$, but according to the low shape factor $H_{12}$ no significant amplification of TS-wave occurs before laminar separation. Inside the bubble, $H_{12}$ grows rapidly and reaches $H_{12max} \cong 6.5$. At $\bar{s}=0.744$ a TS-amplitude of $0.1\%U_\delta$ is observed, which shows that the transition process develops inside the bubble and is therefore dominated by the stability characteristics of the separated shear-layer. Free shear-layers are highly unstable and in the present experiment the amplification from 0.1% to 3% TS-amplitude took place over a distance of one TS-wavelength ($\lambda_{TS} \cong 0.03\bar{s}$). The measured amplification rates are slightly higher inside the bubble in comparison to linear stability calculations. This may be contributed to the parallel flow assumption which is questionable in this case. Nevertheless the calculated n-factor for the onset of transition reaches n=10.5. This is in good agreement with the above mentioned measurement and shows the consistency of this semi-empirical $e^n$-method.

## Transition prediction with envelope-methods

A well known simplification of linear stability calculations for transition prediction are the so-called enevelope-methods. They were first introduced by Gleyzes et al. [12] and later used by Drela in the airfoil design program XFOIL. This method takes advantage from the observation that for a similar boundary layer the amplitude development of different amplified frequencies can be reduced to one envelope covering the amplification curve of all single frequencies (fig.10). This envelope can then be used for the calculation of n-factors. For this purpose, Drela derived an analytic function depending on Falkner-Skan-profiles for the determination of the instability point (1) and a second one for the gradient of the envelope (2). Equation (3) and (4) are derived from the same profiles to enable a direct integration in s.

$$log_{10}(Re_{\delta 2in}) = 2.492\left(1/(H_{12}-1)\right)^{0.43} + 0.7\left(tanh\left(14.0/(H_{12}-1)-9.24\right)+1.0\right) \quad (1)$$

$$dn/d\,Re_{\delta 2} = 0.028(H_{12}-1) - 0.0345 \cdot e^{-(3.87/(H_{12}-1)-2.52)^2} \quad (2)$$

$$m_{H12} = -0.05 + 2.7/(H_{12}-1) - 5.5/(H_{12}-1)^2 + 3.0/(H_{12}-1)^3 \quad (3)$$

$$dn/ds = m_{H12} \cdot dn/d\,Re_{\delta 2}/\delta_2. \quad (4)$$

In the vicinity of the instability point a small circular correction is made to the gradient of the envelope-curve to achieve a smooth start.

The key assumption of this method is that the transition development in a boundary layer with non constant shape factor can be described by stepwise integrating along these envelopes. Dini [13] shows that this is not true and results in an „envelope-error". He analysed a boundary layer development with a sharp increase in the shape factor. In this case, the amplification must be regarded along a single frequency and is usually higher than calculated along the envelope. For example (fig.10), if a boundary layer is stable to all frequencies until $Re\delta 2=200$ is reached and then a jump to $H_{12}=2.8$ occurs in the shape factor, the most amplified single frequency will follow the dotted line. After a short distance, the corresponding „envelope" is crossed and the n-factor of n=9 is reached further upstream.

In practice, boundary layers develop without sharp jumps in the shape factor and for analysing such a continuous change a special series of boundary layers were calculated (fig.11). An inverse boundary layer method was used to prescribe a linear growth in the shape factor, starting with a flat plate solution. The onset of transition is then calculated according to linear stability theory using fixed single frequencies in comparison with the envelope-method. For a constant shape factor the predicted transition points (n=9) agree quite well. For stronger increase of $H_{12}$, linear stability theory predicts transition further upstream. The maximum differences in $\bar{s}_{tr}$ are not observed for the strongest

increase in $H_{12}$ because of the higher instability of the boundary layer resulting in a rapid increase in the n-factor with $\bar{s}$.
Fig.12 shows a stability analysis for the above mentioned SM701 airfoil for a lift coefficent of $c_l$=0.67. The calculated amplitude development for single frequencies is plotted against the corresponding envelope-curve. The experimentally detected transition by a oil and lampblack coating can be seen in fig.13. The increase in wall shear stress caused by the onset of transition results in a bright zone starting at $\bar{s}$=0.4. The lampblack from this zone is transported to the darker zone following downstream. For $\bar{s}$=0.4 a n-factor of n=11.3 can be derived from the linear stability calculations, which is in good agreement with the results from the boundary layer experiments. For the same transition position the envelope-method reaches only n≅6. Since the accuracy of the transition prediction with this method depends strongly on the shape of the boundary layer development, which differs from one airfoil to the other (also for changes in angle of attack, or the upper and lower surface) it is not possible to get consistent n-factors. For practical airfoil design it should also be observed that small changes in the velocity distribution can cause large differences in the transition prediction. A comparison of two slightly different airfoils with this envelope-method may therefore be questionable.

## Conclusions

Boundary layer experiments were performed in the Laminar Windtunnel of the IAG using hot-wire anemometry. The measurements were made under „natural" conditions without introducing a single dominant frequency. According to a careful experimental setup, the development of TS-waves could be studied over a large amplitude range of $A_{max}/A_{min}$≅1800. The comparison with linear stability theory shows good agreement for the amplified frequency-band and the amplification rates in attached boundary layers, whereas for the laminar separation bubble slightly higher amplification rates were measured. Based on the stability calculations, consistent n-factors for the „onset" of transition could be derived. It could be shown that the simplified envelope-method leads to an uncertain transition prediction.

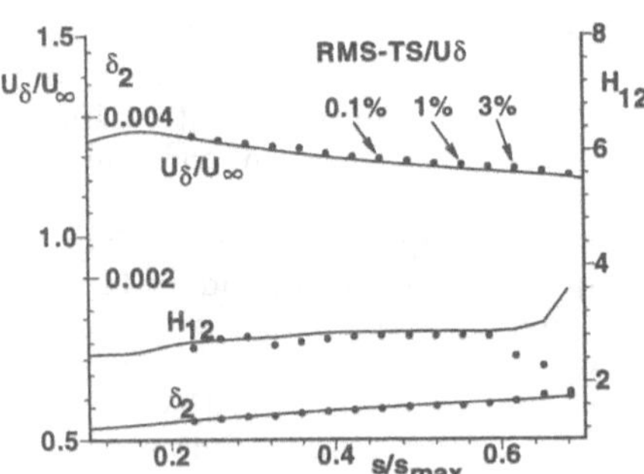


**Fig.4:** Boundary layer parameters XIS40MOD, α=1°, • experiment

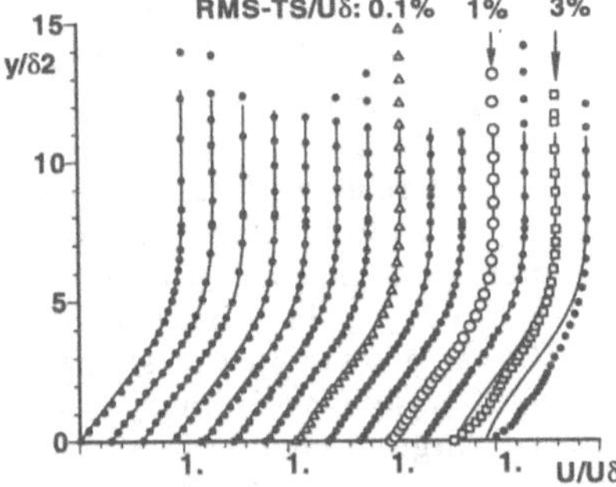


**Fig.5:** Mean velocity profiles • experiment, — theory [11]

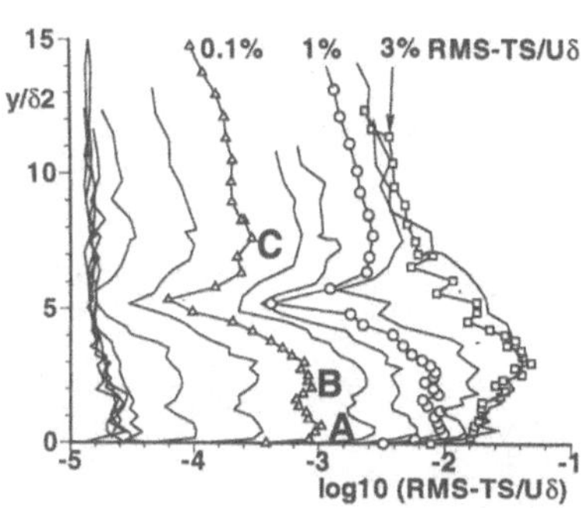

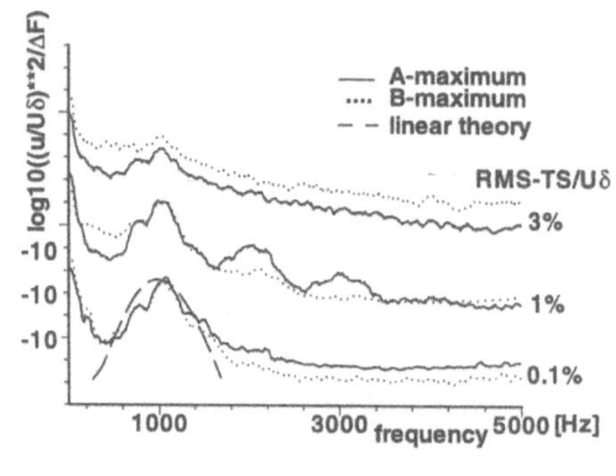

**Fig.6:** Eigenfunctions for the TS-frequency

**Fig.7:** Frequency spectra (shiftet by 2 decades)

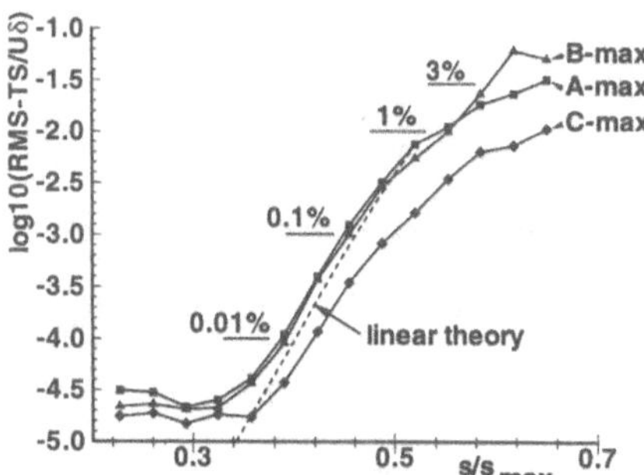

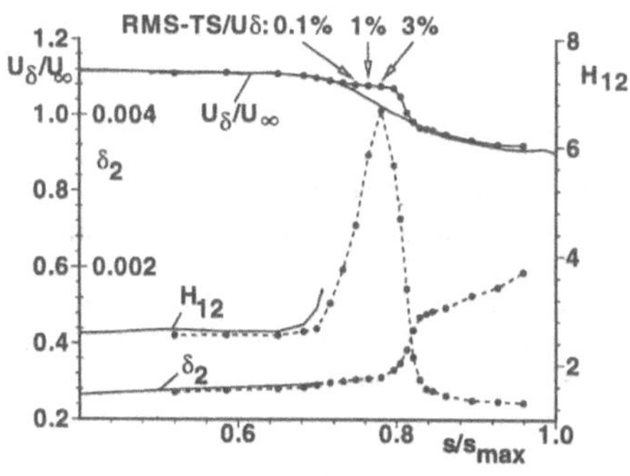

**Fig.8:** Amplitude development for the TS-frequency

**Fig.9:** Boundary layer parameters XIS40MOD, $\alpha$=-3°

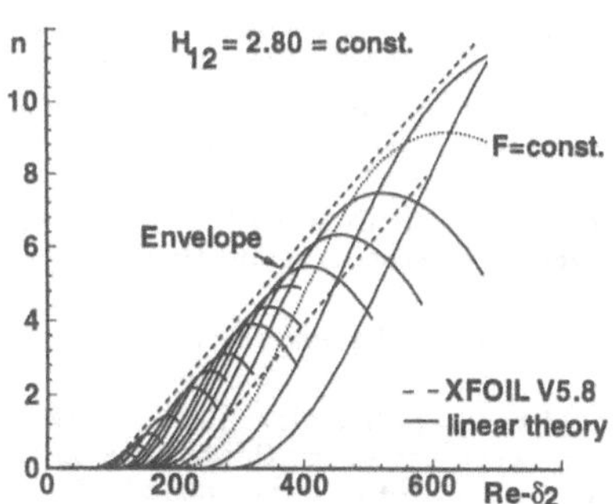

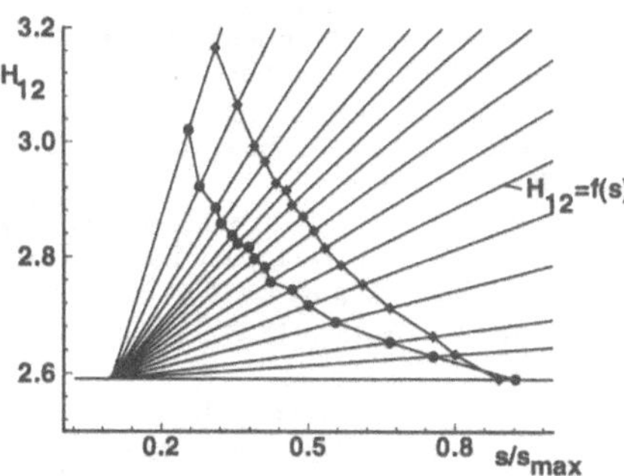

**Fig.10:** Amplitude development for similar boundary layers

**Fig.11:** Transition prediction, n=9, • linear theory, ♦ envelope-method

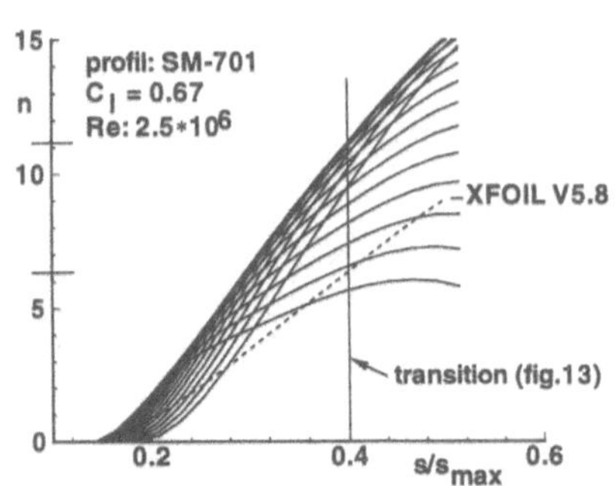


**Fig.12:** Stability analysis for the SM701 airfoil, - - envelope-curve

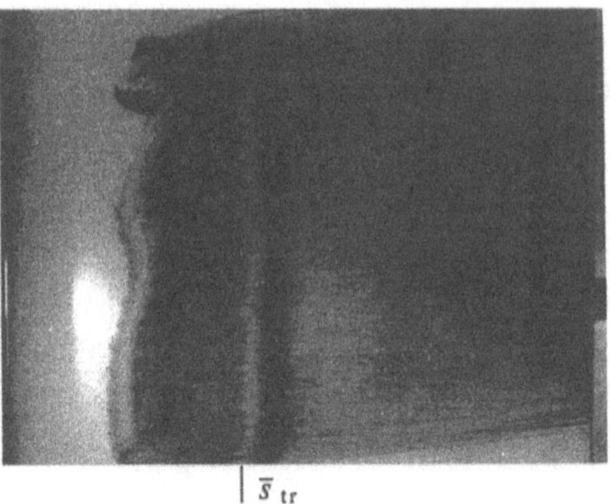

**Fig 13:** Oil and lampblack coating for transition detection, SM701, flow direction from left to right

# References


[1] Arnal,D.: „Description and Prediction of Transition in two-dimensional, incompressible Flow“, AGARD-R-709,pp2.1-2.71 (1984).

[2] Arnal,D.: „Boundary Layer Transition: Prediction, Application to Drag Reduction“, AGARD-R-786, pp5.1-5.59 (1992).

[3] Somers,D.M.; Maughmer,M.D.: „ The SM701 Airfoil: An Airfoil for World Class Sailplanes“, Technical Soaring, Vol.14,No.3 (1992).

[4] Eppler,R.: „Airfoil Design and Data“, Springer Verlag, Berlin-Heidelberg-New York, (1990).

[5] Drela,M.; Giles,M.B.: „Viscous-Inviscid Analysis of Transonic and Low Reynolds Number Airfoils“, AIAA-86-1786-CP (1986).

[6] Althaus,D.; Würz,W.: „Wind Tunnel Tests of the SM701 Airfoil and the UAG88-143/20 Airfoil“, Technical Soaring Vol.17,No.1 (1993).

[7] Horstmann,K.H.; Quast,A.; Redecker,G.: „Flight and Wind-Tunnel Investigations on Boundary-Layer Transition“, Journal of Aircraft, Vol27, No.2, pp146-150 (1989).

[8] Würz,W.: „Hitzdrahtmessungen zum laminar-turbulenten Strömungsumschlag in anliegenden Grenzschichten und Ablöseblasen sowie Vergleich mit der linearen Stabilitätstheorie und empirischen Umschlagskriterien“, Dissertation Universität Stuttgart (1995).

[9] Wortman,F.X.; Althaus,D.: „Der Laminarwindkanal des Instituts für Aerodynamik und Gasdynamik der Technischen Hochschule Stuttgart“ Zeitschrift für Flugwissenschaften, Nr.12, Heft 4.

[10] Wubben,F.J.M.: „Experimental Investigation of Tollmien-Schlichting Instability and Transition in Similar Boundary Layer Flow in an Adverse Pressure Gradient (Hartree $\beta = -0.14$)“, TU Delft, Report LR-604 (1991).

[11] Cebeci,T.; Smith,A.M.O.: „Analysis of Turbulent Boundary Layers“, Academic Press, New York (1974).

[12] Gleyzes,C.; Cousteix,J.; Bonnet,J.L.: „Theoretical and Experimental Study of Low Reynolds Number Transitional Separation Bubbles“, Proceedings of the Conference on Low Reynolds Number Airfoil Aerodynamics, UNDAS-CP-77B123, Notre Dame, pp137-151 (1985).

[13] Dini,P.; Selig,M.S.; Maughmer,M.D.: „A simplified $e^n$-Method for Separated Boundary Layers“, AIAA-91-3285 (1991).

# On-Line Data Processing for the DLR-F9 Windtunnel Experiment

## A Prototype for a Decision Support System

R. Zores, J. Trapp, S. Standfuß, H.-G. Pagendarm
DLR, Bunsenstr. 10, D 37073 Göttingen, Germany

## Summary

Introduction of new windtunnel testing technology including Circulation Control Splitter Blades called for improved methods for the control of the parameters and sequential process of the experiment. This was achieved by the implementation of a new, dedicated on-line data processing and flow feature extraction system which allowed the windtunnel engineers to validate experimental results and take decision on further experiments within a short time frame while the experiment was conducted. These decisions were based on information presented in a higher level of abstraction compared to previous windtunnel data display systems.

## Introduction

Advanced windtunnel testing of modern aircraft configurations calls for fast and effective data processing to allow to control and adjust the experiment while running. The experiments conducted on the DLR-F9 transport aircraft model aimed at investigations in the fuselage area involve a number of new technologies beyond the traditional well-known testing methods [1]. Such new approaches include an adaptive test section of the Transonic Windtunnel of DLR in Göttingen, an extremely large model with a high blockage ratio, and the use of Circulation Control Splitter Blades (CCSB) [2] for compensating for the cut-off wing tips.

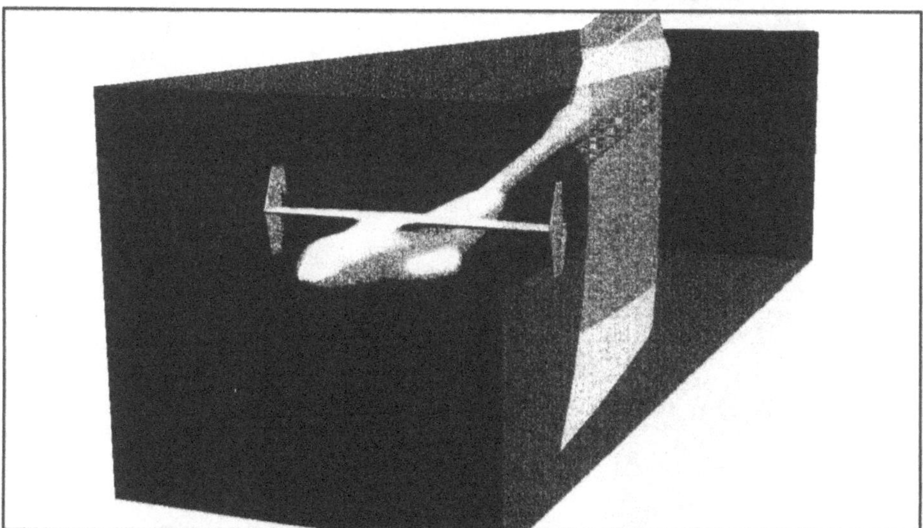

**Fig. 1 F9-Model in the windtunnel test section. CCSBs are fitted to the wings to correct for effects of model scale and clipped wings. The geometrical representation of the windtunnel which is shown here was also used for comparative numerical studies**

This step into new areas of windtunnel testing requires the human in the loop in order to set model and windtunnel parameters in such a way that valid flow testing may be conducted. In order to support the decisions of the testing crew while the experiment is running, a fast run-time data processing and evaluation system was implemented which presents important evaluation of the measured data and feature extraction immediately after a set of data is sampled. This allows to minimize windtunnel testing time and ensure the quality of the experiments. Changes of the testing sequence or testing parameters are suggested while the experiment is running and may be entered on the fly into the windtunnel control system

## The DLR-F9 test case

The DLR-F9 model is a generic high wing transport aircraft. The model is designed to be used for basic flow studies of new geometries and adaptive components in particular near the wing body junction and on the inner parts of the wing. Therefore the model was built in an unusual large scale in order to provide high resolution data in the region of interest and offer sufficient space for instrumentation and adaptive components. The use of such large scale models in a relatively small windtunnel became possible only due to the modern adaptive wall technology of the windtunnel test section. The adaptive ceiling and bottom walls help to avoid windtunnel blockage with large models.

The scale of the model requires the wings outer parts to be clipped in this given windtunnel. In order to correct for the effect of the clipped wing the model is equipped with CCSBs. Fig. 1 shows the model in the windtunnel in a virtual view using the data for accompanying analysis in a virtual windtunnel approach.

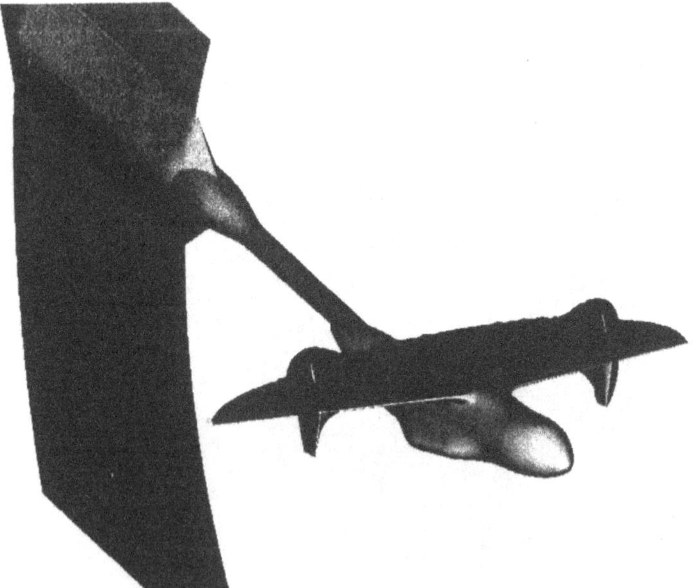

**Fig. 2 Comparison of numerical simulation of windtunnel flow around the DLR-F9 model with CCSBs and Windtunnel walls (not shown) and flow around a full wing model. Sonic surfaces shown for comparison exhibit no interference near the wing body junction**

Several steps are required to set up the CCSB concept. CFD-analysis of the full wing for various lift coefficients verifies a shock location along the wing span which is approximately along a line x/l=const. This is referred to as Isobar Concept. Evaluation of the flow field in a plane at the span position where the CCSB will be placed on the windtunnel model allows to design a set of unsymetrically twisted double sided winglets which work as side blades controlling the circulation distribution along the cut-off wing. A different twist of the winglets results for each desired lift coefficient. Fig. 2 illustrates that the CCSBs may be used to correct the flow field near the wing body junction in the desired way. The figure compares two numerical simulations of the model equipped with full wings and clipped wings with CCSBs and windtunnel walls in place. By comparing the shape of the sonic surface (Ma=1) in the flow field one can obtain a good impression of the local influence of CCSBs.

Ideally these CCSBs would be adjustable. However, mechanical requirements don't allow for adjustable winglets. Therefore, during the actual experiment flow parameters such as free-stream Mach number and angle of attack were varied until the flow meets the design criteria of the CCSBs. This was controlled using the pressure distribution and shock location. Once the optimal set of flow parameters was identified, the desired data of the flow around the aircraft could be sampled with the flow already being corrected for the undesired side effects of model scale. In order to decide whether the concept worked and where the design criteria were met a rapid and effective detailed analysis of the flow field was necessary. The analysis involved the evaluation of pressure distribution and local lift at various cross sections of the wing as well as an automatic detection of shock position along the wing.

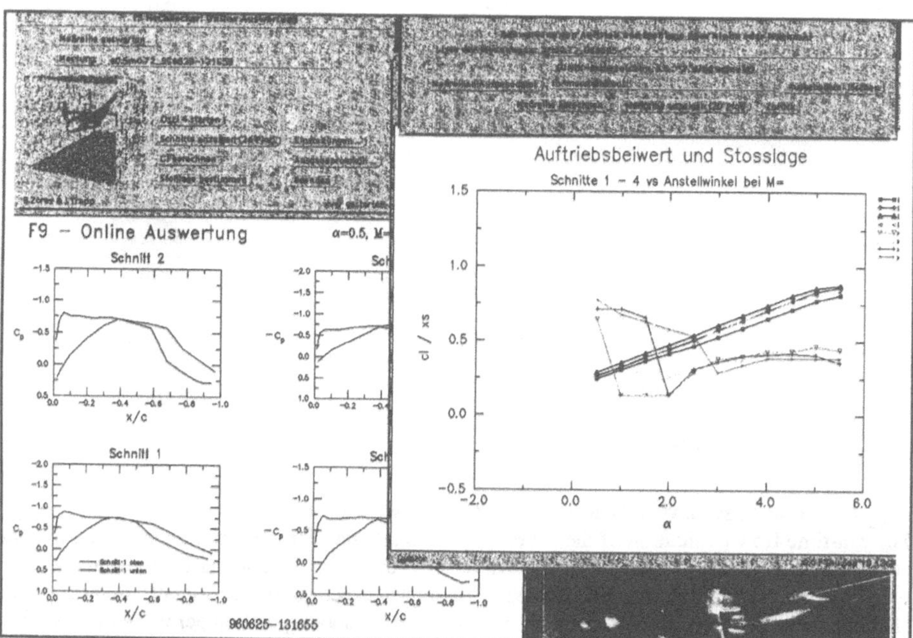


**Fig. 3 Screenshot of the F9-Online data evaluation system: The main control panel positioned top left offers various flow evaluation tools, bottom left a series of pressure distribution plots for various cross sections, on the right integrated lift coefficients and shock position in comparative visualisation, video surveillance of the model in the windtunnel for additional control of vibration.**

## Data acquisition and flow evalution system

The transsonic windtunnel of DLR in Göttingen is equipped with a modern data acquisition and control system named DeAS [3], [4]. One of the features of this system is an interface to add further measurement devices. While such devices are usually data acquisition hardware or mechanical actuators they may as well be so-called virtual devices. Virtual devices are software running on the data acquisition computer which behave like ordinary devices but make use of the data stream from the data acquisition system in some intelligent manner.

For the F9 experiments a virtual device was implemented which builds up a data communication to a second powerful workstation in the DLR network. The virtual device selects a large subset of the measured data and supplies it to the data evaluation system. The selection of this subset is fast enough in order to not slow down the windtunnel experiment.

The actual flow evaluation and feature recognition process will then be performed in parallel on the second computer. The flow evaluation system was developed according to the special demands of the F9 experiments. The system is highly modular which makes all components easily re-usable for further related application. Many modules were taken from a set of well tested tools developed earlier for flow analysis [5]. Some modules and the graphical user interface were tailored toward the particular requirements of the F9-tests.

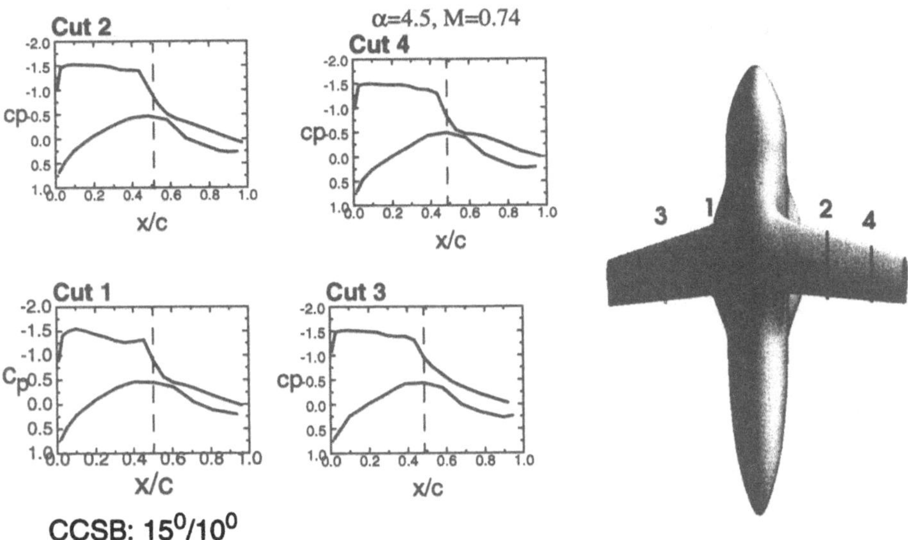


**Fig. 4 Comparative visualization of pressure distribution**

The run-time flow evaluation of the F9-experiments allowed for the interactive evaluation and visualization of pressure distribution in various cross sections, the comparative visualization of such pressure distribution, numerical integration of cross section lift coefficients, determination of shock wave position, and comparative visualization of shock position across wing span. Experience with shock detection and comparative visualization gained in earlier research [6], [7] were taken into account.

All these different post-processing tools may be selected interactively and the results are displayed graphically as shown in Fig. 3. This visualization provides support for the windtunnel crew to select parameters for the following tests. An example of the comparison of pressure profiles is presented in Fig. 4.

## Typical Application of Run-Time Evaluation

The results obtained with the run-time data evaluation system will be briefly illustrated using a typical example. One of the selection criteria for defining a set of flow parameters allowing for valid aerodynamic testing with corrected scale and wing clip influence was the position of the shock wave on the upper surface of the wing. For the design point of the CCSBs the shock should be parallel to the trailing edge of the wing.

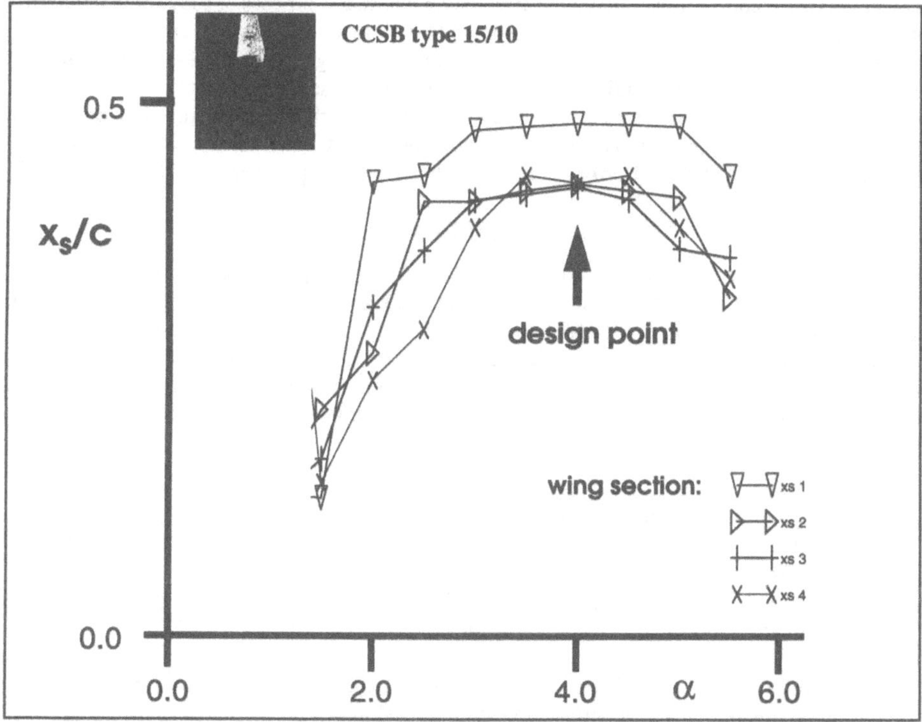


**Fig. 5** Automatic evaluation of shock position for Ma=0.74 and CCSB type 15/10

The software provides extraction of the shock position by analysis of the steepest gradient in the pressure distribution. This is performed for all measured cross sections of the wing. Fig. 5 shows a comparative visualization of the shock position $x_s$ versus chord length c of four different cross sections plotted against angle of attack $\alpha$. The wing sections are numbered 1 to 4, number 1 being the closest to the aircraft body. When the CCSBs fit optimal to the flow conditions three of the shock positions coincide, indicating a shock at x/l=const.or generally speaking a flow on top of the wing equivalent to an unclipped wing. The inner wing section ($xs_1$) is not expected to follow this trend because of the significantly different cross section.
Such evaluation quickly proved the feasibility of the experimental approach. In the meantime a series of tests has been conducted successfully using the F9 model and the software described above. Scientific evaluation of these test will be subject of future publications.

## Conclusions

Presently the human in the loop is required due to the new nature of the CCSB approach. Test engineers have to decide on the fly whether the flow meets the design criteria of the experimental setup. However, the DeAS windtunnel data acquisition and control system is prepared to accept parameter control from virtual devices. Thus, once the selection criteria may be expressed algorithmically or by some sort of an expert system, optimization of flow parameters in such complex experimental settings may well become an automatic process. Today the complexity of the experiment suggests a permanent human interaction. Due to the convenient windtunnel control system as well as the fast and elaborate data processing system decisions are available at a similar time as needed to adjust the windtunnel to a new set of flow parameters. Thus the experiment is not slowed down by human interaction. On the other hand the run-time data evaluation ensures a fast approach to the design point, avoids many unnecessary tests, and provides a valuable, advanced check for errors in an early phase of the experiment.

## Future Goals

The run-time flow data evaluation system is considered to be the prototype of a decision support system for the windtunnel crew. It also provides a feasibility study of advanced data evaluation and parameter optimization needed for expert systems which could help to control the windtunnel experiments in complex setting. The combination of such post-processing tools with the modern object-oriented data acquisition system DeAS opens new opportunities to add complexity to windtunnel tests without the costs of increasing windtunnel running time.

## References


[1] H. Sobieczky, Theoretical Knowledge Base for Accelerated Transsonic Design, AIAA 96-2115, 1 st AIAA Theoretical Fluid Mechanics Meeting, June 17-20, 1996 New Orleans, LA.

[2] R.E. Zores, H. Sobieczky, Using flow control devices in small wind tunnels, abstract submitted to AIAA 28th Fluid Dynamics Conference, June 29- July 2, 1997, Snowmass, Colorado.

[3] H. Geisen, W. Sachs, C++ im Windkanal - DeAs: Ein Programmsystem zur Datenerfassung und Anlagensteuerung, IX-Multiuser-Multitasking Magazin, 8/94, Hannover, 1994.

[4] W. Sachs, H. Geisen, Programmsystem zur Datenerfassung und Anlagensteuerung, Proceedings of Messcomp '94, Wiesbaden 13-15. Sep. 94, Germany.

[5] J. Trapp, GRTools Manual, published on internet June 1996, see: http://www.ts.go.dlr.de/sm-sk_info/library/docs/trapp_GRTools.html.

[6] H.-G. Pagendarm, B. Walter, Feature Detection from Vector Quantities in a Numerically Simulated Hypersonic Flow Field in Combination with Experimental Flow Visualization, in: R. D. Bergeron, A.E. Kaufman, Proceedings of Visualization '94, IEEE Computer Society Press, Los Alamitos, CA, 1994.

[7] H.-G. Pagendarm, B. Walter, Competent, Compact, Comparative Visualization of a Vortical Flow Field, IEEE Transactions on Visualization and Computer Graphics, Vol.1, No.2, 142-150, June, 1995.


## Acknowledgements


CFD analysis was contributed by T. Gerhold using DLR-Tau-Code. The CFD grid used 2.5 million tetrahedral cells. Windtunnel experiments were contributed by A. Heddergott.

**Addresses of the Editors of the Series "Notes on Numerical Fuid Mechanics"**

Prof. Dr. Ernst Heinrich Hirschel (General Editor)
Herzog-Heinrich-Weg 6
D-85604 Zorneding
Federal Republic of Germany

Prof. Dr. Kozo Fujii
High-Speed Aerodynamics Div.
The ISAS
Yoshinodai 3-1-1, Sagamihara
Kanagawa 229
Japan

Prof. Dr. Bram van Leer
Department of Aerospace Engineering
The University of Michigan
3025 FXB Building
1320 Beal Avenue
Ann Arbor, Michigan 48109-2118
USA

Prof. Dr. Michael A. Leschziner
UMIST-Department of Mechanical Engineering
P.O. Box 88
Manchester M60 1QD

Prof. Dr. Maurizio Pandolfi
Dipartimento di Ingegneria Aeronautica e Spaziale
Politecnico di Torino
Corso Duca Degli Abruzzi, 24
I-10129 Torino
Italy

Prof. Dr. Arthur Rizzi
Royal Institute of Technology
Dept. of Aeronautics
Aerodynamics Division
S-10044 Stockholm
Sweden

Dr. Bernard Roux
Institut Recherche sur les Phénomènes Hors
d'Equilibre (IRPHE)
Technopole de Chateau-Gombert
F-13451 Marseille cedex 20
France

**Brief Instruction for Authors**

Manuscripts should have well over 100 pages. As they will be reproduced photomechanically they should be produced with utmost care according to the guidelines, which will be supplied on request. In print, the size will be reduced linearly to approximately 75 per cent. Figures and diagrams should be lettered accordingly so as to produce letters not smaller than 2 mm in print. The same is valid for handwritten formulae. Manuscripts (in English) or proposals should be sent to the general editor, Prof. Dr. E. H. Hirschel, Herzog-Heinrich-Weg 6, D-85604 Zorneding.

Zeitfracht Medien GmbH
Ferdinand-Jühlke-Straße 7
99095 Erfurt, Deutschland
produktsicherheit@kolibri360.de